Mathematics and Archaeology

Mathematics and Archaeology

Editors

Juan A. Barcelo and Igor Bogdanovic

Quantitative Archaeology Lab
Department de Prehistory, Facultat de Lletres
Universitat Autonoma Barcelona
Barcelona, Spain

CRC Press
Taylor & Francis Group
Boca Raton London New York

CRC Press is an imprint of the
Taylor & Francis Group, an **informa** business

A SCIENCE PUBLISHERS BOOK

CRC Press
Taylor & Francis Group
6000 Broken Sound Parkway NW, Suite 300
Boca Raton, FL 33487-2742

First issued in paperback 2020

© 2015 by Taylor & Francis Group, LLC
CRC Press is an imprint of Taylor & Francis Group, an Informa business

No claim to original U.S. Government works

ISBN-13: 978-1-4822-2681-2 (hbk)
ISBN-13: 978-0-367-73818-1 (pbk)

Visit the Taylor & Francis Web site at
http://www.taylorandfrancis.com

and the CRC Press Web site at
http://www.crcpress.com

Foreword

It is sometimes said that "archaeology is too important to be left to archaeologists". Of course, this quip means different things to different people. To me it means that archaeology is not just about treasure-hunting or finding attractive ruins to lure tourists or excavations to entertain mass TV audiences. It is not just about collecting and analysing excavation and other evidence and reconstructing interesting past life-ways. It is not even just about the recovery of the past trajectories of particular human societies as a step towards long-term human social system theory. Ultimately it is about using such access as we have to our own past to guide our choices for our future as a species. It is about using archaeological evidence and interpretation to help understand the long-term dynamics of human society, both in development and in decline, and hence to make beneficial strategic decisions for the whole of humanity: promoting social justice, achieving different kinds of sustainability, and minimising the risk of catastrophes: meteorite strikes, pandemics, nuclear wars, uncontrolled technological growth. Prehistory and history must work together on this fundamental task of global understanding and strategy to ensure the future of our complex and fragile global system that seems all too vulnerable to both endogenous and exogenous stressors. Concern for our future is currently centre stage. See, for example, the website of the recently established University of Cambridge *Centre for Existential Risk* (CSER).

To achieve this overarching goal, mathematical and algorithmic precision is needed at all interpretive levels as we climb the ladder to reliable theories and models of human society and its possible events and trajectories. Furthermore, experience suggests that these global models will have to be individual (or "agent") based—unless we are very lucky and find simple model compressions that can support precise and effective social modelling.

Global agent-based modelling (ABM) to guide collective strategic action may at first seem wildly over-ambitious. Yet national scale agent-based modelling is already being taken seriously by governments, and global work using more traditional mathematical models dates back at least to 1972 when the Club of Rome published "Limits to Growth". Such studies are ongoing now in the contexts of, for example, climate change impact and war gaming.

We can foresee controversies, challenging predictions and daunting conceptual and technical difficulties ahead. Mathematical and computational models challenge unreliable, ambiguous, often merely fashionable, verbal theorising that is elegant but has little precision and little practical value. Not all scholars understand or welcome this challenge or can accept that at our disposal is important "white magic": "mechanical" structures and processes running within a digital computer such that their properties (mathematically provable or experimentally demonstrable) *can* reliably inform our real world choices.

There are great and well-known difficulties in carrying through any such research program. All formal modelling faces problems with the choice and handling of models involving unavoidable issues of abstraction, calibration, and validation. Additional identifiable problem areas for agent-based global modelling are (1) the sheer size and diversity of the system to be modelled, (2) linking archaeological evidence to formal model-based social theory, (3) capturing human general cognition within a model— unavoidable if we are to seriously address the dynamics of *human* society, and (4) the subtle difficulty that our collective action choices are sometimes already determined by a social and physical context potentially itself represented within our models.

The results of model-based studies can themselves trigger disharmony. Consider the potential impact of global warming and the fractious international response to it. What if reliable global models told us, for example, that massive self-imposed social structure backtracks are the best long-term strategy for humankind? What chance of collective implementation would such a strategy have?

But how is the archaeological facet of this great endeavour actually working out? How goes the effort to deploy mathematical and computer-based methods usefully in archaeology? There has been half a century of work dating back, in the UK, to 1962. Early pioneers included Albert Spaulding, David L. Clarke, David Kendall, F. Roy Hodson, Jean-Claude Gardin, Carl-Axel Moberg, William S. Robinson, and Jacob Sher. This list is far from complete. Both David L. Clarke's "Analytical Archaeology" published in 1968 and the collection of articles he edited under the title "Models in Archaeology" (1972) were influential. The text I published with Roy Hodson in 1975 had some impact; its final section entitled "The Archaeologist and Mathematics in the Future" (13.5) makes interesting reading forty years on!

Of course, it was the development of the automatic computer that was the driver. Once digital computers became generally available, a diversity of methods and algorithms that had previously been conceivable, but computationally far too burdensome to use on the available hand calculators, became possible to trial, use, exploit and develop. Thus many new research threads were discovered and followed and new more scientific attitudes to archaeological evidence developed.

So how ARE things going? Unfortunately there has been ebb and flow of fashion and it seems to me that even now the techniques deployed in archaeological work often still feel experimental, unproven, a ragbag rather than a structured toolkit. Of the early ideas and techniques, usually imported into archaeology from other disciplines but sometimes home-grown, some seem to have had little impact, e.g., the "Horse-Shoe" method as a seriation tool, the specifics of catastrophe theory, and heuristic computer programs for hypothesis generation (this last one of my own imports). Other techniques have indeed proved useful and survived, e.g., numerical taxonomy (automatic typology), archaeological databases, shape description methods, direct methods of seriation, and multidimensional scaling (metric and non-metric). Currently prominent in the archaeological research literature (and the social sciences generally) is agent-based modelling (ABM) first trialled by archaeologists in the early 70s. For recent ABM studies see, for example, Kohler and van der Leeuw's "Model-Based Archaeology of Socionatural Systems". As I have emphasised, the foregoing ABM fits well with a drive to large-scale social modelling for long-term human strategy choices.

Why have we not made more solid progress over the past fifty years and more? Is it because social systems are so much more complex than cognitive or physical systems? To expect systems of minds to be no more complex than individual minds is surely optimistic! Or might it be that disciplinary barriers and distrust too often impede the necessary cooperation and training? Computer scientists and sociologists have often been mutually dismissive!

What is certain is that the social sciences and experimental and theoretical archaeological do need mathematical and computational precision. Natural language and the level of thinking that goes with it are pervasively ambiguous and too often engender inferences that are more stylish than sound.

Everything points to the importance of this volume. Professors Barceló and Bogdanovic have gathered together for us an impressive set of international contributors who address a diverse range of applications of formal methods in archaeological work. Formal tools deployed range from mathematical logic, through shape analysis, artefact databases, cluster analysis, artificial neural networks, lattice theory, Bayesian networks to settlement pattern analysis and the uses of mathematics in heritage studies. Several chapters report studies using agent-based modelling which, as I have stressed, agrees well with our current pressing need to understand much more about long-term human social dynamics past and future and the collective survival strategies open to us. The editors are fully aware of these needs. I am confident that the volume they have put together for us containing exciting new ideas, results and trends will be a milestone on our long and dangerous road to self-understanding.

Emeritus Professor James E. Doran
March 2014   University of Essex, UK

Preface

In May 2013 I received an email from a representative of Science Publishers, addressed to Juan Antonio Barceló. The email was an invitation to edit a book on some aspect of Mathematics. Well, I have been working for the last 25 years on "mathematical" subjects, but this work has always been applied to a discipline that is apparently very far away of Mathematics. I am an Archaeologist, considered by many as a "crazy" archaeologist working on esoteric methods like statistics, simulation, machine learning and data mining. That the publisher of a series of books on Mathematics would invite me to edit a book on a subject so specific was something that I did not understand at first. The answer actually is pretty simple. I am not the only Juan Antonio Barceló in the world. There is a famous mathematician in Madrid, and the email was addressed to him, but it arrived to me. I quickly responded to the publisher indicating that they should contact the right "Barceló" at the "Universidad Autónoma de Madrid"—and not me at the "Universitat Autònoma de Barcelona", my University. At the same time I took the initiative of suggesting that I could contribute with a different book, a volume on Mathematics *and* Archaeology. They asked me for a proposal, and I sent an invitation to many archaeologists, mathematicians and computer scientists that I knew had been developing new methods for archaeological research. I received many positive answers, and with this material Igor and I organized a series of subjects that could be of interest. The proposal was accepted by the Publisher, contributing authors sent their chapters, and the result is the book you have before your eyes.

One year after receiving that first misaddressed invitation, the final draft was sent to the publisher. Igor and I have interacted a lot with the contributing authors, insisting that all of us should read most of the chapters, so the Project becomes more cohesive. Therefore, I acknowledge first of all authors and co-authors, the people that interacted with us, and the people who contributed writing the chapters or with the research presented in them. This is their book, and should be read as a collective effort.

Since the inception of the Project I wished that two outstanding scholars addressed some general conclusions from the perspective of a Mathematician and from the point of view of an Archaeologist. Michael Greenacre accepted at the very beginning, and Keith Kintigh joined the project at a later moment. Many thanks also to Jim Doran, one of the pioneers in quantitative archaeology, who has contributed with the foreword to the book.

Some colleagues have worked as anonymous reviewers, and their comments have contributed to the present version. I also acknowledge the comments by Mike Baxter, Ben Ducke, Michael Greenacre, Irmelda Herzog, Keith Kintigh and Dwight Read.

We acknowledge the help from Vijay Primlani, Director of Science Publishers, who made the initial contact and was kind enough to accept the many delays the Project has experimented.

This has been a long Project, and my contribution has been intertwined with research work at the Quantitative Archaeology Lab (Universitat Autònoma de Barcelona). Thanks to all members: Florencia, Vera, Joan, Giacomo, Katia, Berta, Hendaya, Oriol, Laura. My introductory chapter is based on ideas and materials from our collective research, funded by the Spanish Ministry for Economy and Competitiveness (HAR2012-31036 and CSD2010-00034, this last one under the CONSOLIDER-INGENIO Framework).

Juan A. Barcelo
Igor Bogdanovic

Contents

CONCLUSIONS

Introduction

1

Measuring, Counting and Explaining: An Introduction to Mathematics in Archaeology

*Juan A. Barceló**
with contributions from
*Katiu F. Achino, Igor Bogdanovic,
Giacomo Capuzzo, Florencia Del Castillo,
Vera Moitinho de Almeida* and *Joan Negre*

What Are Numbers For?

"What is archaeology about?" asks the professional mathematician shocked by the use of numbers, functions, equations, probabilities, set-theoretic propositions and the like by archaeologists. The answer is pretty simple. "Archaeology is what archaeologists do" (Gardin 1980). We excavate and find stones, pottery sherds, animal and human bones, the remains of ancient buildings, what our ancestors made and discarded at some time, etc. These are our objects of study, but not our objective. The goal of archaeology is to describe the past, that is to say, to find out what people did some time ago and why. Then, where is the place for mathematics? Why a book with such an unusual title: "Mathematics *and* Archaeology"?

We should keep in mind that mathematics is not a property of nature. Although some physicists may disagree, I think that there are some phenomena that are "mathematical" and other phenomena which are not. Whenever we express an idea through order relations among its components, we are expressing it mathematically. The basic meaningful unit of this artificial language is the idea of *quantity*. More than a property or characteristic in itself, it is a *kind* of property: certain entities have "quantities" of

Universitat Autònoma de Barcelona, Dept. de Prehistoria. Facultat de Lletres-Edifici B. 08193 Bellaterra (Barcelona), Spain.
web page: http://gent.uab.cat/barcelo.
* Corresponding author: juanantonio.barcelo@uab.cat

something. These are those properties of entities expressing a gradation or intensity. Therefore, *quantities* will be the opposite of *qualities*: those characteristic features that do not imply gradations, and cannot be expressed in terms of relations of order. The only thing we can say about a *quality* is that it is "present" or "absent". *Quantities* allow the ordering of objects or phenomena in terms of relations of order:

A is greater than B in q
A is equal to B in q
A is less than B in q

By extension, I suggest that a phenomenon that can only be expressed quantitatively, in terms of intensities and orderings, is a mathematical phenomenon, not because it has a different nature, but because if described in verbal terms it would add too much ambiguity and misunderstanding. Mathematics is simply an artificial language used to represent ideas that cannot be expressed in another language. Therefore, there is not a "quantitative" archaeology that may be distinguished from another "qualitative" archaeology. I am just saying that we can do better archaeology using quantification. This is what Nicolucci and colleagues consider in Chapter 3: "data cannot be compared, if the data structures have no clear semantics and the provenance of knowledge is not transparent." Mathematics, as an artificial and formal language should be considered as an attempt to make explicit and well-defined in formal terms the many current archaeological (subjective) implicit terms and concepts. Nicolucci, Hermon and Doerr clarify the archaeological concepts and terms that can be formalized, although they give no details about the way mathematics can be used to disambiguate the description of archaeological primary data, that is, archaeological observables.

It is surprising to realize how scarce is the use of numbers or formal languages for the proper description of archaeological percepts (Kintigh et al. 2014). Nearly 65 years ago, Oliver Myers (1950) "was a forerunner (though apparently without descendant) of what today constitutes the bulk of quantitative and statistical applications in archaeology" (Read 1989:6). The famous dictum by David Clarke "the proper place of archaeology is the faculty of Mathematics" (Clarke 1968) had no impact at that time, and perhaps even today it sounds as a heresy to many archaeologists. Thirty years ago, the use of mathematics was still an exception rather than a norm (Orton 1980). François Djindjian in his chapter at the beginning of this book addresses this difficulty of formal languages to enter in mainstream archaeology. He argues that a major quantitative trend in all the Social and Human Sciences started after the last world war, and it was driven by the development of mathematics for the optimization of war logistics. Since 1960s, the new advances in computer technology allowed the first applications in Archaeology. At that time, the success of Quantitative Archaeology was associated with the revolution in multidimensional data analysis, which occurred with computerization and improvements in the algorithms. From 1975 to 1990, the very idea of a Quantitative Archaeology went a step forward, as exemplified by the transition from preliminary theoretical discussion to the application of quantitative methods to archaeological data, the diffusion of multidimensional data analysis, the development of new algorithms and the possibility of using ready-made software. At around the same time, theoretical developments, usually known as *"New Archaeology"* or *"Processual Archaeology"*, emphasized the use of statistical tests, algebraic models, spatial analysis, system dynamics, linear programming, catastrophe theory, etc. It had an important basis on hypothetic-deductive approaches. Numerous papers in scientific reviews, conference proceedings and books were published during the seventies and the eighties. After 1990, the quantification of archaeological data and the development of new methods experienced a clear decrease, probably due to the new theoretical fashions in Social Science studies and the Humanities, like the "deconstruction" approach, the post-modern paradigms and the collapse of formal rationality. The failure of "classical" Artificial Intelligence was also a factor that may explain the sudden stop in quantifying archaeology twenty-five years ago.

The decrease in popularity of mathematical theories, techniques and tools in archaeology and related disciplines is still a problem, given the lack of a clear consensus of what "numbers" allow when describing archaeological percepts (Kintigh et al. 2014). And dealing with the intrinsic variability of what has been perceived and recorded, as Nicolucci, Hermon and Doerr consider in Chapter 3. It is true that nowadays, mathematics is embedded in software applications (for example Database Management, Teledection, Geographic Information Systems, Visualization, and Virtual Reality) and computing archaeology seems to be in the center of many research interests in Culture Heritage Studies. New mathematics are being tested in archaeology (chaos theory, fractals, neural networks, non-linear systems, multi-agent systems, Bayesian statistics) proving the fundamental role of mathematics in the description and explanation of archaeological complex systems. But much of this effort seems to be out of mainstream professional archaeology.

We have edited this volume to look for solutions to this real problem. Most archaeologists are not aware that the task of understanding the past can be done better with the help of geometry, probabilities, and equations. Most mathematicians do not imagine the possibilities to develop new algorithms to solve archaeological problems because the archaeological problem has not yet been expressed in formal terms (Fig. 1). Dwight Read, in his contribution to this volume (Chapter 4) asserts that the power of mathematics when applied to archaeological problems lies in extending archaeological reasoning (see also Read 1985). To this, there must be concordance between the assumptions of formal methods and the structure of data from past societies recovered through archaeological research. Among the many uses of mathematics, statistical methods for instance, assume homogeneity in underlying data structuring processes, whereas archaeological theorizing begins with process heterogeneity, thereby leading to a double-bind problem that arises before archaeological reasoning can begin: quantitative methods presume the dimensions along which data homogeneity can be delineated are already known whereas archaeological research begins with heterogeneous data and is aimed at determining those dimensions. Solving the double-bind problem does not lie in turning to more complex methods and models, but in integrating formal logics with archaeological theorizing to determine an archaeologically grounded solution to the double-bind problem. This has been Dwight Read long-term interest's (Read 1987, 1989), and this is also the goal of this book

Fig. 1. A General Schema for Archaeological Problems.
Color image of this figure appears in the color plate section at the end of the book.

The Mathematics of Ancient Artifacts

The first domain where archaeologists use "numbers" is in data acquisition and description. Archaeology is assumed to be a quintessentially 'visual' discipline, because visual perception makes us aware of such fundamental properties of objects as their size, orientation, form, color, texture, spatial position, and distance, all at once. Information that should make us aware of how archaeological artifacts and materials were produced and or used in the past is multidimensional in nature: *size*, which makes reference to height, length, depth, weight and mass; *shape* and *form*, which make reference to the geometry of contour, surfaces and volume; *texture* or the visual appearance of what we observe at the archaeological site; and finally *material* and *location properties*: the position of the object in space and time. The real value of archaeological observations comes from our ability to extract meaningful information from them. This is only possible when all relevant information has been captured and coded.

We call *measuring* to the operation of assigning numbers representing the degree or intensity a particular object or phenomenon has a material property. Numbers describe aspects of observables that nouns, adjectives and verbs cannot. This is a very important aspect. As soon as we see the world, we are imposing our personal (subjective) experience on the visual model and, hence on information we may extract from the archaeological observation. By using an instrument able to document a surface without our direct implication—a range-scanner, for instance—we externalize the act of observation, and the observation output, the geometrical model appears to be a correct representation of the empirical world as it is, and not as I believe I see it. However, too many archaeologists tend to consider only the most basic observational properties, like size and a subjective approximation to the idea of shape. Sometimes, a deeper detail of visual appearance is also taken into account, or the mineral/chemical composition. Space and time are usually referred indirectly. The problem is that in too many cases, such properties are not rigorously measured and coded. They are applied as subjective adjectives, expressed in terms of verbal descriptions: round, big, scarce, abundant, far, near, old, older, etc. If the physical description of such visual properties is somewhat vague, then possibilities for understanding what people did sometime and somewhere in the past is compromised; we hardly can infer the object's physical structure nor its location in space and time. The insufficiency and lack of a clear consensus on the traditional methods of archaeological description—mostly visual, narrative, ambiguous, subjective and qualitative—have invariably led to ambiguous and subjective interpretations of the way tools, objects, houses or landscapes were produced and/or used in the past. Nicolucci et al. contribution to this volume (Chapter 3) deals extensively with this issue.

Let us consider the idea of shape measuring, as it is brought forward by Kampel and Zambanini and by Michael Merrill in their contributions to this book (Chapters 6 and 7). The attempts at formally defining the terms "form" or "shape" are often based on the idea of any single, distinct, whole or united visual entity; in other words, it is the structure of a localized field constructed around an object (Small 1996, Costa and Cesar 2001, Barceló 2010a, 2014, Leymarie 2011). Therefore, the shape/form of an object located in some space can be expressed in terms of the geometrical description of the part of that space occupied by the object—abstracting from location and orientation in space, size, and other properties such as color, content, and material composition (Johansson 2008, 2011, Rovetto 2011, Barceló and Moitinho de Almeida 2012). Kampel and Zambanini consider the advantages offered by 3D Vision methods to extract relevant shape information from archaeological observables. They suggest the relevance of shape segmentation based on expert knowledge and general geometrical principles (i.e., curvature). They insist on the necessary three-dimensional nature of archaeological observables (they use the example of ancient pottery and of Roman coins) and the need to go beyond classic 2D methods based on shape from silhouette and using solid volume representation in terms of boundary representation and constructive solid geometry. They discuss how to create such models from instrumented observation (range-scanning). This emphasis on 3D shape extraction and measurement is of topmost relevance for current research in archaeology. There are still archaeologists that think that 3D is just a question of realism in the representation of enhanced aesthetic qualities. As a matter of fact, visual impressions may seem more accurately represented by the two-dimensional projective plane but experience proves the value of the third dimension: the possibilities of understanding *movement* and *dynamics*, that is, "use" and "function" (Barceló 2014).

Michael Merrill (Chapter 7) uses equivalent approaches to create an objective model of shape geometry. His method involves a transformation from a coordinate based elliptical Fourier representation of a complex two-dimensional outline to a coordinate-free representation of the outline using curvature. He investigates prehistoric human bones (the endocranium of Middle Pleistocene (ca. 700,000–150,000) hominines: *Homo erectus*, *Homo heidelbergensis*, and *Homo neanderthalensis*). Beyond mere geometry, the two-dimensional outline being examined is a biological boundary, and the goal is to provide a biologically meaningful representation of the shape. The purpose of the study is to identify segments of the endocranial outline with locally maximal curvature that may result from significantly greater rates of growth in specific sub-regions of the brain.

There are many other quantitative approaches to measure the shape and form of archaeological observables (Özmen and Balcisoy 2006, Rovner and Gyulai 2007, for an overview see the contributions in Elewa 2010, and also Barceló and Moitinho de Almeida 2012, Lycett and von Cramon-Taubadel 2013, Grosman et al. 2014, López et al. 2014, Lucena et al. 2014a, 2014b, Smith et al. 2014, Thulman 2012). It is important to take into account that the need of *measuring* shape is not only applicable to artifacts and objects. The contour of built structures and even macro-spatial patterns (villages, territories, and landscapes) can be quantitatively measured (Conolly and Lake 2006, Hengl and Reuter 2008, Miller and Barton 2008, Oltean 2013).

We need much more than "shape" or "form" for a proper description of archaeological observations. Texture has always been used to describe archaeological materials. It is possible to distinguish between different archaeological materials, because of the appearance of the raw material they are made of. For example, based on textural characteristics, we can identify a variety of materials such as carved lithic tools, stripped bones, polished wood, dry hide, painted pottery, etc. In many ways, "decoration" is a series of texture patterns when studied in terms of its physical nature, and not only stylistically. Engraved, carved or painted, decorative patterns are man-made modifications on the surface of some objects, and they can be considered as an example of induced texture (Maaten et al. 2006).

Texture can be quantitatively defined as those attributes of an object's surface having either visual or tactile variety, and defining the appearance of the surface (Tuceryan and Jain 1998, Fleming 1999, Mirmehdi et al. 2008, Barceló 2009, Engler and Randle 2009, Moitinho de Almeida et al. 2013). It is useful to distinguish between "visual appearance" (color variations, brightness, reflectivity, and transparency), from "tactile appearance", which refers to micro topography (i.e., roughness, waviness, and lay). Color measurement and quantification is a relatively easy task with the appropriate equipment *Colorimeters* measure tri-stimulus data, that is, lightness (value), chromaticity (saturation), and hue (rainbow or spectrum of colors) of a sample color; the color's numeric value is then visually determined using a specific three-dimensional color model or three-valued system. *Spectrophotometers*, for their part, measure spectral data, that is to say, the amount of spectral reflectance, transmitting, and/or emitting properties of a sample color at each wavelength on the visible spectrum continuum, without interpretation by a human. The measured data has the advantage of not being dependent upon light, object micro topography/finishing, and viewer.

In archaeology, tactile information has been traditionally measured in terms of transforming grey-level image information into a map of bumps within a surface, and detecting significant local changes among luminance values in a visually perceived scene and its translation into a geometric language. In other words, texture appears to be a consequence of anisotropic reflection, given the underlying assumption that light waves undergo reflection when they encounter a solid interface (surface), and this reflection is irregular depending on the heterogeneity of the surface. Such an approach has produced good results in archaeology (Barceló 2009), but it is no more tenable because it is still based on the probably wrong assumption that digital pictures (coded in pixels) are surrogates of real surfaces. Luminance variations do not always allow distinguishing differentiated micro topographic variations because they can be an effect of the perceptual acquisition mechanism (the microscope, the eye, the camera, the sensor), and consequently images not only show features of the object being analyzed but they mix this variation with variability coming from the context of observation and the mechanical characteristics of the observation instrument. That means that an image texture not only contains the object surface irregularity data, but additional information which in the best cases is just random noise, and in many other cases makes difficult to distinguish what belongs to the object from what belongs to the observation process. Consequently, the constructed geometric models of a scanned surface can be used for a quantitative representation of the visual appearance of the archaeological object.

Nowadays, the high resolution precision of many non-contact close-range 3D scanners can capture 3D surface data points with less than 50 microns between adjacent points. In addition, confocal, digital holography, focus variation, interferometric, or scanning tunneling microscopy, optical focus sensing, Nomarski differential profiler, phase shifting or coherence scanning interferometry, angle resolved SEM, SEM stereoscopy, just to name some other non-contact instruments, can generate 3D representations of surface irregularities with even higher detail, and thus allowing finer measurements. In this way, instead of using grey-level values measured at pixel resolution, we have proper measurements of depth and height at well localized points within the surface (Stytz and Parrott 1993, Swan and Garraty 1995, Lark 1996, Van Der Sanden and Hoekman 2005), whose geometric irregularities can be quantified in terms of planes variations or curvature variability (angle and distribution), and patterns (Barceló and Moitinho de Almeida 2012, Moitinho de Almeida 2013, Moitinho de Almeida et al. 2013). Kampel and Zambanini (Chapter 6) offer some details on pottery decoration description in terms of texture quantification.

Visual features are not enough for an exhaustive documentation of archaeological evidence. Among non-visual data we can mention compositional data, which are most frequently understood as the enumeration of basic or fundamental elements and properties defining a material. Although in the historical beginnings of the discipline, the enumeration of the substances an archaeological object was made of was regarded as a visual inference based on the scholar's previous experience (in terms of the color or texture of different materials like 'pottery', 'stone', 'bone'). Nowadays, mineralogical and physicochemical compositions are measured objectively using appropriate instruments, such as x-ray and μ-Raman spectrometry, Neutron Activation Analysis (NAA) for elemental composition information, neutron scattering for revealing alloys and organic material; particle accelerator, Laser Induced Breakdown Spectroscopy (LIBS), etc. Archaeometrical analysis provides an unquestionably valuable source of numeric data for inferring some possible ways an ancient or prehistoric artifact may have been produced and/or used in the past. Nevertheless, we should take into account that the material components of any archaeological object can be defined and delimited at a variety of scales (e.g., atomic, molecular, cellular, macroscopically), what prevents taking compositions as magnitudes. Instead, we have different compositions at different analytical scales.

As suggested by Martin et al. (Chapter 8) and Mucha et al. (Chapter 9), a composition can be defined as a D-part row vector $\mathbf{x} = [x_1, x_2,..., x_D]$, $x_j > 0$, $j = 1,...,D$, representing parts of a whole, that only carry relative information. Usually, compositions are represented subject to a constant-sum constraint, $x_1 + x_2 + ... + x_D = k$ (e.g., $k = 1$, 100 or 10^6). Such data have proved difficult to handle statistically, because of the awkward constraint that compositions are not mere lists of substances but multi-component vectors, where the addition of components is a constant in the population under study. Compositional vectors should fulfill two conditions:

a) The components should be "generic", in the sense that all objects can be described as different combinations of the same components. For instance, the material components of a knife can be decomposed in steel and wood; the components (for instance, chemical elements) of a pottery vase can be decomposed into *Al, Mg, Fe, Ti, Mn, Cr, Ca, Na, Ni*.

b) The components should be expressed as a proportion of the total sum of components, which defines the composition of the entity. Compositions should be expressed as vectors of data, which sum up to a constant, usually proportions or percentages. To say that there is steel and wood in this object, is not a true decomposition of the knife. Instead, we have to say that 13% of the object consists in wood for the grip, and the remaining 87% is composed of steel. In this case the components sum a constant (100), and composition is measured against this total.

Shape, texture and non-visual properties of archaeological entities (from artifacts to landscapes) should be regarded as changing not as a result of their input-output relations, but as a consequence of the effect of processes on them. Consequently, reasoning about the material remains of the past depends on the following factors and senses (Bicici and St. Amant 2003):

•   *Form/Texture/Composition:* For many tools, form, texture and composition is a decisive factor in their effectiveness.
•   *Planning:* Appropriate sequences of actions are key to tool use. The function of a tool usually makes it obvious what kinds of plans it takes part in.

- *Physics:* For reasoning about a tool's interactions with other objects and measuring how it affects other physical artifacts, we need to have a basic understanding of the naive physical rules that govern the objects.
- *Dynamics:* The motion and the dynamic relationships between the parts of tools and between the tools and their targets provide cues for proper usage.
- *Causality:* Causal relationships between the parts of tools and their corresponding effects on other physical objects help us understand how we can use them and why they are efficient.
- *Work space environment:* A tool needs enough work space to be effectively applied.
- *Design requirements:* Using a tool to achieve a known task requires close interaction with the general design goal and requirements of the specific task.

This list suggests that reasoning about the archaeological objects recovered at the archaeological site requires a cross-disciplinary investigation ranging from recognition techniques used in computer vision and robotics to reasoning, representation, and learning methods in artificial intelligence. The idea would be to investigate the interaction among planning and reasoning, geometric representation of the visual data, and qualitative and quantitative representations of the dynamics in the artifact world. Since the time of Galileo Gallilei we know that the "use" of an object can be reduced to the application of forces to a solid, which in response to them moves, deforms or vibrates. *Mechanics* is the discipline which investigates the way forces can be applied to solids, and the intensity of consequences. Pioneering work by Brian Cotterell and Johan Kamminga (1992) proved that many remarkable processes of shaping holding, pressing, cutting, heating, etc. are now well known, and can be expressed mathematically in equations. Consequently, we can obtain additional sources of quantitative data in archaeology:

- **Structural analysis** is the determination of the effects of loads on physical structures and their components. Structures subject to this type of analysis include all that must withstand loads, such as buildings, bridges, vehicles, machinery, furniture, containers, cutting instruments, soil strata, and biological tissue. Structural analysis incorporates the fields of applied mechanics, materials science and applied mathematics to compute a structure's deformations, internal forces, stresses, support reactions, accelerations, and stability. The results of the analysis are used to verify a structure's fitness for use, often saving physical tests. Structural analysis is thus a key part of the engineering design of structures. We can distinguish between linear and non-linear models. *Linear models* use simple parameters and assume that the material is not plastically deformed. ***Non-linear models*** consist of stressing the material past its time-variant capabilities. The stresses in the material then vary with the amount of deformation.
- **Fatigue analysis** may help archaeologists to predict the past duration of an object or building by showing the effects of cyclic loading. Such analysis can show the areas where crack propagation is most likely to occur. Failure due to fatigue may also show the damage tolerance of the material.
- **Vibration analysis** can be implemented to test a material against random vibrations, shock, and impact. Each of these incidences may act on the natural vibration frequency of the material which, in turn, may cause resonance and subsequent failure. Vibration can be magnified as a result of load-inertia coupling or amplified by periodic forces as a result of resonance. This type of dynamic

information is critical for controlling vibration and producing a design that runs smoothly. But it's equally important to study the forced vibration characteristics of ancient or prehistoric tools and artifacts where a time-varying load excited a different response in different components. For cases where the load is not deterministic, we should conduct a random vibration analysis, which takes a probabilistic approach to load definition.

- **Heat Transfer analysis** models the conductivity or thermal fluid dynamics of the material or structure. This may consist of a steady-state or transient transfer. Steady-state transfer refers to constant thermo properties in the material that yield linear heat diffusion.
- **Motion analysis (kinematics)** simulates the motion of an artifact or an assembly and tries to determine its past (or future) behavior by incorporating the effects of force and friction. Such analysis allows understanding how a series of artifacts or tools performed in the past—e.g., to analyze the needed force to activate a specific mechanism or to exert mechanical forces to study phenomena and processes such as wear resistance. It can be of interest in the case of relating the use of a tool with the preserved material evidence of its performance: a lithic tool and the Stone stelae the tool contributed to engrave. This kind of analysis needs additional parameters such as center of gravity, type of contact and position relationship between components or assemblies; time-velocity.

Among current applications of this way of quantifying characteristics of ancient materials we can mention: Barham (2013), Bril et al. (2012), Haan (2014), Homsher (2012), Moitinho de Almeida (2013), O'Driscoll and Thompson (2014). There is nothing new, modern techniques of Finite Element Analysis can do the work. Without such information any effort in explaining what people did in the past seems to me impossible. It wonders me the absolute lack of such information, not only in current virtual archaeology projects, but also in culture heritage databases. Archaeologists insist in documenting ancient artifacts, but such documentation never takes into account the physical and mechanical properties of ancient materials.

The Mathematics of Social Action in the Past

We should also take into account the possibility of the direct measuring and quantification of human behavior in the past. The reader may think that we are limited to the physical measuring of observables. Let us consider an example that shows how more social data can be extracted from archaeological observations. In an investigation of the origins of the city in the Etrurian area of Central Italy between 11th and 6th centuries BC (Barceló et al. 2002) research goals were:

- the archaeological correlates for *generators* of capital accumulation,
- the archaeological correlates for *restraints* on capital accumulation.

Such correlates can be calculated by measuring qualitatively the presence/absence of social actions (settlement, resources acquisition, labor action, distributive/exchange activities, and ritual action). Among others, observational inputs can be:

- Presence/absence of colonial import goods
- Presence/absence of indigenous import goods (pottery, metal)

- Presence/absence of locally produced valuable pottery
- Presence/absence of weapons
- Presence/absence of metallurgical activities
- Presence/absence of store buildings and structures
- Presence/absence rich burials
- Presence/absence of complex residential structures (multi-room houses)
- Presence/absence of subsistence activities (farming, husbandry, etc.).

The *spatial density or settlement concentration* can be empirically measured in terms of a spatial probability density measure associated to each location, based on the geographical proximity with neighboring locations. Concentration maps, however, can be misleading. If settlement concentration is a relevant variable, then *distance* is one of the main dynamic factors determining the process of city formation. Spatial Interaction is related to distance, in such a way that the less distance between social agents, the higher the probabilities of social interaction. The definition of such *distances* is very complex:

- Distance produced by the diversity of resources at different locations
- Distance produced by the diversity of production activities at different locations
- Distance produced by the differences in the volume of produced goods at different locations
- Distance produced by the diversity of non productive activities—consumption— at different locations
- Distance produced by the differences in the volume of non productive activities —consumption—at different locations
- Distance produced by the differences of the nature of social agents at different locations
- Distance produced by the differences and diversity of social interactions between different locations.

If a city is the area where social interaction has highest levels, then, any modification of the settlement pattern towards an increasing concentration of sites should be also related with a reduction of inter site distances. However, we have to remember that a city is not only a spatial container of people. It is an *attractor* for all interactions generated in its periphery and directed to the core area. Interaction flows between various places are proportional to the probability of contacts of their residents, and this probability is a function of the size of the places. We need demographic estimates to calculate the theoretical maximum degree of interaction depending on population. Demographic estimates can be obtained through extension measures of settlement areas.

We may also need to estimate the inequality and directionality of interaction flows. In other words, we have to integrate into the model the hierarchy between social cores and peripheries, that is to say, the differentiation between the emergent urban core and the exploited rural periphery, for instance. This allows understanding capital accumulated in the center *as a function of* capital extracted from the periphery. Therefore, if Power and Dominance may be analyzed in terms of spatial attraction, then the inequality of interaction from core to periphery is directly proportional to capital generated in both points, and inversely proportional to the cost of coercion and domination. The attraction force exerted by a city core area may be estimated as directly proportional to the number or intensity of interactions between periphery locations and the center and the squared distance between both entities, and inversely proportional to the attraction force exerted by alternative periphery locations. Local factors may be understood in terms of locally accumulated mobile capital (for instance: quantity of colonial imported goods, presence/absence of metallurgical luxury objects). If the model represents the process of capital accumulation at the city, then the volume of capital or wealth accumulated is directly proportional to the level of dominance or power of the city over the surrounding rural area, plus the frictional effects due to the cost of coercion, and inversely proportional to the total amount of capital accumulated at the periphery. The more productive is a location, and the more independent are their local elites, the most difficult is to ensure dominance and capital transfers from periphery to core areas.

This is just an example of measuring social action in the past from the amount and/or differences among archaeological observables. Consider, alternatively, the case of quantity of labor for the production of a certain tool or built structure, the precise efficiency of tool use, or cost/benefit ratios of hunting or herding at a specific region during a concrete period of time. A measure of the complexity of the organization of production can be approximated through the reconstruction of the technological chain used in the creation/manufacture of the object. The idea of technological chain or sequence of operations needed for the manufacture of a tool was developed by André Leroi-Gourhan and has been successfully applied to the study of lithic technology. In short, it is based into the differentiation of labor activities arranged in manufacture order. For instance, in the case of ceramics: raw material acquisition, modeling, turning, decorating (and the various decorative techniques), drying, cooking or processing. Different objects can be ordered in terms of the length of such a technological chain, and we can use a normalized measure of such ordering as an estimation of the quantity of labor invested in its manufacture (Soressi and Geneste 2011). In some specialized fields within archaeology, like the study of animal bones (archaeozoology) it is usual to refer to the utility index of anatomical parts, total biomass, obtainable biomass, etc. (Lyman 2008). In archaeobotany, the study of grain yields, the caloric efficiency of different woods, etc. can be measured, thanks to theoretical models from biological sciences and ecology (Buxó and Piqué 2009). Some other measurements of the complexity of a production system, of a social system or a settlement pattern can be obtained (Feinman et al. 1981, Morrison 1994, Francfort 1997, Barceló et al. 2006, Barceló 2009, Betts and Friesen 2004, Betts and Friesen 2006, Barton et al. 2004, Bentley et al. 2005, Brantingham 2006, among many others). Rogers et al. (Chapter 23) also suggest some formal measurement of past social action, notably the idea of *wealth*.

The best known example of measuring social action in the past is based on the frequency of grave goods per burial to provide quantitative estimates of social distance and hierarchy. The absolute and relative quantities of different grave goods in different burials, together with estimates of the quantity of labor involved in the burial practice has allowed the quantification of social prestige, and even the identification of complex societies with differentiated hierarchical roles (O'Shea 1984, McHugh 1999, Chapman 2003, Drennan et al. 2010, Sosna et al. 2013, Strauss 2012, Sayer and Wienhold 2013).

The Mathematical Nature of Archaeological Problems

The use of mathematics in archaeology does not end with measuring shape, size, texture and material properties of objects, built structures and landscapes, or with the estimation of the intensity of social activities in the past. We should solve historical "problems" using archaeological quantitative data (measurements).

Archaeological research involves an intricate set of interrelated goals, and therefore, an intricate set of interrelated *problems*. We constantly deal with many different kinds of problems; most of them can be formalized (Table 1).

Table 1. A List of Possible Archaeological Problems (Barceló 2009).

Type	Goal to be Achieved	Formal Problem
Definition	What is society? What is a social class? What is an archaeological site? What is a tool?	Invention of concept or taxonomy
Theory	How do we explain the distribution of this Pottery type? What is the cause of the fact that these objects have this shape?	Invention of Theory
Data	What information is needed to test or build a Theory?	Observation, experiment
Technique	How can we obtain data? How do we analyze it? How may the phenomenon best be displayed?	Invention of instruments and methods of analysis and display
Evaluation	How adequate is a definition, theory, observation or technique? Is something a true anomaly or an artifact?	Invention of criteria for evaluation
Integration	Can two disparate theories or sets of data be integrated? Does Binford contradict Hodder?	Reinterpretation and rethinking of existing concepts and ideas
Extension	How many cases does a theory explain? What are the boundary conditions for applying a theory or a technique?	Prediction and testing
Comparison	Which theory or data set is more useful?	Invention of criteria for Comparison
Application	How can this observation, theory or technique be used?	Knowledge of related unsolved problems
Instrument insoluble	Do these data disprove the theory? Is the technique for data collection appropriate?	Recognition that problem is as stated

Regrettably, most of these questions are not usually considered in mainstream archaeological studies. Current archaeological explanations, like most social science explanations seem addressed to simply tell us what happens *now* at the archaeological site, or *when* something is believed to have happened. They do not tell us *what* happened in the past, nor *who, why* or *how*. A substantial proportion of research effort in archaeology isn't expended directly in explanation tasks; it is expended in the business of discovering and confirming the evidence of social action at some specific moment in the past, without arguing *why* those actions took place there and then. The lion's share of the effort goes into the description and data analysis, while discovering and explaining the causes they uncover rarely stand in the research agenda.

Archaeology should become a problem solving discipline, centered on *historical problems*, whose focus is on explaining existing perceivable phenomena in terms of long past causes:

WHY IS PRESENT OBSERVATION THE WAY IT IS?
WHAT ACTION OR PROCESS **CAUSED** WHAT I'M SEEING NOW?

In other words,

WHY DO THE OBSERVED MATERIAL ENTITIES HAVE SPECIFIC VALUES OF SIZE, SHAPE, TEXTURE, MATERIAL?

By assuming that what we perceive in the present (archaeological excavation) is simply the material effects of human work performed at some time in the past, we should understand 'archaeological percepts' as material things that were products in some moment of their causal history. Archaeology is involved in solving such *why*-questions. In our opinion, the answer to a *why*-question is a *causal affirmation* about the formation process of society (Barceló 2009).

Functional analysis can be defined as the analysis of the object's disposition to contribute causally to the output capacity of a complex system of social actions (Cummins 1975, 2000, 2002). Such a definition includes the use of objects with a material purpose (instruments) and objects used in a metaphorical way with an ideological intention (symbols). The term "function" can be defined as a causal explanation of behaviors the item was involved at some moment. We can argue that archaeological functional statements should provide an answer to the question "how does *S* work?" where *S* is a goal-directed system in which the material entity whose function we are interested in appears. Archaeological observables should be explained by the particular causal structure in which they are supposed to have been participated. The knowledge of the function of some perceived material element should reflect the causal interactions that someone has or can potentially have with needs, goals and products in the course of using such elements. We should consider how size and weight will affect what the object did in the past; its overall form (for holding or halting); the edge angle where cutting, scraping, or holding was important; the duration of its use, how specialized the working parts needed to be; whether it was at all desirable to combine two or more functions in the same tool; how reliable the tool needed to be; and how easily repaired or resharpened it needed to be (Hayden 1998, Read 2007). A possible way to carry out this functional analysis would be through the decomposition of use-behavior processes into chains of single mechanisms or operations, each one represented by some part (or physico-chemical/mineralogical component) of the studied object. In order to know the use of an object, we need to infer its proper usage position, the direction of the action, and the pressure to be applied by a prospective user. These cannot be learned without spatial relations between parts and subparts, which imply that the parts and subparts directly relate to behaviors made with the object. Changing the direction of forces, torques, and impulses and devising plans to transmit forces between parts are two of main problems that arise in this framework. To solve these, we need to integrate causal and functional knowledge to see, understand, and be able to manipulate past use scenarios. Functional analysis appears then as the application of an object in a specific context for the accomplishment of a particular purpose. In this sense, important work is being done in archaeology for the computer simulation of mechanics and/or life-cycle among other forms of functional/productive behavior (Atkins 2009, Dunn and Woolford 2013, Hopkins 2008, Barceló and Moitinho de Almeida 2012, Moitinho de Almeida 2013, Kilikoglou and Vekkinis 2002, Walker and Schiffer 2014).

The main assumption is that some percept (*archaeological description*) is related to a causal affirmation about the causal event (social action) having produced the perceived evidence (*archaeological explanation*). In our case, it implies to *predict* the cause or formation process of some archaeological entity given some *perceived* evidences of the effect of this causal process. In its most basic sense, then, the task may be reduced to the problem of detecting localized key perceptual stimuli or features, which are unambiguous clues to appropriate causal events. For instance, a distinctive use-wear texture on the surface of a lithic tool, and not on others predict that these tools have been used to process fresh wood, and we infer that at some moment a group of people was cutting trees or gathering firewood. Alternatively, we can consider that the shape of some pottery vases predicts their past use as containers for wine when its shape appears to be clearly different from the shape of vases used for other purposes; the distinguished or privileged position, and/or the composition of some graves predicts the social personality of the individual buried there and hence the existence of social classes, in case they are *different* qualitatively and quantitatively, both in the number of grave-goods and in the estimated amount of labor invested in the ritual. Here the output is not the object (stone, pottery, grave), but the causal event is the social action itself: cutting trees or gathering firewood, wine production and trade, social power and political hierarchy.

In a majority of cases, we cannot study the causes of individual action, but we will seek to determine the causes of collective action in the past, i.e., the repetitiveness of labor activities by men and women, and the degree of regularity manifested by the material consequences of such activities. We need "sets of artifacts", instead of mere "artifacts". We should measure observed variability among archaeological objects and remains (shape, size, texture, material, etc.) within a differentiated set and between different sets to see if the action: a) determined the presence of effects (all effects are same for all instances of the action : no variability within the results of a single action and total differentiation between results of different actions); b) conditioned its presence (there is some variation among the results of a single action, but its intensity is lower than what would be expected by chance alone; differences among results of the same action are always lower than differences with the results of another action), or the alleged action had nothing to do with the presence of these effects (very high variability among results of a single action, statistical independence between results of different actions). This is what D. Read, in his contribution to this book (Chapter 4), states when mentioning "patterning in the aggregate case". See also Chapter 5, by Luciana Bordoni, on the need to integrate different observations into relational entities.

Interpretations of this kind typically constitute the solution of an *inverse problem*. The idea of an inverse problem refers to a wide range of problems that are generally described by saying that their answer is known, but not the question. In our case: "Guessing a past event from the variability of its vestiges". It entails determining unknown *causes* based on observation of their *effects*. This is in contrast to the corresponding direct problem, whose solution involves finding effects based on a complete description of their causes. That is to say, we are looking for the way to infer the motivations and goals of social action based on the observed variability of perceived material transformations, which are assumed to be the consequence of such motivations and goals. Nicolucci et al. (Chapter 3) suggest a relevant example of inverse problem, the case of the Varus battle, an historical event inferred from weapons, pieces of armors or shields, helms, skeletons, etc. In archaeology, the main source for inverse problems lies in the fact that archaeologists generally do not know why archaeological observables have the shape, size, texture, material, and spatio-temporal location they have. Instead, we have sparse and noisy observations or measurements of perceptual properties, and an incomplete knowledge of relational contexts and possible causal processes. Consequently, quantification of observation is critical; if we make errors in our descriptions, typologies, etc. then we have already compromised the ability to engage in finding solutions to the inverse problem.

Inverse problems are among the most challenging in computational and applied science and have been studied extensively (Hensel 1991, Kirsch 1996, Woodbury 2002, Sabatier 2000, Pizlo 2001, Kaipio and Somersalo 2004, Tarantola 2005, Bunge 2006). To solve the archaeological problem we need controlled observations, and a theory about the relationship between the nature of observations and the model to infer the values of the parameters representing the model, that is, a true experimental design.

A naïve solution would be to list all possible material consequences of the same action that was performed in the past. This universal knowledge base would contain all the knowledge needed to "guess" in a rational way the most probable cause of newly observed effects. This way of solving archaeological problems would imply a kind of instance-based learning, which would represent knowledge in terms of specific cases or experiences, relying on flexible matching methods to retrieve these cases and apply them to new situations. This way of learning, usually called *case-based learning*, is claimed to be a paradigm of the human way of solving complex diagnostic problems in domains like archaeology. Archaeologists make decisions based on their accumulated experience on successfully solved cases, stored in a knowledge base for later retrieval. When the expert perceives a new case with similar parameters, she tries to recall stored cases with characteristics similar to the new one, finds the closest fit, and applies the solutions of the old case to the new case. Successful solutions are tagged to the new case and both are stored together with the other cases in the knowledge base. Unsuccessful solutions also are appended to the case base along with explanations as to why the solutions did not work. A precise and quantitative way to deal with variability and similarity seems of topmost relevance for applying prior knowledge to new archaeological observations waiting for an explanation. Richards et al. (Chapter 12) offer a detailed account of the mathematics of similarity in the domain of text mining.

This approach relies on an assumption asserting that 'similar problems have similar solutions'. The idea is that once we have a rule that fits past data, if present observations are similar to past ones, we will make correct predictions for novel instances. This mechanism implies the search for maximal explanatory similarity between the situation being explained and some previously explained scenario. The trouble is that in most real cases, there are infinite observations that can be linked to a single social action, making them impossible to list by extension. Even the most systematic and long-term record keeping is unlikely to recover all the possible combinations of values that can have arose as a result of social action in the past. Thus, the learning task becomes one of finding some solution that identifies essential patterns in the samples that are not overly specific to the sample data.

This is exactly what philosophers of science have called *induction* (Holland et al. 1986, Genesareth and Nilsson 1987, Langley and Zytkow 1989, Gibbins 1990, Gillies 1996, Williamson 2004, Tawfik 2004, Bunge 2006). It can be defined as the way of connecting two predicates to each other, based on a number of examples exhibiting the relevant predicates. Virtually all inductive inferences may be regarded in one sense as either generalizations or specializations. Since Aristotle, generalization has been the paradigmatic form of inductive inference. He and many subsequent logicians discussed the structure and legitimacy of inferences from the knowledge that some observed instances of a kind *A* have a property *B* to the conclusion that all *A* have the property *B*. In the past several decades in philosophy of science, the problem has been conceived in terms of the conditions under which observed instances can be said to confirm the generalization that all *A* are *B*. The underlying assumptions were once suggested by Bertrand Russell (1967):

A. When a thing of certain sort *A* has been found to be associated with a thing of a certain other sort *B*, and has never been found dissociated from a thing of the sort *B*, the greater the number of cases in which *A* and *B* have been associated, the greater is the probability that they will be associated in a fresh case in which one of them is known to be present;

B. Under the same circumstances, a sufficient number of cases of association will make the probability of a fresh association nearly a certainty, and will make it approach certainty without limit.

Consequently, one of the most fundamental notions in solving archaeological problems methods is that of similarity: an inverse problem can be solved in case we can find cases of a particular input-output relationship that are enough similar. Two entities are *similar* because they have many *properties* in common. According to this view (Medin 1989):

- similarity between two entities increases as a function of the number of properties they share,
- properties can be treated as independent and additive,
- the properties determining similarity are all roughly the same level of abstractness,
- these similarities are sufficient to describe a conceptual structure: a concept would be then equivalent to a list of the properties shared by most of its instances.

It means that an archaeologist has to be able to identify the common property shared by two or more material effects of the same social action to acquire the ability of explaining similar evidences as generated by the same cause. In any case, the very idea of similarity is insidious. First, we must recognize that similarity is relative and variable. That means that the degree of similarity between two entities must always be determined relative to a particular domain. Things are similar in color or shape, or in any other domain. There is nothing like overall similarity that can be universally measured, but we always have to say in what respects two things are similar. Similarity judgements will thus crucially depend on the context in which they occur.

In our case, the task will be to find the common structure in a given perceptual sequence under the assumption that structure that is common across many individual instances of the same cause-effect relationship must be definitive of that group. This implies that we may learn explanatory concepts such as "15th century", "cutting", "killing", "social elite", provided we have enough known instances for the underlying event, and a general background knowledge about how in this situation a human action has generated the observed modification of visual appearances that it is using as perceptual information. That is, archaeologists will learn a mapping from the cause to the effect provided some instances of such a mapping are already known or can be provided by direct experience in the world. When subsequently asked to determine whether novel instances belong to the same causal event, those instances that are similar to instances characteristic of a single event of a single class of events will tend to be accepted. Here, the *givens*—the archaeological description of material consequences of social action—are the condition, and reverse engineering can be used here as a process to help finding a *generalization* of input-output mappings connecting cause and effect (Fig. 2).

Each instance of a solved problem may consist of: 1) a set of related observations, and 2) the corresponding social activity that caused them. For instance, suppose we are asking about the formation process of an animal bone assemblage. These quantitative properties constitute the inverse problem we want to solve. The fact that these bones are the consequence of a hunting event would be the solution to the problem. This solution will be possible only when our automated archaeologist learn what "hunting" is, or more precisely how to relate the action with its material consequences. This learning has been possible because an archaeologist has been trained on a variety of correlated sets of hunting strategies features (supposed to correspond to known instances of human hunting). These descriptions should have been obtained by selecting one set of features and stipulating that these describe each prototypical hunting strategy. In case of opportunistic hunting, for instance, carcasses will be presumably butchered unsystematically, and this fact will be preserved in the number and kind of animal

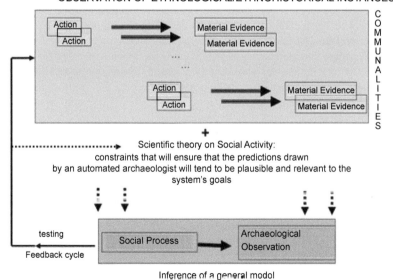

Fig. 2. From Observed Instance to Inferred causes. A schema for archaeological problem-solving.

Color image of this figure appears in the color plate section at the end of the book.

bones, in their fractures, cut marks, and butchery traces. We assume the archaeologist never sees a single prototypical "opportunistic hunting strategy", but many instances of approximate examples of opportunistic hunting. She learns to extract the general pattern exemplified in the overall set of instances.

When an archaeologist attempts to use a body of observed evidences to discover a way to reconstruct the social process or processes which generated them, he or she exploits certain properties in the data, which can be referred as trends, regularities, similarities and so on. This is the very basics of archaeological reasoning. The presence of communalities implies a high level of *regularity* in the data, which means that certain characteristics or properties are more probable that others (Zytkow and Baker 1991: 34). Regularity has the advantage of increasing useful redundancy into the mechanism. When we introduce useful redundancy into an encoding scheme, we want to provide the means whereby a receiver agent can *predict* properties of parts of the message from properties of other parts, and hence generalize how most similar inputs are related with a single output.

Chris Thornton, in a thought provoking essay, suggests that generalization inferences should be presented in the form of *prediction tasks* (Thornton 2000). Solving an archaeological problem is a predictive task because its aim is to extract a decision rule from known data that will be applicable to new data. Prediction and learning are associated according to the following general principle:

Given a collection of examples of f, *learn a function* h *that* predicts *the values of* f.

Archaeologist should be able to generalize the set of known examples of *f* in such a way that *h* can be considered as an agreement of what is common to all *f*, leaving out the apparently irrelevant distinguishing features. The function *h* is called a hypothesis, and it will be used to *predict* the most appropriate explanation for new archaeological data not previously seen. The idea is that any wrong hypothesis will almost certainly be "found out" after a small number of examples, because its low associative strength does not allow making a correct prediction. Problem solving can be seen then as the enhancement of "predictability" (or associative strength) to a previous experienced state of the problem. According to this view, an inductive hypothesis is a universally quantified sentence that summarizes the conditions under which archaeological evidence has been generated by a social action (or by a natural disturbation process!). In such cases, the problem of induction reduces to that of forming the group of all effects caused by the same action.

We may call this approach to archaeological explanation *category-based* because explanatory elements are accessed through a process of categorization. It implies that the perceived input is sorted out into discrete, distinct categories whose members somehow come to resemble one another more than they resemble members of other categories. In categorization, two operations are involved. First, we decide an observable is a member of a large number of known categories according to its input properties. Second, this identification allows access to a large body of stored information about this type of object, including its function and various forms of expectations about its past behavior. This two-step schema has the advantage that any explanatory property can be associated with any object, because the relation between the physical properties of the object and the information stored about its function, history, and use can be purely arbitrary, owing to its mediation by the process of categorization. That means that our responses to an incoming input are not dependent of any particular attribute of the input. Rather, the solution to the archaeological problem will make sense only when considered as one component in a causal chain that generates responses entirely according to the probability distribution of the past significance of the same or related input. The final answer to the question exemplifies not the stimulus or its sources as such, but the accumulated interactions with all the possible sources of the same or similar stimuli in proportion to the frequency with which they have been experienced.

This procedure is analogous to classical experimentation. Experimental analysis is the process whereby the antecedents of a phenomenon are manipulated or controlled and their effects are measured. Hypotheses investigated in classical experimental science postulate regularities among different repetitions of the same event. A test condition C is inferred from the hypothesis to predict what should happen if C is performed (and the hypothesis is true). In many of these experiments, C is held constant (repeated) while other experimental conditions are varied. Experimentation does not stop with a successful experiment. Scientists continue to fiddle with introducing variations in the conditions while repeating C. They also try removing C while holding conditions constant. When one considers that vast number of additional conditions (known and unknown) that might affect the outcome of an experiment independently of the truth of the hypothesis, all three of these activities make good sense.

The same should be true in the case of archaeology (Dibble and Rezek 2009, Domínguez-Rodrigo 2008, Hopkins 2013, Schiffer 2013, Seetah 2008, Griffitts 2011). However, here the meaning of the word "experiment" should change a little. Archaeologists need observed situations in which a causal relationship is present. Our only chances are by rigorous experimentation or through "controlled" observation. In the first case, the cause is replicated in laboratory conditions in order to generate the material effect as the result of a single action, all other actions being controlled. An obvious example is modern use-wear analysis (Grace 1996, Odell 2001, Stemp 2013, Van Gijn 2013, Hurcombe 2014). By replicating lithic tools and using them over a determined period of time performing some activity—i.e., cutting fresh wood—we will be able to test the relationship between kinematics, worked material and observed use-wear on the surface of the tool. It is the archaeologist who makes the tool and who performs the activity. In this way, the material consequences of cutting fresh wood can be make explicit, and used to discriminate other activity also performed by the archaeologist, for instance, cutting dry bone. Provenience studies, where the raw material source is empirically known, would be another example of proper experimental design in archaeology.

Regrettably, not all social activities performed in the past can be replicated in the present. We cannot replicate human groups, social reproduction processes or coercive actions, among many others. What cannot be replicated, in many occasions can be observed. Ethnoarchaeology has been defined as the observation in the present of actions that *were probably performed* in the past. That is to say, it is not a source of analogies, because the observed action is not like the action to be inferred, but a source for hypothesis (Roux 2007, Cunningham 2009, Gavua 2012, Gifford-González 2010, Schiffer 2013). In this book, Rogers et al. (Chapter 23) and Barceló et al. (Chapter 25) use ethnoarchaeological data for testing purposes.

If we use modern ethnographic data to directly explain by analogy transfer ancient archaeological evidences, we are forgetting change, and social dynamics. This is a serious mistake. However, ethnographic and historically preserved ancient written sources can be used as observational situations in which some causal events took place and were described. They are instances of the general event we want to learn. Remember that our task is to find perceptual properties that are coherent across *different realizations* of the causal process. Therefore, the basic problem is distinguishing the social invariants from historical and contextual variability. Here the invariance is a predicate about social action that assigns probabilities (including 1 or 0) to observable outcomes in a series (the more exhaustive possible) of historical situations.

In any case, historical predictability critically depends on the ability to accommodate variability in the social domain. If the observed cases for a social event are highly variable with respect to some well-defined features, then predictions whose strength is extreme, and which provide evidence for the causing social action, will acquire more strength than they would if the experimental replications or the observed cases in well-described historical situations were less variable. As a result, archaeologically observed evidence with an extreme strength will provide more evidence for its hypothesized cause (reflected in higher support level) in the former case. The greater the overlap in the features of the material evidences generated by other causal processes, the more difficult it is to generalize positive instances of the relationship. The number of alternative causes also affects the quality of the generalization. Both category overlap and the number of categories will contribute directly to the degree of competition between alternative possible categorizations of instances (Barceló 2009).

What can we do when we do not have experimental data or ethnoarchaeological analogies, but just the quantitatively measured list of unearthed artifacts and ancient garbage? When there is no prior causal knowledge to oversee the generalization process, machine learning textbooks refer to a different way of acquiring knowledge, the so called *unsupervised* or *self-organized learning methods*. In such an unsupervised or self-organized task, the goal is to identify clusters of patterns that are similar, and using them to generate potential generalizations. The learner agent will suppose there is a structure to the input space such that certain patterns occur more often than others do, and it would look for what generally happens and what does not. That is to say, a set of explanations will be modeled by first describing a set of prototypes, then describing the objects using these prototypical descriptions. The prototype descriptions are chosen so that the information required to describe objects in the class is greatly reduced because they are "close" to the prototype. This information reduction arises because only the differences between the observed and expected (prototypical) values need to be described. This approach is popular within statistics: Principal Component Analysis, Cluster Analysis, etc., are good examples (Djindjian 1991, Shennan 1997, Baxter 1994, 2003, Drennan 2010, VanPool and Leonard 2010). Such methods are based on some distance measure. Each object is represented as an ordered set (vector) of features. "Similar" objects are those that have nearly the same values for different features. Thus, one would like to group samples to minimize intra-cluster distances while maximizing inter-cluster distances, subject to the constraints on the number of clusters that can be formed. Another approach to unsupervised learning, beyond classical statistical procedures are *vector quantization* methods, a general term used to describe the process of dividing up space into several connected regions, using spatial neighborhood as an analogue of similarity (Delicado 1998, Schilkopf and Smola 2001, Hamel 2011, Hastie et al. 2011, Murphy 2012). Every point in the input space belongs to one of these regions, and it is mapped to the corresponding nearest vector. For example, the attributes for "object *A*" are mapped to a particular output unit or region, such that it yields the highest result value and is associated with that object, while the attributes for "object *B*", etc. are mapped to different regions.

The trouble with unsupervised learning is that we are not discovering how to instantiate a specific input-output function. Whereas supervised learning involves learning some mapping between input and output patterns, unsupervised learning can be viewed as describing a mapping between input patterns and themselves—i.e., input and output are identical. Once the archaeologist is aware of the statistical regularities of the input data, he or she may develop the ability to form internal representations for encoding features of the input and thereby to create new categories automatically. Prediction ability is dictated solely by the statistical properties of the set of inputs, and the kind of the statistical rule that governs clustering.

This distinction between supervised and unsupervised or self-organized learning leads us directly to the concepts of Classification and Clustering. Both mechanisms can be defined as the partition of a series of observations according to the similarity criterion and generating class descriptions from these partitions. The classification problem is just like the supervised learning problem: known cases illustrate what sort of object belongs in which class. The goal in a classification problem is to develop an algorithm which will assign any artifact, represented by a vector x, to one of c classes (chronology, function, provenance, etc.), that is, to find the best mapping from the input patterns (descriptive features) to the desired response (classes). That explains why classification and prediction are frequently interrelated. A prediction of an historical event is equivalent to a classification within a given set of events. A prediction of flint knives implies a distinction between longitudinal and transversal use-wear traces, for instance. Conversely, a classification of flint tools also means a prediction of their past function. There are exceptions, of course, where such a relation does not exist. Unsupervised learning, that is, clustering, does not have to assume that the clusters it develops at each stage are meaningful in themselves, just that they are going to be useful in deriving a yet-more-accurate cluster at the next step. A new object would then be categorized as an *A* or *B* as a function of the average distance from the central tendency of each of the dimensions underlying category *A* and category *B*, and the dispersion of each of the dimensions around their central tendencies. In addition, knowledge of the dispersion of *A* and *B* can be used to decide whether a novel instance is so unlikely to belong to either known category that a new category concept should be formed to accommodate it. Therefore, the acquisition of explanatory knowledge cannot be reduced to clustering, because such methods are limited by the natural grouping of the input data, and they are based on restricting knowledge production to finding regularities in the input. If we want to go beyond the usual archaeological template matching, we should make emphasis on learning and categorizing, and on how meaning can be generalized from known examples of a given concept. It is the non-linear and adaptive way of learning what allows for the formation of discriminative classes. A computational inductive approach should share three characteristics with explanatory concepts formed by humans: (a) the boundaries of learned classes should be fuzzy in that no single feature is required to distinguish one class from another, (b) the formation of learned classes should be path-dependent in that the final properties of the class vary with the details of the learner's selection history, and (c) the cumulative effect of selection should cause the mechanical system to function *as if* a general category has been formed. In this way, the explanatory knowledge we learnt is not a store of associations constituted as a large database that can be explicitly addressed (Typology), but concepts (input-output functions) created from a variety of experiences. There is no limit to the complexity or non-linearity of the relationship.

The discussion is well covered in this book in the Chapters by Nicolucci et al. (Chapter 3), Read (Chapter 4), Bordoni (Chapter 5), Mucha et al. (Chapter 9), Esquivel et al. (Chapter 10), O'Brien et al. (Chapter 11), Richards et al. (Chapter 12), Nicolucci and Hermon (Chapter 13) and Ducke (Chapter 18). Archaeologists have been doing clustering to achieve some kind of classification for years, instead of a real conceptual learning. It is important to understand the difference between clustering and classification (Baxter 2006, Bevan et al. 2014, Hörr et al. 2014, Östborn and Gerding 2014, Gansell et al. 2014, Kovarovic et al. 2011, Papaodysseus 2012). A good classification should both impose structure and reveal the structure already present within the data. With the exception of data reduction tasks, classification techniques are generally favored in analysis since the user retains control over the classes imposed. The outcome from a clustering of a set of data may have little meaning since the resulting clusters are not associated (by design) with any concept arising from the domain of study (although they may be because of inherent structure in the data).

Dwight Read (Chapter 4) analyzes clustering techniques (notably cluster analysis) from the perspective of why it does not work for the goal of determining culturally salient artifact types. More precisely, neither weighting of variables by themselves, nor weighting of cases by themselves, is adequate and one essentially needs to simultaneously weight by cases and variables, but to do that we would need to know the types in advance! The assumption of numerical taxonomy, which has been carried forward to cluster analysis, that increasing the number of variables so as to provide more information about the entities that are to be grouped through a clustering algorithm, does not converge on the inherent data structure that the cluster analysis is supposed to recover. Instead, we get divergence when variables are included that are not sensitive to the inherent data structure, but we don't know in advance which variables these might be.

Mucha et al. (Chapter 9) also advocate for clustering in a bottom up approach. They suggest how we can begin our analysis with a single variable that best finds a cluster structure in the dataset, then adding an additional variable, and so on. Read also points out the need not only for weighting variables, but simultaneously weighting variables and subdividing the dataset according to its cluster structure as the latter is revealed, and even before doing that, subdividing the dataset by qualitative criteria that have cultural relevance.

Esquivel et al. (Chapter 10) discuss available methods of numerical classification, and consider that discriminant function analysis (DFA) has important use limitations, especially in archaeology. The first limitation is clearly shown by the sensitivity of this technique regarding the composition of the sample. This can prove problematic in archaeological samples composed of a low observation number, which makes them especially sensitive to outliers. Another problem archaeologists must face is the fragmentary character of archaeological observables. For the above reasons, it is important for archaeology to be equipped with classification tools that allow us to work with a low or moderate number of observations while enabling the use of the fewest possible number of variables and/or parameters. Among the techniques of numerical classification, the authors present the Lubischew test, which has remained forgotten for decades. They characterize the use of the Lubischew test on archaeological samples to determine the conditions under which this method is most effective, the type of data that can provide the best results and, therefore, contribute to the resolution of archaeological problems where the objective is to characterize dichotomous samples from single variables in many fields of Archaeology, specifically when we only know the number of observations, average and standard deviation.

Most archaeological classifications are restricted to shape measurements. The purpose is to find functional regularities in the shape differences and similarities. The same can be true in the case of texture measurements, like in classic use-wear analysis. However, when considering compositional data, inductive methods and classification techniques should be based on different mathematical assumptions. Martin et al. (Chapter 8) and Mucha et al. (Chapter 9) deal with this subject. The special characteristic of compositional data means that the variables involved in the study occur in constrained space defined by the simplex, a restricted part of a mathematical space, implying dangers that may befall the analyst who attempts to interpret correlations between ratios whose numerators and denominators contain common parts. It is important for archaeologists to be aware that the usual multivariate statistical techniques are not applicable to constrained data (Aitchison 1986, Pawlowsky-Glahn and Buccianti 2011, van den Boogaart and Tolosana Delgado 2013). Martin et al. (Chapter 9) suggest the use of Log-ratio analysis to solve these problems. According to them, the statistical analysis of compositional data should be governed by two main principles: scale invariance and subcompositional coherence. To satisfy these principles a particular geometry in S^D is required. This geometry is based on three basic elements: two operations, perturbation and powering, and an inner product. These elements provide an Euclidean structure for the simplex, called *Aitchison (simplicial) geometry*. With these tools at hand one can exploit the following *Euclidean* elements, among others: angles, norm and distance, orthonormal coordinate representation, and orthogonal projections. Consequently, one can apply properly any multivariate method to analyze such data: Principal Component Analysis, Cluster Analysis, Discriminant Analysis, etc.

Mucha et al. (Chapter 10) take a more usual approach to understanding compositional variability in terms of differentiated classes of artifacts. They look for sub-populations (clusters, groups) of compositional vectors which can be related to different proveniences of pottery or other artifacts (in terms of raw material of a differentiated workmanship). They compare hierarchical clustering and adaptive methods of clustering to evaluate the number of possible differentiated groups of compositions. A logarithmic transformation is applied in order to handle both the scaling problem and the existence of outliers. Authors make the very important remark that cluster analysis usually presents always clusters—even in the case of no structure in the data. Moreover, hierarchical clustering presents nice dendrograms in any case containing all the clusters that are established during the agglomeration or the division process. A very important part of this chapter deals with the methods of cluster validation.

Although statistical reasoning is still giving its support to all these inductive approaches to classification and clustering, we are not restricted to classical statistical inference. O'Brien et al. (Chapter 11) suggest that the history of cultural changes is recorded in the similarities and differences in characters (attributes of phenomena) as they are modified over time by subsequent additions, losses, and transformations. Therefore they use similarity measurements not to simply "classify" or "cluster" archaeological assemblages, but to create phylogenetic trees that shows the kinds of changes that occur over generations of, say, spear-point or ceramic-vessel manufacture that usually arise through modification of existing structures and functions. Authors describe the basics of the cladistic method, focusing first on distinguishing between homologous and analogous characters and, in the case of the former, distinguishing between derived and ancestral characters. They explain how phylogenetic trees are different from standard classification or clustering, dividing the analytic process into four steps: (1) generating a character-state matrix; (2) establishing the direction of evolutionary change in character states; (3) constructing branching diagrams of taxa; and (4) generating an ensemble tree.

Nicolucci and Hermon (Chapter 13) take the discussion even further and consider the need of "fuzzy classifications", that is categories (explanations) that although emerged from an inductive reasoning are characterized in terms of a continuum of truth values ranging from 0 (false) to 1 (true), and assuming intermediate values for statements on which the value is 'Maybe true'. In these cases the truth value corresponds to the subjective degree of belief in the membership of an instance to a category or "type". Such an approach, besides reflecting closer the archaeological reality, also provides a method to determine and quantify qualitative matters of data under investigation. The authors describe a 'reliability index' which is assigned to each classification by the researcher itself. Such an index reflects the level of confidence of the research in the assignment of a particular item to a typological class. At assemblage level, the reliability index gives an indication to the 'quality' of the classification process, i.e., how well artifacts fit into selected categories.

Consequently, more than the classical view of an archaeological typology as a classification or clustering, I advocate the analogy with the idea of an *associative memory*. This is a device storing not only associations among individual perceptual representations, but organizing "conceptual" information not directly derived from the senses. Richards et al. (Chapter 12) develop this idea.

Computer scientists are intensively exploring this subject and there are many new mechanisms and technologies for knowledge expansion through iterative and recursive revision. Artificial Intelligence offers us powerful methods and techniques to bring about this inferential task. Modern computational inductive algorithms avoid some of these drawbacks, specially the sensitivity to noise, by not performing an exhaustive search, but a heuristic one. Heuristic approaches to logical concept induction carry out a partial search through the description space, using an evaluation function that measures the fit to the training data, and selects the best hypotheses at each level. The resulting algorithm is more robust in the presence of noise and target concepts that violate the conjunctive assumption. Fuzzy logic, rough sets, genetic algorithms, neural networks and Bayesian networks are among the directions we have to explore.

Richards et al. (Chapter 12) discuss some of these new methods of discrimination and clustering to classify archaeological texts, and to build semantic spaces. They present a probabilistic framework based upon the assumption that there is an 'ideal answer set' which contains exactly the relevant category for a given classification. Knowing the description and attributes of such an 'ideal answer set' will lead us to successful results. The authors' develop the idea of induction in terms of extracting several types of information from a corpus of over 1000 unstructured archaeological grey literature reports. The project employs a combination of a rule-based (KE) and an

Automatic Training (AT) approach. The rule-based approach was applied to information that matched simple patterns, or occurred in regular contexts. The AT approach was applied to information that occurred in irregular contexts and could not be captured by simple rules, such as place names, temporal information, event dates, and subjects. In addition, both approaches were combined to identify report title, creator, publisher, publication dates and publisher contacts. The authors also explore rule-based techniques that combined different types of entities in a meaningful way, while retaining the semantics of each entity in the ontological output, what can be seen as complementary to the broader classifications of the preliminary method.

We can use neural networks as a non-linear fitting mechanism to find form-and-function regularities in a set of experimental data or ethnographical observations (Baxter 2006, Deravignone and Macchi 2006, Barceló 2009, 2010b, Di Ludovico and Pieri 2011, Lu et al. 2013, Negre 2014). An Artificial Neural Network (ANN) is an information processing paradigm that is inspired by the way biological nervous systems, such as the brain, process information. In analogy to the way chemical transmitters transport signals across neurons in the human brain, a mathematical function (learning algorithm) controls the transport of numerical values through the connections of the artificial neural network. If the strength of the signal arriving in a neuron exceeds a certain threshold value, the neuron will itself become active and "fired", i.e., pass on the signal through its outgoing connection. The power of neural computation comes from the massive interconnection among the artificial neurons, which share the load of the overall processing task, and from the adaptive nature of the parameters (weights) that interconnect the neurons. The nodes of the network are either input variables, computational elements, or output variables. It is the pattern of interconnections which is represented mathematically as a weighted, directed graph in which the vertices or nodes represent basic computing elements (neurons), the links or edges represent the connections between elements, the weights represent the strengths of these connections, and the directions establish the flow of information and more specifically define inputs and outputs of nodes and of the network. Each node's activation is based on the activations of the nodes that have connections directed at it, and the weights on those connections.

The role of neural networks is to provide general parameterized non-linear mappings between a set of input variables and a set of output variables. The neural network builds discriminant functions from its neurons or processing elements. The network topology determines the number and shape of the different classifiers. The shapes of the discriminant functions change with the topology, so the networks may be considered semi parametric classifiers.

There are two modes in neural information processing:

- Training mode,
- Using mode.

During training, the network "learns" a non-linear function associating outputs (expected functions) with input patterns (known instances, experimentally based of distinct objects having performed a particular function). The network modifies selectively its parameters so that application of a set of inputs produces the desired (or at least consistent) set of outputs. Essential to this learning process is the repeated presentation of the input-output patterns. Training is accomplished by sequentially applying input vectors, while adjusting network weights according to a predetermined procedure. The more data are presented to the network, the better its performance will be. As the network is trained, the weights of the system are continually adjusted to incrementally reduce the difference between the output of the system and the desired response. Once trained, a network's response becomes insensitive to minor variations in its input. Then, when the network is used, it identifies the input pattern and tries to output the associated output pattern. In the using mode, the presentation of an input sample should trigger the generation of a specific prototype. It is clear that a single prototype represents a wide range of quite different possible inputs: it represents the extended family of relevant features that collectively unite the relevant class of stimuli into a single category. Any member of that diverse class of stimuli will activate the entire prototype. In addition, any other input stimulus that is *similar* to the members of that class, in part or completely, will activate a pattern that is fairly close to the prototype. Consequently, a prototype vector activated by any given visual stimulus will exemplify the accumulated interactions with all the possible sources of the same or similar stimuli in proportion to the frequency with which they have been experienced. A prototype as formed within a neural network is by definition "general", in the same sense in which a property is general: it has many instances, and it can represent a wide range of diverse examples. However, this property does not mean that prototypes are universal generalizations. No prototype feature needs to be universal, or even nearly universal, to all examples in the class. Furthermore, prototypes allow us a welcome degree of looseness precluded by the strict logic of the universal quantifier: not all Fs need to be Gs, but the standard or normal ones are, and the non-standard ones must be related by a relevant similarity relationship to these that properly are G.

Deravignone et al. (Chapter 17) offer an application of neural network techniques to the solution of a classical archaeological problem, and they develop the idea that classification should be considered within a prediction framework. I will comment further on this investigation when considering the relevance of spatial data analysis in archaeology.

The mathematics of similarity and inductive inferences are one of the main domains of application of mathematics in archaeology. It is so relevant to this domain that in many cases "mathematics" in archaeology is confounded with "statistics". It should be no surprise then that in this book statistical applications are predominant. However, although classificatory and clustering approaches seem quite appropriate in the case of archaeological research, we should take into account that inductive techniques that rely 'only on the input' are of limited utility. Explanation is an inference process whose very nature is beyond a mere mapping out of the statistical correlation present in the descriptive features of material evidences. Rather they involve identifying and disentangling the *relationships* that exists in the data, utilizing available knowledge about the process that generated those effects. Nicolucci et al. (Chapter 3) Bordoni (Chapter 5) and Richards et al. (Chapter 12) deal with the idea of semantic relationships beyond mere visual resemblance.

Beyond Literal Similarity
The Mathematics of Archaeological Space and Time

Just aggregating over many observations is not enough to learn an appropriate explanation, because of the large input space and the ambiguities implicit in it.

Consequently, induction can be dangerous as a general explanatory mechanism because it leaps from a few experiences to general rules of explanation. If the experiences happen to be untypical, or the archaeologist misidentifies the relevant conditions, predicted behavior may be permanently warped. We can be victims of self-reinforcing phobias or obsessions, instilled by a few experiences. To avoid these difficulties, we should go beyond explicit commonalities between observed inputs and consider also implicit relationships between them.

We say that two objects may be "implicitly" related when they are not similar, but there is something connecting them: the paper and the pencil are very dissimilar, but they are related when we use them to write a letter. For instance, consider an arrow point and a human skeleton. These are objects with different shape, size, texture and material. But there is something *relating* them: they have appeared at the same location: in a burial, the arrow point was found inside one of the bones. Both elements are related because they constitute part of the same event, the death of this individual. Bordoni (Chapter 5) offers an exhaustive presentation of mathematical tools to describe such relationships (networks and graphs). The use of graphs and networks is necessary in order to create models, or simplified representation systems of networks of relationship. Over the past decade *network* has become a "buzzword" in many disciplines across the Humanities and Sciences. Researchers in archaeology and history, in particular, are increasingly exploring network-based theory and methodologies drawn from complex network models as a means of understanding dynamic social relationships in the past, as well as technical relationships in their data.

The number of potential implicit relationships in a given scenario is generally unbounded, implying that the number of possible relational regularities is infinite. Given the fact that *everything may be related with everything*, this is, in principle, an infinitely hard operation. Does it mean that relational learning is out of the archaeologist range? To identify and disentangle the non-explicit *relationships*, we should use available knowledge about the processes that generated those effects, because they are not always apparent. If learning proceeds constructively on the identification of certain relationships, then those relationships presumably need to be explicitly identified.

"Space" and "Time" are among the most interesting implicit relationships to be investigated. Nicolucci et al. (Chapter 3) insist in the fact that all archaeological entities are embedded in space-time, and the connection with their spatio-temporal location is a key passage in archaeological interpretation. Clarifying every detail of this process and the related concepts in light of knowledge organization theory is fundamental to enable a better use of accumulated data and to organize archaeological knowledge on a sound basis. Last decades have witnessed a proliferation and diversification of theoretical discussions about time, space and space-time and its impact on archaeological interpretation (Murray 2004, Bailey 2005, 2007, Buck and Millard 2004, Lock and Molyneaux 2006, Lucas 2008, Holdaway and Wandsnider 2008, Barceló 2008, Conolly and Lake 2006, Kamermans et al. 2009, Verhagen 2013, Bevan and Lake 2013, Alberti 2014, among many others). Many chapters of this book are dedicated to this kind of relational explanation. The reader is referred to Nicolucci and Hermon (Chapter 13), and Carver (Chapter 15), for an interesting definition of the basic concepts of duration and time. Bronk Ramsey (Chapter 14) explains in detail the mathematics of time measurement in archaeology. Deravignone et al. (Chapter 17), Ducke (Chapter 18), and Negre (Chapter 19) cover different aspects of spatial analysis. Crema (Chapter 16) deals with space *and* time.

However, "space", "time" and "space-time" have not yet been formally defined in Archaeology. The notion of space is ambiguous in current speaking, but also between different scientific disciplines. We can refer to "abstract" spaces, "physical" spaces, and "social" spaces or even to "archaeological" spaces. Abstract space is governed by the principles of mathematical logic. Physical space is the consequence of the localization of objects in the real world and the time I need to reach them from the point of observation. Movement is what imposes "distances" among observables. Social space is a framework in which the entities are social agents which carry on different activities and social actions of production, consumption, distribution and reproduction. Archaeological space deals with the localizations of archaeological remains which are the material evidences of past social actions.

A similar ambiguity is usual for defining the concept of *Time*. What we learn from this debate is that time does not exist as an autonomous physical entity, which can be observed, described and measured. What exists is the evidence of change. We understand by it, a characteristic of a concrete event that defines how the quality of the event has changed from state O_1 to state O_2 at two different places E_1 and E_2, and at two different moments of time T_1 and T_2. The operational concept that can help us integrating both kinds of relationships is the idea of *distance*. It has been defined as the difference between the values of any property at two (or more) spatial/temporal locations (Gatrell 1983). The main organizing principle for understanding how space and time are related is expressed in Tobler's law, according to which "everything is related to everything else, but near things are more related than distant things" (Tobler 1970). Here we should take the idea of "distance" both in its spatial and temporal sense. This principle constitutes a key-concept in order to explain the archaeological record.

Spatial and temporal location is characteristically quantitative information. Instruments for measuring spatial coordinates are well known and widely used in archaeology (GPS, Theodolite, Electronic distance meter (EDM) and other electronic surveying tools); the measure of time is also one of the main achievements of archaeology in the last 60 years, through the dating of samples based on the measurements of various types of isotopes (^{14}C is the most used measure for events having occurred in the last 40000 years). When available, absolute dating can be used as a strong indicator of the time-span of the archaeological context where the analyzed object has been found, and extended to all the other objects found in its immediate spatial neighborhood and with the same sedimentological formation process (i.e., taphonomy; see Carver, Chapter 15). Nevertheless, as suggested by Nicolucci and Hermon (Chapter 13) and Bronk Ramsey (Chapter 14), absolute dating leaves ample margins of intrinsic uncertainty related to the technique used, and a coarse time granularity implied by assimilating all the objects in a given context to the same time-span. Furthermore, current measures of archaeological time attach a value to a context (a layer, an occupation floor, a depositional unit, etc.) and not a time-span, as suggested by Nicolucci and Hermon. For this purpose a 'relative dating' based on the topological ordering of depositional units within a stratigraphic trajectory is more useful, as it places archaeological observables and their micro-spatial context 'before' or 'after' other distinct time slices. So dating often implies coarsening the time granularity to make time-span assignments feasible or adopting approximations.

Nicolucci and Hermon (Chapter 13) introduce the formal definition of *chronological relation* between events, of *dating functions* as a mapping *d* of *E* on the set *I* of intervals of real numbers, associating to an event *e* an interval (a, b), the time-span of *e*, i.e., the set of the real numbers (here, times) during which *e* is happening, and the *duration of an event* is another mapping *f* from *E* to, which assigns a real number to an event. The duration measures the time-span of the event. Bronk Ramsey (Chapter 14) develops those ideas suggesting the need to integrate all sources of information on chronology. This includes stratigraphic information, information on deposition processes, isotope frequency and other constraints, which can be applied relative to the passage of calendar time. According to him, Bayesian statistics provide the ideal way to combine the information from these different sources and it has been highly successful in improving the precision and accuracy of archaeological chronologies. That said, the aim of such mathematical models is to provide a degree of impartiality between absolute date estimates and relative date estimates; both of these allow us to understand processes of change. Chronological models therefore aim to be unbiased in the estimates of age, and duration. What Bronk Ramsey means by unbiased is critical in formulating our models and is discussed in depth in his contribution.

An additional trouble comes from the fact that although space and time must be considered different dimensions of the same aspect, it is usual to analyze spatial variability within particular fixed chunks of time called *periods*. As explained by Nicolucci and Hermon (Chapter 13), assigning archaeological observables to predetermined time-frames requires clustering them with some inevitable approximation and deciding if they belong, or do not belong, to a concrete time-frame or another, despite the sometimes problematic assignment of an observation within a crisp category. The time-frames that we choose to define are a product of our understanding of the human activity reflected in the archaeological record. Furthermore, due to the fact that events extend in time, there may be overlaps which complicate the temporal sequence and the number of possible relations that may exist between two events: the first ends before the second begins; they have some overlap; one starts later and ends earlier than the other, and so on (Nicolucci and Hermon, Chapter 13, Bronk Ramsey, Chapter 14, Carver, Chapter 15). If a dating function exists for all the events considered, i.e., if it is known, for every event, 'when' it starts and 'when' it ends, the mutual chronological relation between two events may be represented by the relations between the corresponding time-spans, the intervals assigned by this omnipotent dating function to each event. Nicolucci and Hermon (Chapter 13) suggest that it may correspond to the Allen algebra of time intervals as defined in (Allen 1983). Allen algebra deals with time intervals, with instants considered as time intervals of atomic granularity. It defines thirteen possible relations between time intervals, in fact seven, i.e., equality and six mutual relations with their inverses. Given two time intervals *A* and *B*, the possible relations between them are: *A equals B*; *A before B*, i.e., the end of *A* occurs before the beginning of *B*; *A meets B*, i.e., the end of *A* corresponds to the beginning of *B*; *A overlaps B*, i.e., the end of *A* occurs after the beginning, but before the end, of *B*; *A during B*, i.e., *A* is fully contained in *B*, the beginning of *A* being after the beginning it *B*, and the end of *A* being before the end of *B*; *A starts B*, i.e., the beginning if *A* is the same of *B*, and the end of *A* is before the end of *B*; and finally *A ends B*, i.e., the beginning of *A* is after the beginning of *B* and the two ends of *A* and *B* coincide. In archaeology, this kind of temporal reasoning is based on original ideas by E. Harris (1975, 1979, 1989) and it has been developed recently by Cattani and Fiorini 2004, Desachy 2008, Herzog 2002, 2004, 2014, Holst 2001, 2004, Traxler and Neubauer 2008).

Unfortunately, Allen's model is only an abstraction of reality that does not fit with archaeological use. Actually, no such thing as a universal dating function exists: or, at least, to make it practical it is necessary to deal with the granularity of time and consider for our reference time a discrete set instead of a continuous one. In other words, when measuring time in archaeology it is impossible to consider it as continuous flow. Consequently, two events may appear as contemporary even if they start, or end, with a difference in time so small that the clock cannot measure it, as it is easily experienced in everyday life. In this book, two different solutions to this problem are discussed. On the one hand, Nicolucci and Hermon (Chapter 13) build "non-crisp" versions of time-frames or periods, based on rough sets and fuzzy logic. When applied to dating, a fuzzy set X is defined using a membership function f_X from time (modeled as the real line) to the interval [0, 1]. The value of $f_X(t)$ represents, in a scale from 0 to 1, the likelihood of the event X taking place at time t. The extreme values correspond to the traditional truth value: when $f_X(t) = 0$, the event X does not exist; when $f_X(t) = 1$, the event X does exist. The membership function may assume any shape, but for convention it is assumed to be continuous (events are not intermittent) and zero outside some—possibly large—interval (no eternity). On the other hand, Enrico Crema (Chapter 16) and Christopher Bronk Ramsey (Chapter 14) adopt a more probabilistic approach. Crema discuss the limits imposed by temporal uncertainty and suggests a solution based on probabilistic reasoning and Monte-Carlo simulation. A case study from prehistoric Japan will then illustrate an example of this approach. Bronk Ramsey (Chapter 14) introduces Bayesian chronological analysis to express temporal information in probabilistic terms. The author uses the Bayes theorem to relate a set of parameters t expressing information about time to a series of observations y that refer to measurements on isotope ratios, for example, and thus the likelihood terms are the way we relate these observations to the age parameters in the model; radiocarbon dates would thus be expressed as likelihoods in an overall scheme.

The way to reconstruct a temporal ordering (historical trajectory) and its specific metric is of the topmost relevance in archaeology. Not only because we are interested in positioning past events in space *and* time, but because temporal variation should be used to approach the very idea of change and evolution. Michael Leyton (1992) has argued that a trajectory of changes (a history) can be described as a discontinuous sequence composed of a minimal set of distinguishable actions. The key idea is that what appears to be different in the present speaks about some action in the past that generated such a difference. Variability in the present is understood as having arisen from variability in the formation processes. In an archaeological dataset with no variability, nor any differences among its elements, the best hypothesis is that the corresponding causal process has the least amount of variation. If a property is invariant (unchanged) under an action, then one cannot infer from the property that the action has taken place. We cannot explain the history of water in a lake, because water is spatially and temporally undifferentiated. However, if we can distinguish topographic variation along the basin lake perimeter, we can follow the geological transformation of this landscape. Therefore, we may regard the complexity of a temporal trajectory of visually apparent differences as a measure of the amount of change needed to produce perceivable variation.

O'Brien et al. (Chapter 11) argue that if visually similar character states are ordered correctly, a historical sequence of forms is created, although independent evidence is needed to root the sequence—that is, to determine which end of the sequence is older.

Such evidence could come, for example, from chronological dating, stratigraphic topology or historical sources. Over the past several years, archaeologists have grown to appreciate that the methods biologists have developed to reconstruct the evolutionary, or phylogenetic, relationships of species can help them create not only historical sequences of artifact forms—what came before or after what—but sequences based on heritable continuity—what produced what. As with DNA, the history of cultural changes is recorded in the temporal ordering of similarities and differences in characters (attributes of phenomena) as they are modified over time by subsequent additions, losses, and transformations.

But the effect of time is not always teleological. Although any temporal trajectory has an implicit directivity, the "arrow of time" is not necessary linear. Bronk Ramsey (Chapter 14), Carver (Chapter 15) and Crema (Chapter 16) develop the non-linearity of time. Nevertheless, we can still use perceived and quantified "variation" in shape, size, texture, material and spatio-temporal location values to "run time backwards" and explaining the mechanism that may have generated those differences. Distinctions between successive stages of an archaeological trajectory of changes and modifications point to a past event where variation did not exist; sufficiently far back, no difference existed. Archaeologists may explain a trajectory of changes by imposing a temporal *slicing* on archaeologically perceived *discontinuities* (Barceló et al. 2006). The intention for such a slicing will be to represent visually the transitions between events. In this way, we would reproduce the past occurrences of the events in an historical sequence. Such a trajectory of events would be "explanatory" because the same occurrence of an event within the trajectory, and its spatio-temporal relationship with the preceding and successive event would serve as the *explanandum* of what happened in the past.

Besides temporal discontinuities, we need to distinguish *spatial discontinuities*. This assumption leads us to the methods of classification and clustering referred before. Of course, it is not a class of "visually" resembling objects what we should build, but a class of objects with a similar location in space *and* time that is, aggregated patterns of locations with similar *distances*. The question that arises is whether the spatial and temporal coordinates displayed any systematic pattern or departure from randomness. Questions of interest to archaeologists include:

- Is the observed neighborhood due mainly to natural background, topography, for instance?
- Over what spatial scale does any aggregation or concentration occur?
- Are concentrations merely a result of some obvious a priori heterogeneity in the region studied?
- Are they associated with proximity to other specific features of interest, such the location of some other social action or possible point sources of important resources?
- Are observables that aggregate in space also clustered in time?

The key idea is that effective discontinuities in the spatial probabilities of a social action often coincide with important limits in causal process having modified physical space. Archaeological space, as any other kind of socially modified physical space, will be understood when interfacial discontinuities between successive events can be detected. Its study is then a matter of reporting at what spatial locations a change in the observable frequency of some archaeological feature leads to a change in the probability of its causing action or process.

Ben Ducke (Chapter 18) looks for local concentrations of archaeological sites in terms of point sets in archaeological distribution maps. Through statistical clustering techniques "hot spots" of social activity in the past can be discovered. The author focuses on a few selected methods for spatial cluster detection, examining how far current mathematical approaches have evolved since the comparative study by Blankholm (1991). Practical advice on the potentials and limits of each discussed method is given, as well as a general outline for exploring the clustering properties of spatial data. Joan Negre (Chapter 19) argues that many usual methods of estimating aggregation in space are dependent on a wrong assumption: the homogeneous behavior of the Space. The surface of the Earth is irregular; therefore, Euclidean distances are the wrong way to quantify spatial relationships. When we consider cost-weighted distances among the location of archaeological sites, based on topographic irregularity and/or the influence of social attraction/repulsion points or areas, the Euclidean assumptions become misleading and can misguide the clustering of such "hot spots" of social activity in the past. The movements of people and the transfer of goods and ideas present anisotropy and directionality, from a social point of view. For example, it is assumed that is not as difficult to travel between these two points when they are in a flat and firm area as when the path must pass through a very hilly mountainous area. Under geometric anisotropy, the variance of any spatial process is the same in all directions, but the strength of the spatial autocorrelation is not. The author proposes to address this issue presenting a non-Euclidean adaptation of the well-known Ripley's K-function in order to implement a more reliable tool for detecting deviations from spatial homogeneity.

Enrico Crema (Chapter 16) offers a spatio-temporal equivalent of this measure of spatial homogeneity and defines spatial heterogeneity in the temporal variation of residential density in terms of a variant of the Spatio-Temporal Join-Count Statistic. The rationale of the method can be summarized as follows. For a given transition from one event to *another*, we first compute the number of spatial locations that are occupied by at least one residential unit in both temporal time-frames. We then permute the occupied locations of both periods for n times, and generate a distribution of expected number of joins given a null hypothesis of random settlement locations.

Michael Merrill, in his contribution to this volume (Chapter 20) looks for structural discontinuities by investigating the nature of spatially cohesive assemblages in terms of lattice theory. Lattice theory in mathematics is the study of partially ordered sets of objects (called lattices) in which any two elements of the lattice have a least upper bound and a greatest lower bound. This contribution is related with the more general approach adopted by Bordoni in Chapter 5. Merrill suggests that lattices that are useful for discovering and representing a broad range of patterning in spatially cohesive sets of artifacts are lattices whose operations are provided by set union and intersection. These kinds of lattices are often called distributive lattices. He uses a specific kind of distributive lattice, called a Galois lattice, to identify spatially cohesive sets of artifacts in an archaeological case study from southern California. When used to depict the complete spatial structure of artifacts detected by an appropriate combination of statistical and graph theoretical methods, these kinds of lattices also reveal overlapping and hierarchical structure within and among these sets that may otherwise not be easily observed. Revealing hierarchical and overlapping structure in and among spatially cohesive sets of artifacts is necessary for interpreting fundamental properties of the artifacts in these sets, such as their degree of specialization and the variability of their use over space and time.

We take for granted that it was social action what varied across *space* and *time*. It was social action that produced the precise location and accumulation of material evidences at specific places and specific moments, and the local circumstances at the moment of the action. For instance, human settlements may be tied together by proximity (a measure of distance), but the causal mechanism is not physical proximity. The real cause should be explained in terms of the "influence" an action performed at a location has over all locations in the proximity and the future use of the same spot. From this assumption we cannot deduce that social actions are necessary adapted to the environment but productive actions (hunting, fishing, and gathering) determine the location of residential actions (settlement). The suggestion is that a social action can generate the reproduction of similar actions around it, or it can prevent any other similar action in the same vicinity. Consequently spatio-temporal analysis implies trying to solve the following problem:

WHY OBSERVED MATERIAL ENTITIES WITH THOSE SPECIFIC VALUES OF SIZE, SHAPE, TEXTURE, OR MATERIAL, HAVE BEEN ARCHAEOLOGICALLY OBSERVED *HERE* AND WERE DEPOSITED *THEN*?

Clearly, nothing is gained if the archaeologist introduces the archaeological evidence that some *y* occurred *there* and *then* as an explanation of why some *x* occurred *there* and *then*, an indicator that some *y* occurred (where *x* and *y* refer to different acts, events or processes). Such spatio-temporal descriptions, even if quantified, are not explanations but are themselves something to be explained. Spatio-temporal regularities do not explain, but require explanation by appeal to the activities of individual entities and collections of entities.

In a way, solving the spatial and temporal problem implies that we are trying to discover whether what happens (or *happened*) in one location has been the cause of what happens (or happened or *will* happen) at neighboring locations (Barceló 2002). Social actions are performed in specific locations because of their position relative to another location used for any another action or a reproduction of the same action. Consequently, actions performed at one location may have some relationship with actions performed at a neighbor location. For instance, when an action is performed *here*, it may increase the probability of some posterior action and decrease the probability of some others. Therefore, if the location of action influences, conditions or determines the location of other actions, we can calculate a general model of spatial dependence. For instance, humans have always decided the location of their settlements according to many reasons, but more specifically we make "placement" decisions based on the social strategies for resources management in the area around the settlement. This means a sophisticated effort balance between social needs, available techniques and resources.

To sum up, this is the domain of application for a spatio-temporal analysis in archaeology: to infer the location (in space and time) of what cannot be seen based on observed things that are causally related to the action to be explained. Knowing where someone made something based on what she did, is an inverse problem with multiple solutions, which can be solved using some of the methods and technologies already presented. A remaining question is how the location differences among the effects of cause C have determined or conditioned the differences among the effects of cause B. This property has also been called *location inertia*: it is a time-lag effect that activities experience in the adjustment to new spatial influences (Wheeler et al. 1998). In other words, changes in the probability of a social action at some location and at some distance of the place of another action should determine changes in the probability value of the spatial variability of material effects (archaeological record), not only of the same action, but also of other actions performed at the same place at different times. The spatio-temporal variability of such an "influence" is what we call the dynamics of the socially configured relational space.

Luca Deravignone et al. (Chapter 17) can be read as a specific example of this formalization of the archaeological spatial problem. The basic assumption of their approach is that archaeological sites may be conceived as entities located in a territory, characterized by the variables of the place where they are located. This is especially true for settlement sites since the reasons involved in the choice of one place instead of another are usually far from being random. The variables can be environmentally, socially or economically oriented, or related to many other aspects. In the investigation, such variables are used as input features of a neural network based predictive model, and presence/absence of an archaeological site as the final output. We can conceive the output of the net as a prediction model for that settlement, which is important not only for using it to produce predictive maps, but also for the analysis of the settlement pattern itself. In this way what is common in the environmental, social or economic characteristics of some sites can be used to predict the reasons for human settlement as evidenced by the archaeological record. Moreover, this method allows visualizing the evolution of settlement systems by observing and comparing results diachronically in time-line or even synchronically in different territories. The choice of variables to be included initially included altitude, slope, aspect, linear distance and cost distance from rivers, sea, and dioceses. The purpose was to reach a wider description and understanding using variables that represent in a much comprehensive way the quality and features of each archaeological site.

A related way of predicting the spatio-temporal location of unknown observations based on a sample of spatio-temporally positioned archaeological data can be obtained through interpolation methods. Given the location of archaeological evidences, the purpose of any spatial interpolation method is to predict the value of the dependent variable at any imaginable spatio-temporal coordinate. The result is an interpolated manifold (usually a surface), which can be understood as a probabilistic map for the placement of social actions performed in the past. In such a map, nearer things appear to be more related than distant things (Tobler's law), because the synchronicity of social actions states that, all else being equal, activities that occur at the same time will tend to increase the joint frequency of their effects. In the same sense, elements that are located within some interfacial boundary are spatially related, and configure a *common region*. For example, suppose that we know the presence or absence of archaeological evidences for settlement at different places at different historical periods. Then spatial-temporal interpolation would estimate the probability of human settlement at unsampled locations and times. Ideally, instead of predicting for each location (x,y) a single value $f'(x,y)$ or this value plus a variance, it would be interesting to estimate the probability density of the value $f(x,y)$. Given a sample of observed frequencies at known locations, what the archaeologist needs to generalize is a non-linear function that represents the probability density function of the place and time the social action was performed. In a simpler way, we are trying to calculate the probability density function of finding archaeological evidences $Pr[f(x,y) \geq th]$ for a fixed critical threshold *th* and for any location (x,y) of any archaeological evidence. Obviously, the ability to interpolate accurately depends ultimately on the availability of data commensurate with the particular target scale of output.

Interpolation and related methods only maximize available information from often sparsely distributed data. As such, it does not solve the archaeological problem of inferring the original location of the social action, nor why the materials are where they are, but gives us some information about the spatial modality of the social action based on the spread and density of its materials effects. Discontinuity detection is essentially the operation of detecting significant local changes among spatially sampled values of some physical properties. A good rule of thumb is to divide the data array into components at *maximal concavities*, which mathematically speaking, are the local minima of curvature. Formally, such a discontinuity in the spatial probabilities of the social action is defined as an observable edge in the first derivative of the mathematical function that describes the archaeological frequencies over space. This task can be approached by calculating the *spatial gradient* in the data array—that is, the direction of maximum rate of change of the perceived size of the dependent values, and a scalar measurement of this rate. This spatial gradient describes the modification of the density and the size of archaeologically measured values and so regularity patterns in spatial variation can be determined. It is calculated by finding the position of maximum slope in its intensity function (a graph of the value of the dependent variable as a function of space). Thus, the intensity profile of spatial frequencies can be graphed as a curve in which the *x*-axis is the spatial dimension and the *y* axis corresponds to the dependent variable (for instance, the quantity of some archaeological material at each sampled location).

Spatio-temporal gradients constitute a tool for measuring and visualizing expansive historical phenomena in space *and* time. The idea of *expansion* is usually correlated with the notion of directivity. In dynamical terms, we may explain the presence of some degree of directivity in spatio-temporal data in terms of movement, and hence of "expansion", in the mathematical and physical senses of the word. In physics, expansion refers to a system in which a gradient of a scalar field can be detected. In fact, three types of gradient should be detected to consider the *expansivity* of a phenomenon, a spatial gradient, a temporal gradient and a spatio-temporal one. They are closely connected; we cannot consider the spatial gradient without the time dimension within it. The variation in space or in time of any quantity can be represented graphically by a slope.

Expansive phenomena in historical research have been traditionally related with the movement of people through space: invasions, migrations, colonization, and conquests what gives the appearance of an expanding population of men and women moving through space (and time). In recent times, however, expansive phenomena in historical research are not limited to the assumption of population movement but can imply also the movements of goods and/or ideas. Therefore, "historical expansions" are not always a consequence of movement of people (a demic diffusion) but can be caused also by phenomena of cultural diffusion (acculturation) dealing with the "migration" of ideas, knowledge or goods. As soon as time passes, farther places begin to use previously unknown goods or ideas, increasing the distance between the place where the good or idea appeared for the first time, and the place where it is used anew.

The notion of *spatial diffusion through time* covers all processes that contribute to the changes in the location of some phenomenon and the reactive effects generated by those changes. The more complex the diffused or transmitted innovation, the more influence its diffusion process will have on transformation of its propagation environment, as effects induced by its adoption will be all the more increased. Therefore, the expansive nature of the historical phenomenon should be analyzed as an increase in the spatial distance between social agents resulting from some transformation in social ties (social fission), or a growth in the absolute number of agents. Contraction would be the reverse process; for instance, a decrease in distance between social agents as a result of an increase in social ties (social aggregation, social fusion). It brings about the intrinsic dynamic nature of the phenomenon, which refers to the idea of spatial change in a determined period of time (Barceló et al. 2014a).

Many attempts have been made to model the diffusion dynamics of expansive phenomena in particular by geographers, epidemiologists, demographists and botanists, but also by archaeologists and historians. Early results were obtained using diffusion or difference equation models (reviewed in Okubo and Levin 2001). A variety of other classes of models have subsequently been studied (e.g., individual-based models), showing that rates of expansion can be either linear or accelerating and that movement through space and time can be smooth or patchy depending on assumptions about individual movements, demography, adaptation and environmental structure (reviewed in Hasting et al. 2005). The goal is to analyze where, when and why the chronology of the first occurrence of an event "varies from one location to another". In other words:

- how the spatial distribution of the values of some property *depends* (or "has an influence") over the spatial distribution of other(s) value(s) or properties,
- how the temporal displacement of the values of some property *depends* (or "has an influence") over the spatial distribution of other(s) value(s) or properties,
- how the temporal displacement of the values of some property *depends* (or "has an influence") over the temporal displacement of other(s) value(s) or properties,
- how the spatial distribution of the values of some property *depends* (or "has an influence") over the temporal displacement of other(s) value(s) or properties.

The classic study of spatio-temporal dynamics in Archaeology is Ammerman and Cavalli Sforza's paper "Measuring the rate of spread of early farming in Europe" (Ammerman and Cavalli-Sforza 1971). Analyzing a wide dataset of georeferenced radiocarbon dates the authors suggested a model of demic diffusion to understand the sudden apparition of early farmers at different moments and at different places according to a relatively regular gradient. According to their model from a point of origin located in the area of Jericho, in the Middle East, the agriculture would have expanded to Eastern and the North-Eastern territories through several waves of advance. The authors calculated a constant isotropic expansion rate of 1km/year. The main cause for explaining such a movement traditionally was traced in an episode of demographic growth that would have led to an excessive stress on the available resources. Therefore, this increase in the demographic pressure would have produced a sort of migration toward territories with a lower degree of exploitation. Modern developments of such an approach do not equate exactly demic diffusion with migration (Ammerman and Cavalli-Sforza 1984, Gkiasta et al. 2003, Russell 2004, Pinhasi et al. 2005, Dolukhanov et al. 2005, Bocquet-Appel et al. 2009, Isern et al. 2012, Isern et al. 2014).

In addition to tracking expansive phenomena from georeferenced databases of radiocarbon estimates, archaeologists have recently begun to estimate temporal variations in activity and investigate the probabilities of local population growth rates and subsequent dynamics (including episodes of rapid depopulation, sometimes related to extreme climatic events). A number of recent studies have addressed this problem by graphing summed calibrated probability distributions (SCPDs) for all the radiocarbon dates in their datasets. The peaks and troughs in these SCPDs, where these are sufficiently robust to be more than mere artifacts of the inflections of the calibration curve, are interpreted as evidence of population fluctuations. The proxy data on population numbers provided by radiocarbon dating can be combined with estimates of fertility and migration in the construction of colonization models. There are many applications of those models (Gkiasta et al. 2003, Fort et al. 2004, Gamble et al. 2005, Mellars 2006, Shennan and Edinborough 2007, Hamilton and Buchanan 2007, Collard et al. 2010, Hinz et al. 2012, Shennan et al. 2013, Capuzzo 2014). The rationale of the method assumes that the number of dated archaeological contexts in a given time period can be expected to be monotonically related to population size. Consequently observed peaks in the summed calibrated probability distribution are taken as evidence of higher populations and troughs as evidence of lower populations, with the steepness of the slope of an increase or decrease showing the rapidity of the population rise or fall. It would obviously be possible to examine patterns like these mathematically, but archaeological practice has generally been simply to examine probability distributions visually. Nevertheless, there is no method without shortcomings. Chiverrell et al. (2011) warn of the fact that georeferenced radiocarbon databases incorporate multiple types of dated contexts with differing chronological relationships between the ^{14}C measurements and the dated events, with pre-dating, dating or post-dating chronological control, each displaying variable length temporal lags all mixed together in the same analysis. Ambiguities inherent in the process of radiocarbon dating can thus both create patterns in sets of radiocarbon dates that mimic those that archaeologists often interpret as evidence of long-term demographic trends and obscure real demographic patterns. Distributions of radiocarbon dates are potentially affected by sample selection and by a range of factors that condition the probability that archaeological remains are deposited, survive, and are then discovered (Chamberlain 2009, Baggaley et al. 2012). A note of caution relating to the uncritical use of temporal distributions of radiocarbon dates has been raised by Hazelwood and Steele (2004) who have argued, with the aid of a simple model of population expansion, that such a pattern will only survive in the modern archaeological record when some rather narrow conditions are met for the demographic parameters that determined the original population expansion. Surovell and Brantingham (2007) have also pointed out that a monotonic increase in the frequency of dates through time can be generated by a systematic taphonomic bias if (as may often be the case) the probability of archaeological site survival is negatively correlated with the age of the site. Furthermore, large numbers of dates from individual sites might skew the overall data-set. More serious, however, is the tendency of archaeologists to obtain disproportionate numbers of dates from particular site types at the expense of other less distinctive types (Armit et al. 2013). Williams (2012) has recently summarized the latest developments of the method, pointed out different problems and proposed strategies to handle them. In our study we tried to deal with them in different ways according to his suggestions.

Carsten Lemmen (Chapter 21) uses this approach to test archaeological and climatological hypotheses about the reciprocal relationships between society and nature, and to quantify these interactions. The purpose is to simulate the dynamics of population density and three population-averaged characteristic sociocultural traits x = $\{T_A, T_B, C\}$: technology T_A, share of agro pastoral activities C, and economic diversity T_B. The author uses Gradient adaptive dynamics to describe the co-evolutionary development of population density p and growth-rate influencing traits x from an aggregated community-average perspective over time t. These simulations help to (1) interpolate and extrapolate the archaeological record; (2) test archaeological hypotheses about endogenous and exogenous factors governing the transition to agriculture; and (3) provide consistent estimates of past regional and global population, as well as land use and carbon emissions.

Fabio Silva and James Steele (Chapter 22), present an alternative approach to the study of dispersal episodes in human prehistory based on level set methods, in which fronts are tracked on a discrete lattice by assigning to each point the time at which the front reaches it. To model heterogeneous surfaces, grid points on the two-dimensional domain can be assigned friction terms—effectively reducing or boosting the dispersal speed locally. The distribution of friction terms can be estimated from any empirical or semi-empirical ecological, geographical or cultural variable. These methods can be applied to cases of multiple competing fronts originating at different spatial and temporal locations, as well as to the classic single origin, uniform wave speed case. Such an approach can be used to test alternative dispersal hypotheses using archaeological data. The authors illustrate this from their own work on the spread of farming in Neolithic Europe and in the Bantu-speaking regions of sub-Saharan Africa.

The Mathematics of Accumulative Process

Archaeologists usually draw their inferences about past behavior from dense, spatio-temporally discrete aggregations of artifacts, bones, features, and debris. We have traditionally assumed that the main agent responsible for creating such aggregates was *only* human behavior. Even though nowadays most archaeologists are aware of natural disturbance process and the complexities of archaeological formation processes, archaeological assemblages are still usually viewed as a *deposit* or aggregate of items.

When we are in face of a spatio-temporally determined accumulation of archaeological observables it seems obvious to think in quantitative terms, and hence of the apparent "intensity", "importance" or "abundance" of the actions that generated such accumulation in the past. Depositional events are then usually referred as accumulation or aggregation episodes, in which the probability that a social action occurs is related to its dimensions. In other words, the more frequent the refuse materials at a specific place (location), the higher the probability that a social action was performed in the vicinity of that place. For example, in archaeozoological studies, the largest number of bones is often interpreted in terms of a larger number of preys captured. In archaeobotanical research, a greater number of preserved seeds, charcoal fragments or pollen evidence of some taxa than others is often interpreted in terms of the degree of dominance of these taxa in the prehistoric environment, or in terms of key economic interest (or symbolic) to human populations of the most abundant. For Archaeometrical studies, the greatest abundance of artifacts made from some type of raw material is usually interpreted in terms of a frequency of artifacts taken from the place where the raw material was more abundant. Trade routes and even commercial systems have been reconstructed based on quantitative variation of artifacts with different physicochemical compositions from diverse geographical features.

In general, the intuition of archaeologists led to the belief that if the abundance of similar archaeological observations is different in different places, there is a chance that in the past the intensity or frequency of the action was greater in certain places or at certain times. This characterization of archaeological formation processes in terms of *accumulation* has led to the idea of a direct relationship among the population size at a site, the site occupation time span, and the amount of material discarded by its inhabitants, as if the number of artifacts increase directly with a settlement's occupation span (see discussion in Varien and Mills 1997, Varien and Ortman 2005). In this volume we review such assumptions and propose a formal framework to investigate the quantitative variation of archaeological events in space and in time.

Once we define the aggregate as the unit of analysis, the scale of measuring changes. To describe the variability between different aggregates we should refer to *frequencies*, but archaeological data are recorded as *counts*. Lindsey (1995) defined the term *count* as the enumeration of evidence, that is, the number of distinct observations. Frequency is an estimation not only of the number of times the event was repeated in the past and produced the observed *count* but also an estimation of the length of the time interval over which the event was repeated (see Nicolucci and Hermon, Chapter 13; Carver, Chapter 15; see also Premo 2014). Hence if archaeologists record five vessels of a particular type in four different graves situated at different spatial locations, their deposition may be related to one single event (the simultaneous burial of 5 individuals), or to five repetitions of a single event (a burial every year, for instance). Only in the case these five events occurred independently in different space-time locations (or at the same location, but the archaeologist has sedimentary or stratigraphic information to distinguish between five successive depositions), we can calculate the frequency of the event in the past. What will give us an idea of the social organization in the past is the frequency of the event, and not necessary the difference in the absolute number of vessels in different locations. In case of burials, it is relatively easy to differentiate discrete events, then, the number of *k* independent tombs with pottery vessels decorated with the same pattern of geometric motifs can be considered a measure of the frequency with which the funerary ritual took place in the past, and we can compare these values with those from other cemeteries to have an idea of social differences. However, the observation of 5 vessels at an occupation layer, and 15 in another, cannot be compared because the occupation layer does not represent a single event. Frequencies *per time and/or spatial unit* can be compared, *counts* from different sampled units cannot be compared and its spatial and temporal variation cannot be explained in functional or causal terms (see discussion in Carver, Chapter 15). A count of 1350 flint tools may seem very high, but the accumulation frequency in the past might be very low considering that this archaeological context has a minimum duration of 300 years during which the material consequences of many independent actions of carving could have accumulated. The frequency is calculated by dividing the count by the length of the time interval; in this case 1350/300 = 4.5 tools per year. Similarly, the archaeological finding of 71 bovid left humeri in a spatiotemporal context constitutes a mere count. To estimate the frequency with which the historical event "herding cows" took place, we must count the number of animals that were produced and/or consumed during the period of time the archaeological deposit was formed. If 15 years were needed for the formation of what the archaeologist has distinguished as a depositional unit, the process of formation had a frequency of 4.7 animals per year. It is this frequency that we need to investigate the spatio-temporal variation of that event.

In any case, to infer the cause (social action performed in the past) from the effect (the frequency of material evidences observed in the present—the archaeological site—), we have to rebuild the real frequency that was generated in the past by the social action. The major problem is whether the absolute number of accumulated items can be attributed to the number of times the social action was repeated in the past. Nevertheless, most of the times, it is impossible to give a simple expression to characterize the spatio-temporal modality of the social (or natural) process. We should take into account that the spatial trend contains both the process that generated the original frequencies *prima facie*, and all post-depositional process that altered the original values. What happened at the specific location after the social action was performed—post-depositional processes—have the effect of disordering the artifact patterning in the archaeological record, and increasing entropy. Loss, discard, reuse, decay, and archaeological recovery are numbered among the diverse formation processes that in a sense, mediate between the past behaviors of interest and their surviving traces. They make archaeological assemblages more amorphous, lower in artifact density, more homogeneous in their internal density, less distinct in their boundaries, and more similar (or at least skewed) in composition. Furthermore, some post-depositional disturbance process may increase the degree of patterning of artifact disturbances, but towards natural arrangements. Carver (Chapter 15) and Crema (Chapter 16) develop this problem.

There is a considerable source of error when describing social action in the past in terms of accumulation processes, because in many cases different parts of the material record may be lost at varying rates. The consequences are the dispersal or differential accumulation of items and the selection of some kinds of things and not others, leading to biases in assemblage composition. Consequently, determining whether the various frequencies of items in an assemblage or deposit have resulted from differential distribution, differential preservation, or both is the problem. In faunal analysis, the prediction that original and disturbed archaeological deposits will vary in element spatial distribution is based on the premise that in paleontological settings, carcasses tend to be deposited whole and elements undergo limited post-depositional dispersal. Elements representing a single individual are likely to occur together.

Regardless of how much evidence is present, the archaeologist cannot read the frequency of past events directly from the counts in the archaeological record. Interpreting the content and frequency of an archaeological assemblage must be grounded in an understanding of both the social and natural events that have influenced the presence/absence, alteration, and displacement (relative to it as a primary site of production, use or discard) of its individual components and of the assemblage as a whole (Hassan 1987).

Cowgill (1970) proposed a preliminary solution: we have to recognize three basic populations (in the statistical sense):

1) events in the past,
2) material consequences created and deposited by those events, and
3) artifacts that remain and are found by the archaeologist ("physical finds").

By stressing the discontinuities, Cowgill asserted the nature of archaeological formation process as agents of bias within a sampling framework. Each population is a potentially biased sample drawn from the previous population that was itself a potentially biased sample. We may view these discontinuities as sampling biases in the sense that what we recover and observe does not proportionately represent each aspect of the antecedent behavior.

At the beginning, material remains are organized in the archaeological record in a way coherent with the social practices that generated it. Once the site of social action was abandoned, those remains are subjected to bio-geologic forces, which introduce a new material organization. This new patterning of social material remains is opposite to the original pattern, and consequently increases entropy (des-organization, chaos, ambiguity), until the original patterning become unrecognizable. Carr (1984, p. 114) developed Cowgill's idea, by distinguishing activity sets from depositional sets. The tool and debris types that are repeatedly found together in the archaeological record may be termed *depositional set*. In contrast, the sets of tools used repeatedly in the past to perform a particular task may be called *activity set*. The distinction of activity sets from depositional sets is necessary because they may differ internally in their defining attributes and organization, and externally in their relations with entities. An area in which several kinds of tools and debris cluster together does not necessarily correspond to an "activity area".

Therefore, changes and transformations in the original patterning of activity sets are not a simple accumulation process from low entropy sets (identity between depositional and activity set) to higher entropy patterns (disturbed deposits), but a non-linear sum of quantitative changes, which beyond a threshold, produce a qualitative transformation. A depositional set may be thought of as a mathematical set, the organization of which is the end product of structural transformations operating upon a previously structured set. In this sense, the occurrence of specific formation process is determined by specific causative variables. That means, that depending on the degree of entropy, the transformed archaeological set is not necessary a random sample of the original population. That is, the difference between a depositional and an activity set is based on a deep discontinuity generated from the irregular addition of minor quantitative modifications (Estévez 2000).

During the past decades, several researchers have emphasized the problems involved in identifying the effects of formation processes. As suggested by Carver (Chapter 15), the theory behind is that formation processes can be inferred in principle because they have predictable physical effects which are related to the laws that govern them. Furthermore, the archaeological record has in large part been shaped by the same processes (in particular those which involve the post-depositional phase) that have modeled the landscape. A general model for understanding the frequency of observables in the archaeological record as result of the accumulation of discard objects after use has been suggested:

$$T_D = \frac{St}{L}$$

Where:

T_D = total number of artifacts discarded
S = number of artifacts normally in use

t = total period of use of the artifact type (expressed in units of time, such as months or years)

L = use-life of the artifact (expressed in the same units of time as t).

This is the "discard equation" or Cook's Law (Schiffer 1987, 2010). It allows estimating systemic quantities such as the number of cooking pots in a site at a given point in time and, from these data, population size or occupation span. The largest source of error in applications of Cook's Law is usually in estimating T_D from archaeological samples (Sullivan 2008, Schiffer 2010). This can be made in terms of a standard Poisson, or a non-uniform Poisson process (Orton 1982, De Meo et al. 2000, Bevan and Conolly 2009, Crema et al. 2010, Achino and Capuzzo, forthcoming).

Suppose some action of production or consumption was repeated in the past numerous times, and at each repetition certain material evidence specific to that action was materialized. In those events where the action was not performed, the archaeological observable did not materialize. We refer here to events in the past (hunting, manufacturing of ceramics, jewelry or weapons, building a house, etc.) in which goods or refuse material was generated as a result of human activity. Let E denote the event in which the action took place and something was generated (a shape, a texture feature, a particular mineralogical composition, etc.). A represents the material evidence, so that the specific event denoted E_A represents the action that produced that characteristic evidence. What we are looking for is then the probability of this event ($P(E_A)$). Events in which only two outcomes are possible ("the action took place"/"action did not take place", "presence of material evidence"/"material evidence is absent") are known as *Bernoulli events*, and its distribution is a *Binomial distribution.*

Doran and Hodson (1975, see also Buck et al. 1996, Banning 2000) showed some examples of the basic application of the binomial distribution to explain its main features; they considered that this distribution can adequately describe some archaeological issues, such as:

a) the process by which vessels of two different types are discarded into a rubbish pit

b) the process which caused the occupation or not of locations suitable for settlement in a region

c) determining the proportion among blades with one-sided retouch (with retouch on the left or on the right side).

It has been highlighted that for particular combinations of parameter values the binomial closely approximates to other standard distributions; for example, as said above, if n is very large and p correspondingly small, then the binomial distribution approximates more closely to the Poisson distribution.

Nevertheless the kind of *binomial* distribution (and its extension, the *hypergeometric* distribution) that such description seems to assume is impractical in archeology, because it would force us to work with fixed sets of events whose number must be known in advance. Unfortunately, in archeology, we do not always know the total number of relatives who should have died in a single family during Bronze Age, or the number of swords that a certain artisan could have produced all along his/her life. What we would like to find out is the probability of occurrence of a certain event within a continuous framework (knowing the rate of occurrence per unit weight, volume, time, etc.), even without knowing the total number of repetitions of the action in the past. The assumption is different: once we know the number of artifacts an artisan was able to produce per year, for instance, what is the probability that the archaeologists discover x artifacts in an archaeological depot that needed 30 years to accumulate? Here we do not know the number of repetitions in the past, but the frequency of occurrence per unit time (or space).

In the book, Bronk Ramsey (Chapter 14) and Carver (Chapter 15) develop the apparent Stochasticity of archaeological accumulation processes. Carver suggests the use of Markov models. A discrete-time Markov chain is a mathematical system that undergoes transitions from one state to another on a state space. It is a random process usually characterized as memory less: the next layer of archaeological deposition is assumed to depend only on the current state and not on the sequence of previous depositions that preceded it. This specific kind of "memorylessness" is called the Markov property. Bronk Ramsey (Chapter 14) takes the opposite approach, that is, the case in which each event is not independent, and hence influenced by the knowledge of previous depositions event, there are some situations where this condition is not met. When depositional units exhibit a stratigraphic relationship, we can infer the temporal topology between events, and this additional layer of knowledge can considerably improve the accuracy of the explanation.

Consequently, when

- we are aware of the rate at which the value of the property appeared in the past,
- the probability that a simple event occuring in a short interval was proportional to the spatial extension or temporal duration of that interval,
- the probability that an event occurring in an brief interval was independent of the events that occurred outside that interval,
- the probability of more than one event in a sufficiently small interval is negligible (not simultaneous events occur).

We can explain the observed accumulation in terms of a continuous-time counting process $\{N(t), t \geq 0\}$ that possesses the following properties:

- $N(0) = 0$
- Independent increments (the numbers of occurrences counted in disjoint intervals are independent from each other)
- Stationary increments (the probability distribution of the number of occurrences counted in any time interval only depends on the length of the interval)
- No counted occurrences are simultaneous.

Consequences of this definition include:

- The probability distribution of all archaeological observables produced by the same and unique process can be a Poisson distribution.
- The probability distribution of the time of occurrence of a single archaeological event (that is, the time until the next occurrence of an archaeological observable of the same kind) is an exponential distribution.
- The occurrences of archaeological observables should be distributed uniformly on any interval of time. Note that $N(t)$, the total number of occurrences, has a Poisson distribution over $(0, t]$, whereas the location of an individual occurrence on $t \in (a, b]$ is uniform.

Thus, if the assumptions are met, in particular the independence of the events and the regularity of the expected number of occurrences per interval, we may predict the likelihood that something happened in the past: that someone hunted two deer, that an iron dagger was placed in a female grave, that five vessels were manufactured for a specific type with specific dimensions, etc.

Archaeological assemblages may be regarded as aggregates of individual elements, which interact with various agents of modification in statistical fashion; with considerable potential for variation in the traces they ultimately may show (Mameli et al. 2002). Therefore, the effects of depositional accumulations of material remains through time can be modeled using stochastic methods, which follow some random probability distribution or pattern, thereby allowing behavior to be analyzed statistically but not predicted precisely. In case we have *counts* of specific types of archaeological observables with similar physical properties, *and* we know the most probable frequency of their production in the past, it should be possible to infer the duration of the archaeological event. Alternatively, if we know the duration of the event and the total quantity of material evidence produced in that period, it would be possible to deduce how many people were working on that. In this sense, and based on the observed mean of counts of artifacts with the same quality, at different locations, and the number of locations where the action occurred, we may estimate the probability that the action produced the observed and counted effect as the ratio of the observed mean of counts and the number of locations.

The Mathematics of Social Models

History only runs once, but inside a computer a virtual model of the historical past would run infinite times. In the computer, we would explore (by altering the variables) the entire *possible* range of outcomes for different past behaviors (Barceló 2012, Cioffi-Rivilla 2014, Edmonds 2010, Epstein 2006, Macy and Willer 2002, Sokolowski and Banks 2009, Squazzoni 2012, Zacharias et al. 2008). The idea is then simulating inside a computer what we know about actions having been performed in the past and experimenting with the effects they may produce in such a virtual world. Archaeology appears then as a discipline dealing with *events* (Sewell 2005, Lucas 2008) instead of mere objects. That means that, inside a computer, the Past would be seen in the Present as a sequence of finite states of a temporal trajectory. Such a simulation would not "see" the past as it once was but as potentialities for action, that is, explanations that can take place when it encounters a situation of some sort.

The starting point of the explanation of past social systems by means of computer simulation is not the simulation of one particular system but the investigation of the mathematically possible development of specific classes of model systems (potentialities). This way of explaining what happened in the past requires the problem solver (human being or machine) to simulate in the present, perhaps in very sketchy terms, a *mechanism*, which, given the properties of the constituent components and of the environment, gives rise to the phenomena of interest. "Mechanisms are entities and activities organized such that they are productive of regular changes from start or set-up to finish or termination conditions" (Machamer et al. 2000, p. 3, see also Darden 2002). Obviously, the word "mechanism" is here a parable of how social intentions, goals and behaviors are causally connected. It should explain how social activity worked, rather than why the traits contributing to these activities or workings are there (Demeulenaere 2011, Galam 2012, Hedström and Swedberg 1998, Naldi et al. 2010). In this way, the building blocks for explanations in the social domain are products (people, goods, information), production (human labor, social action), and events (the context in which production took place) organized such that they are productive of some changes—regular or irregular—from start-up to finish or termination conditions. The termination conditions are the effects explained by an account of the workings of the social mechanisms producing them. Such explanatory mechanisms are more than static beginning and end-points. The stages are dynamically connected via intermediate operators. It is the ability of such operators to *produce* the subsequent changes in the social mechanism that keep the process going. The simulation happens when we execute this social mechanism in a controlled way. Running such a model simply amounts to instantiate agent populations, letting the agents interact, and monitoring what emerges. That is, executing the model—spinning it forward in time—is all that is necessary in order to "solve" it. Since the model is "solved" merely by executing it, there results an entire dynamical history of the process under study.

A computer simulation should allow us to understand archaeological observables in terms of a priori relationships between observed properties and the inferred properties/ abilities of people having generated those properties. It involves establishing and exploiting constraints (between the user/producer and the material evidence of his/her action, the user/producer and the natural environment, and the material evidence and the natural environment). For this sort of explanatory task to work, the archaeologist, as programmer, should know what precipitating conditions generate an increase in the probability of occurrence of an effect. Beyond a simple addition of individual random decisions, social activity should be defined in terms of social *dispositions* or *capacities* within a system of subjects, intentions, activities, actions and operations, some of them rational, others clearly indeterminate, impulsive or unconscious. The fact that the performance of some social action A, in circumstances T, has a probability P of causing a change Y in some entity N (social agent, community of social agents or the nature itself), is a property of the social action A (Barceló 2009).

In this way, the simulation can be understood as operating at different levels: the individual agent behavior, norms emergence and the institutionalization of some ways of interaction, what constitutes the historical event we want to explain. According to R. Conte (2009), it implies the ontological *dependence* of simulated institutionalized collective action on lower levels but endowed with autonomous *causal power.* All higher levels of reality (not only the macro-social) would be then ontologically dependent on the lower levels, in that they are implemented on lower levels until the very lowest: social order, political organization and the like are not ontologically autonomous, since each level of reality is implemented or realized on top of a lower level—for instance, labor, subsistence, reproduction, until the (so far) lowest, i.e., the biological constraints or the influence of environmental features (topography, resources, climate, etc.). At any given level of reality, entities may have the power to cause effects at the other levels. Hence, they are necessary for explaining the occurrence of these effects. To say that a higher level property is implemented on a lower-level mechanism, even a specific one, does not mean to deny causal autonomy to the higher levels.

There are many ways of simulating computationally social events in the past. From top-down approaches to bottom-up models, from equation systems to agent-based artificial societies, the dynamics of social mechanisms can be expressed in functional terms. Many chapters of this book can be considered as examples of social simulation. Martin et al. (Chapter 8) simulate the proportion of each component in a compositional dataset in terms of some probability distribution. Bronk Ramsey (Chapter 14) introduces stochastic simulations of accumulative processes, in the same way as Carver (Chapter 15) or Crema (Chapter 16) simulate the original state of a population of material effects. There are many kinds of social mechanisms, and nearly all historical phenomena can be analyzed in mechanical terms. What can be difficult is to formalize the Physics of the underlying processes, that is, the functions that relate some input to a simulated output.

Four contributions in this book can be considered as proper examples of social simulation. Carsten Lemmen (Chapter 21) simulates the foraging-farming transition that occurred during the last 10,000 years; there, the relationship between humans and their environment underwent a radical change from mobile and small groups of foraging people to sedentary extensive cultivators and on to high-density intensive agriculture modern society; these transitions fundamentally turned the formerly predominantly passive human user of the environment into an active component of the Earth system. The author assumes the adaptive co evolution of the food production system and population.

The close relationship which exists today between population density and food production system is the result of two long-existing processes of adaptation. On the one hand, population density has adapted to the natural conditions for food production (…); on the other hand, food supply systems have adapted to changes in population density (Boserup 1981, p. 15).

Fabio Silva and James Steele (Chapter 22) deal with the same historical subject, the transition from foraging to agriculture, and consider the dispersal of farming in Neolithic Europe as an innovation whose transmission would have followed the form of an advancing front. The authors estimate the mean speed of dispersal and the proportion best-fitting speeds of dispersal in different directions as a function of habitat, with coasts, rivers, and major ecoregions all being given individual values for their possible effects on rates of spread. Cost-distances from the origin to each dated site retained for our analysis are calculated using a specific algorithm that takes into account variable costs for traversing the various geographical features, with friction-cost parameter sets estimated both from exting scenarios in the literature and by systematically sampling the wider parameter space using a Genetic Algorithm.

J. Daniel Rogers et al. (Chapter 23), and Federico Cecconi et al. (Chapter 23), use Multi-Agent Systems to simulate prehistoric events. An Artificial Society is a synthetic representation of a society modeled as a Multi-Agent System (MAS) where autonomous agents imitate society's real actors as well as their interactions (Railsback and Grimm 2011, Helbing 2012, Rogers et al. Chapter 23). Agents are then computational objects with attitude. They should be implemented as discrete entities, autonomous, but with their own goals and behaviors and with their own thread of control. They should be capable to adapt and modify their behavior. A MAS consists of a set of autonomous software entities (the agents) having autonomy to act without direct human or other agents' intervention, thus taking their own decisions based on their goals and the state of the world in which they are. The goals of the agent must be inherent to the agent, rather than being assigned according to a pragmatic 'stance' of an observer. A goal-directed action is under an agent's control if (1) the goal normally comes about as the result of the agent's attempt to perform the action, (2) the goal does not normally come about except as the result of the agent's action, and (3) the agent could have not performed the action. Agents interact among them and with their environment, to solve problems that cannot be resolved individually. In any case the system is not directly modeled as a globally integrated entity. Systemic patterns emerge from the bottom up, coordinated not by centralized authorities or institutions (although these may exist as environmental constraints) but by local interactions among autonomous decision-makers. Emergence is the process of higher-level phenomena resulting from the interactions of lower-level rules, where the lower-level rules do not contain any direct description of the higher-level phenomena. This process is known as "self-organization".

As MAS explicitly attempts to model specific behaviors of specific *individuals*, it may be contrasted to *macro* simulation techniques that are typically based on mathematical models where the characteristics of a *population* are averaged together and the model attempts to simulate changes in these averaged characteristics for the whole population. Thus, in macro simulations, the set of individuals is viewed as a structure that can be characterized by a number of variables, whereas in micro simulations the structure is viewed as emergent from the interactions between the individuals. These models as well as real phenomena, for example, the societies, are dynamic because they change in time; therefore, a model will consist not only of structure but also of behavior. To observe a models' behavior it is necessary to know/find/measure the passage of time on it and it is here where computer simulation functionality is required.

An artificial historical society should have the ability to react to emergent properties, a trait called by some "second order emergence". First order emergent properties arise directly from interactions between parts of the system. Second order emergence occurs when the system reacts to its own first order emergent properties and often changes the structure of the system itself. How do the ways in which a complex system reacts to its own emergent properties influence the further emergent aspects of the system? By allowing parts of the system to react in different ways by forming different abstractions, can we influence what properties emerge and the manner in which they do so?

There are many modern applications of Agent-based modeling in Archaeology (Barceló et al. 2014b, Beekman and Baden 2005, Christiansen and Altaweel 2006, Costopoulos and Lake 2010, Doran and Palmer 1995, Kohler and Van der Leeuw 2007, Kohler et al. 2012, Kowarik 2012, Lake 2014, Madella et al. 2014, McGlade 2014, Wurzer et al. 2013). In this book, Rogers et al. (Chapter 23) implement an empirically calibrated, spatial agent-based model (ABM) as a tool for studying why and how wealth differentials and associated social inequalities are generated and sustained over multiple generations. Following standard research in ethnology, authors formalize the idea of "wealth" among nomadic populations based on the Gini coefficient, and the Pareto power-law of wealth distribution. The simulation is designed to replicate a broad range of behaviors pertaining to pastoralist households living in a kinship structured social and economic landscape referenced to that of Inner Asia during the Early Bronze Age. The model includes pastoralist households and their herds, situated in a biophysical landscape affected by weather. The landscape is endowed with biomass for herds to subsist. The overall dynamics are as follows: Weather affects biomass vegetation, which affects herding throughout the annual seasons, which, in turn, affects the movement of households across the steppe. In time, households' herds change in size and location as households congregate in camps as they undergo nomadic migrations in interaction with their herds. Three simulation experiments were conducted to address the central question—what are the factors that most affect wealth maintenance over the course of generations? Considering the length of time involved, the relative longevity of kinship lineages was the best measure of wealth maintenance.

Cecconi et al. (Chapter 24) study the phenomenon of the birth of proto-urban centers, on the threshold of the first millennium BC, in Central Italy. At this time, settlements increased their defensibility and the strict control of territory also increased). Previous villages were abandoned, and population concentrated on a few centers with the same "military" prerogatives, and at considerably higher dimensional scale than before. The settlements and village communities are represented by software agents, whose number grows in function of time. Each settlement controls an area of the surrounding territory, more or less wide depending on N, i.e., in function of the number of inhabitants. The sequence of operations affecting settlements takes place in "cycles" of two years. Starting at *2300* BC, every two years the villages grow according to available resources and their capacity of controlling the surrounding territory, and the possibility to store the food and shelter the cattle in places safe from any raids. Applying the simulation, only some of the social dynamics inferred from archaeological data are replicated. Among the results that differ from those expected, the authors mention the fast decrease of the total number of settlements. In fact, the system seems to undergo a collapse, rising from 250 villages starting in about sixty, and in [....] the following centuries remains roughly stable, with limited fluctuations. This diverges from traditional historical explanation: after an increase between the beginning of the Early Bronze Age and Middle Bronze Age, the total number settlement appears to be pretty stable up to the radical change in the First Iron Age, when the average size of occupied sites increased. This phenomenon is interpreted as the evidence of a gradual but steady population growth. At the same time, the total number of the population shows a slight decrease in the simulation; on closer inspection, one cannot exclude that the effect of war and conflict has led to temporary reductions in the number of persons in the community what coincides with the tendency to settle in positions of high defensive potential.

J.A. Barceló et al. (Chapter 25) is a different approach to modeling methods in archaeological theory. Authors use Bayesian Network Models to simulate the way hunter-gatherers took the decision to cooperate with other human groups, provided their respective "cultures" were similar enough. It is not an agent-based model, but the approach can be considered as the mechanism that an agent model could use to represent heuristic decisions based on an idea of bounded limited rationality, and where optimal economic decisions were constrained by cultural circumstances and social subjective evaluations. Here the rational for using an explicitly probabilistic approach comes from the incomplete nature of archaeological evidence. Is there any way of estimating the truth likeness of hypothesis about social action of the past, given the partial evidence provided by archaeological observations?

A Final Word Regarding "Truth" and "Likelihood"

Absolute truth is not to be found in any empirical science, most especially in historical disciplines. What distinguishes the Archaeology from purely laboratory sciences is its inability to know with near certainty the sequence of events that led to the complexity we see today in the present at the archaeological site. Although uncertainty has long been recognized as being important in archaeology, it is still a problematic issue. It is not always clear what the term means, how it should be discussed, used, and measured, what its implications are. In general, the assumption seems to be that our knowledge or dataset on the phenomena under discussion is in some way imperfect. In this book, Chapter 3 and 4, Chapter 13 to 17 and Chapter 25 deal explicitly with such topic.

Archaeological explanation is always uncertain because some indeterminacy may appear between actions of human activity and the visual and material properties of the preserved remains of such an activity. With only a partial knowledge of the past, historical science provides a plausible account of the present, but the observed present may not be the only outcome of hypothesized events and processes. Sometimes a social action happens, but the expected material consequence does not take place as expected. Other times, the entity we study does not seem to have experienced any perceptible change allowing the automated archaeologist to know if some social action or sequence of social actions had any causal influence. Even more,

- diverse actions can determine the existence of similar material effects
- the same action not always will determine the same material effects.

There is always room for individual variation within collective action. For instance, social agents who elaborate pottery to contain liquids will not always produce vessels with the same shape, size, material, and texture; neither use these containers in the same place nor waste them in the same way. Furthermore, the consequences of single action may have been altered by the consequences of successive actions, in such a way that effects may *seem* unrelated with causes.

The challenge is to derive a consistent mapping from a potentially infinite set of social actions through time to a relatively small number of observable outcomes in the present. As a result, causal explanations generated by archaeologists are necessarily as ambiguous as the stimulus (description/measurement of archaeologically observable features) they presumably encode. As suggested by Read in Chapter 4, "as ambiguous as the stimulus" can also refer to ambiguity introduced by measurement and statistical methods that are not in concordance with underlying structuring processes (the "double blind problem"). It is difficult to decide among the large number of possible explanations that could fit the variation of input data equally well, or the perceived ambiguity of archaeological observations. The problem is that an infinite number of possible visual features can correspond to a particular material effect of a social action, and even this material outcome can be related in many different ways with many causing social actions. There are millions of reasons why a knife is so long, why this wall is made of adobe and not of stones, why this decorative element was painted in red, etc. Determining whether the various frequencies of items in an assemblage or deposit have resulted from differential distribution, differential preservation, or both is the problem. The actual combination of processes that could have given rise to specific physical properties of the archaeological record is nearly infinite, and so one cannot expect to find many simple correspondences between lists of evidences and the characteristics of their formation processes. One can hardly argue that uniformitarian principles may be formulated concerning the social scope of human communities, given the profoundly varied nature of social action. In fact, even recent methodological advances provide little or no basis for connecting such inferences to other than non-human process or differential representation and damage morphology.

The fact that we cannot *predict* the precise material outcome of a single causal action, does not mean that an archaeological feature cannot be analyzed as caused by a series of social actions and altered by other series (or the same). While we may not be able to predict the precise outcome, when we have a good functional understanding of the causal action, we can make plausible predictions. We can often eliminate what will not occur, thus reducing the "design space" to a tractably small size. Thus, although we do not know what single actions have produced precise material consequences, we can relate the variability of observable features (shape, size, content, material, and texture) with the variability of social actions through time and space. Consequently, we should infer the variability of social action from the variability of the archaeological record, and we must infer social organization from the variability of inferred social actions.

Archaeologists, as most social scientist, should adopt a *probabilistic view* to explanation. The idea of defining necessary and sufficient conditions may be replaced with that of the probable properties for a member of a given hypothesis. A probabilistic view accounts for graded categorization, since the "better" members will be those exhibiting more of the characteristic properties. Instead of representing several concrete instances in memory, we judge category membership by degree of connection to an abstract model or *prototype*. By treating perceptual recognition as a form of probabilistic inference, various conclusions may be assigned subjective probabilities of correctness on the basis of given evidence. Archaeologically perceived evidences are not determined univocally by human labor; there is only certain probability that this specific material entity had been produced when a concrete action or series of actions have been performed, among many others. If and only if the perceived trace could not have produced in absence of that action (Probability = 0), then the automated archaeologist will be reasonably sure that the percept can be recognized as having been determined by that action. Consequently, probabilities can be used to infer the most appropriate solution to the problem at hand, which may not even be based on the highest probability. If we accept that archaeological recognition can never be made absolutely reliable, it is necessary to describe the goal of recognition as maximizing the probabilities of a correct identification, and also providing a confidence measure. On the other hand, where material objects are functionally constrained by the effectiveness of doing a task in the phenomenal domain, we have a much better chance of making these probabilities. Where the material objects are constrained by their "meaning" in a social context, then predictability is much more difficult (Dwight Read, personal communication).

In agreement with the most habitual definition of probability, we could affirm, then, that a causal event would exhibit some degree of regularity when the more characteristics are "frequent", and the fewer characteristics are "infrequent" in the known series of observed events. In the same way, we could define the "regularity" of the social action when the material elements used to produce and to reproduce the social group show the same shape, when they have the same size, when their material and its texture are similar, and when we found them in the same place. Associations are likely to be learned if they involve properties that are important by virtue of their relevance to the goals of the system. The propensity, inclination or tendency of certain states or events to appear together is, then, what we need to learn how the world is (Cartwright 2004).

Therefore, understanding those elements of the past that have been seen in the present assumes that the perceived strength of causes is directly stored in memory under the form of mental connections between the *potential cause* and the observed effect (Van

Overwalle and Van Rooy 1998). An automated archaeologist (a so-called "intelligent" robot) would only learn when perceived events violate their previous expectations, and it assumes that an increasing number of comparison cases with a similar outcome will cause an increase in the perceived influence of the context (Barceló 2009).

Computer simulation has allowed to us thinking about social activity in the past in terms of social *dispositions* or *capacities* within a system of subjects, intentions, activities, actions and operations, some of them rational, others clearly indeterminate, impulsive or unconscious. The fact that the performance of some social action A, in circumstances T, has a probability P of causing a change Y in some entity N (social agent, community of social agents or the nature itself), is a property of the social action A. It is a measurement of the intensity of the propensity, tendency, or inclination of certain events to appear in determined causal circumstances (Popper 1957, Bunge 1959, Cartwright 2004, Salmon 1989, Eells 1991, Rivadulla 1993). Thus, the primary *explanandum* of archaeological theory is social capacities: the capacity to work, to produce, to exchange, to interact, to obey, to impose something or someone.

This approach is based on the idea that the probability of an event is the inversely proportional relation between the occurrence of the effect and the capacity of the cause to produce the effect (see Barceló et al. Chapter 25). Then the probability would not be more than a measurement of the regularity; that is, of the frequency whereupon the performance of a social action is associated with a material outcome. More specifically, the probability is usually identified with the limit towards which it tends the relative frequency of the effect as long as the cause has acted. If in a finite number of cases in which cause C was present the automated archaeologist has observed some effect E with a relative frequency h_n, it can be postulated that in a greater number of cases not yet observed, the frequency of observation of the effect will tend to a value limit around h_n. Therefore, when the number of observed cases increases, the probability that the next case be effect of the most probable cause will converge towards the relative frequency of cases with the characteristics produced by that cause and not by another one. What allows the automated archaeologist to assure that C causes Y in circumstances T is not the increase of probability of Y with C in T, but the fact that in T some Cs regularly cause E. This leads to the homogeneity assumption; that is to refinement of T to T^* so that the likelihood of C regularly causing E is greater in T^* than in T. As I mention in my own contribution—Chapter 25—(or at least I think I mention it!) this is done implicitly by all archaeologists, otherwise we would simply make a single, giant, undiferentiated database consisting of all universal observations. All archaeologists implicity (and sometimes explicitly) "know" that T needs to be refined to T^* to T^{**} in order to be able to identify when "some C's regularly cause E" in a meaningful manner that can lead to explanatory arguments.

Capacities are best understood as the causal disposition to contribute to something (Cummins 1975, 2000, 2002, Treur 2005). Aristotle did introduce such a concept; he called it *potentiality* (to move), or movable (see discussion in Treur 2005). The difference between the arrow at rest and the snapshot of the moving arrow at time t at position P is that the former has no potentiality to be at P', whereas the latter has. This explains why at a next instant t' the former arrow is still where it was, at P, while the latter arrow is at a different position P': Why is the arrow at t' at position P'? The arrow is at position P' at t' because at t it was at position P, and at t it had the *potentiality* to be at P', and at t nothing in the world excluded it to be at P'. Then, in the general case of the archaeological problem-type, for each of kind of observable change in the archaeological record, a specific kind of potentiality should be considered, e.g., the potentiality of a flint pebble with shape S to become a scrapper with shape S'. This differs from Aristotle's potentiality in that the arrow in motion has the potential to be at position P' by virtue of a property of the arrow, namely that it is in motion. The potentiality of a flint pebble becoming a scrapper depends on the perception of an artisan who connects the property of the pebble, with the desired outcome of the artisan. S is the consequence of a geologic process responsible for the material origin of the stone, S' is the consequence of a work operation which transformed a raw material into an instrument. In general, if the potentiality (occurring in a state S) to have state property X has led to a state S' where indeed X holds, then this state property X of state S' is called the fulfillment or actualization of the potentiality for X occurring in state S.

Probabilities measure the capacity a cause can produce an effect in some specific context. It is a measure of the regularity a cause produces an effect, and therefore a measure of the robustness of the cause, or a measure of the degree a context can modify the causal capabilities of a process. What we are suggesting is that archeologists will be able to explain how a behavior, social act or work operation causally produces an observable effect (the shape, size, texture, material or spatio-temporal location of some perceivable entity) when we are able to generalize probabilistically a series of known examples of the input-output pairs. A social action or sequence of social actions will be causally related with a state change if and only if the probability for the new state is higher in presence of that action than in its absence. That is to say, the cause is not "necessary" for the production of the effect, but "simply probable". Dwight Read has suggested the following example: "if the artisan is motivated to make an end scraper, then with high probability he or she will select a flint nodule of such and such dimensions and make an end scraper of such and such form. What we may not be able to say is why at that point in time and location, he/she was motivated to make an end scraper, though we can say that it is likely that there will be a location and time when he/she is so motivated."

What we need is inverse reasoning methods that allow predicting a cause even when it is not universally and directly tied with its effect. Rather than assuming that data are generated by a single underlying event, it should be assumed that the environment could be modeled as a collection of idiosyncratic "processes", where a process is characterized by a particular probabilistic rule that maps input vectors to output vectors (Jordan and Jacobs 1992). In other words, we should be able to generate abstractions that may not exist explicitly in the sensor input, but which capture the salient, invariant visual properties of a generic causal model used as a solution to the perceptual problem. Therefore, what we need is a heuristic classification machine (Clancey 1984), a classifier which has the smallest probability of making a mistake.

References Cited

Aitchison, J. 1986. The Statistical Analysis of Compositional Data. Chapman and Hall, London.

Alberti, G. 2014. Modeling Group Size and Scalar Stress by Logistic Regression from an Archaeological Perspective. PLOS-One 9(3): e91510.

Allen, J.F. 1983. Maintaining knowledge about temporal intervals. Communications of the ACM 26(11): 832–843.

Ammerman, A.J. and Ll. Cavalli-Sforza. 1971. Measuring the rate of spread of early farming in Europe. Man 6: 674–88.

Ammerman, A.J. and Ll. Cavalli-Sforza. 1984. The Neolithic Transition and the Genetics of Populations in Europe. Princeton, New Jersey.

Armit, I., G.T. Swindles and K. Becker. 2013. From dates to demography in later prehistoric Ireland? Experimental approaches to the meta-analysis of large 14C data-sets. Journal of Archaeological Science 40: 433–438.

Atkins, T. 2009. The Science and Engineering of Cutting: The Mechanics and Processes of Separating, Scratching and Puncturing Biomaterials, Metals and Non-Metals. Butterworth, Oxford.

Baggaley, A.W., R.J. Boys, A. Golightly, G.R. Sarson and A. Shukurov. 2012. Inference for population dynamics in the Neolithic period. Annals of Applied Statistics 6(4): 1352–1376.

Bailey, G.N. 2005. Concepts of time. pp. 268–273. *In:* C. Renfrew and P. Bahn (eds.). Archaeology: The Key Concepts. Thames and Hudson, London.

Bailey, G.N. 2007. Time perspectives, palimpsests and the archaeology of time. Journal of Anthropological Archaeology 26: 198–223.

Banning, E.B. 2000. The Archaeologist's Laboratory: The Analysis of Archaeological Data. Springer, New York (Interdisciplinary Contributions to Archaeology).

Barceló, J.A. 2002. Archaeological Thinking: between space and time. Archeologia e calcolatori 13: 237–256.

Barceló, J.A. 2007. Automatic Archaeology: Bridging the gap between Virtual Reality, Artificial Intelligence, and Archaeology. pp. 437–456. *In:* F. Cameron and S. Kenderdine (eds.). Theorizing Digital Cultural Heritage. A critical Discourse. The MIT Press, Cambridge (MA).

Barceló, J.A. 2008. La incertesa de les cronologies absolutes en Arqueologia. Probabilitat i Estadística. *Cypsela* 17: 23–34.

Barceló, J.A. 2009. *C*omputational Intelligence in Archaeology. The IGI Group, Hershey (NY).

Barceló, J.A. 2010a. Visual analysis in archaeology. An artificial intelligence Approach. In: A.M.T. Elewa (ed.). Morphometrics for Nonmorphometricians. Springer Verlag, Berlin. Lecture Notes in Earth Sciences 124: 51–101.

Barceló, J.A. 2010b. Computational Intelligence in Archaeology. State of the art. pp. 11–22. *In:* B. Frischer, J. Webb and D. Koller (eds.). Making History Interactive. ArcheoPress (BAR Int. Series, S2079), Oxford.

Barceló, J.A. 2012. Computer Simulation in Archaeology. Art, Science or Nightmare? Virtual Archaeology Review 3(5): 8–12.

Barceló, J.A. 2014. 3D Modeling and Shape Analysis in Archaeology. pp. 15–23. *In:* F. Remondino and S. Campana (eds.). 3D Recording and Modelling in Archaeology and Cultural Heritage—Theory and Best Practices. Archaeopress BAR Publication Series 2598, Oxford.

Barceló, J.A. and V. Moitinho de Almeida. 2012. Functional Analysis from Visual and Non-visual Data. An Artificial Intelligence Approach. Mediterranean Archaeology & Archaeometry 12(2): 273–321.

Barceló, J.A., G. Pelfer and A. Mandolesi. 2002. The origins of the city. From social theory to archaeological description. Archeologia e Calcolatori 13: 41–64.

Barceló, J.A., I. Briz, I. Clemente, J. Estévez, L. Mameli, A. Maximiano, F. Moreno, J. Pijoan, R. Pique, X. Terradas, A. Toselli, E. Verdún, A. Vila and D. Zurro. 2006. Análisis etnoarqueológico del valor social del producto en sociedades cazadoras-recolectoras. pp. 189–207. *In*: Etnoarqueología de la Prehistoria más allá de la analogía. CSIC, Madrid.

Barceló, J.A., G. Capuzzo and I. Bogdanović. 2014a. Modeling expansive phenomena in early complex societies: the Transition from Bronze to Iron Age in Prehistoric Europe. Journal of Archaeological Method and Theory 21(2): 486–510.

Barceló, J.A., F. Del Castillo, R. Del Olmo, L. Mameli, F.J. Miguel Quesada, D. Poza and X. Vilà. 2014b. Social Interaction in Hunter-Gatherer Societies: Simulating the Consequences of Cooperation and Social Aggregation. Social Science Computer Review 32(3): 417–436.

Barham, L. 2013. From hand to handle: the first industrial revolution. Oxford University Press.

Barton, C.M., J. Bernabeu Auban, O. Garcia Puchol, S. Schmich and L. Molina Balaguer. 2004. Long-term socioecology and contingent landscapes. Journal of Archaeological Method and Theory 11(3): 253–295.

Baxter, M.J. 1994. Exploratory Multivariate Analysis in Archaeology. Edinburgh University Press.

Baxter, M.J. 2003. Statistics in Archaeology. Wiley, London.

Baxter, M.J. 2006. A Review of Supervised and Unsupervised Pattern Recognition in Archaeometry. Archaeometry 48(4): 671–694.

Beekman, C.S. and W.W. Baden (eds.). 2005. Nonlinear models for archaeology and anthropology: continuing the revolution. Ashgate Publishing, Hampshire.

Bentley, R.A.E., M.W. Lake and S.J. Shennan. 2005. Specialisation and wealth inequality in a model of a clustered economic network. Journal of Archaeological Science 32: 1346–1356.

Betts, M.W. and T.M. Friesen. 2004. Quantifying hunter-gatherer intensification: a zooarchaeological case study from Arctic Canada. Journal of Anthropological Archaeology 23: 357–384.

Betts, M.W. and T.M. Friesen. 2006. Declining foraging returns from an inexhaustible resource? Abundance indices and beluga whaling in the western Canadian Arctic. Journal of Anthropological Archaeology 25: 59–81.

Bevan, A. and J. Conolly. 2009. Modelling spatial heterogeneity and nonstationarity in artifact-rich landscapes. Journal of Archaeological Science 36(4): 956–964.

Bevan, A. and M. Lake (eds.). 2013. Computational Approaches to Archaeological Spaces. Left Coast Press, Walnut Creek.

Bevan, A., X. Li, M. Martinón-Torres, S. Green, Y. Xia, K. Zhao, Z. Zhao, S. Ma, W. Kao and T. Rehren. 2014. Computer Vision, Archaeological Classification and China's Terracotta Warriors. Journal of Archaeological Science 49: 249–254.

Bicici, E. and R. St. Amant. 2003. Reasoning about the functionality of tools and physical artefacts. Technical Report TR-2003:22, Department of Computer Science, North Carolina State University.

Blankholm, H. 1991. Illtrasite spatial allalysis in theory and practice.Aarhus University Press. Aarhus.

Bocquet-Appel, J.P., S. Naji, M. Vander Linden and J.K. Kozlowski. 2009. Detection of diffusion and contact zones of early farming in Europe from the space-time distribution of 14C dates. Journal of Archaeological Science 36: 807–820.

Boserup, E. 1981. Population and Technological Change: A Study of Long Term Trends. University of Chicago Press.

Brantingham, P.J. 2006. Measuring Forager Mobility. Current Anthropology 47(3): 435–459.

Bril, B., J. Smaers, J. Steele, R. Rein, T. Nonaka, G. Dietrich and V. Roux. 2012. Functional mastery of percussive technology in nut-cracking and stone-flaking actions: experimental comparison and implications for the evolution of the human brain. Philosophical Transactions of the Royal Society B: Biological Sciences 367(1585): 59–74.

Buck, C.E., W.C. Cavanaghwg and R. Litton. 1996. Bayesian Approach to Interpreting Archaeological Data. Wiley, Chichester (UK).

Buck, C.E. and A. Millard (eds.). 2004. Tools for constructing chronologies: crossing disciplinary boundaries (Vol. 177). Springer, Berlin.

Bunge, M. 1959. Causality. The Place of Causal Principle in Modern Science. Harvard University Press, Cambridge.

Bunge, M. 2006. Chasing Reality: Strife over Realism. University of Toronto Press.

Buxó, R. and R. Piqué. 2009. Arqueobotánica: Los usos de las plantas en la Peninsula Ibérica. Editorial Ariel, Barcelona.

Capuzzo, G. 2014. Space-temporal analysis of radiocarbon evidence and associated archaeological record: from danube to ebro rivers and from bronze to iron ages. PhD. Dissertation. Universitat Autònoma de Barcelona, Bellaterra.

Carr, C. 1984. The Nature of Organization of Intrasite Archaeological Records and Spatial Analytic Approaches to their Investigation. pp. 103–222. *In*: M.B. Schiffer (ed.). Advances in Archaeological Method and Theory 7. Academic Press, New York.

Cartwright, N. 2004. Causation: One word, many things. Philosophy of Science 71(5): 805–820.

Cattani, M. and A. Fiorini. 2004. Topologia: identificazione, significato e valenza nella ricerca archeologica. Archeologia e Calcolatori XV: 317–340.

Chamberlain, A. 2009. Archaeological Demography. Human Biology 81(2–3): 275–86.

Chapman, R. 2003. Death, society and archaeology: the social dimensions of mortuary practices. Mortality 8(3): 305–312.

Chiverrell, R.C., V.C. Thorndycraft and T.O. Hoffmann. 2011. Cumulative probability functions and their role in evaluating the chronology of geomorphological events during the Holocene. Journal of Quaternary Science 26(1): 76–85.

Christiansen, J. and M. Altaweel. 2006. Simulation of natural and social process interactions: An example from Bronze Age Mesopotamia. Social Science Computer Review 24(2): 209–226.

Cioffi-Rivilla, C. 2014. Introduction to Computational Social Science: Principles and Applications. Springer, Berlin-New York (Texts in Computer Science).

Clancey, W. 1984. Heuristic Classification. Artificial Intelligence 27: 289–350.

Clarke, D. 1968. Analytic Archaeology. Cambridge University Press.

Collard, M., B. Buchanan, M.J. Hamilton and M.J. O'Brien. 2010. Spatiotemporal dynamics of the Clovis-Folsom transition. Journal of Archaeological Science 37(10): 2513–2519.

Conolly, J. and M. Lake. 2006. Geographical Information Systems in Archaeology. Cambridge University Press.

Conte, R. 2009. From Simulation to Theory (and Backward). pp. 29–47. In: F. Squazzoni (ed.). Epistemological Aspects of Computer Simulation in the Social Sciences. Springer, Berlin.

Costa, L.F. and R.M. Cesar. 2001. Shape Analysis and Classification: Theory and Practice. CRC Press, Boca Raton (FL).

Costopoulos, A. and M.W. Lake (eds.). 2010. Simulating Change: Archaeology into the Twenty-First Century. University of Utah Press, Salt Lake City.

Cotterell, B. and J. Kamminga. 1992. Mechanics of Pre-Industrial Technology. An introduction to the mechanics of ancient and traditional material culture. Cambridge University Press.

Cowgill, G.L. 1970. Some sampling and reliability problems in archaeology. pp. 161–175. In: Archéologie et Calculateurs. Problèmes Semiologiques et Mathematiques. Colloque International du CNRS. Editions du CNRS, Paris.

Crema, E.R., A. Bevan and M.W. Lake. 2010. A probabilistic framework for assessing spatio-temporal point patterns in the archaeological record. Journal of Archaeological Science 37(5): 1118–1130.

Cummins, R. 1975. Functional Analysis. Journal of Philosophy 72(20): 741–765

Cummins, R. 2000. How does it work? vs. What are the laws? Two conceptions of psychological explanation. pp. 117–145. In: F. Keil and R. Wilson (eds.). Explanation and Cognition. The MIT Press, Cambridge (MA).

Cummins, R. 2002. Neo-Teleology. In: A. Ariew, R. Cummins and M. Perlman (eds.). Functions. New Essays in the Philosophy of Psychology and Biology. Oxford University Press.

Cunningham, J. 2009. Ethnoarchaeology beyond correlates. Ethnoarchaeology 1(2): 115–136.

Darden, L. 2002. Strategies for Discovering Mechanisms: Schema Instantiation, Modular Subassembly, Forward/Backward Chaining. Philosophy of Science 69: 354–365.

De Meo, A., G. Espa, S. Espa, A. Pifferi and U. Ricci. 2000. Study of archaeological areas by means of advanced software technology and statistical methods. Journal of Cultural Heritage 1(3): 233–245.

Delicado, P. 1998. Statistics in Archaeology: New Directions. pp. 29–37. In: J.A. Barceló, I. Briz and A. Vila (eds.). New Techniques for Old Times. Computer Applications in Archaeology. Archaeopress, Oxford (BAR International Series S757).

Demeulenaere, P. 2011. Analytical Sociology and Social Mechanisms. Cambridge University Press.

Deravignone, L. and G. Macchi. 2006. Artificial Neural Networks in Archaeology. Archeologia e Calcolatori 17: 121–136.

Desachy, B. 2008. De la formalisation du traitement des données stratigraphiques en archéologie de terrain. PhD Dissertation. Université Panthéon-Sorbonne-Paris I.

Di Ludovico, A. and G. Pieri. 2011. Artificial Neural Networks and ancient artefacts: justifications for a multiform integrated approach using PST and Auto-CM models. Archeologia e calcolatori 22: 91–128.

Dibble, H.L. and Z. Rezek. 2009. Introducing a new experimental design for controlled studies of flake formation: results for exterior platform angle, platform depth, angle of blow, velocity, and force. Journal of Archaeological Science 36(9): 1945–1954.

Djindjian, F. 1991. Les Méthodes de l'Archéologie. Armand Colin, Paris.

Dolukhanov, P., A. Shurukov, D. Gronenborn, D. Sokoloff, V. Timofeev and G. Zaitseva. 2005. The chronology of Neolithic dispersal in Central and Eastern Europe. Journal of Archaeological Science 32: 1441–1458.

Domínguez-Rodrigo, M. 2008. Conceptual premises in experimental design and their bearing on the use of analogy: an example from experiments on cut marks. World Archaeology 40(1): 67–82.

Doran, J. and F.R. Hodson. 1975. Mathematics and Computers in Archaeology. Harvard University Press.

Doran, J.E. and M. Palmer. 1995. The EOS project: Integrating two models of Palaeolithic social change. pp. 103–125. In: N. Gilbert and R. Conte (eds.). Artificial Societies: The Computer Simulation of Social Life. UCL Press, London.

Drennan, R. 2010. Statistics for Archaeologists. 2nd Edition: A Common Sense Approach, Springer, Berlin-New York.

Drennan, R.D., C.E. Peterson and J.R. Fox. 2010. Degrees and kinds of inequality. pp. 45–76. *In*: T.D. Price and G.M. Feinman (eds.). Pathways to Power. Springer, Berlin-New York.

Dunn, S. and K. Woolford. 2013. Reconfiguring Experimental Archaeology Using 3D Movement Reconstruction. pp. 277–291. *In*: Electronic Visualisation in Arts and Culture. Springer, Berlin-New York (Series on Cultural Computing).

Edmonds, B. 2010. Computational modelling and social theory—the dangers of numerical representation. pp. 36–68. *In*: E. Mollona (ed.). Computational Analysis of Firm Organisations and Strategic Behaviour. Routledge, London.

Eells, E. 1991. Probabilistic Causality. Cambridge University Press.

Elewa, E.M.T. (ed.). 2010. Morphometrics for Non-Morphometricians. Lecture Notes in Earth Sciences 124. Springer, Berlin.

Engler, O. and V. Randle. 2009. Introduction to Texture Analysis: Macrotexture, Microtexture, and Orientation Mapping. CRC Press, Boca Raton (FL).

Epstein, J.M. 2006. Generative social science: Studies in agent-based computational modeling. Princeton University Press.

Estévez, J. 2000. Aproximación dialéctica a la Arqueología. Revista Atlántica-Mediterránea de Prehistoria y Arqueología Social 3. Cádiz (Spain).

Feinman, G.M., S. Upham and K.G. Lightfoot. 1981. The production step measure. An ordinal index of labor input in ceramic manufacture. American Antiquity 46(4): 871–884.

Fleming, B. 1999. 3D Modeling and Surfacing. Morgan Kaufmann Publishers, San Francisco (CA).

Fort, J., T. Pujol and Ll. Cavalli-Sforza. 2004. Paleolithic population waves of advance. Cambridge Archaeological Journal 14: 53–61.

Francfort, H.P. 1997. Archaeological interpretation and non-linear dynamic modelling: between metaphor and simulation. *In*: S.E. van der Leeuw and J. McGlade (eds.). Time, Process and Structured Transformation in Archaeology. Routledge, London.

Galam, S. 2012. Sociophysics: A Physicist's Modeling of Psycho-political Phenomena. Springer, Berlin (Understanding Complex Systems Series).

Gamble, C., W. Davies, P. Pettitt, L. Hazelwood and M. Richards. 2005. The Archaeological and genetic foundation of the European population during the Late Glacial: implications for "agricultural thinking". Cambridge Archaeological Journal 15(2): 193–223.

Gansell, A.R., J.W.V.D. Meent, S. Zairis and C.H. Wiggins. 2014. Stylistic clusters and the Syrian/South Syrian tradition of first-millennium BCE Levantine ivory carving: a machine learning approach. Journal of Archaeological Science 44: 194–205.

Gardin, J.C. 1980. Archaeological Constructs. An Aspect of Theoretical. Archaeology. Cambridge University Press.

Gatrell, A.C. 1983. Distance and Space: a geographical perspective. Oxford University Press.

Gavua, K. 2012. Ethnoarchaeology. Oxford University Press.

Genesareth, M.R. and N.J. Nilsson. 1987. Logical Foundations of Artificial Intelligence. Palo Alto (CA): Morgan Kaufmann.

Gibbins, P. 1990. BACON Bytes Back. *In*: J.E. Tiles, G.T. McKee and G.C. Dean (eds.). Evolving Knowledge in Natural Science and Artificial Intelligence. Pitman, London.

Gifford-González, D. 2010. Ethnoarchaeology-looking back, looking forward. The SAA Archaeological Record 10(1): 22–25.

Gillies, D. 1996. Artificial Intelligence and the Scientific Method. Oxford University Press.

Gkiasta, M., T. Russell, S. Shennan and J. Steele. 2003. Neolithic transition in Europe: the radiocarbon record revisited. Antiquity 77: 45–62.

Grace, R. 1996. Review article use-wear analysis: the state of the art. Archaeometry 38(2): 209–229.

Griffitts, J. 2011. Designing Experimental Research in Archaeology: Examining Technology through Production and Use. Ethnoarchaeology 3(2): 221–225.

Grosman, L., A. Karasik, O. Harush and U. Smilansky. 2014. Archaeology in Three Dimensions: Computer-Based Methods in Archaeological Research. Journal of Eastern Mediterranean Archaeology and Heritage Studies 2(1): 48–64.

Haan, H.J. 2014. More Insight from Physics into the Construction of the Egyptian Pyramids. Archaeometry 56(1): 145–174.

Hamel, L.H. 2011. Knowledge Discovery with Support Vector Machines. Wiley, London (Series on Methods and Applications in Data Mining).

Hamilton, M.J. and B. Buchanan. 2007. Spatial gradients in Clovis-age radiocarbon dates across North America suggest rapid colonization from the north. Proceeding of the National Academy of Sciences of the United States of America (PNAS) 104(40): 15625–15630.

Harris, E.C. 1975. The Stratigraphic Sequence: A Question of Time. World Archaeology 7(1): 109–121.

Harris, E.C. 1979. The Laws of Archaeological Stratigraphy. World Archaeology 11(1): 111–117.

Harris, E.C. 1989. Principles of archaeological stratigraphy. 2nd edition, Academic Press, London.

Hassan, F.A. 1987. Re-Forming Archaeology: A Foreword to Natural Formation Processes and the archaeological Record. pp. 1–9. *In*: D.T. Nash and M.D. Petraglia (eds.). Natural Formation Processes and the archaeological Record. ArcheoPress, Oxford (British Archaeological Series S352).

Hastie, T., R. Tibshirani and J. Fiedman. 2011. The Elements of Statistical Learning: Data Mining, Inference, and Prediction. Second Edition. Springer, Berlin.

Hasting, A., K. Cuddington, K.F. Davies, C.J. Dugaw, S. Elmendorf, A. Freestone, S. Harrison, M. Holland, J. Lambrinos, U. Malvadkar, B.A. Melbourne, K. Moore, C. Taylor and D. Thomson. 2005. The spatial spread of invasions, new developments in theory and evidence. Ecology Letters 8(1): 91–101.

Hayden, B. 1998. Practical and Prestige Technologies: The Evolution of Material Systems. Journal of Archaeological Method and Theory 5: 1–55.

Hazelwood, L. and J. Steele. 2004. Spatial dynamics of human dispersals: Constraints on modelling and archaeological validation. Journal of Archaeological Science 31(6): 669–679.

Hedström, P. and R. Swedberg (eds.). 1998. Social Mechanisms: An Analytical Approach to Social Theory. Cambridge University Press.

Helbing, D. 2012. Social Self-Organization: Agent-Based Simulations and Experiments to Study Emergent Social Behavior. Springer, Berlin.

Hengl, T. and H.I. Reuter (eds.). 2008. Geomorphometry: Concepts, Software, Applications. Elsevier, Amsterdam.

Hensel, E. 1991. Inverse Theory and Applications for Engineers. Prentice Hall, Englewood Cliffs, NJ.

Herzog, I. 2002. Possibilities for Analysing Stratigraphic Data. CD of the Workshop Archäologie und Computer, Vienna.

Herzog, I. 2004. Group and Conquer—A Method for Displaying Large Stratigraphic Data Sets. pp. 423–426. *In*: Procedings of the 31th Conference of CAA 2003, BAR Int. Ser. 1227, Oxford.

Herzog, I. 2014. Datenstrukturen zur Analyse archäologischer Schichten. Archäologische Informationen 26(2): 457–461.

Hinz, M., I. Feeser, K.G. Sjögren and J. Müller. 2012. Demography and the intensity of cultural activities: an evaluation of Funnel Beaker Societies (4200–2800 cal BC). Journal of Archaeological Science 39(10): 3331–3340.

Holdaway, S. and L. Wandsnider (eds.). 2008. Time in Archaeology: time perpectivism revisited. University of Utah Press. Salt Lake City.

Holland, J.H., K.J. Holyoak, R.E. Nisbett and P.R. Thagard. 1986. Induction. Processes of Inference, Learning, and Discovery. The MIT Press, Cambridge (MA).

Holst, M.K. 2001. Formalizing Fact and Fiction in Four Dimensions: A Relational Description of Temporal Structures in Settlements. pp. 159–163. *In*: Z. Stančič and T. Veljanovski (eds.). Computing Archaeology for Understanding the Past, CAA 2000, ArcheoPress, Oxford (BAR Int. Ser. 931).

Holst, M.K. 2004. Complicated Relations and Blind Dating: Formal Analysis of Relative Chronological Structures. pp. 129–147. *In*: C.E. Buck and A.R. Millard (eds.). Tools for Constructing Chronologies. Springer, Berlin. Lecture Notes in Statistics 177.

Homsher, R.S. 2012. Mud bricks and the process of construction in the Middle Bronze Age Southern Levant. Bulletin of the American Schools of Oriental Research 368: 1–27.

Hopkins, H. 2008. Using Experimental Archaeology to Answer the Unanswerable: A case study using Roman Dyeing. pp. 103–118. *In*: P. Cunningham, P.J. Heeb and R.P. Paardekooper (eds.). Experiencing Archaeology by Experiment. Oxbow Books, Oxford.

Hopkins, H. 2013. The importance to archaeology of undertaking and presenting experimental research: a case study using dyeing in Pompeii. Archaeological and Anthropological Sciences. DOI 10.1007/s12520–013–0159–y.

Hörr, Ch., E. Lindinger and G. Brunnet. 2014. Machine Learning based Typology Development in Archaeology. Journal on Computing and Cultural Heritage (JOCCH) 7(1). DOI 10.1145/2533988.

Hurcombe, L. 2014. Archaeological Artefacts as Material Culture. Routledge, London.

Isern, N., J. Fort and M. Vander Linden. 2012. Space competition and time delays in human range expansions. Application to the Neolithic transition. PLOS One 7(12): e51106.

Isern, N., J. Fort, A.F. Carvalho, J.F. Gibaja and J.J. Ibañez. 2014. The Neolithic transition in the Iberian Peninsula: data analysis and modeling. Journal of Archaeological Method and Theory 21(2): 447–460.

Johansson, I. 2008. Functions and Shapes in the Light of the International System of Units Metaphysica. International Journal for Ontology & Metaphysics 9: 93–117.

Johansson, I. 2011. Shape is a Non-Quantifiable Physical Dimension. *In*: J. Hastings, O. Kutz, M. Bhatt and S. Borgo (eds.). SHAPES 1.0. The Shape of Things. Proceedings of the 1st Interdisciplinary Workshop on SHAPES, Karlsruhe.

Jordan, M.I. and R.A. Jacobs. 1992. Modularity, Unsupervised Learning and Supervised Learning. pp. 21–29. *In*: S. Davis (ed.). Connectionism: Theory and Practice. Oxford University Press.

Kaipio, J. and E. Somersalo. 2004. Statistical and Computational Inverse Problems. Springer, Berlin.

Kamermans, H., M. van Leusen and Ph Verhagen (eds.). 2009. Archaeological Prediction and Risk Management. Alternatives to current practice. Leiden University Press.

Kilikoglou, V. and G. Vekkinis. 2002. Failure Prediction and Function Determination of Archaeological Pottery by Finite Element Analysis. Journal of Archaeological Science 29: 1317–1325.

Kintigh, K.W., J.H. Altschul, M.C. Beaudry, R.D. Drennan, A.P. Kinzig, T.A. Kohler, W.F. Limp, H.D.G. Maschner, W.K. Michener, T.R. Pauketat, P. Peregrine, J.A. Sabloff, T.J. Wilkinson, H.T. Wright and M.A. Zeder. 2014. Grand challenges for archaeology. American Antiquity 79(1): 5–24.

Kirsch, A. 1996. An Introduction to the Mathematical Theory of Inverse Problems. Springer, Berlin.

Kohler, T.A. and S.E. van der Leeuw. 2007. The Model-Based Archaeology of Socionatural Systems. SAR Press, Santa Fe (NM).

Kohler, T.A., D. Cockburn, P.L. Hooper, R.K. Bocinsky and Z. Kobti. 2012. The coevolution of group size and leadership: an agent-based public goods model for prehispanic Pueblo societies. Advances in Complex Systems 15(1&2). DOI 1150007–1–1150007–29.

Kovarovic, K., L.C. Aiello, A. Cardini and C.A. Lockwood. 2011. Discriminant function analyses in archaeology: are classification rates too good to be true? Journal of archaeological science 38(11): 3006–3018.

Kowarik, K. 2012. Agents in Archaeology-Agent-Based Modelling (ABM) in Archaeological Research. Beiträge zum Geoinformationssysteme 17: 238–251.

Lake, M.W. 2014. Trends in archaeological simulation. Journal of Archaeological Method and Theory 21(2): 258–287.

Langley, P. and J. Zytkow. 1989. Data-Driven Approaches to Empirical Discovery. Artificial Intelligence 40: 283–312.

Lark, R.S. 1996. Geostatistical description of texture on an aerial photograph for discriminating classes of land cover. International Journal of Remote Sensing 17(11): 2115–2133.

Leymarie, F.F. 2011. On the Visual Perception of Shape—Analysis and Genesis through Information Models. *In*: J. Hastings, O. Kutz, M.L. Bhatt and S. Borgo (eds.). SHAPES 1.0. The Shape of Things. Proceedings of the 1st Interdisciplinary Workshop on SHAPES, Karlsruhe.

Leyton, M. 1992. Symmetry. Causality, Mind. The MIT Press, Cambridge (MA).

Lindsey, J.K. 1995. Modelling Frequency and Count Data. Clarendon Press, Oxford.

Lock, G. and B.L. Molyneaux (eds.). 2006. Confronting scale in archaeology. Issues of theory and practice. Springer, Berlin.

López, P., J. Lira and I. Hein. 2014. Discrimination of Ceramic Types Using Digital Image Processing by Means of Morphological Filters. Archaeometry. DOI: 10.1111/arcm.12083.

Lu, P., Y. Tian and R. Yang. 2013. The study of size-grade of prehistoric settlements in the Circum-Songshan area based on SOFM network. Journal of Geographical Sciences 23(3): 538–548.

Lucas, G. 2008. Time and archaeological event. Cambridge Archaeological Journal 18(1): 59–65.

Lucena, M., A.L. Martínez-Carrillo, J.M. Fuertes, F. Carrascosa and A. Ruiz. 2014a. Applying Mathematical Morphology for the Classification of Iberian Ceramics from the Upper Valley of Guadalquivir River. pp. 341–350. *In*: Pattern Recognition. Springer International Publishing.

Lucena, M., A.L. Martínez-Carrillo, J.M. Fuertes, F. Carrascosa and A. Ruiz. 2014b. Decision support system for classifying archaeological pottery profiles based on Mathematical Morphology. Multimedia Tools and Applications: 1–15.

Lyman, R.L. 2008. Quantitative Paleozoology. Cambridge University Press.

Lycett, S.J. and N. von Cramon-Taubadel. 2013. Toward a "Quantitative Genetic" Approach to Lithic Variation. Journal of Archaeological Method and Theory. DOI 10.1007/s10816–013-9200–9.

Maaten, L.J.P. van der, P. Boom, G. Lange, H. Paijmans and E. Postma. 2006. Computer Vision and Machine Learning for Archaeology. Proceedings of the Computer Applications and Quantitative Methods in Archaeology Conference (CAA'06), Fargo (ND).

Machamer, P., L. Darden and C. Craver. 2000. Thinking about Mechanisms. Philosophy of Science 67: 1–25.

Macy, M.W. and R. Willer 2002. From factors to actors: Computational sociology and agent-based modeling. Annual review of sociology 28: 143–166.

Madella, M. and B. Rondelli (eds.). 2014. Simulating the Past: Exploring Change Through Computer Simulation in Archaeology. Journal of Archaeological Method and Theory 21(2): Special Issue.

Mameli, L., J.A. Barceló and J. Estévez. 2002. The Statistics of Archeological Deformation Process. A zooarchaeological experiment. pp. 221–230. *In*: G. Burenhult (ed.). Archaeological informatics: pushing the envelope. ArcheoPress, Oxford (BAR International Series 1016).

McGlade, J. 2014. Simulation as narrative: contingency, dialogics, and the modeling conundrum. Journal of Archaeological Method and Theory 21(2): 288–305.

McHugh, F. 1999. Theoretical and Quantitative Approaches to the Study of Mortuary Practice. Archeo Press, Oxford (BAR Int. Series, S785).

Medin, D.L. 1989. Concepts and conceptual structure. American psychologist 44(12): 1469–1481.

Mellars, P. 2006. A new radiocarbon revolution and the dispersal of modern humans in Europe. Nature 439: 931–935.

Miller, A. and C.M. Barton. 2008. Exploring the land: a comparison of land-use patterns in the Middle and Upper Paleolithic of the western Mediterranean. Journal of Archaeological Science 35: 1427–1437.

Mirmehdi, M., X. Xie and J. Suri (eds.). 2008. Handbook of texture analysis. Imperial College Press, London.

Moitinho de Almeida, V. 2013. Towards Functional Analysis of Archaeological Objects through Reverse Engineering Processes. PhD. Dissertation. Universitat Autònoma de Barcelona.

Moitinho de Almeida, V., J.A. Barceló, R. Rosillo and A. Palomo. 2013. Linking 3D Digital Surface Texture Data with Ancient Manufacturing Procedures. IEEE Digital Heritage International Congress, Marseille. A.C. Addison, G. Guidi, L. De Luca and S. Pescarin (eds.). pp. 735–738.

Morrison, K. 1994. The intensification of production: archaeological approaches. Journal of Archaeological Method and Theory 1 (2): 111–159.

Murphy, K.P. 2012. Machine Learning: A Probabilistic Perspective. Springer, Berlin (Adaptive Computation and Machine Learning series).

Murray, T. (ed.). 2004. Time and archaeology. Routledge, London.

Myers, O.H. 1950. Some Applications of Statistics to Archaeology. Government Press, Cairo.

Naldi, G., L. Pareschi and G. Toscani. 2010. Mathematical Modeling of Collective Behavior in Socio-Economic and Life Sciences. Birkhäuser, Basel.

Negre, J. 2014. Implementation of Artificial Neural Networks in the design of archaeological expectation predictive models. Mapping 23(165): 4–16.

O'Driscoll, C.A. and J.C. Thompson. 2014. Experimental projectile impact marks on bone: implications for identifying the origins of projectile technology. Journal of Archaeological Science 49: 398–413.

O'Shea, J. 1984. Mortuary Variability. Academic Press, Orlando (FL).

Odell, G.H. 2001. Stone tool research at the end of the millennium: classification, function, and behavior. Journal of Archaeological Research 9(1): 45–100.

Okubo, A. and S.A. Levin. 2001. Diffusion and ecological problems, Modern perspectives. Springer, New York, Berlin.

Oltean, I.A. 2013. Burial mounds and settlement patterns: a quantitative approach to their identification from the air and interpretation. Antiquity 87(335): 202–219.

Orton, C.R. 1980. Mathematics in archaeology. Cambridge University Press.

Orton, C.R. 1982. Stochastic process and archaeological mechanism in spatial analysis. Journal of Archaeological Science 9(1): 1–23.

Östborn, P. and H. Gerding. 2014. Network analysis of archaeological data: a systematic approach. Journal of Archaeological Science 46: 75–88.

Özmen, Y.C. and S. Balcisoy. 2006. A framework for working with digitized cultural heritage artefacts. Springer. Lecture Notes in Computer Science 4263: 394–400.

Papaodysseus, C. 2012. Pattern Recognition and Signal Processing in Archaeometry: Mathematical and Computational Solutions for Archaeology. The IGI Group, Henshey (NY). Information Science Reference.

Pawlowsky-Glahn, V. and A. Buccianti (eds.). 2011. Compositional data analysis: Theory and applications. John Wiley & Sons, London.

Pinhasi, R., J. Fort and A.J. Ammerman. 2005. Tracing the origin and spread of agriculture in Europe. PLoS Biol 3(12): e410.

Pizlo, Z. 2001. Perception viewed as an inverse problem. Vision Research 41(24): 3145–3161.

Popper, K. 1957. The Poverty of Historicism. Boston. Beacon press, Boston.

Premo, L.S. 2014. Cultural Transmission and Diversity in Time-Averaged Assemblages. Current Anthropology 55(1): 105–114.

Railsback, S. and V. Grimm. 2011. Agent-Based and Individual-Based Modeling: A Practical Introduction. Princeton University Press.

Read, D. 1985. The Substance of Archaeological Analysis and the Mold of Statistical Method: Enlightenment Out of Discordance? pp. 45–86. *In*: C. Car (ed.). For Concordance in Archaeological Analysis. Bridging Data Structure Quantitative Technique and Theory. Westport Publishing Co, Boulder (CO).

Read, D. 1987. Archaeological Theory and Statistical Methods: Discordance, Resolution and New Directions. pp. 151–184. *In*: M. Aldenderfer (ed.). Quantitative Research in Archaeology: Progress and Prospects. Sage Publications.

Read, D.W. 1989. Intuitive typology and automatic classification: Divergence or full circle? Journal of Anthropological Archaeology 8(2): 158–188.

Read, D.W. 2007. Kinship theory: A paradigm shift. Ethnology 46(4): 329–364.

Rivadulla, A. 1993. Probabilidad e inferencia estadística. Anthropos, Barcelona.

Roux, V. 2007. Ethnoarchaeology: a non historical science of reference necessary for interpreting the past. Journal of Archaeological Method and Theory 14(2): 153–178.

Rovetto, R. 2011. The Shape of Shapes: An Ontological Exploration. *In*: J. Hastings, O. Kutz, M. Bhatt and S. Borgo (eds.). SHAPES 1.0. The Shape Of Things. Proceedings of the 1st Interdisciplinary Workshop on SHAPES. Karlsruhe, Germany.

Rovner, I and F. Gyulai. 2007. Computer-Assisted Morphometry: A New Method for Assessing and Distinguishing Morphological Variation in Wild and Domestic Seed Populations. Economic Botany 61(2): 1–19.

Russell, B. 1967. The Problems of Philosophy. Oxford University Press (1st ed. 1912, Home University Library).

Russell, T.M. 2004. The spatial analysis of radiocarbon databases: the spread of the first farmers in Europe and of the fat-tailed sheep in Southern Africa. ArcheoPress, Oxford. British Archaeological Reports.

Sabatier, P.C. 2000. Past and Future of Inverse Problems. Journal of Mathematical Physics 41: 4082–4124.

Salmon, W.C. 1989. Four decades of scientific explanation. *In*: P. Kitcher and W.C. Salmon (eds.). Scientific Explanation Minnesota Studies in the History of Science 13. University of Minnesota Press, Minneapolis.

Sayer, D. and M. Wienhold. 2013. A GIS-Investigation of Four Early Anglo-Saxon Cemeteries Ripley's K-function Analysis of Spatial Groupings Amongst Graves. Social Science Computer Review 31(1): 71–89.

Schiffer, M.B. 1987. Formation Processes of the Archaeological Record. University of New Mexico Press, Albuquerque.

Schiffer, M.B. 2010. Behavioral Archaeology: Principles and Practice. Equinox, London.

Schiffer, M B. 2013. Contributions of Ethnoarchaeology. pp. 53–63. *In*: The Archaeology of Science. Springer, Berlin.

Schilkopf, B. and A.J. Smola. 2001. Learning with Kernels: Support Vector Machines, Regularization, Optimization, and Beyond Spriger, Berlin (Adaptive Computation Series).

Sewell, Jr., W.H. 2005. Logics of history: Social theory and social transformation. University of Chicago Press.

Seetah, K. 2008. Modern analogy, cultural theory and experimental replication: a merging point at the cutting edge of archaeology. World Archaeology 40(1): 135–150.

Shennan, S. 1997. Quantifying Archaeology, Edinburgh University Press, Edinburgh.

Shennan, S. and K. Edinborough. 2007. Prehistoric population history: from the Late Glacial to the Late Neolithic in Central and Northern Europe. Journal of Archaeological Science 34(8): 1339–45.

Shennan, S., S.S. Downey, A. Timpson, K. Edinborough, S. Colledge, T. Kerig, K. Manning and M.G. Thomas. 2013. Regional population collapse followed initial agriculture booms in mid-Holocene Europe. Nature Communications 4: article number 2486. DOI. 10.1038/ncomms3486.

Small, C.G. 1996. The Statistical Theory of Shape. Springer, Berlin.

Smith, N.G., A. Karasik, T. Narayanan, E.S. Olson, U. Smilansky and T.E. Levy. 2014. The pottery informatics query database: a new method for mathematic and quantitative analyses of large regional ceramic datasets. Journal of Archaeological Method and Theory 21(1): 212–250.

Sokolowski, J.A. and C.M. Banks. 2009. Modeling and simulation for analyzing global events. John Wiley & Sons, London.

Soressi, M. and J.M. Geneste. 2011. The History and Efficacy of the Chaîne Opératoire Approach to Lithic Analysis: Studying Techniques to Reveal Past Societies in an Evolutionary Perspective. pp. 344–350. *In*: G. Tostedin (ed.). Reduction Sequence, Chaîne Opératoire and Other Methods: The Epistemologies of Different Approaches to Lithic Analysis. PaleoAnthropology: Special Issue

Sosna, D., P. Galeta, L. Šmejda, V. Sladek and J. Bruzek. 2013. Burials and Graphs Relational Approach to Mortuary Analysis. Social Science Computer Review 31(1): 56–70.

Squazzoni, F. 2012. Agent-Based Computational Sociology. Wiley, London.

Stemp, W.J. 2013. A review of quantification of lithic use-wear using laser profilometry: a method based on metrology and fractal analysis. Journal of Archaeological Science 48: 15–25.

Strauss, A. 2012. Interpretative possibilities and limitations of Saxe/Goldstein hypothesis. Boletim do Museu Paraense Emílio Goeldi. Ciências Humanas 7(2): 525–546.

Stytz, M.R. and R.W. Parrott. 1993. Using Kriging for 3D Medical Imaging. Computerized Medical Imaging and Graphics 17(6): 421–442.

Sullivan, A.P. 2008. Ethnoarchaeological and archaeological perspectives on ceramic vessels and annual accumulation rates of sherds. American Antiquity 73(1): 121–135.

Surovell, T.A. and P.J. Brantingham. 2007. A note on the use of temporal frequency distributions in studies of prehistoric demography. Journal of Archaeological Science 34(11): 1868–1877.

Swan, A.R.H. and J.A. Garraty. 1995. Image analysis of petrographic textures and fabrics using semivariance. Mineralogical Magazine 59: 189–196.

Tarantola, A. 2005. Inverse Problem Theory. Society for Industrial and Applied Mathematics. Philadelphia (PA).

Tawfik, A.Y. 2004. Inductive Reasoning and Chance Discovery. Minds and Machines 14: 441–451.

Thornton, C. 2000. Truth from Trash. How Learning Makes Sense. The MIT Press. Cambridge (MA).

Thulman, D.K. 2012. Discriminating Paleoindian point types from Florida using landmark geometric morphometrics. Journal of Archaeological Science 39(5): 1599–1607.

Tobler, W. 1970. A computer movie simulating urban growth in the Detroit region. Economic Geography 46(2): 234–240.

Traxler, C. and W. Neubauer. 2008. The Harris Matrix Composer, A New Tool To Manage Archaeological Stratigraphy. pp. 3–5. *In*: Archäologie und Computer-Kulturelles Erbe und Neue Technologien-Workshop.

Treur, J. 2005. A Unified Perspective on Explaining Dynamics by Anticipatory State Properties. pp. 27–37. *In*: J. Mira and J.R. Álvarez (eds.). Mechanisms, Symbols, and Models Underlying Cognition: First International Work-Conference on the Interplay Between Natural and Artificial Computation, IWINAC 2005, Las Palmas, Canary Islands, Spain. Proceedings, Part I. Berlin, Springer-Verlag, Lecture Notes in Computer Science 3561.

Tuceryan, A. and K. Jain. 1998. Texture Analysis. pp. 207–248. *In*: C.H. Chen, L.F. Pau and P.S.P. Wang (eds.). The Handbook of Pattern Recognition and Computer Vision. World Scientific Publishing.

Van den Boogaart, K.G. and R. Tolosana-Delgado. 2013. Analyzing compositional data with R. Springer, Berlin.

Van Der Sanden, J.J. and D.H. Hoekman. 2005. Review of relationships between grey-tone co-occurrence, semivariance and autocorrelation based image texture analysis approaches. Canadian Journal of Remote Sensing 31(3): 207–213.

Van Gijn, A.L. 2013. Science and interpretation in microwear studies. Journal of Archaeological Science 48: 166–169.

Van Overwalle, F. and D. Van Rooy. 1998. A connectionist approach to causal attribution. Connectionist models of social reasoning and social behavior 143–171.

VanPool, T.L. and R.D. Leonard. 2010. Quantitative Analysis in Archaeology. Wiley-Blackwell, London/Oxford.

Varien, M.D. and B.J. Mills. 1997. Accumulations research: problems and prospects for estimating site occupation span. Journal of Archaeological Method and Theory 4(2): 141–191.

Varien, M.D. and S.G. Ortman. 2005. Accumulations research in the Southwest United States: middle-range theory for big-picture problems. World Archaeology 37(1): 132–155.

Verhagen, P. 2013. Site Discovery and Evaluation Through Minimal Interventions: Core Sampling, Test Pits and Trial Trenches. pp. 209–225. *In*: Good Practice in Archaeological Diagnostics, Springer, Berlin.

Walker, W.H. and M.B. Schiffer. 2014. Behavioral Archaeology. pp. 837–845. *In*: Encyclopedia of Global Archaeology. Springer New York.

Wheeler, J.O., P.O. Muller and G.I. Thrall. 1998. Economic Geography. John Wiley and Sons, London.

Williams, A.N. 2012. The use of summed radiocarbon probability distributions in archaeology: a review of methods. Journal of Archaeological Science 39(3): 578–589.

Williamson, J. 2004. A dynamic interaction between machine learning and the philosophy of science. Minds and Machines 14(4): 539–549.

Woodbury, K.A. 2002. Inverse Engineering Handbook. CRC Press, Boca Raton (FL).

Wurzer, G., K. Kowarik and H. Reschreiter (eds.). 2013. Agent-based Modeling and Simulation in Archaeology. Berlin, Springer.

Zacharias, G.L., M. Macmillan and S.B. Van Hemel. 2008. Behavioral Modeling and Simulation: From Individuals to Societies. National Academies Press, Washington.

Zytkow, J.M. and J. Baker. 1991. Interactive Mining of Regularities in Databases. *In*: G. Piatetsky-Shapiro and W.J. Frawley (eds.). Knowledge Discovery in Databases. AAAI Press/The MIT Press, Menlo Park (CA).

2

A Short History of the Beginnings of Mathematics in Archaeology

François Djindjian

Introduction

In the years following 1950, like most scientific disciplines, Archaeology became the field of major research in the application of Mathematics and Computers. At the same time Archaeometry was applied to Physics and Earth Sciences with environmental reconstitution and palaeoclimatology. In this chapter we attempt, in brief, to rebuild the history of this period, the main actors and their contribution to the evolution of Archaeology.

1945–1950: A Worldwide Context for the Development of Scientific Research

The period between the two world wars saw the emergence of a quantitative movement in Anthropology, Sociology and Psychology that imperceptibly influenced Archaeology. But, after 1945, when scientists servicing the war effort (operational research) returned to basic research, they began to apply the methods and tools developed during the war, promoting quantitative approaches and the use of Mathematics.

Many European countries, inspired by the model of the USSR Academy of Sciences, founded large research organisations: CNRS in France, CNR in Italy, and of course in all the Eastern and Central Europe countries located behind the iron curtain. Other countries applied the classical model of lecturer-researchers in the University (USA, UK, Netherlands and Germany). Whatever the case, the very large number of researchers recruited between 1945 and 1965 in those institutions drove forward the

Université Paris I-Panthéon-Sorbonne France.
Email: francois.djindjian@wanadoo.fr

fundamental and applied research for thirty years until the end of the sixties when the recruitment of new researchers significantly decreased.

1950–1965: The General Development of a Quantitative Movement in Social and Human Sciences and in Archarology: Statistics and Graphics before Computers

From 1950 to 1965, Archaeology registered an impulse to quantification, using elementary statistics and graphics, which constitute of the actual methods for processing archaeological data. Many pioneers from USA and Europe contributed to the movement towards quantification:

Brainerd (1951), Robinson (1951)	Seriation
Spaulding (1953)	Typology and statistics (X^2)
Bordes (1953)	Cumulative diagram for assemblage identification
Clark and Evans (1954)	Nearest Neighbour Analysis and spatial analysis
Bohmers (1956)	Graphics and statistics for typology
Meighan (1959)	Seriation
De Heinzelin (1960)	Typology and statistics
Vescelius (1960)	Sampling
Ford (1962)	Graphics for seriation
Vertes (1964)	Statistics and graphics
Laplace (1966)	X^2 test and "Synthetotype" for cultural facies
Angel (1969)	Palaeodemography

The first book dedicated to Quantitative Archaeology was edited by Heizer and Cooke (1960). During the same period pioneering works concerned with the formalisation, the description and the recording of archaeological data for the purpose of archaeological data banks arose, which initially used punched card machines (Gardin 1956 for a punched card archaeological data bank and Gardin 1976, second edition, for the code of pottery description, one of the numerous set of artefact codes realised in the 1955–1965 periods). This was also the period when archaeologists used data matrices by reorganisation of rows and columns for seriation or typology (Clarke 1962).

1965: The Computer Liberates the Archaeologist from Manual Computations

After the first experimental machines (1946–1950), the fifties saw the deployment of the first computers: 1951, Univac 1 (Remington Rand); 1952, Gamma 2 (Machines Bull); 1952, IBM 701 (IBM). In 1955, FORTRAN computer language was developed on the IBM 704, the first scientific computer. In 1959, the first business computer, IBM 1401 was introduced. This was followed in 1964 by the famous IBM 360, the first number of the fully compatible upgraded IBM machines.

From 1960 onwards, computers for academic research were installed in computer centres at the universities for basic research. It was a heroic time for scientists who were obliged to develop their own software in the assembly or FORTRAN language. Fortunately, the computer departments in the universities had started developing the first software packages offering statistical tools (SPSS, Osiris, and BMDP). Not limiting themselves to elementary statistics, statistical tests or graphics, these packages were also offering tools to develop sophisticated algorithms like numerical taxonomy, factor analysis, and quantitative spatial analysis. This period also saw the very beginnings of mapping (Bertin 1967 for the origin of graphic semiotics), and maps were printed by special costly color graphic plotters until the end of the seventies.

During the same period, many methodological advances influenced the field of archaeological data analysis:

- Numerical taxonomy (Sokal and Sneath 1963, Sneath 1957),
- Quantitative Geography (Haggett 1965, Berry 1967, Chorley and Haggett 1967, Chisholm 1962),
- Mathematical Ecology (Pielou 1969), for the first applications in intrasite spatial analysis,
- System dynamics by J.W. Forrester (1968, Industrial Dynamics), (1969, Urban Dynamics) (1970, World dynamics),
- Catastrophe theory (Thom 1972),
- Mathematical modelling (algebraic and exponential models, linear programming, game theory, etc.), most of them announcing multi-agent system (Doran 1981),
- Sampling (Cochran 1977, Desabie 1966),
- Quantitative environmental studies and paleoclimatology,
- Physical and chemical characterization (Archaeometry, name introduced in 1958), using quantitative analysis,
- Expert systems (first ones are Dendral in 1965 and Mycin in 1972), and first archaeological applications at the end of the seventies (Gardin 1987).

1966–1976: The Quantitative Revolution in Archaeology

Around 1966, several papers marked the beginnings of a quantitative revolution in Archaeology:

- Hodson et al. (1966): a cluster analysis on Münsingen fibulae,
- Doran and Hodson (1966): a multidimensional scaling on upper Paleolithic assemblages,
- Binford and Binford (1966): a Factor analysis on Mousterian assemblages,
- Ascher and Ascher (1963), Kuzara et al. (1966), Hole and Shaw (1967), Craytor and Johnson (1968), Elisseeff (1968), Renfrew and Sterud 1969, etc.: Seriation algorithms,
- Renfrew, Dixon and Cann (1968): Characterization and exchange of Obsidian through Mediterranean sea (archeometry).

In 1970, an international conference at Mamaia (Romania), "Mathematics in the Archaeological and Historical Sciences", became the venue where statisticians met archaeologists: Rao, Kruskal, Kendall, Sibson, La Vega, Lerman, Wilkinson, Solomon, Doran, Ihm, Borillo, Gower, and where archaeologists showed how they could use statistics: Moberg, Spaulding, Cavalli-Sforza, Hodson, Orton, Hesse, Ammerman, and Goldmann. The first use of multidimensional scaling and cluster data analysis in Archaeology and historical texts was presented at this conference.

The period 1966–1976 belonged to the pioneers, scientists that were most often educated both in Archaeology, Natural Science and computers or mathematics:

- USA: R.G. Chenhall (1967), A.C. Spaulding (1953), G.L. Cowgill (1977), A.J. Ammermann (1979), F. Limp (1969), R. Whallon (1973), E. Zubrow (1975), D.W. Read (1989), D. Snow (1969),
- UK: J.D. Wilcock (1975), J. Doran (1970), Cl. Orton (1976), I. Hodder (1976), I. Graham (1980), D.G. Kendall (1963,1971),
- Italy: A. Bietti (1982),
- Germany: I. Scollar (1978), P. Ihm (1978), A. Zimmermann (1978),
- Netherlands: A. Voorrips (1973),
- Russia: P. Dolukhanov (1979),
- France: F. Djindjian (1976, 1977), M. Borillo (1977, 1978), J. Cl. Gardin (1970),
- Australia: L. Johnson (1977),
- Danemark: T. Madsen (1988).

The next generation of researchers, mostly archaeologists, arrived in the period 1976–1986: P. Moscati (1987), A. Guidi (1985), F. Giligny (1995), S. Shennan (1988), J.A. Barcelo (1996), H. Hietala (1984), K. Kintigh (1982), K. Kvamme (1980), J.M. O'Shea (1978), S. Parker (1985), S. Van der Leeuw (1981), M. Aldenderfer (1998), R. Drennan (2009), M. Baxter (1994), C. Buck (1996), D. Snow (1996), H. Kammermans (1995), R. Laxton (1989), etc.

1966–1976 saw the greatest growth in Quantitative Archaeology.

- Mathematics and computers (Doran and Hodson 1975 , World Archaeology 1982),
- « Spatial » Archaeology (Clarke 1977, Hodder and Orton 1976),
- Simulation in Archaeology (Clarke 1972, Hodder 1978, Sabloff 1981),
- Image processing in Archaeology (Scollar 1978),
- Harris matrix (Harris 1979),
- Demography (Hassan 1973, 1981, Masset 1973),
- Site catchment analysis (Vita-Finzi and Higgs 1970, Higgs 1975, Zubrow 1975),
- Sampling in archaeological surveys and excavations (Mueller 1975, Cherry et al. 1978),
- Mathematical models (Doran 1970, 1981),
- Exchange models (Wilmsen 1972, Renfrew 1975, 1977, Sabloff and Lamberg-Karlovsky 1975, Earle and Ericson 1977, Renfrew and Shennan 1982),
- Physical and chemical characterization in Archaeometry (Barrandon and Irigoin 1979),
- Environmental studies and paleoclimatology (Imbrie and Kipp 1971).

1970: The Revolution of Multidimensional Data Analysis

The mathematical foundations of multidimensional data analysis are well known since the beginning of 20th century (Principal Component Analysis by K. Pearson 1901). But the computations for obtaining eigenvalues during the process of diagonalizing the matrix were too long to be used without computers. This is the reason why their application were limited around 1930 to psychometrics (Spearmann, Thurstone, Guttman, Burt), with the design of special questionnaires that simplifying the computations.

From 1960 to 1970, the first computerised algorithms which were at the origin of a new revolution in Statistics appeared:

- Cluster analysis techniques, which allowed the graphical (dendrograms) representation of similarities between objects described by numerous variables, and production of clusters of objects. 1963 is the year of first publication of the famous book of R.R. Sokal and P.H.A. Sneath: "Numerical Taxonomy".
- The scaling techniques that reduced multidimensional space of data in to a one or two dimension scale, most famous of them being the "Non metric Multidimensional Scaling" of J.B. Kruskal (Kruskal, 1971) from Bell labs—first published in 1964.
- The "Factor" analysis techniques, a family of techniques based on the diagonalization of a matrix of correlation or association between individuals or variables, including the Principal Components Analysis (Pearson 1901), the Factor analysis (Spearman 1904), the Discriminant analysis (Mahalanobis 1927, Fisher 1936) and the Correspondence analysis (Benzecri 1973).

The seventies and eighties saw dynamic research in multidimensional data analysis methods, with thousands of new algorithms added every year. The leading researchers of that period were polarized around the English speaking school (Hodson et al. 1971) that emphasised the use of the multidimensional scaling method, and the French school of data analysis which inspired numerous archaeological applications from 1974 onwards in France (Djindjian 1989 for a review). Correspondence Analysis, developed by French mathematicians found its way in to the English speaking world from 1985, first in the Scandinavian countries and later in United-Kingdom and USA (Baxter 1994).

During the eighties, the capabilities of multidimensional data analysis techniques to solve many archaeological methods formed the roots for its later success. Among the numerous classic or prototypal techniques of data analysis, Correspondence analysis and PCA, associated with an appropriate cluster analysis, appeared to be robust and easy to use techniques, even to non mathematician researchers. Archaeology played a major role among all the Human and Social Sciences, in showing how to integrate Statistics into archaeological methods. Quantification, statistics, and data analysis are in the tool box of archaeologists as a result of their integration in easy to use computer packages:

- Surveying (artifact surface collecting studies),
- Stratigraphic analysis (Harris matrix computerized reorganization),
- Artifact analysis (see under Chapter 8),
- Stylistic analysis,
- Taxonomy and Classification (Anthropology, Paleontology, Genetics),
- Identification of cultural systems (see under Chapter 10),
- Seriation (see under Chapter 9),
- Intrasite spatial analysis of dwellings and funerary structures (see under Chapter 11),
- Paleoenvironmental studies,
- Raw material procurement and craft manufacturing sources,
- Mathematical modeling (see under Chapter 12),
- Intersite spatial analysis and landscape studies,
- Any intrinsic and extrinsic structuring (general case).

Prehistory, Classical Archaeology and the New Archaeology

The development of mathematics and computers in Archaeology, have more or less been welcomed, depending on the nature of the different archaeological sub disciplines. Prehistoric Archaeology has preferred multidimensional data analysis for typometry, assemblage identification, intrasite spatial analysis and environmental studies. Classical Archaeology has been the field for data base management, notably for indexing Epigraphic remains, and in Cultural Resource Management. In all fields within Archaeology, G.I.S. software and related methodologies have been used for intersite spatial analysis and the study of ancient exchange systems. Theoretical developments since the 60s, like New Archaeology, oriented cultural anthropology, functionalism and cultural ecology, explain the preference for deductive models implementing statistical tests and mathematical modelling. Archeometry and environmental studies (geoarchaeology, palaeoclimatology, archaeobotany, etc.) have emphasized the use of multidimensional data analysis methods. Nevertheless, since 1980, changes in the dominant theoretical background, hermeneutic post-processual Archaeology, founded as a reaction to the scientifism of New Archaeology, has rejected the use of formalization and logical proofs and, as such, it does not approve of the need for statistics or mathematics.

Processual Archaeology, in USA (Binford, Flannery, Hill, Plog and many others) and in UK (Clarke, Renfrew and Hodder, this scholar just before his "conversion" to an alternative post-processualist archaeology), and all the numerous archaeologists who followed the theoretical movement, promoted the use of statistical tests in a deductive framework approach (see Cowgill 1977 for a discussion). It is certainly the place where parametric and non-parametric statistical tests have been used most intensively in archaeology. A review of the production of these archaeologists before 1990 shows how difficult it has been to have representative samples of archaeological data, and then to have the adequate conditions to apply a statistical test (see for example Soffer 1985 for a disciplined but without "common sense" application in a PhD thesis). It is probably the reason why Schiffer (1976) developed his "Behavioural Archaeology" as a divergent evolution from mainstream processualist Archaeology, to distinguish "systemic culture" (the reality) from "archaeological culture" (the archaeological data) revealing the major role of the taphonomic analysis.

A First Case Study: Typometry

Artefact classification or typology is one of the basic methods of Archaeology since Montelius. Until 1950, the classification was the result of a subjective visual inspection of artefacts. The analogy with numerical taxonomy methods in the Earth Sciences allowed the formalization of artefact description (attributes), and created the basis for formal classifications on the basis of similarities between artefacts, quantified from descriptive measurement.

Numerous statistical approaches have been proposed, some of them are cited below:

- *Attribute analysis* (Spaulding 1953) is based on the use of X^2 tests for measuring the association between attributes, extended logically to multidimensional contingency table analysis (Spaulding 1977).

- *Matrix analysis* (Tugby 1958, Clarke 1962) is based on the reorganization of the rows and columns of a matrix of presence-absence or percentages to reveal a partition inside the matrix, demonstrating the evidence of several types.
- *Biometry* (Bohmers 1956, Heinzelin 1960) is based on the use of Laplace-Gauss elementary statistics to reveal the existence of multi modal peaks in histograms or separated point clouds in diagrams, to isolate types.
- *Numerical taxonomy* (Hodson et al. 1966) is based on the use of techniques of cluster analysis to identify archaeological types.
- *Typological Analysis* (Djindjian 1976) is an improvement of numerical taxonomy techniques. The Typological Analysis is based on R + Q Correspondence Analysis and/or PCA associated with a Cluster analysis. The "*Multiple Typological Analysis*" (Djindjian 1991) is based on a several Typological Analysis applied on each homogeneous intrinsic variables (morphology, technology, decoration, gripping, raw material, etc.) and a final one applied to the matrix of the clusters resulting in the previous analyses.
- *Morphology analysis* (pattern recognition) is based on multidimensional data analysis techniques, applied on the profile digitalization of artefacts. Different coding of profile measures have been tested and proposed: Sliced method (Wilcock and Shennan 1975), Tangent-profile technique (Main 1986), Extended sliced method (Djindjian 1985c), B-spline curve (Hall and Laflin 1984), Fourier series (Gero and Mazzula 1984), Centroïd and cyclical curve (Tyldesley et al. 1985), Two-curves system (Hagstrum and Hildebrand 1990), etc.

A Second Case Study: Seriation

Seriation is certainly the most original method in Archaeology, in the sense that it was created by archaeologists for archaeologists (with the help of mathematicians). It is used for discovering a chronological order between artefacts (based in their quantitative description) and mainly between closed sets, particularly burials in a cemetery (from a previous classification of diagnostic types). Many different algorithms have been proposed to solve the problem of the seriation:

- Similarity matrix ordering ((Brainerd 1951, Robinson 1951, Landau and de La Vega 1971),
- Graphical comparative methods (Ford 1962, Meighan 1959),
- Matrix reorganization (Clarke 1962, Bertin 1977),
- Direct ordering of the incidence matrix (Kendall 1963, Regnier 1977),
- Computerised similarity matrix ordering (Asher and Asher 1963, Kuzara et al. 1966, Hole and Shaw 1967, Craytor and Johnson 1968),
- Rapid methods on similarity matrix (Dempsey and Daumhoff 1963, Elisseeff 1968, Renfrew and Sterud 1969, Gelfand 1971),
- Multidimensional scaling (Kendall 1971),
- Travelling salesman problem (Wilkinson 1971),
- Reciprocal averaging method (Goldmann 1973, Wilkinson 1974, Legoux 1980),
- Correspondence analysis (Djindjian 1976),
- PCA (Marquardt 1978),
- Rapid method for incidence data matrix (Ester 1981),

- Toposeriation (Djindjian 1985a),
- Other (Ihm 1981, Laxton and Restorick 1989, Baxter 1994, Barcelo and Faura 1997, etc.). More recently, we should mention Bayesian applications (Buck and Sahu 2000).

Today, Correspondence analysis is the most popular and easy to use technique of seriation, delivering a double parabola (the Guttmann effect), ordering chronologically both objects (burials) and types. The technique is very robust, and is able to reveal errors of recording or excavation and inaccuracies of typology, and allows for the separating of time scales from other non-time parasite scales. Many applications of seriation algorithms were published during 1970 and 1990 (see Djindjian 1991 for a more exhaustive bibliography).

A Third Case Study: Typology, Assemblage, Culture and «System» in the Culture-Historical Archaeology

When confronted with closed sets of artifacts, the archaeologist usually compares them to each other, as well as to some reference sets, and links the similarities between assemblages within time and space. This is the origin of the concept of "culture", which has many analogues in the archaeological literature (techno-complexes, assemblages, factories, industries, and cultures).

Traditionally, specific methods have related artifacts to types (typology), and then types to cultures (or reference sets), by partitioning a matrix of percentages of types in stratigraphically ordered archaeological layers. The approach started with simple statistical techniques, such as the cumulative diagram of Bordes (1953) or the histograms and chi-squared tests of Laplace (1966) who defined his approach as the "*synthétotype*" method.

The real multidimensional scale of the problem was understood by Binford and Binford (1966) in their famous revisiting study of European Mousterian assemblages, refuted unfortunately by a wrong use of their particular implementation of factor analysis. At the same time, the first use of a Multidimensional Scaling algorithm by Doran and Hodson (1966) proved the potential of the multidimensional approach to solve a similar problem.

The main technical difficulty at that time was to process both individuals (Q method) and variables (R method). Several techniques were proposed for such tasks around the 1970s (Christenson and Read 1977). Later, the use of Correspondence analysis on contingency tables of types (Djindjian 1976) or Burt tables of attributes, avoided the use of subjectively built typologies (Djindjian 1980, 1989). Many applications using multidimensional data analysis have been published on assemblage tables of artefact counts. We can mention studies of Middle Palaeolithic times (Callow and Webb 1981), the Upper Palaeolithic time period (many contributions producing a new classification and chronology of the European upper Palaeolithic), Mesolithic and Neolithic (see Djindjian 1991 and Djindjian et al. 1999 for a more complete bibliography).

A Fourth Case Study: Intrasite Spatial Analysis in Archaeology

The beginning of spatial analysis in Archaeology is associated with the influence from Quantitative Ecology (Pielou 1969). Among these early methods we can mention Nearest

Neighbor Analysis and tests on grid counting. Whallon (1973, 1974) and Dacey (1973) were the first to apply such techniques to artifact distribution on occupation floors, to discover the existence of material concentrations. Ever since those preliminary investigations, it has been evident that archaeological artifact distributions have a multidimensional component, especially due to the different nature of archaeological materials involved: lithics, ceramics, stones, bones, etc.

Tests for spatial associations were proposed in the seventies as a technical improvement: Hietala and Stevens (1977), Hodder and Orton (1976), Hodder and Okell (1978), Clarke (1977), Berry et al. (1980), Hietala (1984), Carr (1985), etc. The most successful method was the multidimensional intrasite spatial data analysis:

- Local density analysis by Johnson (1977),
- Spectral analysis by Graham (1980),
- (x, y) clustering by Kintigh and Ammerman (1982),
- Unconstrained clustering by Whallon (1984),
- Spatial structure analysis with topographical constraints by Djindjian (1988),
- Spatial structure analysis on refitting artifacts by Djindjian (1999).

These methods today constitute the State of the Art in spatial analysis. But the rapid development of G.I.S. software suggests the need to integrate all these techniques in an easy to use computational framework.

A Fifth Case Study: Mathematical Modeling

In the seventies, mathematical modeling techniques were put to various uses in Archaeology, for example, algebraic and exponential equations, gravity models, linear programming, stochastic processes (including the use of Monte-Carlo techniques), etc. We can mention the following applications:

- Population models of hunter-gatherer groups (Wobst 1974),
- Fitting logistic curves for demographic estimation of cities or regions (Ammerman et al. 1976): $Y = A/(1 + B \exp(-kT))$,
- Population estimation of hunter-gatherer groups from surface and structure of dwelling areas (Hassan 1975): $A = 0{,}7105\ P \exp 1{,}76$,
- Boundary models (Renfrew and Level 1979): $I = C \exp(a) - kd$ with $a = 0{,}5$, $k = 0{,}01$,
- Gravity models (Fulford and Hodder 1974, Hodder and Orton 1976),
- Subsistence models (Jochim 1976, Keene 1979),
- Transition model from hunter-gatherer economy to farming and breeding economy (Reynolds and Ziegler 1979, Reynolds 1981),
- Stochastic models for random walk process (Hodder and Orton 1976).

Beyond elementary mathematical models, the modeling of complex systems has been a major research target since the seventies, using more and more elaborate methods (Doran 1990, Djindjian 2011).

The study of (linear) **System Dynamics** to understand complex interactions and feedback among social, environmental, cultural and economic factors has been the domain of System theory (Forrester 1968, Principles of Systems 1969, Urban Dynamics 1970 World Dynamics). In this approach, the model is committed to study the interactions between the main components of the system, from a set of linear and differential equations, defining a matrix of interaction, whose properties allow highlighting equilibrium solutions, independent of time. This technique has been used in the hypothetical models of peopling (Thomas 1972, Zubrow 1975, Reynolds 1986, Ammermann and Cavalli-Sforza 1973, 1984, O'Shea 1978, Black 1978) and population (Ward et al. 1973, Wobst 1974, Martin 1972, McArthur et al. 1976, Snow 1993). Lowe (1985) uses it to study the Mayan collapse.

At the same time, **Fractal theory** proposed by B. Mandelbrot (1983), which has ties to Chaos theory, has some interesting applications in Archaeology (Brown et al. 2005), especially in the following fields: use-wear traces (mathematical characterization of the traces), artifact fragmentation (stone knapping, ceramic breaking), size distribution of sites by the Zipf-Mandelbrot law (Guidi 1985) and non-linear dynamic systems particularly for the study of transitions (Nottale et al. 2000).

Similarly, **the catastrophe theory** of R. Thom (1972) offers applicable modeling when gradually changing forces produce sudden effects, called collapses, due to their discontinued nature, not only in time, but also in the form, that is morphogenesis. Renfrew (1978), Renfrew and Cooke (1979) introduced this modeling tool in archaeology for cultural change explanatory models.

The underlying linear assumption of system dynamics according to J.W. Forrester, has considerably limited its applications in human and social sciences. The works of I. Prigogine (1968) concerning **non-linear systems** and of T.Y. Li and J.A. Yorke (1975) concerning **chaos theory** (system dynamics which are sensitive to initial conditions) led to a revival of complex systems modeling in Archaeology at the end of the 1990s (Van der Leeuw 1981, Van der Leeuw and Mc Glade 1997, Beekman and Baden 2005). McGlade and Allen (1986) have studied the horticulture economic system of the Huron people (North America) through non-linear differential equations with stochastic variables. They emphasize the unstable equilibrium of the system and the constant stress of the populations, which a more simple functional approach could not detect. However, these applications have remained very limited, although the nature of the non-linearity of the socio-cultural processes has always been admitted.

Pioneering work on **multi-agent systems** by J. Doran concerning the Mayan collapse (1981), allowed the modern development of this new modeling approach. In the last twenty years, numerous archaeological applications of multi-agent systems have been published (Gilbert and Doran 1994, Kohler and Gumerman 2000, Kohler and Van der Leeuw 2007, Gilbert 2008):

- The EOS project (Doran et al. 1994) for the social organization of Magdalenian hunter-gatherers in Perigord,
- The Anasazi peopling in Arizona between 800 and 1350 AD (Dean et al. 2000, Axtell et al. 2002, Janssen 2009),
- The Pueblos peopling in the Mesa Verde region between 600 and 1300 AD with the Village Ecodynamics Projects VEP1 (2002–2008) and VEP2 (2009–2013) in (Kohler and Van der Leeuw 2007),
- The Enkimdu project for Mesopotamian archaeology (Wilkinson et al. 2007),
- A kinship based demographic simulation of societal processes (Read 1998).

Criticism of Quantitative Archaeology

Towards the end of the eighties, Quantitative Archaeology and Statistics were no longer the most favoured. Many reasons seem to have contributed to explain such a situation:

- Quantitative Archaeology was passing from the field of research to the field of current use, corresponding to the edition of general textbooks (see bibliography),
- The available archaeological data had been exhausted by the quantitative movement and it was necessary to come back to a new data acquisition phase,
- The ambitious objectives at the end of the 60s (the so called "New" Archaeology) were often applied with a naïve approach and without sufficient mathematical and methodological rigour, and were considered more as fashion than as the right epistemological basis of the discipline,
- The development of microcomputers has allowed the real development of computing archaeology and consequently new methods research has shifted from mathematics and statistics to computing applications (data banks, G.I.S., Archaeological Information System, CRM, etc.), as in other fields of scientific research, notably in Human and Social Sciences.

But a strong criticism about the need for quantification in the Human and Social Sciences was also emerging, as exemplified by the enormous academic success in USA and UK of French deconstructivism (Derrida, Foucault, etc.). It supported a new theoretical background (the so called "Post-processual Archaeology"). Among the main criticisms, it may be necessary to focus on the following aspects:

- A measure is not knowledge (S.J. Gould, *"the mismeasure of man"*, 1981);
- A structure is not a system (deconstruction of the concept of archaeological "culture");
- Any theoretical models can be fitted to any data set (poverty of models, complexity of civilizations, data failure);
- The bias of the archaeological record does not allow any reliable quantitative or statistic process (Behavioural Archaeology).

Nevertheless, such criticism has been useful for quantitative archaeology to distinguish the useful from the good and from the bad, and to develop new cognitive approaches based on new methodological and technical grounds.

New Recent Applications

Since the nineties, archaeology has always been the field of new developments in applied mathematics but more so in the context of computing than as abstract solutions. Geographical Information Systems are a major development in archaeology. This software solution to integrate mapping, data and data query mechanisms has found its way in CRM (Cultural Resource Management), Rescue Archaeology and large surveys but it has allowed developments in intersite spatial archaeology, towards a new Landscape archaeology. Classical applications of Thiessen polygons, frontier and gravity models, Central Place theory (Johnson 1975), etc., have been superseded by the recent progress in quantitative geography, as integrated in modern GIS software. The most popular techniques are now intervisibility or viewshed analysis (Wheatley and Gillings 2002) and the measuring of least cost distances (Herzog 2013). In any case, too many archaeologists do not know the algorithms behind the tools and we have the risk of using these as a new theoretical hammer.

The emergence of an "Archeo-geography" whose research objective would be the study of territory in the past and not only the archaeological sites is a real challenge (Chouquer 2008). The concepts of resilience and self-organization may have echoes of old mathematics (catastrophe theory, chaos theory), but they offer new insights to the study of historical dynamics. Also groundbreaking might be the possibility of a dynamic analysis of spatial data (Durand-Dastes et al. 1998, Archaedyn 2008), based on quantitative ratios of geographical statistics (hierarchy ratios, intensity and stability ratios, etc.). Unfortunately, most archaeological data comes from old excavations, from the 19th century, or from modern rescue interventions where time and money are of paramount relevance, and not the exhaustiveness of data and the reliability of sampling methods. As a consequence, what we have actually measured from the past is not representative enough of the original population in the past. Statistics in Geography, working on contemporaneous data, rely on well extracted samples of a known population. This is not the case in Archaeology. Therefore, the use of robust statistics and careful validated conclusions is an obvious need. The criticism of Schiffer (1976) is particularly significant in this case.

More recently, since 2010, the applications of graph-theory based models or network analysis has arrived in Archaeology (Brughmans 2010, Knappett 2013). It is also important to note the relevance of Bayesian statistical inference, which has been introduced in Archaeology (Buck et al. 1996, Cowgill 2002) with particular success in ^{14}C date calibration programs.

1990–2000: Standardization, Embedding and Theorization

Since 1990, Quantitative Archaeology seems to have a few enthusiastic archaeologists and their number continues to diminish. The decline of more "scientific" influences (Physics, Mathematics, etc.) in Social and Human Sciences becomes more and more noticeable. And the academic success of Earth Sciences, boosted by the considerable fear of climatic change on the earth's environmental catastrophes, neglects archaeological data for more suitable non-anthropic records.

Use of computers in Archaeology can be limited in a majority of cases to word processing, spreadsheets, drawing, picture processing and presentation software. But the use of DBMS by National Heritage Agencies, the huge interest in G.I.S. and multimedia data banks, the possibilities offered by Virtual Reality are transforming archaeology into a computer addicted discipline. Computers are then not only a strategic weapon for the most dynamic archaeologists but also the professional tool for mainstream work with Cultural Heritage. Moreover, the Internet has also revolutionized the practice of Archaeology, in communication, on-line libraries queries, Google assisted retrieval systems, etc. Statistical techniques are still present in typometry, spatial analysis, raw material provenance analysis, seriation, Harris matrix management, archaeological surveys, etc., but they are now more and more embedded into software applications (statistical packages, GIS, Virtual Reality, etc.) and archaeological methods—"*Techniques are changing, methods are going on*" (Djindjian 1991, Djindjan and Ducasse 1987).

A further step forward has been the attempt to integrate quantitative methods with archaeological reasoning in an integrated construct, with the ambition to edify a general theory of archaeological knowledge independent of traditional theoretical paradigms (Gardin 1979, Read 1989, Djindjian 2002, Djindjian and Moscati 2002). It opens the door for cognitive approaches. To advance in this direction we would need to reduce the artificial opposition between quantification and semiotics. In this way we could converge towards a three step cognitive model inspired by the logic of Charles Peirce (1992):

Step 1	Acquisition	Qualitative and quantitative acquisition
Step 2	Structuring	Structuring by data Analysis
Step 3	Reconstitution	Modeling by logic and/or mathematical formalized, demonstrative and standardized discourse

Conclusions

The national and regional development of Cultural Resource Management, the specialization of the archaeological research domains, and the fast development of Rescue Archaeology are contributing largely to professionalize Archaeology, given the availability of funding to recruit young archaeologists. These modern circumstances impose the need of pursuing elaborate "State of the Art" archaeological techniques and methods to facilitate the access to computerized tools. Further, there is also need of an emergence of new standards (recording, reporting, thesaurus, etc.), codes for good practices and increased productivity.

Even though Quantitative Archaeology is present everywhere now-a-days, and is embedded into application software, statistical packages and current methods of surveying, data acquisition and processing, but also new advances are needed for increasing the conceptualization and the formalization of archaeological projects. Quantitative Archaeology will be always there, because Archaeology is more and more a multidisciplinary science, integrating exact sciences, geosciences, social and human disciplines and computer engineering, where quantitative approaches are natural.

References Cited

To avoid giving too long a list of references, only the papers cited in this chapter are listed here. It is possible to read more complete references in:

Djindjian, F. 1991. Méthodes pour l'Archéologie, Paris, Armand Colin.

Ryan, N.S. 1988. A bibliography of computer applications and quantitative methods in archaeology. *In*: S.P.Q. Rahtz (ed.). Computer and Quantitative Methods in Archaeology. BAR International Series 446(1): 1–27.

Wilcock, J.D. 1999. Getting the best fit? 25 years of statistical techniques in archaeology. pp. 35–52. *In*: L. Dingwall, S. Exon, V. Gaffney, S. Laflin and M. van Leusen (eds.). Archaeology in the Age of the Internet. CAA97. Computer Applications and Quantitative Methods in Archaeology. Proceedings of the 25th Anniversary Conference (Birmingham 1997), BAR International Series 750, Oxford, Archaeopress.

It is also possible to find references in reviews or proceedings dedicated to computing archaeology:

Review *"Archéologia E Calcolatori"* since 1990.

Proceedings of the annual conference "Computer Applications and mathematical methods in Archaeology" (CAA), since 1973.

Proceedings of the conferences of the commission 4 of the International Union for Prehistoric and protohistoric Sciences (UISPP), since 1976 interligne.

Aldenderfer, M.S. 1998. Quantitative methods in archaeology: A review of recent trends and developments. Journal of Archaeological Research 6(2): 91–120.

Ammerman, A.J. and L.L. Cavalli-Sforza. 1973. A Population model for the diffusion of early farming in Europe. pp. 43–358. *In*: C. Renfrew (ed.). The Explanation of Culture Change. London: Duckworth.

Ammermann, A.J., L.L. Cavalli-Sforza and D.K. Wagener. 1976. Towards the estimation of population growth in Old world Prehistory. In: Demographic Anthropology: Quantitative approaches. E. Zubrow (ed.). Albuquerque, University of Mexico Press.

Ammermann, A.J. and L.L. Cavalli-Sforza. 1984. The Neolithic transition and the genetics of populations in Europe. Princeton: Princeton University Press.

Angel, J.L. 1969. The Bases of paleodemography. American Journal of Physical Anthropology 30: 427–437.

Archaedyn. 2008. 7 millennia of territorial dynamics, final conference Archaedyn, June 2008, preprints, pp. 232.

Ascher, R. and M. Ascher. 1963. Chronological ordering by computer. American Anthropologist 65: 1045–1052.

Axtell, R.L., J.M. Epstein, J.S. Dean, G.J. Gumerman, A.C. Swedlund, J. Harburger, S. Chakravarty, R. Hammond, J. Parker and M. Parker. 2002. Population Growth and Collapse in a Multi-Agent Model of the Kayenta Anasazi in Long House Valley, Proceedings of the National Academy of Sciences 99(3): 7275–7279.

Barcelo, J.A. 1996. Arqueología Automática. El uso de la Inteligencia Artificial en Arqueología. Editorial Ausa, Barcelona.

Barcelo J.A. and J.M. Faura. 1997. Time series and neural networks in archaeological seriation. *In*: Archaeology in the Age of Internet. L. Dingwall, S. Exon, V. Gaffney, S. Laflin and M. Leusen (eds.). Oxford, Archeopress, BAR S750, pp. 91–102.

Barrandon, J.N. and J. Irigoin. 1979. Papiers de Hollande et papiers d'Angoumois de 1650 à 1810. Archaeometry 8: 101–106.

Baxter, M.J. 1994. Exploratory multivariate Analysis in Archaeology, Edinburgh University Press, Edinburgh.

Beekman, C.S. and W.W. Baden (eds.). 2005. Nonlinear Models for Archaeology and Anthropology, Continuing the Revolution, Hampshire-Burlington, Ashgate.

Benzecri, J.-P. 1973. L'Analyse des données. Paris: Dunod, vol. 2.

Berry, B.J.L. 1967. The Geography of market centres and retail distribution. Englewood Cliffs, N.J.: Prentice Hall.

Berry, K.J., K.L. Kvamme and P.W. Mielke. 1980. A Permutation technique for the spatial analysis of the distribution of artefacts into classes. American Antiquity 45: 55–59.

Bertin, J. 1967. Sémiologie graphique: les diagrammes, les réseaux, les cartes. Paris, La Haye: Mouton; Paris: Gauthier-Villars, cop.

Bertin, J. 1977. Le Graphique et le traitement graphique de l'information. Paris, Flammarion.

Bietti, A. 1982. Techniche matematiche nell'analisi dei dati archaeologici. (Contributi del Centro Linceo interdisciplinare di scienze matematichi e lero applicazioni, 61.) Roma, Academia Nationale dei Lincei.

Binford, L.R. and S.R. Binford. 1966. A Preliminary analysis of functional variability in the Mousterian of Levallois facies. American Anthropologist 68: 238–295.

Black, S. 1978. Polynesian outliers: a study in the survival of small populations. pp. 63–76. *In*: I. Hodder (ed.). Simulation Studies in Archaeology. Cambridge: Cambridge University Press.

Bohmers, A. 1956. Statistiques et graphiques dans l'étude des industries lithiques préhistoriques. Palaeohistoria, 5.

Bordes, F. 1953. Essai de classification des industries moustériennes. Bulletin de la Société Préhistorique Française 50: 457–466.

Borillo, M. 1977. Raisonnement et méthodes mathématiques en archéologie/ed. M. Borillo. Paris, Éditions du C.N.R.S.

Borillo, M. 1978. Archéologie et calcul: textes recueillis/M. Borillo. Paris: Union générale d'éditions (Collection 10/18: ser. 7; 1215).

Brainerd, G.W. 1951. The Place of chronological ordering in archaeological analysis. American Antiquity 15: 293–301.

Brown, C.T., W.R.J. Witschey and L.S. Liebovitch. 2005. The broken past: Fractals in Archaeology, Journal of Archaeological methods and theory 12(1): 37–78. Simuler une "artificial society": organisation sociale, gouvernance et attitudes societales 31.

Brughmans, T. 2010. Connecting the dots: towards archaeological network analysis. Oxford Journal of Archaeology 29(3): 277–303.

Buck, C.E., W.G. Cavanagh and C.D. Litton. 1996. Bayesian approach to interpreting archaeological data. New-York, Wiley.

Buck, C.E. and S.K. Sahu. 2000. Bayesian models for relative archeological chronology buildings. Applied Statistics 49(4): 423–440.

Callow, P. and E. Webb. 1981. The application of multivariate statistical techniques to middle palaeolithic assemblages from southwestern France. Revue d'Archéométrie 5: 330–338.

Carr, C. 1985. Alternative models, alternative techniques: variable approaches to intrasite spatial analysis. pp. 302–473. *In*: Christopher Carr (ed.). For Concordance in Archaeological Analysis. Kansas City, Missouri: Westport Publishers.

Chenhall, R.G. 1967. The description of archaeological data in computer language. American Antiquity. 32: 161–167.

Cherry, F., C. Gamble and S. Shennan (eds.). 1978. Sampling in contemporary British archaeology. Oxford: British Archaeological Reports (British Archaeological Reports: British ser., 50).

Chisholm, M. 1962. Rural settlement and land use. London, Hutchinson. Chorley, R. and P. Haggett (ed.). 1967. Models in Geography. London, Methuen la référence Chorley et Hagett a disparu.

Chorley, R. and P. Haggett (ed.). 1967. Models in Geography. London, Methuen.

Chouquer, G. 2008. Traité d'archéogéographie. Paris, Errance.

Christenson, A.L. and D.W. Read. 1977. Numerical taxonomy, R-mode factor analysis and archaeological classification. American Antiquity 42: 163–179.

Clark, P.J. and F.C. Evans. 1954. Distance to nearest neighbour as a measure of spatial relationships in populations. Ecology 55: 445–453.

Clarke, D.L. 1962. Matrix analysis and archaeology with particuliar reference to British Beaker Pottery. Proceedings of the Prehistorical Society 28: 371–382.

Clarke, D.L. 1972. Models in archaeology/ed. D.L. Clarke. London: Methuen.

Clarke, D.L. 1977. Spatial archaeology/ed. D.L. Clarke. 2ed. London, Methuen & Co.

Cochran, W.G. 1977. Sampling techniques. New York, Wiley.

Cowgill, G.L. 1977. The Trouble with signifiance tests and what we can do about it. American antiquity 42: 350–368.

Cowgill, G.L. 2002. Getting Bayesian ideas across a wide audience. Archaeologia E Calcolatori 13: 191–196.

Craytor, W.B. and L. Johnson. 1968. Refinements in computerized item seriation. Bulletin/Museum of Natural History, University of Oregon, 1968, 10.

Dacey, M.F. 1973. Statistical tests of spatial association in the location of tool types. American antiquity 38: 320–328.

Dean, J.S., G.J. Gumerman, J.M. Epstein, R.L. Axtell, A.C. Swedlund, M.T. Parker and S. McCarroll. 2000. Understanding Anasazi Culture Change Through Agent-Based Modeling pp. 179–207. *In*: A. Timothy Kohler and G.J. Gumnerman (eds.). Dynamics in Human and Primate Societies, Agent-Based Modeling of social and spatial processes, New York-Oxford, Oxford University Press.

Dempsey, P. and M. Baumhoff. 1963. The Statistical use of artefact distributions to establish chronological sequence. American Antiquity 28(4): 496–509.

Desabie, J. 1966. Théorie et pratique des sondages. Paris, Dunod. (Statistique et programmes économiques; 10.)

Djindjian, F. 1976. Contributions de l'Analyse des Données à l'étude de l'outillage de pierre taillée. Mémoire de Maîtrise, Paris 1: 1976.

Djindjian, F. 1977. Burins de Noailles, burins sur troncature et sur cassure: statistique descriptive appliquée à l'analyse typologique. Bulletin de la Société Préhistorique Française, 72, Comptes rendus des séances mensuelles 5: 145–154.

Djindjian, F. 1980. Faciès chronologiques aurignacien et périgordien à La Ferrassie (Dordogne). Dijon, Archéologia. 1980. (Dossiers de l'archéologie; 42), pp. 70–74.

Djindjian, F. 1985a. Seriation and toposeriation by correspondence analysis. *In*: A. Voorrips and S.H. Loving (eds.). To Pattern the Past. PACT 11: 119–136.

Djindjian, F. 1985b. Typologie et culture: l'exemple de l'aurignacien. pp. 15–38. *In*: M. Otte (ed.). La Signification Culturelle des Industries Lithiques. British Archaeological Reports (British Archaeological Reports: International ser.; 239).

Djindjian, F. (ed.). 1985c. Rapport d'activités du séminaire Informatique et mathématiques appliquées en Archéologie, 1984–1985. Paris: G.R.A.Q.

Djindjian, F. 1988. Improvements in intrasite spatial analysis techniques. *In*: C.L.N. Ruggles and S.P.Q. Rahtz (eds.). Computer and Quantitative Methods in Archaeology 1988. Oxford: B.A.R., Intern. series 446: 95–106.

Djindjian, F. 1989. Fifteen years of contributions of the French school of Data Analysis to quantitative Archaeology. *In*: S.P.Q. Rahtz and J. Richards (eds.). Computer and Quantitative Methods in Archaeology 1989. Oxford: B.A.R., Intern. series 446(i): 95–106.

Djindjian, F. 1991. Méthodes pour l'Archéologie. Paris: Armand Colin.

Djindjian, F. 1999. L'analyse spatiale de l'habitat: un état de l'art. Archeologia E Calcolatori, n°10, pp. 17–32.

Djindjian, F. 2002. Pour une théorie générale de la connaissance en archéologie. XIV° Congrès International UISPP, Liège Septembre 2001. Colloque 1.3. Archeologia e Calcolatori, n°13, pp. 101–117.

Djindjian, N.F. 2011. Manuel d'Archéologie. Paris: Armand Colin.

Djindjian, F. and H. Ducasse (eds.). 1987. Data Processing and Mathematics Applied to Archaeology. Ecole d'été Valbonne-Montpellier: Mathématiques et Informatique appliquées à l'Archéologie, 1983. Pact n°16.

Djindjian, F. and P. Moscati (eds.). 2002. Proceedings of the commission 4 Symposia, XIV° UISPP congress, Liege, 2001. Archeologia E Calcolatori, vol. 13.

Dolukhanov, P.M. 1979. Ecology and Economy in Neolithic eastern Europe. London, Duckworth.

Doran, J. 1970. Systems theory, computer simulations and archaeology. World archaeology 1: 289–298.

Doran, J. 1981. Multi-actor systems and the Maya collapse. pp. 191–200. *In*: Congrès de l'Union Internationale des Sciences Préhistoriques et Protohistoriques. 10. 1981. Mexico. Colloquium 5: Data management and mathematical methods in archaeology (preprint).

Doran, J. 1990. Computer-based simulation and formal modelling in Archaeology. pp. 93–114. *In*: A. Voorips (ed.). Mathematics and Information Science in Archaeology: a flexible framework Bonn: Holos, Studies in Modern Archaeology, vol. 3.

Doran, J.E. and F.R. Hodson. 1966. A Digital computer analysis of Palaeolithic flint assemblages. Nature 210: 688–689.

Doran, J.E. and F.R. Hodson. 1975. Mathematics and computers in Archaeology. Edinburgh, Edinburgh University Press.

Doran, J., M. Palmer, N. Gilbert and P. Mellars. 1994. The EOS Project: Modeling Upper Palaeolithic Social Change. pp. 195–221. *In*: N. Gilbert and J. Doran (eds.). Artificial Societies, London, UCL Press.

Drennan, R.D. 2009. Statistics for Archaeologists: A Commonsense Approach. Second Edition. New York: Springer.

Durand-Dastès, F., F. Favory, J.-L. Fiches, H. Mathian, D. Pumain, C. Raynaud, L. Sanders and L. Van Der Leeuw. 1998. Des oppida aux métropoles. Archéologues et géographes en vallée du Rhône. Paris: Anthropos.

Earle, T.K. and J.E. Ericson (eds.). 1977. Exchange systems in Prehistory. New-York, Academic Press.

Elisseeff, W. 1968. De l'application des propriétés du scalogramme à l'étude des objets. pp. 107–120. *In*: J.-Cl. Gardin (ed.). Calcul et Formalisation dans les Sciences de l'Homme. Paris: Éditions du C.N.R.S.

Ester, M. 1981. A Columm-wise approach to seriation. pp. 125–156 (preprint). *In*: Congrès de l'Union Internationale des Sciences Préhistoriques et Protohistoriques, 10. Mexico. Colloquium 5 on Data management and mathematical methods in archaeology.

Fisher R. 1936. The use of multiple measurements in taxonomic problems. Annals of Eugenics 7: 179–188.

Ford, J.A. 1962. A Quantitative method for deriving cultural chronology. Washington: Department of social affairs, Pan-American Union (Technical manual; 1.).

Forrester, J.W. 1968. Principles of systems. Cambridge, Wright Allen Press.

Forrester, J.W. 1969. Urban dynamics. Cambridge: MIT Press.

Forrester, J.W. 1970. World dynamics. Cambridge: Wright Allen Press.

Fulford, M.G. and I. Hodder. 1974. A Regression analysis of some Late Romano-british fine pottery: a case study. Oxoniensia 39: 26–33.

Gardin, J.-C. 1956. Le fichier mécanographique de l'outillage. Outils en métal de l'âge du bronze, des Balkans à l'Indus, Beyrouth, Institut Français d'Archéologie de Beyrouth.

Gardin, J.-Cl. (dir.). 1970. Archéologie et calculateurs: problèmes sémiologiques et mathématiques/ dir. J.-Cl. Gardin. Paris : Éditions du C.N.R.S. (Colloques internationaux du C.N.R.S.: Sciences humaines.)

Gardin, J.-Cl. 1976. Code pour l'analyse des formes de poteries, *établi en* 1956, révisé en 1974. Paris, Éditions du C.N.R.S.

Gardin, J.-Cl. 1979. Une Archéologie théorique. Paris, Hachette.

Gardin, J.-Cl. (dir.). 1987. Systèmes experts et sciences humaines: le cas de l'archéologie. Paris, Eyrolles.

Gelfand, A.E. 1971. Rapid seriation methods with archaeological applications. pp. 185–201. *In*: F.R. Hodson, D.G. Kendall and P. Tautu (eds.). Mathematics in the archaeological and historical sciences. Edinburgh: Edinburgh University Press.

Gero, J. and J. Mazzulla. 1984. Analysis of artifact shapes using Fourier series in closed form. Journal of Field Archaeology 11: 315–322.

Gilbert, N. 2008. Agent-Based Models, Quantitative Applications in the Social Sciences, 153, Sage Publications, Thousand Oaks, CA.

Gilbert, N. and J. Doran. 1994. Artificial Societies: The Computer simulation of social phenomena, London, UCL Press.

Giligny, F. 1995. Evolution des styles céramiques au Néolithique final dans le Jura, Revue Archéologique de l'Ouest, supplt n°7, pp. 191–212.

Goldmann, K. 1973. Zwei Methoden chronologischer Gruppierung. Acta prehist et archaeol 3: 1–34.

Gould, S.J. 1981. The mismeasure of man. New-York, Norton.

Graham, I.D. 1980. Spectral analysis and distance methods in the study of archaeological distributions. Journal of Archaeological Science 7: 105–130.

Guidi, A. 1985. An Application of the rank size rule to Protohistorie settlements in the Middle Tyrrhenien area. pp. 217–242. *In*: Papers in Italian Archaeology: Cambridge conférence 4. Oxford: British Archaeological Reports (British Archeological Reports: International ser.; 245) vol. 3: Patterns in Protohistory.

Haggett, P. 1965. Locational analysis in human Geography. London: Arnold, Trad. en français par H. Frechou. Paris: A. Colin, 1973.

Hagstrum, M.B. and J.A. Hildebrand. 1990. The two-curve method for reconstructing ceramic morphology. American Antiquity 55(2): 388–403.

Hall, N.S. and S. Laflin. 1984. A computer aided design technique for pottery profiles, database. pp. 177–188. *In*: S. Laflin (ed.). CAA 1984 Birmingham: University of Birmingham Computer Center.

Harris, E.C. 1979. Principles of archaeological stratigraphy. London, s.n.

Hassan, F.A. 1973. On mechanisms of population growth during the Neolithic. Current Anthropology 14: 535–540.

Hassan, F.A. 1975. Determination of the size, density and growth rate of hunting-gathering people pp. 27–52. *In*: S. Polgar (ed.). Population ecology and social evolution. The Hague: Mouton.

Hassan, F.A. 1981. Demographic archaeology. New York, Academic Press.

Heinzelin, J. de. 1960. Principes de diagnose numérique en typologie. Mémoire de l'Académie Royale de Belgique 14, 6.

Heizer, R.F. and S.F. Cook (eds.). 1960. The Application of quantitative methods in archaeology. Chicago, Quadrangle Books (Viking Fund publications in anthropology; 28).

Herzog, I. 2013. Least-cost networks pp. 240–51. *In*: G. Earl, T. Sly, A. Chrysanthi, P. Murrieta-Flores, C. Papadopoulos, I. Romanowska and D. Wheatley (eds.). *CAA* 2012. Proceedings of the 40th Annual Conference of Computer Applications and Quantitative Methods in Archaeology Southampton, 26–30 March 2012, Amsterdam: Pallas Publications.

Hietala, H.J. (ed.). 1984. Intrasite Spatial Analysis in Archaeology. Cambridge, Cambridge University Press.

Hietala, H.J. and D.S. Stevens. 1977. Spatial analysis : multiple procedures in pattern recognition studies. American antiquity 42: 539–559.

Higgs, E.S. 1975. Palaeoeconomy. Cambridge, Cambridge University Press.

Hodder, I. (ed.). 1978. Simulation studies in archaeology/ed. I. Hodder. Cambridge, London, New York, University Press (New directions in archaeology).

Hodder, I. and E. Okell. 1978. An Index for assessing the association between distributions of points in archaeology. pp. 97–107. *In*: I. Hodder (ed.). Simulation Studies in Archaeology. Cambridge, London, New York, Cambridge University Press (New directions in archaeology).

Hodder, I. and Cl. Orton. 1976. Spatial analysis in archaeology. Cambridge, Cambridge University Press (New studies in archaeology; 1).

Hodson, F.R., D.G. Kendall and P. Tautu (ed.). 1971. Mathematics in the archaeological and historical sciences. Proceedings of the Anglo-Romanian conference, Mamaia 1970/ed. F.R. Hodson, D.G. Kendall, P. Tautu. Edinburgh: Edinburgh University Press.

Hodson, F.R., P.H.A. Sneath and J.E. Doran. 1966. Some experiments in the numerical analysis of archaeological data. Biometrika 53: 311–324.

Hole, F. and M. Shaw. 1967. Computer analysis of chronological seriation. Houston: Rice University (Rice University studies 53(3): 1–166.

Ihm, P., J. Lüning and A. Zimmermann. 1978. Statistik in der Archeologie. Archaeophysica, n°9.

Ihm, P. 1981. The Gaussian model in chronological seriation. pp. 108–124 (preprint). *In*: Congrès de l'Union Internationale des Sciences Préhistoriques et Protohistoriques. 10. Mexico. Colloquium 5 on Data management and mathematical methods in archaeology.

Imbrie, J. and N.G. Kipp. 1971. A New micropalaeontological method for quantitative palaeoclimatology: application to a Late Pleistocene Carribean core. pp. 71–181. *In*: K.K. Turrekian (ed.). The Late Cenozoïc glacial ages. New Haven: Yale University Press.

Janssen, M.A. 2009. Understanding Artificial Anasazi. Journal of Artificial Societies and Social Simulation, 12(4)13 (http://jasss.soc.surrey.ac.uk/12/4/13.html).

Jochim, M.A. 1976. Hunter-gatherer subsistence and settlement: a predictive model. New York: Academic Press.

Johnson, G.A. 1975. Locational analysis and the investigation or Uruk local exchange systems. pp. 285–339. *In*: J.A. Sabloff and C.C. Lamberg Karlovsky (eds.). Ancient Civilisation and Trade. Albuquerque: University of New Mexico.

Johnson, J. 1977. Local density analysis: a new method for quantitative spatial analysis. pp. 90–98. *In*: Computer Applications in Archaeology. Birmingham, University of Birmingham.

Kammermans, H. 1995. Survey sampling, right or wrong ? *In*: CAA 1994. J. Huggett and N. Ryan (eds.). Tempus reparatum, BAR S600, pp. 123–126.

Keene, A.S. 1979. Economic optimization models and the study of hunter-gatherer subsistence settlement systems. pp. 369–404. *In*: C. Renfrew and K.L. Cooke (eds.). Transformations: Mathematical Approaches to Culture Change. New York: Academic Press.

Kendall, D.G. 1963. A Statistical approach to Flinders Petrie's sequence dating. Bulletin int. stat. 40: 657–680.

Kendall, D.G. 1971. Seriation from abundance matrices. pp. 215–252. *In*: F.R. Hodson, D.G. Kendall and P. Tautu (eds.). Mathematics in the Archaeological and Historical Sciences. Edinburgh: Edinburgh University Press.

Kintigh, K.W. and A.J. Ammerman. 1982. Heuristic approaches to spatial analysis in archaeology. American antiquity 47: 31–63.

Kohler, T.A. and G.J. Gumerman. 2000. Dynamics in Human and Primate Societies, Agent-Based Modeling of social and spatial processes, New York-Oxford, Oxford University Press.

Kohler, T.A. and S.E. Van Der Leeuw. 2007. The Model-Based Archaeology of Socio-natural Systems, Santa Fe, School for Advance Research Press.

Knappett, C. (ed.) 2013. Network Analysis in Archaeology: New Approaches to Regional Interaction. Oxford: Oxford University Press.

Kruskal, J.B. 1971. Multidimensional scaling in archaeology: time is not the only dimension. pp. 119 –132. *In*: F.R. Hodson, D.G. Kendall and P. Tautu (eds.). Mathematics in the Archaeological and historical sciences. Edinburgh: Edinburgh University Press.

Kuzara, R.S., G.R. Mead and K.A. Dixon. 1966. Seriation of anthropological data : a computer program for matrix ordering. American Anthropologist 68: 1442–1455.

Landau, J. and W.F. Vega, de la. 1971. A New seriation algorithm applied to European Protohistoric anthropomorphic statuary. pp. 255–262. *In*: F.R. Hodson, D.G. Kendall and P. Tautu (eds.). Mathematics in the archaeological and historical sciences. Edinburgh: Edinburgh University Press.

Laplace, G. 1966. Recherches sur l'origine et l'évolution des complexes leptolithiques. Paris: de Boccard. Mélanges d'archéologie et d'histoire/École Française de Rome: Suppléments; 4.

Laxton, R.R. and J. Restorick. 1989. Seriation by similarity and consistency. pp. 215–224. *In*: S. Rahtz and J. Richards (eds.). CAA 89. B.A.R. Intern. ser. n°548.

Legoux, R. 1980. *In* "Perrin, P, La datation des tombes mérovingiennes. Historique, méthodes, applications", Genève, Droz, 1980 (IVe section de l'Ecole pratique des Hautes Etudes), Hautes Etudes médiévales et modernes, V, 39.

Li, T.Y. and J.A. Yorke. 1975. Period three implies chaos. American Mathematical Monthly 82: 985–992.

Limp, W.F. 1989. The Use of Multispectral Digital Imagery in Archeology. Arkansas Archeological Survey Research Series No. 34. 122p.

Lowe, J.W.G. 1985. The Dynamics of Apocalypse: a system simulation of the classic Maya Collapse, Albuquerque, The University of New Mexico Press.

McArthur, N., I.W. Saunders and R.L. Tweedie. 1976. Small population isolates : a micro-simulation study. Journal of the Polynesian Society 85: 307–326.

Madsen, T. 1988. Multivariate statistics and Archaeology. pp. 7–27. *In*: T. Madsen (ed.). Multivariate Archaeology. Numerical approaches in Scandinavian archaeology. Jutland Archaeological Publication, 21.

Main, P.L. 1986. Accessing outline shape information efficiently within a large database. pp. 73–82. *In*: S. Laflin (ed.). CAA 1986. Birmingham: University of Birmingham Computer center.

Mahalanobis, P.C. 1927. Analysis of race mixture in Bengal. Journal of proceedings of the Asiatic Society of Bengal 23: 301–333.

Mandelbrot, B.B. 1983. The Fractal Geometry of Nature, New York, Freeman and Co.

Marquardt, W.H. 1978. Advances in archaeological seriatio. pp. 257–314. *In*: M.B. Schiffer (ed.). Advances in Archaeological Method and Theory. London, New York: Academic Press, 1978, 1.

Martin, J.F. 1972. On the estimation of the sizes of local groups in a hunting-gathering environment. American Anthropologist 75: 1448–1468.

Masset, C. 1973. La Démographie des populations inhumées: essai de paléodémographie. L'Homme 13(4): 95–131.

McGlade, J. and P.M. Allen. 1986. Fluctuation, instability and stress: understanding the evolution of a Swidden horticultural system. Science and Archaeology 28: 44–50.

Meighan, C.W. 1959. A New method for seriation of archaeological collections. American antiquity 25: 203–211.

Moscati, P. 1987. Archeologia e calcolatori. Firenze: Giunti.

Mueller, J.W. (ed.). 1975. Sampling in archaeology/ed. J.W. Mueller. Tucson: The University of Arizona Press.

Nottale, L., J. Chaline and P. Grou. 2000. Les arbres de l'évolution, Paris, Hachette.

O'Shea, J.M. 1978. A Simulation of Pawnee site development. pp. 39–40. *In*: I. Hodder (ed.). Simulation Studies in Archaeology. Cambridge: Cambridge University Press.

Pearson, K. 1901. On lines and planes of closest fit to systems of points in space, Philosophical Magazine, Series 6, 2(11): 559–572.

Parker, S. 1985. Predictive modelling of site settlement systems usingmultivariate logistics, in Carr, C. (ed.). For Concordance in Archaeological Analysis. Bridging Data Structure, Quantitative Technique, and Theory. KansasCity: Westport Publishers, pp. 173–207.

Peirce, C.S. 1992. Reasoning and The Logic of Things. The Cambridge conferences. Lectures of 1898. Cambridge: Harvard University Press.

Pielou, E.C. 1969. An Introduction to mathematical ecology. Chichester: J. Wiley.

Prigogine, I. 1968. Introduction à la thermodynamique des processus irréversibles, Paris, Dunod. Simuler une "artificial society": organisation sociale, gouvernance et attitudes sociétales 33.

Read, D.W. 1989. Statistical methods and reasoning in archaeological research: A review of praxis and promise. Journal of Quantitative Archaeology, pp. 15–78.

Read, D.W. 1998. Kinship based demographic simulation of societal processes. Journal of Artificial Societies and Social Simulation, 1 (http://jasss.soc.surrey.ac.uk/1/1/1.html).

Regnier, S. 1977. Sériation des niveaux de plusieurs tranches de fouilles dans une zone archéologique homogène. pp. 146–155. *In*: M. Borillo, W.F. de La Vega and A. Guenoche (eds.). Raisonnement et méthodes mathématiques en archéologie. Paris: Éditions du C.N.R.S.

Renfrew, C. 1975. Trade as action at distance. pp. 1–59. *In*: J.A. Sabloff and C.C. Lamberg-Karlovsky (eds.). Ancient Civilization and Cade. Albuquerque: University of New Mexico Press.

Renfrew, C. 1977. Alternative models for exchange and spatial distribution. pp. 71–90. *In*: T.K. Earle and J.E. Ericson (eds.). Exchange Systems in Prehistory. New York: Academic Press.

Renfrew, C. 1978. Trajectory discontinuity and morphogenesis, the implications of catastrophe theory for archaeology. American Antiquity 43: 203–244.

Renfrew, C., J.E. Dixon and J.R. Cann. 1968. Further analyses of Near Eastern obsidians. Proceedings of the Prehistoric Society 34: 319–331.

Renfrew, C. and G. Sterud. 1969. Close-proximity analysis: a rapid method for the ordering of archaeological materials. American Antiquity 34: 265–277.

Renfrew, C. and K.L. Cooke (eds.). 1979. Transformations: mathematical approaches to culture change/ ed. C. Renfrew, K. L. Cooke. New York, San Francisco, London: Academic Press.

Renfrew, C. and E.V. Level. 1979. Exploring dominance: predicting polities from centers. pp. 145–162. *In*: C. Renfrew and K.L. Cooke (eds.). Transformations: Mathematical Approaches to Culture Change. New York: Academic Press.

Renfrew, C. and S. Shennan (eds.). 1982. Ranking, resource and exchange/ed. C. Renfrew, S. Shennan. Cambridge : Cambridge University Press.

Reynolds, R.G.D. 1981. An Adaptative computer simulation model of acquisition of incipient agriculture in prehistoric Oaxaca, Mexico. pp. 202–216 (preprint). *In*: Congrès de l'Union International des Sciences Préhistoriques et Protohistoriques. 10. 1981. Mexico. Colloquium 5 : Data management and mathematical methods in archaeology.

Reynolds, R.G. 1986. An adaptative computer model for the evolution of plant collecting and early agriculture in the eastern valley of Oaxaca. pp. 439–500. *In*: K.V. Flannery (ed.). Guila Naquitz: Archaic Foraging and early Agriculture in Oaxaca, Mexico, Orlando, Academic Press.

Reynolds, R.G.D. and B.P. Zeigler. 1979. A Formal mathematical model for the opération of consensous-based hunting-gathering bands. pp. 405–418. *In*: C. Renfrew and K.L. Cooke (eds.). Transformations: Mathematical Approaches to Culture Change. New York: Academic Press.

Robinson, N.S. 1951. A Method for chronologically ordering archaeological deposits. American Anthropologist 15: 301–313.

Sabloff, J.A. 1981. Simulations in archaeology/ed. J.A. Sabloff. Albuquerque: University of New Mexico Press (Advanced seminar scr./School of American Research).

Sabloff, J.A. and C.C. Lambert-Karlovsky (eds.). 1975. Anciens civilization and trade/ed. J.A. Sabloff, C.C. Lambert-Karlovsky. 1975. Albuquerque: University of New Mexico.

Schiffer, M.B. 1976. Behavioral Archeology. New York: Academic Press.

Scollar, I. 1978. Progress in aerial photography in Germany and computer methods. Aerial Archaeology 2: 8–18.

Shennan, S. 1988. Quantifying archaeology. Edinburgh: Edinburgh University Press.

Sneath, P.H.A. 1957. The Application of computers to taxonomy. Journal of general microbiology 17: 201–226.

Snow, D. 1969. Ceramic Sequence and Settlement Location in Pre-Hispanic Tlaxcala. American Antiquity 34(2): 131–145.

Snow, D. 1993. Archaeology and Computer Simulation, The Case of Iroquois Demography. In Actes du XIIe Congrès International des Sciences Prèhistoriques et Protohistoriques, Bratislava, 1–7 Septembre 1991, 4 vols., edited by Juraj Pavúk, 1: 394–399. Institut archéologique de l'Académie Slovaque des Sciences, Bratislava, Slovakia.

Snow, D. 1996. GIS Applications in North America. In The Coloquia of the XIII Congrès International des Sciences Prèhistoriques et Protohistoriques, Vol. 1 Theoretical and Methodological Problems, pp. 159–168, edited by Ian Johnson. A.B.A.C.O. Edizioni, Forlì, Italy.

Soffer, O. 1985. The upper Paleolithic of the central Russian plain. Orlando, Florida : Academic Press.

Sokal, R.R. and P.H.A. Sneath. 1963. Principles of numerical taxonomy. San Francisco, London: Freeman.

Spaulding, A.C. 1953. Statistical techniques for the discovery of artefact types. American antiquity 18: 305–313.

Spaulding, A.C. 1977. On Growth and Form in Archaeology: Multivariate Analysis. Journal of Anthropological Research 33(1): 1–15.

Spearman, C. 1904. General Intelligence "objectively Determined and Mesured". The American Journal of Psychology, 15, 2, pp. 201–292.

Thom, R. 1972. Modèles mathématiques de la morphogénèse. Paris: Bourgeois.

Thomas, D.H. 1972. A Computer simulation model of Great Basin Shoshonean subsistence and settlement patterns. pp. 671–704. *In*: D.L. Clarke (ed.). Models in Archaeology. London: Methuen.

Tugby, D.L. 1958. A Typological analysis of axes and choppers from Southeast Australia. American Antiquity 24: 24–33.

Tyldesley, J.A., J.G. Johnson and S.R. Snape. 1985. Shape in archaeological artefacts : two case studies using a new analytic method. Oxford Journal of Archaeology 4(1): 19–30.

Van Der Leeuw, S.E. 1981. Archaeological approaches to the study of complexity/ed. S.E. Van der Leeuw. Amsterdam: Universiteit van Amsterdam (Cingula; 6).

Van Der Leeuw, S. and J. McGlade (éds.). 1997. Archaeology: Time, process and structural transformations, London, Routledge.

Vertes, L. 1964. Statistiques et graphiques dans l'étude des industries paléolithiques. Palaeohistoria, 10.

Vescelius, G.S. 1960. Archaeological sampling : a problem of statistical inference. pp. 457–470. *In*: G.A. Dole and R.L. Cameiro (eds.). Essays in the Science of Culture in Honor of L.A. White. New York: T.Y. Crowell Co.

Vita-Finzi, C. and E.S. Higgs. 1970. Prehistoric economy in the Mount Carmel area of Palestine site-catchment analysis. Proceeding of the Prehistorical Society 36: 1–37.

Voorrips, A. 1973. An Algol-60 program for computation and graphical representation of pollen analytical data. I.P.P. Publications, 155.

Voorrips, A. and S.H. Loving (eds.). 1986. The pattern of the past. Pact n°11.

Voorrips, A. (ed.). 1990. Mathematics and Information Science in Archaeology : A flexible framework. Bonn: Studies in Modern Archaeology, vol. 3.

Ward, R.H., J.W. Webb and M. Levison. 1973. The Settlement of the Polynesian outliers : a computer simulation. Journal of the Polynesian Society 82: 330–342.

Whallon, R. 1973. Spatial analysis of occupation floors: application of dimensional analysis of variance. American Antiquity 38: 266–278.

Whallon, R. 1974. Spatial analysis of occupation floors: the application of nearest neighbour analysis. American Antiquity 39: 16–34.

Whallon, R. 1984. Unconstrained clustering for the analysis of spatial distributions in archaeology. pp. 242–277. *In*: H.J. Hietala (ed.). Intrasite Spatial Analysis in Archaeology. Cambridge, London, New York: Cambridge University Press (New directions in archaeology).

Wheatley, D. and M. Gillings. 2002. Spatial Technology and Archaeology: The Archaeological Applications of GIS. New York, Taylor and Francis.

Wilcock, J.D. and S.J. Shennan. 1975. Shape and style variation in central german Bell Beaker: a computer assisted study. Science and Archaeology 15: 17–31.

Wilkinson, E.M. 1971. Archaeological seriation and the travelling salesman problem. pp. 276–283. *In*: F.R. Hodson, D.G. Kendall and P. Tautu (eds.). Mathematics in the Archaeological and Historical Sciences. Edinburgh: Edinburgh University Press.

Wilkinson, E.M. 1974. Techniques of data analysis-seriation theory. Archaeo-Physica 5: 7–142.

Wilkinson, T.J., J.H. Christiansen, J. Ur, M. Widell and M. Altaweel. 2007. Modeling Settlement Systems in a dynamic Environment, Case Studies from Mesopotamia. pp. 175–208. *In*: T.A. Kohler and E.S. van der Leeuw (eds.). The Model-Based Archaeology of Socio-natural Systems, Santa Fe, School for Advance Research Press.

Wilmsen, F.N. (ed.). 1972. Social exchanges and Interaction/ed. E.N. Wilmsen. Ann Arbor: University of Michigan.

Wobst, H.M. 1974. Boundary conditions for Palaeolithic social systems: a simulation approach. American Antiquity 39(2): 147–170.

World Archaeology. 1982. Quantitative methods in Archaeology.

Zubrow, E.B.W. 1975. Prehistoric carrying capacity: a model. Menlo Park, Calif: Cummings.

3

The Formal Logical Foundations of Archaeological Ontologies

Franco Niccolucci,[1,*] *Sorin Hermon*[2] and *Martin Doerr*[3]

What's in a name? That which we call a rose
by any other name would smell as sweet

William Shakespeare
Romeo and Juliet, Act II, Scene 2

Introduction

Archaeology is a multi-disciplinary science, which, based on the systematic analysis of material remains of past human activity in their archaeological context, aims at enriching our knowledge on past cultures and at understanding past and modern societies. Through the years, archaeology has developed various methods to document these remains and to analyze and interpret them within various methodological, conceptual and theoretical frameworks, eventually finding itself in a sort of epistemological crisis these past decades. Now archaeological research is facing a dramatic challenge vis-à-vis the growth of new scientific paradigms enabled by Information and Communication Technology (ICT).

The Theoretical Crisis of Archaeology

The theoretical and epistemological foundations of archaeology have been debated for a long time, questioning the nature of archaeological 'data' and the way they are

[1] PIN, scrl., Piazza Giovanni Ciardi, 25, 59100 Prato, Italy.
    Email: franco.niccolucci@gmail.com
[2] The Cyprus Institute–20 Konstantinou Kavafi Street, 2121 Aglantzia, Nicosia, Cyprus.
    Email: hermon@cyi.ac.cy
[3] Foundation for Research & Technology–Hellas, Leoforos Plastira 100, Iraklio 700 13, Greece.
    Email: martin@ics.forth.gr
* Corresponding author

used to investigate the past, in a way similar to the trend that has affected all social and human sciences. The debate on the theoretical foundations of archaeology is too vast to be summarized here. We just refer to a few recent books: Johnson 2010, Hodder 1999, Bintliff and Pearce 2011, Praetzellis 2011, Abramiuk 2012, Costopoulos and Lake 2010, and Bentley et al. 2008. Also relevant to the topic is Hodder 2012a, 2012b.

'Post-processualism' and other theories highlighted the dependency of research results on the subjectivity of choices, working hypotheses, observation areas and parameters. Even though this subjectivity of choice of object of investigation and method is in a way a fundamental, inevitable feature common to all sciences, it was taken as an argument to question overall the epistemological foundations of archaeological research, rather than explicating these dependencies and subjective choices as usual in other sciences. Consequently "...the current fragmentation and specialization (in archaeology) has led to a proliferation of multiple and incommensurable agendas. Such a 'hyper-pluralism' actually inhibits debate and discussion, by breaking up research traditions into multiple groups with shrinking scopes" (Olsen et al. 2008).

This fragmentation hampers all notions of an overall evolution of the science towards a better understanding of its subject, as it is expected by the society. Most recent approaches add a 'Defense of the things' (Olsen 2010), recalling that archaeology is based on material culture and giving things the stewardship: "follow them where they lead"; or advocate 'practice' as the pillar that supports research. Nevertheless, archaeological investigations continue, sometimes relying on interim theories that reveal not completely satisfactory even to their authors, or simply producing results that just rely on academic and professional best practice.

The Data Challenge

In recent years there has been an increasing amount of accumulated archaeological records and digital data, and a pressure for sharing and re-use (Niccolucci and Richards 2013). Also thanks to the increased use of analytical methods borrowed from physics and chemistry, producing a vast amount of born-digital archaeological data, the amount of digital data stored seems to be posing a 'Big Data' issue also in the archaeological domain. A new research paradigm could consist of a condition where researchers can collaborate on the same set of data from different perspectives and achieve a 'fourth paradigm' of science besides observation, theory and simulation. It seems, however, that archaeology is not ready to profit from this opportunity for the lack of a theoretical framework supporting the use of somebody else's raw 'data', especially when considering the high level of fragmentation of archaeological datasets. To address this issue, several integrating initiatives are in place, mostly led or coordinated by archaeologists, confirming the awareness of the archaeological community about the issue of digital dataset integration, use and re-use.

However, such integration activities are addressing only the technical aspects of dataset integration. There is a concrete risk that through an enormous effort archaeological data accumulated in years and those coming from new investigations are made available to other researchers with the easiness of present-day communication networks, but they fail to be shared because of the lack of a common well-defined ontology, i.e., a structured system of fundamental concepts and relationships (Smith 2003, Guarino et al. 2009, Doerr and Iorizzo 2008), and of an agreed epistemology, i.e., clearly defined rules of evidence and reasoning, which do not privilege individual experiences or beliefs that cannot be argued against, and which at the same time include clear evaluation mechanisms for the credibility of research conclusions. As argued by (Doerr et al. 2011), such rules of evidence and reasoning need by no means be restricted to formal logic. They can quite well be estimates of probabilities or plausibility arguments based on parallels, etc. By "clearly identified" we mean conformability, i.e., that the principle can be understood by others, its reliability and effect on premises can be verified or falsified or be compared to results obtained by other rules applied to the same premises.

At the semantic level, the ARENA project (Archaeological Records of Europe - Networked Access) already demonstrated in 2005 that technical interoperability may coexist with the impossibility of comparing data due to the difference of underlying data structures.

In conclusion, we claim that data cannot be compared, if the data structures have no clear semantics (identity of concepts employed for data structures), and the provenance of knowledge is not transparent. 'Big Data' need some automated processing to be manageable, posing much stronger requirements on the identity conditions of the concepts used to encode and classify.

A disaster scenario would be one in which archaeological data and discoveries are fed into an overarching, well-functioning integrated system, but nobody cares, because data are assumed by potential users to be theory-laden and biased by preconceptions and subjectivity. They would therefore be untrustworthy and hence unusable by others who don't know the conceptual system used by the authors to organize them, and would ignore the implicit reasoning process that led to the authors' synthesis. This not only would waste the huge resources spent to set up integrated datasets system. Worst, it would lead to a solipsistic approach to data re-use, where people would only trust and access their own datasets, contrasting the 1829 forerunning statement of Gerhard who rightfully claimed (Gerhard 1829) that "in archaeology … the efforts of one person alone may never succeed without the continuous and mutual assistance of many others who mutually share their discoveries and knowledge", which nowadays also include digital data.

The Main Research Questions

This analysis of the problem points out first of all the urgent need of re-considering the system(s) of concepts used to document archaeological remains, and to organize what is distilled from excavations findings, turning 'data' into comparable data with robust notions of identity of meaning. This is not a normative approach, superimposing an abstractly defined conceptual structure on any archaeological documentation; nor do we believe that it is merely a matter of language translation of terms, creation of thesauri or definitions lists. It is, instead, an attempt to make explicit and well-defined in formal terms the many current archaeological (subjective) implicit ontologies, at least and in particular the core notions naturally present in and mediating between most disciplinary specializations. This will prevent a Babel where people call with the same name fundamentally different things, and, aware (or afraid) of this, they don't digitally talk to each other.

As a starting point, it is necessary to consider the 'fundamentals': space/place, time/period and things, and to analyse their identity criteria (which thing is one archaeological remain and how to recognize it again?); extensions (who/what can be considered an actor?); specifications (what is an archaeological site?); and interrelations with the other entities used in the archaeological discourse(s) in light of the techniques of ontology engineering.

The re-organization of the basics of the archaeological discourse from the perspective of logic and of knowledge organization theory starts from the matching of two strands. One, which we might call the 'offer', is internal to mathematical logic, philosophy and computer science, and analyses the basic concepts on which knowledge is structured, regardless of archaeological applications. As such, these clarifications concern the organization of information rather than its processing.

The second, which we might call the 'demand', comes from the use of information technology applied to archaeology and from the consideration that uncontrolled use of IT may not be conformant to the methodology of the discipline. Objections in this second strand initially concerned the domain of GIS applications, questioning the implicit assumptions behind the inference process based on spatial information. They may be grouped under the label of "criticism to environmental determinism" (see, among others, Lock and Stancic 1995). Also a naïve use of databases was questioned (Niccolucci 2002), but this discussion attracted little or no interest. Other arguments advocated the use of multimedia to convey a reflexive approach to archaeological investigations that could be lost in an impersonal collection of data pretending to be neutral (Hodder 1999). This position did not produce much effect on IT applications and eventually remained an unrealized 'great expectation'.

Our goal here is not to contribute to the above archaeological debate. We aim, instead, to clarify how the discussion about the fundamentals of archaeological knowledge organization has a paramount importance for the organization of archaeological data enabling their use in research. So we will start from very general issues indicating how they apply within the archaeological domain.

The Fundamental Concepts and their Logical Organization

There are several important studies dealing with the organization of knowledge starting from the basic concepts of space and time. Their contribution has influenced the definition of CIDOC CRM, the well-known ontology for the organization of cultural heritage documentation (Doerr 2003, 2009). CIDOC CRM (henceforth 'the' CRM) is extending its application domain to various sub-sectors beyond its original realm of museums, and with this revision it is targeting other documentation activities such as field documentation, laboratory analyses and so on. As new fields are considered, the ontology is being extended maintaining compatibility with the general framework and introducing a closer fit to concepts specific to the new domain. Examples of such extensions are CRM dig (Theodoridou et al. 2010, Doerr and Theodoridou 2011), an extension created to manage digital acquisition operations; and CRM-EH (May 2006), an extension/specification created to manage the archaeological documentation created by English Heritage. Work in progress concerns an extension dealing with space, in order to better manage spatial data as those used in GIS, code-named CRM-GEO (Doerr and Hiebel 2013).

The CRM lists all the fundamental concepts calling them **classes** or **entities**, as well as the relationships between classes, called **properties**. Classes are concepts, while individual elements identifiable as manifestations of the concept are called **instances** of the class. The contrast between classes and their instances is similar to the one between 'common noun' and 'proper noun' in the grammar. The properties are ultimately the predicates that allow for formulating propositions, the elements and essence of anything that can be called knowledge or information. Properties however can only be defined if the classes they relate and are applicable to are sufficiently defined as well.

Both entities and properties are identified in the CRM with a label formed by a letter and a number. Letters are specific to the CRM extension: the concepts included in the general version and its revisions are marked with E (entities) or P (properties); those pertaining to a specialization or a refinement use different letters.

A basic distinction of entities organizes them distinguishing between Temporal Entities (E2) and Persistent Items (E77). This distinction matches the one in DOLCE (Gangemi et al. 2002) between what the latter calls **perdurants**, i.e., entities/phenomena that typically extend in time, like processes; and **endurants**, entities/things, either physical/tangible or conceptual/intangible, having an enduring identity. Perdurants are observed diachronically, i.e., they can be said to be "still ongoing", endurants are observed synchronically, i.e., they can be said to be "still existing". With some imprecision, one might say that perdurants are associated with time, so we may say that they occupy a portion of time. Their substance is change itself, the phenomenon by which we perceive the flow of time; endurants exist in a sense transcending time by preserving features over time. Instances of physical/tangible endurants (E18 Physical thing, in CRM terminology) occupy a portion of space: they are constituted of matter. Of course, also things change in time, but in this case when the change is 'substantial' the thing becomes a different one; otherwise, change is regarded as only partial or marginal. This affects especially the classification of objects in archaeology. There are indeed no problems in describing and identifying objects as they are now. Problems arise when trying to identify the original objects of which the current one is a remnant: "This potsherd is a fragment of an amphora that contained wine"; "This piece of stone was a flint tool, a scraper". Of course the identity of the original has not been fully preserved. The archaeologist tries to travel back through the changes undergone by the object: for example, amphora → broken amphora → fragment separated from the others → fragment incorporated in a layer → archaeological find; flint tool → flint discarded (lost, broken, etc.) → remnant incorporated in a layer → archaeological find. In the first transition of each of these processes, these objects lost their function (in the examples: of storing wine; of scraping leather) and with it a great deal of their characteristics, possibly even their whole identity. Every transition is modelled by a relation in the ontology, for example in the CRM there are the properties "P124 transformed", which applies to the original, and "P123 resulted in", which applies to the outcome, both related to the transformation event represented by an arrow in the above diagrams. Reconstructing the modification backwards is however far from simple and clean: it often involves cultural prejudices of the researcher, which may be unrelated to the object but are embedded in the interpretation.

There are perdurants that seem to be space-less. They are generally concepts like the Neolithic Revolution used by researchers to categorize, systematize and interpret fragmented knowledge, and being conceptual constructs may have no specific spatial connotation. On the other hand, most of the perdurants have a clear spatial dimension too beside their main feature, the flow of time. However, all events take place somewhere (and not somewhere else), especially within the archaeological discourse where events are deduced from their physical traces. This link between what happened and the effects it produced on things, eventually discovered by archaeologists, implies a spatial footprint for all archaeological events and introduces the category of space also for such entities, which are intrinsically time-qualified. In conclusion, any entity of both branches, endurants and perdurants, is embedded in space and in time.

On this regard, there are two schools of thought in philosophy. One considers time as a dimension of a different nature compared with the three usual dimensions of space. A consequence of this approach affects things, which in this vision are present with all their parts ('wholly') at any time they exist. The other approach (Sider 1997) considers everything as extending in a 4-dimensional spatio-temporal space, with ordinary objects considered as a time snapshot of the corresponding 4-dimensional object at the given instant in which it is observed. John Locke, anyway, argues that the identity of endurants is based on their behaviour across time—"*considering anything as existing at any determined time and place, we compare it with itself existing at another time, and thereon form the ideas of identity and diversity*" (Locke 1999: 311)—so that regarding things as being determined within a particular instance in time appears rather as an illusion.

It appears rather that the separation of time and space is closer to everyday intuition, whereas the 4 dimensional view is more consistent to describe reality, but much more complex to apply. Therefore we will freely draw upon both theories, under the impression that there are mathematical models encompassing both, and reducing 3-dimensionalism to a limit case of 4-dimensionalism (see also Doerr and Hiebel 2013).

Classification

Classification of real world objects as instances ('particulars') of a general entity requires defining the criteria (Wiggins 2001, Guarino and Welty 2001) that differentiate instances of that entity from instances of another one, or, as it is called, the **substance** criteria of the entity. To characterize a class, also **identity** criteria must be stated, i.e., the criteria that enable to identify or distinguish two instances of the same class. Although this may appear as a trivial task, it is not such when identity in time is considered. Already John Locke (Locke 1999) observed that identity criteria differ between categories, without engaging in an exhaustive analysis. I am the same person through time even if I undergo surgery and implants; but it is less clear how this applies to inanimate objects: if I replace some components of my car, does it remain the same car in any case? If not, which components (e.g., the number plate) are essential and which ones (e.g., the tires) are not? The answer is not easy: for example, if one moves abroad and the car gets a new number plate for the new residence country, is it another car? Identity criteria are tightly related with the **existence** criteria of an instance. If the task of precising existence seems trivial for its coming into existence (birth, creation, etc.) and for its destruction, for transformation it is not the same, as the previous example demonstrates. For archaeological layers the possible notions of coming into existence can be quite confusing: Is it the creation events of its components, its loss, abandonment, destruction, and deposition, the final consolidation of matter into a "layer" or the declaration by an archaeologist? Being able to sort out what makes some substance be part or section of an instance, i.e., defining the **unity** criteria of a class, is also related to identity, but is not the same concept: a car is more than the heap of all its parts and molecules, as the latter cannot be used for transportation. All these criteria are indispensable to make an ontology useful in practice and to organize the knowledge about the real world in a way that we can decide if a thing described in some publication or data set is the same or not as in another. As we will see below, the above-mentioned criteria are often approximate or lacking in the archaeological domain, introducing further difficulty in managing knowledge about the past. The problem is not to replace existing archaeological concepts by better ones or to reduce them to one "correct form", but to reveal ("reengineer") their—often multiple—natures in an explicit form one can refer to according to their utility in the restective context of use.

Space, Time and Space-time

A recently proposed extension to the CRM (Doerr and Hiebel 2013) aims at dealing with difficult cases such as the one exemplified by the location of the battle of the Teutoburg Forest, fought in 9 AD between the Romans, led by Quinctilius Varus (what generated the appellation of "Varus battle"), and allied German tribes led by Arminius. The battle is famous because it led to the annihilation of the Roman troops and was widely popularized by German nationalists in the 19th century. The battle location is still a matter of discussion, although recent investigations have discovered finds hinting towards a precise place; however, the 'battle' took place in various sites and consisted of several fights, according to ancient historians as Tacitus and Velleius Paterculus. The issue is that discussing 'the' spatial location of the battle is confusing: the 'battle' evolved in time through a continuous sequence of fights while the Romans unsuccessfully tried to escape the German ambushes for several days. So there is a prolonged event, which happened continuously throughout an extent of time but during this time took place at any moment in a different place. Additional problems come when we try to identify the modern locations where the battle took place. These locations may be furthermore related to some geometrically identifiable position in a coordinate system. Past locations are usually identified by the presence of objects (the 'finds'), which can be considered as the product of transformations of things of the past (the 'ancestor') related to the event. Historical sources may also help in identifying locations, but we will momentarily ignore this aspect and consider only material sources.

In conclusion, an event, in the example the Varus battle, may be thought of as split into instantaneous time slices, in the example the fighting happening at time t. Each of these instantaneous events took place in space at some location. This location corresponds to a modern place that has some coordinates in a coordinate reference system.

The approach adopted by CRM in its CRM-CRMgeo (Doerr and Hiebel 2013) refinement embeds events (or better Periods, in the CRM terminology) in a spatio-temporal 4-dimensional framework. Here they occupy a region, called **Phenomenal Spacetime Volume**, which has a time projection, the **time-span** of the event, and a space projection, called **Phenomenal Place**. The latter corresponds to the part of space occupied by the event through its existence, in the spatial universe of discourse called **Reference Space**. The Reference Space becomes a non-trivial concern, if movements of soil, slopes or tectonic plates become non-negligible, if things happen on moving platforms such as vessels, or if the precision of determination of the reference space changes, such as between national grids and GPS. The Phenomenal Place may be approximated by a region of the Reference Space called **Declarative Space**, described via coordinates or in other ways, for instance by name. The full body of ontological instruments used by the CRM to develop the organization of spatio-temporal knowledge is described in (Doerr and Hiebel 2013).

Discovery, as usual in archaeology, marches backwards: the finds mark the extent of (modern) space where the event happened. The corresponding ancient space is reconstructed, identifying it as approximately corresponding to an extent of modern space. This process places the objects that generated the finds in the past space they (presumably) belong to, if necessary reverse engineering the effect of movement of artefacts or of phenomena altering the space. Dating the finds places them, or better their ancestors, in the correct time slice, which together with the past position (coinciding with the present one or reconstructed) defines their original location in space-time.

Then the instantaneous spatial footprint of the event at a given instant is reconstructed as the convex hull of the spatial locations of the related finds referable to that precise instant, and the entire space-time volume of the event is the hull of all these time slices. There is a time granularity in dating that makes this reconstruction a stepwise, rather than continuous, process. As already mentioned, sometimes the reconstruction process may meet obstacles, for example when natural or human-made events changed the nature of the space involved and their effects make the identification of past space with the present one impossible or unreliable. Floods, landslides, bradyseisms and other natural phenomena may be responsible for the change and break the above-described conceptual pipeline. This is the case of Baia, near Naples, Italy, a Roman site currently submersed by a bradyseism. Volcanic eruptions may not only destroy a town, as in Pompeii, but also alter the shape of the involved place, as in Akrotiri. At a smaller scale, objects may have been moved or be in fact extraneous to the event 'scene'. In what follows we will assume none of these alterations has happened.

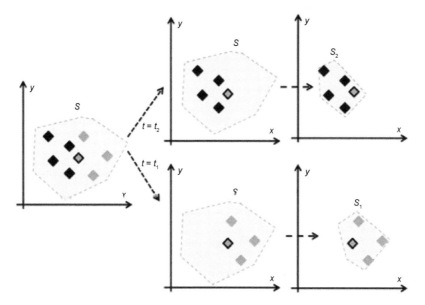

Fig. 1. Identifying the spatial footprint of instantaneous events through the corresponding finds.

Let us consider a (present) space S^* where some finds are discovered. The objects are the remnants produced by an event E in the past: for example, in the Varus battle event, they may consist of weapons, pieces of armours or shields, helms, skeletons, and so on. More precisely, each of these present things is the outcome of a transformation affecting a past thing that participated in the event. For the sake of simplicity, let us assume that the past space S and the present one S^* coincide, and that past and present things have not been moved. The leftmost part of Fig. 1 depicts the spaces $S^* = S$ with the finds (represented by small lozenges) at their location. They are assumed to have the same spatial distribution in S^* as the objects from which the finds derive had in the past space S. If this assumption is not valid, it is necessary to reconstruct the movement of the objects from their original location.

By analysing the finds, they are dated, i.e., the past object each of them derives from is identified and assigned to the time instant in which it existed: those represented in a darker colour belong to time t_2 and the light-colour ones to time t_1; there is furthermore one case (the light grey symbol with dark sides) belonging to both. This is of course a simplification for the sake of explanation, as the things will have an existence extending over a time interval, so t_1 and t_2 are more properly time intervals rather than instants.

Then, as represented in the central part of the figure, the set of finds is split according to their existence time. Each one of the resulting subsets is the collection of the objects participating in the event at different times, still embedded in the overall space S. Finally (right) the two instantaneous event footprints at the different times are identified, basing on the dated finds, represented in the picture by S_1 and S_2, respectively at times t_1 and t_2. The instantaneous event footprint at instant t_i ($i = 1, 2$) is evaluated using the location of the finds participating at the event, using the convex hull of their locations.

Note that all the figures include the spatial axes x and y. They are used to indicate that the reasoning develops in the space: in fact, there may be a reference coordinate system, but it is not relevant in this reasoning. Actually, space will be three-dimensional, only two dimensions being depicted in the figure to enable drawing also the time dimension when necessary

In conclusion, S_1 and S_2 represent two time slices of the spatial footprint of the event E.

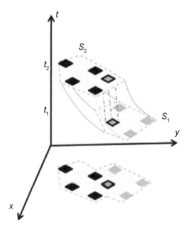

Fig. 2. The time-space volume (Phenomenal Volume) related to an event and its relation with related finds.

The time slices of E form a 4-dimensional volume, called a 'worm' (or, if you prefer, a 'sausage'). Each thing is located somewhere within it with its own worm, which starts at the thing's beginning of existence or beginning of involvement ('participation') in the event and ends with the thing's end of existence or participation in the event. For the sake of simplicity, participation is depicted as instantaneous for all things except for one that continues its participation throughout the time interval from t_1 to t_2. The corresponding 'worm' of this object is also represented in the figure. The projection of the overall worm on the 3D-dimensional space (actually 2-dimensional in the figure) is represented on the x-y plane.

A final explanation concerns the identity of the objects throughout the process. The above process is described with reference to locations and presence/absence of things. No implication is made about the identity of the objects at every instant: the object is assumed to remain the same except at 'major' changes although there may be a continuous modification of its characteristics. For example, a metallic artefact may rust gradually changing its matter continuously, but a change of identity is assumed only when this degradation determines 'substantial' changes. What is most important for the archaeological process is that the object or remains preserves enough of its relevant original form or substance in order to be used as evidence. For instance, a petrified skeleton may chemically be completely changed, but structurally nearly identical to the original. A metal object may have lost its shape, but the chemical fingerprint may be still close enough to the original.

In conclusion, in the CRM vision time and space locations are characteristics of both events and things. Events happen in a spatio-temporal 4-dimensional region. Things occupy space and remain the same during time, until a transformation event happens, altering their identity and turning that thing into another one.

Transformations need attentive consideration. Imagine a snowman, which gradually melts under the sun: the transformation of the snowman into water is a (slow) process that takes considerable time, so the transition is not instantaneous in the observer's eye. Actually, it is a matter of time granularity: it depends on the time-scale of the observations if the time extent of the transformation, during which both the original and the transformed object exist, can be considered instantaneous or not. The situation may become problematic if in the same analysis it is necessary to mix observations with different time granularity.

Ill-defined Concepts

Some concepts used intuitively in the previous sections are more complicated than it may appear at first sight. The first case comes up with the partitioning of space in which everything is embedded. All things occupy space, but of course not all of it: they occupy a portion usually called a region. There have been a number of studies on algebras built on regions, both by ontologists and geographers. Unfortunately the latter seem to talk little to each other, so no unifying theory is still available.

On the ontology side, Varzi (2007) has developed a thorough study of regions in his work on mereology and locations (see also Casati and Varzi 1999). Among other concepts, he argues that although in practice there is an issue of vagueness in some concepts, this does not affect his theory. In his opinion, vagueness derives from semantics rather than mereology, and can be handled within his theory by using a three-valued logic, with values True, False and Indeterminate. As concerns regions, the possible vagueness of region *r* (the mountain Everest, in Varzi's example) sometimes does not allow to decide whether a region *x* is, or is not, included in *r*. According to Varzi, the reason for this is not due to an intrinsic indeterminacy of the region *r*, rather to a poor definition of the terms: it pertains to semantics, and not to mereology. He suggests that these cases might be dealt with so-called supervaluational semantics, which adopt as a truth criterion the following: a statement including vague terms is true if it is true under any precisification; is false if it is false under any precisification; it is indeterminate otherwise. Precisification is a term from logic meaning the process of making a statement more precise. In other words, if the truth of a statement resists to any refinement of the terms it contains, the statement is true; if this holds for its negation, it is false; otherwise, it is indeterminate. This approach may be read in archaeological term as follows: belongingness to or inclusion in an apparently indeterminate region is resolved through archaeological precisification of the latter, i.e., through all known archaeological refinements of the region definition. If during this process belongingness/inclusion holds, the indeterminacy is resolved, otherwise it remains indeterminate. Notwithstanding its theoretical fascination, this approach may fail when applied in practice. There is no practical way of determining the outcome of an algorithm verifying something based on 'any', like the 'any precisification' required above. Furthermore, knowing that the truth of some statement is indeterminate—what would probably be the case for most archaeological statements on location—would put researchers at a dead end.

In another study, Smith and Mark (2003) hypothesize that the spatial vagueness of geographical terms like mountain, river, and lake, depends on being based on a common-sense approach rather than on a scientific definition, which eventually could resolve such uncertainty. They argue that in most cases such terms are used loosely, and this introduces indeterminacy that might be removed with a semantic clarification of the terminology, possibly based on geomorphology and other scientific theories. Although this may be true, there is no evidence that it is feasible.

On the other hand, geographers have tried to handle vagueness since several years (Burrough and Frank 1996, Pauly and Schneider 2008, 2009, Schneider and Pauly 2007, Schneider 2008). There are a number of studies dealing with geographical objects with indeterminate boundaries, fuzzy regions and so on, which aim at providing a theoretical background to dealing with indeterminacy in practice. Such studies assume that vague regions exist in reality, and process them basing on various mathematical theories such as fuzzy sets, rough sets or probability. This approach is very much oriented to practice, introducing only the strictly necessary theory and the related tools.

A similar difference affects a related entity, the **border**, i.e., what separates two contiguous regions. It is a concept that is relevant also when dealing with parts/whole analysis. Smith and Varzi (1997, 2000) distinguish between **bona fide** borders, i.e., real-life borders between objects/regions, and **fiat** borders, i.e., borders precisely and formally defined; only fiat borders are surely crisp, while bona fide ones may result in vagueness. In the authors' opinion, such a dual framework is necessary to deal with knowledge about spatial location and parts/whole relationships. On the contrary, other authors argue that "the distinction between bona fide and fiat boundaries is overdrawn" (Montello 2003: 195). Neither of the above approaches has received much attention in archaeology, notwithstanding the necessity of dealing with the vagueness embedded for example in the concept of archaeological site, a basic concept in the archaeological discourse. We defend rather the position that anything "phenomenal" is vague or fuzzy by nature. It is only a question of scale to reveal the fuzziness. This is also in agreement with the physics of surfaces and measurement in natural sciences in general. This provides on the other side the hint to the general solution: Getting away from the vague areas makes them appear more and more solid. "Encapsulating" vague areas in sufficiently distant outer bounds and inner bounds allows for making absolutely true statement about within which limits things are to be found.

It seems reasonable to expect that the approach presented here is capable of dealing with the various meanings embedded in such a concept as discussed in the Teutoburg battle example in section 4: the (modern) place corresponding to the space-time region where past events took place, the 3D regions that approximate the spatial projections of the spatio-temporal volumes corresponding to past events, and so on. Embedding spatial concepts in a 4-dimensional universe seems to provide a unifying perspective on vagueness and on how to deal with it. Keeping entities separate from the spatio-temporal regions they belong to is also a key factor to avoid the so-called 'reductionist' fallacy, i.e., confusing things with the space they occupy, an error to which geographers seem to be attracted as a professional bias.

As discussed elsewhere in this volume, the classification criteria stated in section 3 are not exempt from vagueness. They should balance comprehensiveness (i.e., reasonable outer bounds), to create significant aggregations of instances, with precision, to provide 'good' definitions and avoid vagueness. Any mixture of these two contrasting features incorporates fuzziness in the criterion and makes classification and typization an uncertain adventure.

Conclusions

As anticipated in the introduction, we have outlined here how to characterize some fundamental concepts in archaeology from a semantic perspective. Firstly, all archaeological entities are embedded in space-time, and the connection with their spatio-temporal location is a key passage in archaeological interpretation. Clarifying every detail of this process and the related concepts in light of knowledge organization theory is key to enable a better use of accumulated data and to organize archaeological knowledge on a sound basis. The apparently complicated and detailed description involved is indeed much simpler than it appears at a first look. It locates usual concepts in a better organized perspective and supports archaeological reasoning, inference and interpretation. It makes the best use of advances in other disciplines as philosophy, information technology and geography. Even when not indispensable, for example when common-sense-based reasoning seems to avoid misunderstandings and pitfalls, it helps in establishing a rigorous and reproducible way for building theories based on factual evidence as is or should be—typical of a science.

References Cited

Abramiuk, M.A. 2012. The Foundations of Cognitive Archaeology MIT Press.
ARENA. http://ads.ahds.ac.uk/arena/.
Bentley, R.A., H.D.G. Maschner and C. Chippindale (eds.). 2008. Handbook of Archaeological Theories AltaMira Press.
Bintliff, J. and M. Pearce (eds.). 2011. The Death of Archaeological Theory? Oxbow Books.

Burrough, P.A. and A. Frank (eds.). 1996. Geographic Objects with Indeterminate Boundaries. CRC.

Casati, R. and A.C. Varzi. 1999. I trabocchetti della rappresentazione spaziale. Sistemi Intelligenti 11: 1, 7–28.

Costopoulos, A. and M.W. Lake (eds.). 2010. Simulating change: archaeology into the twenty-first century. University of Utah Press, Salt Lake City.

Doerr, M. 2003. The CIDOC CRM—An Ontological Approach to Semantic Interoperability of Metadata, AI Magazine 24(3): 75–92.

Doerr, M. 2009. Ontologies for Cultural Heritage. pp. 463–486. *In*: S. Staab and R. Studer (eds.). Handbook on Ontologies, Springer-Verlag.

Doerr, M. and D. Iorizzo. 2008. The dream of a global knowledge network—A new approach. J. Comput. Cult. Herit. 1(1): 1–23.

Doerr, M., A. Kritsotaki and A. Boutsika. 2011. Factual argumentation—a core model for assertions making. Journal on Computing and Cultural Heritage (JOCCH). New York, NY, USA: ACM 3(3): 34.

Doerr, M. and G. Hiebel. 2013. CRMGeo: Linking the CIDOC CRM to GeoSPARQL through a Spatiotemporal Refinement. Technical Report ICS-FORTH/TR-435.

Doerr, M. and M. Theodoridou. 2011. CRMdig: A generic digital provenance model for scientific observation. *In*: Proceedings of TaPP'11: 3rd, USENIX Workshop on the Theory and Practice of Provenance.

Gangemi, A., N. Guarino, C. Masolo, A. Oltramari and L. Schneider. 2002. Sweetening Ontologies with DOLCE. pp. 166–181. *In*: A. Gómez-Pérez and V. Richard Benjamins (eds.). Proceedings of the 13th International Conference on Knowledge Engineering and Knowledge Management. Ontologies and the Semantic Web (EKAW '02). Springer-Verlag.

Gerhard, F.W.E. 1829. Annali dell'Instituto di Corrispondenza Archeologica 1,3.

Guarino, N. and C. Welty. 2001. Identity and subsumption. pp. 111–125. *In*: Rebecca Green, Carol A. Bean and Sung Hyon Myaeng (eds.). The Semantics of Relationships: An Interdisciplinary Perspective. Kluwer.

Guarino, N., D. Oberle and S. Staab. 2009. What is an Ontology? pp. 1–17. *In*: S. Staab and R. Studer (eds.). Handbook on Ontologies, Springer.

Hodder, I. 1999. The Archaeological Process. An introduction. Wiley.

Hodder, I. 2012a. Entanglement. An Archaeology of the Relationships between Humans and Things Wiley-Blackwell, Chichester.

Hodder, I. (ed.). 2012b. Archaeological Theory Today. Second Edition, Polity Press.

Johnson, M. 2010. Archaeological Theory. An Introduction (2nd edition), Wiley-Blackwell.

Lock, G.R. and G. Stancic (eds.). 1995. Archaeology and Geographic Information Systems: A European Perspective. CRC Press.

Locke, J. An Essay Concerning Human Understanding. The Pennsylvania State University 1999. Eletronic Classic Series, Jim Manis, Faulty Editor, Hazleton, PA.

May, K. 2006. CRM-EH. Available at: www.cidoc-crm.org/workshops/heraklion_october_2006/may.pdf.

Montello, Daniel R. 2003. Regions in Geography: Process and Content. *In*: Matt Duckham, Michael F. Goodchild and Michael Worboys (eds.). Foundations of Geographic Information Science. Taylor & Francis.

Niccolucci, F. 2002. XML and the Future of Humanities Computing. Applied Computing Review, 10–1, 43–47.

Niccolucci, F. and J.D. Richards. 2013. ARIADNE: Advanced Research Infrastructures for Archaeological Dataset Networking in Europe. International Journal of Humanities and Arts Computing, 7.1–2, 70–78.

Olsen, B. 2010. In Defense of Things. Archaeology and the Ontology of Objects, Altamira Press, Plymouth.

Olsen, B., M. Shanks, T. Webmoor and C. Witmore. 2008. Presentation of the book Archaeology: the Discipline of Things, U. of California Press, available on http://humanitieslab.stanford. edu/23/1572.

Pauly, A. and M. Schneider. 2008. Vague Spatial Data Types. pp. 1213–1217. *In*: S. Shekhar and H. Xiong (eds.). Encyclopedia of GIS. Springer-Verlag.

Pauly, A. and M. Schneider. 2009. An algebra for vague spatial data in databases. Information Systems 35: 111–138.

Praetzellis, A. 2011. Death by theory: a tale of mystery and archaeological theory, AltaMira Press.

Schneider, M. 2008. Fuzzy Spatial Data Types for Spatial Uncertainty Management in Databases. pp. 490–515. *In*: J. Galindo (ed.). Handbook of Research on Fuzzy Information Processing in Databases. Information Science Reference.

Schneider, M. and A. Pauly. 2007. ROSA: An Algebra for Rough Spatial Objects in Databases. pp. 411–418. *In*: J.T. Yao (ed.). Proceedings of the 2nd Int. Conf. on Rough Sets and Knowledge Technology (RSKT), LNAI 4481, Springer-Verlag.

Sider, T. 1997. Four Dimensionalism. Philosophical Review 106: 197–231.

Smith, B. 2003. Ontology. pp. 155–156. *In*: L. Floridi (ed.). Blackwell Guide to the Philosophy of Computing and Information. Blackwell.

Smith, B. and D.M. Mark. 2003. Do Mountains Exist? Towards an Ontology of Landforms. Environment and Planning B (Planning and Design) 30: 411–427.

Smith, B. and A.C. Varzi. 1997. Fiat and Bona Fide Boundaries. Towards an Ontology of Spatially Extended Objects. pp. 103–119. *In*: S.C. Hirtle and A.U. Frank (eds.). Spatial Information Theory: A Theoretical Basis for GIS. Proceedings of the Third International Conference, Springer-Verlag.

Smith, B. and A.C. Varzi. 2000. Fiat and Bona Fide Boundaries. Philosophy and Phenomenological Research 60: 2, 401–420.

Theodoridou, M., Y. Tzitzikas, M. Doerr, Y. Marketakis and V. Melessanakis. 2010. Modeling and Querying Provenance by Extending CIDOC CRM. Distributed and Parallel Databases Journal 27(2): 169–210.

Varzi, A.C. 2007. Spatial Reasoning and Ontology: Parts, Wholes and Locations. pp. 945–1038. *In*: M. Aiello, L. Pratt-Hartmann and J. Van Benthem (eds.). Handobook of Spatial Logics. Springer-Verlag.

Wiggins, D. 2001. Sameness and Substance Renewed. Cambridge University Press.

4

Statistical Reasoning and Archaeological Theorizing: The Double-Bind Problem

Dwight Read

Introduction

The theme of this chapter is the use of statistical reasoning to extend archaeological reasoning about past social and cultural systems according to their material traces that have survived to the present. Archaeology, though, by its nature as a discipline, deals directly with the material objects formed and produced through deliberate human activity and only indirectly with the underlying cultural framework for which the material objects are the instantiation through culturally mediated behavior: "Culture does not consist of artifacts. The latter are merely the results of culturally conditioned behavior performed by the artisan (Rouse 1939: 15)". This distinction expresses, even if implicitly, the ontological basis both for using statistical ideas to expand upon, and extend, archaeological reasoning and for theorizing aimed at forming a connection between what is directly observed by the archaeologist and the underlying "culturally conditioned behavior." From a mathematical viewpoint, the distinction implies that we need to distinguish between mathematical reasoning used to address the organization and structure of the idea systems making up culture (Lane et al. 2009, Leaf and Read 2012, Read 2012), and the statistical methods used to study patterning determined at the phenomenal level of the material objects produced through culturally conditioned behavior.

Abductive Arguments

The material domain that is the province of archaeology cannot be fully understood without taking into account how patterning in the material domain relates to the

Department of Anthropology, University of California, Los Angeles, CA 90095 USA.
  Email: dread@anthro.ucla.edu

cultural domain. In order to access the latter indirectly through the material remains of the past that have survived to the present, explanatory arguments linking cultural idea systems with the patterning observed in the material consequences of human behavior are formulated using what Charles Peirce referred to as the process of abduction: "[a]bduction is the process of forming explanatory hypotheses" (Peirce 1935: 172, as quoted in Douven 2011). Archaeology as a discipline is dependent upon working abductively from consequence to explanation in order to be more than a descriptive science.

Theory Models and Data Models

To link present-day observations to past behavior and the mediating effect of culture, archaeological research begins with patterning found in the material domain and then abductively connects that patterning to the ideational domain of culture through formulating (using current knowledge) likely explanatory arguments that causally connect the ideational domain to the material domain. Thus the abductive process reverses the inductive research ontology that begins with patterning delineated in the material domain to a deductive ontology that begins with the properties in the ideational domain that would, through the mediating effect of culture, cause the observed patterning in the material domain. The reversal also involves switching between two different kinds of models, hence two different kinds of formal representations. On the one hand, models formulated as a way to represent the logic of cultural idea systems are likely to be in the form of *theory models* (Read 2008) that express properties arrived at deductively from assumed structuring processes, such as a model of the structure of a kinship terminology derived deductively from posited, primary kinship concepts that are part of the cultural domain for the members of a particular society (see Read 1984, 2007b, 2010, 2011, Bennardo and Read 2007, Leaf and Read 2012 [among others] for examples of deductive reasoning applied to the cultural domain). Models like this lend themselves to formal representation using mathematical ideas (Read 2011).

On the other hand, *data models* (Read 2008; see also Chapter 17 for data models in the form of neural nets calibrated to Medieval castle locations in Tuscany and to pit-dwellings on Senja, a Norwegian Island) contrast with theory models and are used to descriptively represent patterning in data observations that reference the phenomenal level of the remnants of material objects produced (in their various modalities) by behavior mediated through the cultural domain. These representations sometimes use qualitative descriptions (such as a ceramic typology based on qualitative attributes) and sometimes quantitative descriptions, with the latter often expressed using summary statistical measures. Statistics is widely accepted in archaeology (see review by Read 1989a) as providing the mathematically based representations of choice for descriptive, quantitative models since statistics has to do with discerning patterning found in the aggregate, as opposed to patterning that may be found on individual cases, and quantitative models typically deal with patterns in the aggregate.

Patterning on Individual Cases versus Patterning in the Aggregate

Patterning on individual cases is exemplified by ceramic typologies based on qualitative patterning of attributes since the attribute pattern on a ceramic object either does or does not satisfy the criteria for belonging to a qualitatively defined ceramic type and so no statistical analysis is needed for this assignment. In contrast, patterning in the aggregate arises when it is not the attribute values for an object that determines the pattern in question, but the patterning among the individual values measured over an aggregate of entities. The values obtained for quantitative variables typically measure the outcome of processes employed to produce an artifact and these are subject to production error between intent and resulting artifact. The measured value will not represent precisely the magnitude of a dimension intended by the artisan, hence it is not the individual value, *per se*, but the pattern expressed through the individual values considered 'in the aggregate' that is of interest. For example, the maximum widths for concave-base projectile points from 4 VEN 39 (see top two rows of Fig. 1 and discussion in Read 1987, 1989b, 2007a), a Paleo Indian site in Ventura County, California, have a frequency distribution with two non-overlapping modes (see Fig. 2), hence we can characterize the concave-base points as being composed of two types: narrow concave-base points and wide concave-base points (see Fig. 1, top two rows).

In addition, there is no reason to expect or to assume that the 'missing' widths in the frequency distribution (missing in the sense of the width values that would have to be added to the actual frequency distribution for it to have a unimodal, rather than a bimodal, distribution) are contingent, say, on the flint knapping technique. The two kinds of points were made in a similar way and so widths that would lie between the two modes were feasible. Hence they are 'missing' in the recovered data because the artisans who made the concave-base projectile points 'chose' not to make those particular widths; that is, the artisans made the points in accordance with two normative (i.e., cultural) values for the maximum width of the concave-base points that are functionally equivalent (Read 2007a), thereby giving rise to what we (etically) label as *narrow* versus *wide* point types, and their functional equivalence suggests groups organized matrilineally (Read 2007a). Thus we abductively explain the observed bimodal pattern

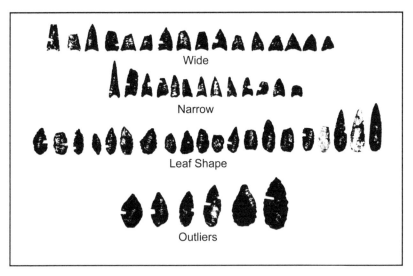

Fig. 1. Projectile points from 4 VEN 39. The top two rows are concave-base triangular points divided into two groups, Narrow and Wide, based on a bimodal distribution for the maximum width of the points. Notches are from sections removed for hydration dating. Some projectile points are missing (Reprinted from Read 2007a: Fig. 5.1 with permission of publisher).

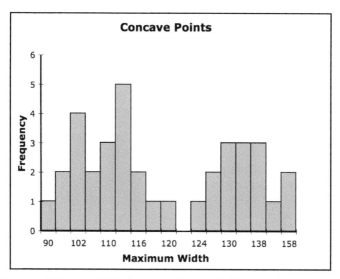

Fig. 2. Histogram of Maximum Width for concave-base projectile points from 4 VEN 39. The left mode can be characterized as *narrow* and the right mode as *wide* (Reprinted from Read 2007a: Fig. 8.7 with permission of publisher).

for the frequency distribution of these point widths through positing that there were two normative values for the concave-base projectile points as part of the cultural repertoire of the inhabitants of 4 VEN 39.

Now let us return to the statistical part of the abductive argument. A concave-base point, in isolation, cannot be assigned to the narrow type or the wide type based on its width measurement except by using that value to determine the mode in the frequency distribution of width values for the concave-base points to which the point in question should be assigned. Critically, by the nature of patterning in the aggregate, if the aggregate changes then the pattern of the frequency distribution may change as well. For the points from 4 VEN 39, if we change the aggregate from 'concave-base projectile points' to 'projectile points' (that is, to all of the points shown in Fig. 1), the frequency distribution of widths no longer even suggests a bimodal distribution, and so if "projectile points" were the aggregate, we would erroneously conclude that only one type of projectile point is present. Our abductive argument would now change and we would now assert that the artisans were operating in a cultural context in which there was a single normative value for the width of projectile points.

Observe that these two different conclusions are not the consequence of obtaining additional projectile points whose properties then changed our previous argument, but the consequence of *our* choice regarding the aggregate for discerning patterning. Both arguments cannot be correct since the database is the same in both cases, and if we have knowledge of the bimodal distribution, we would obviously conclude that the second abductive argument leads to an incorrect explanation. However, it is possible that we may not yet have examined just the concave-base projectile points and if we only have the patterning of all the projectile points considered as a single ensemble, with a unimodal statistical distribution of the width values, then we would have no statistical reason for recognizing that abduction has led us to an invalid explanatory argument.

Abductive arguments for linking the ideational, cultural domain in which past societal members acted to the phenomenal domain of what they produced by reference to patterning found in the surviving materials observed by the archaeologist are dependent, then, on analytical methods being able to discern patterning that accurately reflects the structure and organization of that material that is the consequence of its production and use. Hence quantitative patterning derived through statistical analysis must agree with our understanding of the way the cultural domain affects what is expressed in the material domain through behavior mediated by that cultural domain. For our projectile points, we expect, as a first approximation and by way of the Central Limit Theorem (see Read 2007a: 209–212 for details), that a quantitative dimension will likely have a unimodal, possibly normal, distribution when all artisans agree on a single, desired value for that dimension and will have any one of a number of other possible distributions otherwise. However, the statistical analysis, here in the form of the shape of a frequency distribution, may not provide any indication that we have used the wrong aggregate for discerning pattern or, even more problematic, it will not indicate what variable(s) and aggregate combination(s) will display the relevant patterning. Whether the aggregate should be 'all projectile points', or 'all concave-base projectile points' measured with maximum width, depends on archaeological, not statistical reasoning. As archaeologists, though, we do not know *a priori* what the appropriate aggregate-variable combinations should be. Determining the later is precisely a major goal to be achieved through analysis of the artifacts.

Double-Bind Problem

We have then, a "double-bind" problem (Read 1989b, 2007a: 199–200), as shown in Fig. 3. To know in advance what are the appropriate aggregate-variable combinations (that is, the well-identified data and the well-identified variables in Fig. 3), we need precisely the information we are trying to derive through our analyses, especially statistical analyses. When the initial data set is, for example, all projectile points from 4 VEN 39 and all the variables measured over these points, there is no obvious statistical solution for identifying that the aggregate 'concave-base projectile points', along with the maximum point width measure, is the aggregate-variable combination that leads to division of the concave-base projectile points into narrow versus wide concave-base points. Instead, discovering this combination depends upon conducting data analysis in an exploratory manner that takes advantage of previous, tentative explanatory arguments, which may involve assumptions of a different order than those upon which statistical analyses are based (Carr 1989). The iterative procedure introduced below as a way to resolve the double-bind problem depends upon first discerning the patterning that would occur in the data, given the various factors that structured the production, use and discard of artifacts. Thus it depends upon correctly hypothesizing the connection between the ideational domain and patterning found in the production and use of artifacts.

Archaeological Reasoning

Statistical analysis is brought into archaeological research so as to objectively make inferences going from the data-at-hand to a larger context. By its nature, the material remains of past societies—the objects themselves and the relations among those

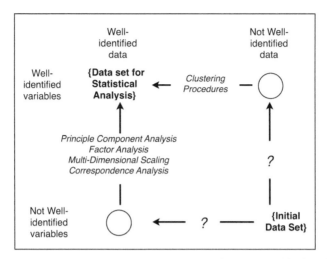

Fig. 3. Double-bind problem. *Well-identified data* correspond to data aggregates identified for analytical purposes that match the inherent data structure. *Well-identified variables* have frequency distributions according to the inherent data structure. Data as recovered by the archaeologist are in the lower right corner: neither well-identified data nor well-identified variables. Data required when statistical analysis is used in abductive reasoning is shown in the upper left: well-identified data and well-identified variables. The double-bind problem arises from needing well-identified data to establish well-identified variables and needing well-identified variables to establish well-identified data. Variable reduction procedures can be used on not well-identified variables defined over well-identified data (arrow from lower left to upper left) and clustering procedures can be used on not well-defined data using well-identified variables (arrow from upper right to upper left). Statistical methods neither provide direct procedures to transform the initial data set into a data set with well-identified data (arrow with a '?' from lower right to lower left), nor into a data set with well-identified variables (arrow with a '?' from lower right to upper right).

objects—typically display patterning in the aggregate rather than on individual cases, whether the patterning is at the micro level of individual artifacts produced by group members or at the macro level of social groups and their forms of organization. Any data model we posit as representing the patterning in the data we analyze and/or theory model we posit as representing the processes responsible for that patterning will be incorrect if we include invalid inferences drawn from observations made on the data-at-hand and/or irrelevant processes arrived at through incorrect abductive arguments. Of necessity, we engage in statistical analyses as a way to objectify inferential arguments when the data-in-hand are not the totality of the data relevant to the phenomena we are trying to incorporate into our arguments. Consequently, we make inferences that have been objectivized using statistical reasoning for the inferential process of going from observed sample measures to unobserved population parameter values

While the rationale for the use of statistical methods for inferential purposes is clear, the conceptual framework of statistics and the conceptual framework we formulate as archaeologists are not necessarily in concordance with each other (see Carr 1989) and, as the double-bind problem indicates, they will seldom be in concordance initially. The goal of inferential statistics applied to data analysis lies in objective inference when generalizing from sample to the population from which the sample was obtained. The statistical methods developed for so doing are mathematically sound and conceptually rich, yet are not central to the concerns of the archaeologist as archaeologists. Statistical inference exists for a simple, pragmatic reason: the inability to have access to populations as a whole. Were we able to access populations as a whole, then there would be no need for inferential statistics.

Inferential Statistics and the Definition of a Population

Consider the notion of a population in more detail. From a statistical viewpoint, a population only needs to be a collection of entities that is well-defined; that is, we can (in principle) determine whether a given entity belongs to that population or not. We can, for example, define the excavated projectile points from 4 VEN 39 as the population since that is a well-defined collection of entities. When we have all the entities making up a population, we can compute the exact value (ignoring measurement error) of statistical parameters such as the population mean, μ, and the population standard deviation, σ. If we have another archaeological site and we consider the projectile points from that site to be a different population, then we can determine, for example, whether the respective means, μ_1 and μ_2, of the two populations are the same simply by comparing their numerical values. No inferential statistical test such as a t-test for comparison of means is needed.

From a statistical viewpoint, then, conceptually there is no reason why we cannot proceed without the machinery of statistical inference so long as our data are in the form of complete populations. From an archaeological viewpoint, however, most likely we would object to a population defined as 'all projectile points recovered by excavation from 4 VEN 39' since such a population is of limited archaeological interest. We want, instead, a description not of what we have recovered through excavation, but also of what we have not recovered. This requires that we sample the totality of the artifacts in the site, e.g., we grid the site and excavate a random selection of grid squares. Now our population becomes 'all projectile points that would be obtained by complete excavation of the site' and our data-in-hand are a (random) sample from that population.[1]

Statistical Null Hypothesis versus Archaeological Null Hypothesis

We compute sample statistics and confidence intervals to estimate the population parameters corresponding to the expanded definition of the population. Instead of comparing parameter values directly, we now form null and alternative hypotheses such as $H_0: \mu_1 = \mu_2$ and $H_1: \mu_1 \neq \mu_2$ and determine the probability with which the observed difference in the magnitude for the sample means for the samples from the two sites would occur if H_0 is true, assuming the sample means follow a known statistical distribution, or use bootstrapping methods when the distribution is unknown. Though a valid way to proceed from a statistical viewpoint, what we have done is an illusion from an archaeological viewpoint since we are dealing with a finite population.

Suppose we had the resources to excavate the two sites in their entirety, recovered all the projectile points from the two sites and computed the exact means, μ_1 and μ_2, for the two sites. Because we are dealing with finite populations, almost surely the two

---

[1] More precisely, we have formed a cluster sample. However important may be the different ways of sampling for carrying out valid statistical inferential analysis, the details of different sample procedures are not of concern here.

parameters are not the same numbers, hence $\mu_1 \neq \mu_2$. Actually, we know this to be the case even without excavating the two sites and without a single measurement. Our use of statistical inference has given us the illusion that we have made a rational decision, as archaeologists, regarding whether we should accept or reject the null hypothesis, whereas we know in advance that the null hypothesis is wrong. The statistical response to this conclusion might be that statistical inference has to do with a hypothetical state of affairs, not whether that hypothetical state of affairs has been empirically observed. However, as archaeologists, we are not dealing with a hypothetical state of affairs, but with what is empirically valid, and empirically it is almost surely the case that two finite populations produced by events in the real world will not have identical population parameters, even if produced by the same processes.

It follows that the null hypothesis of interest to the archaeologist is not H_0: $\mu_1 = \mu_2$, but H_0: $|\mu_1 - \mu_2| \leq \delta$, where δ represents the archeologist's decision regarding how much of a difference, δ, between the two means would be required before the difference would have archaeological significance. The latter is not a statistical question, but is the question of concern to archaeological research, not whether the two means are numerically exactly the same.

Statistical Population versus Archaeological Population

Yet even this expanded definition of a population is not adequate, for we want to know about not only the projectile points left in the site, but also the projectile points that were lost through usage. In addition, since groups of people do not exist in isolation, we want to know about points made by the inhabitants of the site that were given or traded to other groups and points in the site that originated with other groups, and so on. In effect, we want to know about all projectile points made by this particular group and their "life-history", not just about those that survived to the present. A population defined in this manner poses, of course, complex sampling problems that have been discussed extensively by archaeologists (e.g., Schiffer 1987; see also Chapters 3 and 15, this volume). Yet even if we resolve the sampling problem, another problem arises. While we could define the population to be the set of all projectile points ever made by the inhabitants of the site, and while this satisfies statistical criteria for what constitutes a population, from an archaeological viewpoint even this definition is not adequate as it is a population that depends upon happenstance events. Assume that somehow we did recover this population and we computed exact values for population parameters such as the mean and standard deviation. As archeologists, we know that the set of particular points that were made while the site was occupied is happenstance. Under slightly different conditions a flint knapper might have made more or fewer points, fewer or more points might have been lost when used for hunting animals, and so on.

Statistical Parameters and Cultural Norms

Further, consider the archaeological interpretation we might want to make of the population parameters. Following the arguments of Rouse (1939), if length happens to be a mode—that is, a dimension under cultural control—then we may want to interpret parameters such as the mean length as a normative value, hence characteristic of a group, and not simply as the average of individual flint knapper decisions about projectile point lengths. Consider the metal projectile points made by different san groups in Botswana (see Wiessner 1983), where the members of a single group made their points within a narrow size range, but one group may have used a size range distinct from the size range for the points made by another group. We can account for these observations by abductively hypothesizing that group A has μ_A as its normative value for projectile point size, meaning that this is the target size each person in that group has 'in mind' when making projectile points, hence we expect that the points made by members of group A will, on average, have mean μ_A regardless of the particular artisan making projectile points, whereas group B has a different target size 'in mind', hence $\mu_B \neq \mu_A$, when they make projectile points.

We find support for this abductively derived explanation for the regularity in the san metal points through an experiment (Gandon et al. 2013) in which potters in India and potters in France were both asked to duplicate a spherical-shaped pottery object. The target object plays the role of a cultural norm and the use of potters from both India and France introduced not only individual variability in pottery making skills into the experiment, but cultural variation in the technique of pottery making. The experimenters found that the pottery makers made statistically indistinguishable shaped pottery objects independent of whether they were French or Indian potters; in other words, normative values—in this case, the target shape provided by the experimenters—translate into similar objects, independent of the individual artisan.[2]

The points made by the flint knappers at 4 VEN 39 are, then, but a sample of what could have been produced had projectile points been made indefinitely. Consequently, the population we want is the hypothetical population of all possible points that could be made with a normative value as the target, thus implying that the collection of all points that the flint knappers made is but a sample from the infinite population of all points that could ever have been made with the same normative value as the target.[3] Thus, none of the populations—projectile points excavated, all projectile points at 4 VEN 39, all projectile points made by the flint knappers at that site—is the population of interest for reasoning by abduction. From a statistical viewpoint, any one of these is a *valid* population, but what constitutes a *meaningful* population is not a statistical, but an archaeological, question, hence to be answered through archaeological and not statistical reasoning. The latter simply provides the relevant boundaries when making a connection between archaeological reasoning and statistical method.

---

[2] The experimenters interpreted their results as showing equivalent function may arise from culturally distinct motor skills. In other words, artisans with the same cultural norm guiding their artifact production will not necessarily use the same fashioning gestures, yet their outcomes (i.e., shape artifact) will be similar. Their experiment shows that even cultural differences in the background of the potters (French potters versus Indian potters) can be 'overridden' when they all have the same simple target 'in mind' when making a pottery object. In more recent experiments, they have found that potters, when given a complex, unfamiliar type, tend to make potter specific variants, thus suggesting that convergence to a cultural norm for a pottery shape is a dynamic process involving interaction among the potters (Enora Gandon, Personal Communication).

[3] A type defined as the finite population P consisting of 'all artifacts from the site satisfying condition C' is extensively defined since the magnitude of a population parameter defined over P will vary with change in that population's content, whereas the statement 'all objects that could ever be produced, with fixed normative values' provides an intensive definition for a type, hence the latter definition eliminates the problems that arise (see Dunnell 1971, O'Brien and Lyman 2003; Chapter 11, this volume) when types are defined extensively.

That the abductive framework implies a statistical population of artifacts should include not only all points that were made, but those that could have been made keeping fixed normative values (if any) for the dimensions of the artifacts is in keeping with statistical reasoning: "Even in the simplest of cases the values ... before us are interpreted as a random sample of a *hypothetical infinite population of such values as might have arisen in the same circumstances*" (Fisher 1954: 6–7, emphasis added). That is, finite populations such as 'all excavated projectile points' or even 'all projectile points made by the inhabitants of 4 VEN 39' are not the appropriate reference populations regarding the constraints—cultural and otherwise—affecting, for example, the artifacts and their attributes produced by artisans at 4 VEN 39.

These constraints include (but not exclusively) culturally relevant dimensions for which a normative value provides the target for the artisan. Dimensions like this are what Rouse distinguished as modes; other dimensions of the artifacts may be constrained by, for example, the inclinations of the individual artisan, hence for those dimensions parameter values would be artisan specific; yet other dimensions are neither constrained at the cultural level nor at the level of the individual artisan, hence are analogous to neutral traits in genetics (Read 2007a: Chapter 10). Read (1982) provides an example, using Scottsbluff points found in the Midwest, showing how the point dimensions may be divided analytically into functionally constrained and culturally constrained modes, versus those that are neither functionally nor culturally constrained.

Statistical Aggregates versus Archaeological Aggregates

Viewing what constitutes a population of artifacts brings us back to reconsideration of the aggregate over which quantitative patterning is to be discerned. One way archaeologists have addressed the aggregation question is through forming artifact typologies under the notion that the members of the same type are similar to each other and differ from members of other types, hence types are, presumably, the appropriate aggregations. For the most part, typologies, especially pottery typologies, have relied on qualitative measures for their definition, but whether the type differences are expressed qualitatively or quantitatively is not critical for the argument being developed here. Instead, we need a more formal representation of what we mean by a type.

Archaeological Types and Mathematical Relations

The idea of similarity of artifacts as a basis for forming a type can be expressed more formally using the concept of a (mathematical) *relation* defined over a set *S*. By a *relation*, call it *R*, defined over a set *S* is meant a subset of $S \times S$, where $S \times S$ is the collection of all ordered pairs (s, t) of elements s and t from *S* (with *ordered pair* meaning that the pair (s, t) is not the same as the pair (t, s) when $s \neq t$). We will use the expression, sRt (read: "*s* is *R* related to *t*"), to indicate that the ordered pair (s, t) is a member of the subset *R* of $S \times S$.

As an example of a relation, we can represent object similarity by the relation, call it $R_{similar}$, defined over *S*, the collection of artifacts, by including the ordered pair of artifacts (s, t) in the relation $R_{similar}$ when, and only when *s* is similar to *t*. For the set of projectile points with the (incomplete) typology, Projectile Points → Concave-based points + Convex-base points (see Fig. 1, top 3 rows), the concavity dimension determines a similarity relation $R_{concave}$ by including (s, t) in $R_{concave}$ when, and only when, both s and t are concave.

Equivalence Relations

The particular kind of relation we need for connecting relations with types is that of an *equivalence relation*. A relation R defined over the set S is said to be an *equivalence relation over S* when the following three conditions are satisfied by R: (1) R is reflexive: for all s in S, sRs; (2) R is symmetric: if sRt, then tRs for all s, t in S; and (3) R is transitive: if sRt and tRu, then sRu for all s, t and u in S. It can easily be shown, for example, that the relation $R_{concave}$ is an equivalence relation.[4] Corresponding to an equivalence relation R defined over the set S we may define the *equivalence class* R_t corresponding to an element t in S by letting R_t be the set of all s in S for which sRt is true. The analytical power of an equivalence relation stems from the following two theorems:

> *Theorem 1*: An equivalence relation R defined over a finite set S partitions S exhaustively into $n \geq 1$ disjoint subsets $S_1, S_2, ..., S_n$; that is, (1) $S_1 \cup S_2 \cup ... \cup S_n = S$ (read: "the union of the subsets S_i is the set S"), where "$\cup$" stands for set union and (2) $S_i \cap S_j = \varnothing$, the empty set (read: "the intersection of the subsets S_i and S_j is the empty set"), for all i, j, $1 \leq i \neq j \leq n$, where "$\cap$" stands for set intersection. Conversely, a partition of S into n disjoint subsets determines an equivalence relation R over S defined by (s, t) is in R if, and only if, each of s and t is in the same subset from the partition of S.
>
> *Theorem 2*: If R is an equivalence relation defined over the set S and s and t are in the same subset S_i of the partition determined by R, then $R_s = R_t$.

Theorem 2 implies that any member of the subset S_i may be used equally to represent an equivalence class.

Typologies and Equivalence Relations

These two theorems express the logic underlying a typology. The assumption of a typology is that it is exhaustive—all artifacts may be assigned to one of the types—and types are disjoint—no artifact may be assigned to more than one type (Theorem 1). Further, it is assumed that any instance of a type may be used as an exemplar of that type (Theorem 2), hence the practice of presenting ceramic typologies with a picture or drawing of a ceramic object belonging to each of the types in the typology.

The similarity measures typically used in forming artifact typologies determine equivalence relations; hence we can consider an artifact typology as a partition made of a set of artifacts by using an equivalence relation defined over that collection of

---

[4] It is an equivalence relation since: (1) $sR_{concave}s$, for when s is concave then both s and t (= s) are concave and so $sR_{concave}s$; if $sR_{concave}s$, then obviously s and s (= t) are concave; (2) if $sR_{concave}t$, then $tR_{concave}s$, for $sR_{concave}t$ is true when, and only when, both s and t are concave and both s and t are concave when, and only when, both t and s are concave, hence $tR_{concave}s$ is true; and (3) $sR_{concave}t$ is true when, and only when, both s and t are concave, and $tR_{concave}u$ is true when, and only when both t and u are concave, so both are true when, and only when both s and t are concave and both t and u are concave, hence both s and u are concave, thus $sR_{concave}u$ is true.

artifacts. The typology will vary, though, as does the relationship between the numerical value of statistical parameters and aggregates, with a different choice of a similarity measure. For example, Adams and Adams (1991) formed a typology for the seriation of Nubian pottery by basing the types on pottery attributes whose values were sensitive to time periods. Defining similarity by artifacts belonging to the same time period meant excluding other possible types defined by other criteria, such as time independent morphological similarity.

Distance Similarity and Single Linkage Clustering

With quantitative variables measured over artifacts, similarity is often defined by the distance between points in the n-dimensional measurement space determined by n dimensions measured over the artifacts, with each artifact represented as a point in that n-dimensional measurement space located according to the n values obtained for that artifact. For quantitative variables, a simple similarity definition is that two artifacts are similar if, and only if, $d(s, t) \leq \delta$, where d is a distance measure, such as Euclidean distance, for the two artifacts represented by the points s and t in the n-dimensional measurement space and δ is the degree of closeness required for two artifacts to be considered similar.

In the cluster analysis literature, the process of forming a partition determined by a distance similarity measure in the manner just described is referred to as *single linkage clustering* and the sets making up the partition are referred to as *clusters*. The intuitive notion that clusters should be internally cohesive and externally isolated is satisfied by single linkage clustering (Gregorius 2006). However, the clusters formed using single linkage clustering may lead, as is well-known, to chaining when there are a few points located between presumed clusters in the measurement space, hence single linkage clustering is often excluded as a way to determine clusters in a data set. Rejection of clusters formed through chaining depends, though, on using criteria other then similarity measured by distance between points in the n-dimensional measurement space, such as the concern that single linkage clustering can lead to "drawn-out clusters, some members of which are very far from each other" (Kaufman and Rousseeuw 1990: 48, as quoted in Gregorius 2006: 229). What is seen as undesirable chaining may be attributed to either choice of a distance measure, presumably because "the chosen measure of difference cannot detect the presumed structure" (Gregorius 2006: 229) or it "does not appropriately mirror the differences in those properties of objects that are considered to determine groups or classes" (Gregorius 2006: 229). These quotes, however, mainly repeat the fact that the single linkage analysis did not recover what is believed to be the structure for the measurements made over the entities under investigation, and the quotes do not indicate why the distance measure fails to do so. Further, the failure of a cluster analysis to recover even obvious structure in the data measurement space is not peculiar to single linkage clustering.

Invalid Assumption of Numerical Taxonomy and Clustering Algorithms

As will be shown below, the failure stems from the assumption that, in the absence of any well-founded criterion for excluding a measurement, all measures should be included, under the presumption that including additional measurements ensures a better representation of the properties of the objects being studied, hence should, it is assumed, lead to convergence on an underlying data structure composed of clusters with 'internal coherence and external isolation' (see Sokal and Sneath 1963). It is further presumed that possibly needing different variables for some clusters in comparison to other clusters is adequately encompassed within the idea of polythetic, rather than monothetic, groupings.

Both of these assumptions underlie all current clustering methods (see Chapter 9, this volume) and comparisons of clustering methods focus on other topics such as the choice of a distance measure and the characteristics of a clustering algorithm, e.g., in addition to single linkage, algorithms may be based on "complete link, group average, weighted average, centroid, median, or minimum variance" (Gregorius 2006: 229), or on methods such as k-means clustering in which the number of clusters is specified and the procedure then allows for reallocation of cases to clusters as the clustering proceeds. As a consequence, neither the fact that the choice of variables to be used in clustering methods needs to vary according to which cluster is being identified nor the implications this has for the adequacy of current clustering algorithms has been adequately considered.

Empirically, Adding Variables May Lead to Divergence From, Rather than Convergence On, Inherent Data Structure

We can illustrate the problem using the projectile points from 4 VEN 39. As shown in Fig. 4, the points were measured with a suite of nine variables, of which eight are quantitative measures related to the shape of the points and one relates to the

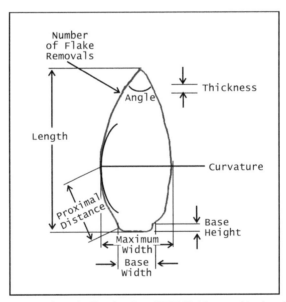

Fig. 4. Measurement system for projectile points from 4 VEN 39. All, except Number of Flake Removals, refer to the form of the projectile point and relate to functional dimensions of the points (Reprinted from Read 2007a: Fig. 6.5 with permission of the publisher).

flint knapping process by which the points were made. We will focus here on the 8 morphological measurements. *A priori*, there is no particular reason to exclude any of these variables as all are based on prior experience of archaeologists regarding relevant projectile point dimensions. The eight variables also measure aspects of the morphology that had functional consequences for the use of these points.

Figure 5 shows a scattergram plot of base height (negative values were used for concave-base points, positive values for convex-base points and zero for points with a flat base) versus the maximum thickness of a point. Two clusters can easily be distinguished by eye.

Black squares and gray circles distinguish the results of a k-means cluster analysis with $k = 2$, using the two variables of the scattergram plot. Not surprisingly, k-means correctly identifies the two, visually obvious clusters with 100% accuracy; the only surprise would be if it had failed to do so. Next, we add the variable, length, to the list of variables used in the k-means analysis, keeping $k = 2$. With the three variables, base height, thickness and length, the clusters found by k-means analysis are as shown in Fig. 6.

Clearly, the addition of the third variable has severely compromised the ability of k-means analysis to find the two, obvious clusters that even naive subjects have no problem discovering when comparing photos of the points.[5]

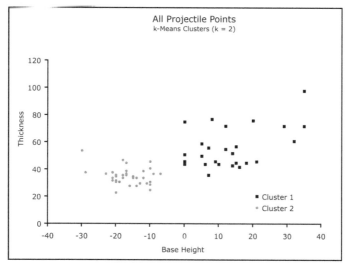

Fig. 5. Scattergram of Base Height versus Thickness (Reprinted from Read 2007a: Fig. 8.6 with permission of publisher).

Formally, Inclusion of Additional Variables Leads to Divergence From, not Convergence on, Inherent Data Structure

The failure of k-means (and other clustering methods) to find the two obvious clusters when three variables are included stems from the fact that length is a dimension

---

[5] As an experiment, pictures of the points were placed on a table and naive subjects (students in a lower division course unrelated to archaeology) were asked to sort the points into as many groups as they thought were reasonable. In less than 15 min, they sorted the points, just by inspection of the photos, into the two groups shown in Fig. 5.

with values whose frequency distribution is independent of the two clusters; the clusters in Fig. 5 have very similar frequency distributions for the length variable. The problem introduced by including a variable whose frequency distribution is independent, or largely independent, of structure in the data is illustrated in Fig. 7.

The addition of a variable whose frequency distribution over the data set is independent of the cluster structure cause some points in one cluster, and some points in the other cluster, to both be moved vertically in the same direction along the added dimension, and for some points in the one cluster to be moved upward and yet other points in the same cluster to be moved downward. Consequently, the points from the two clusters that were moved vertically in the same direction will have a small distance measure between each pair of these points in comparison to a pair of points from the

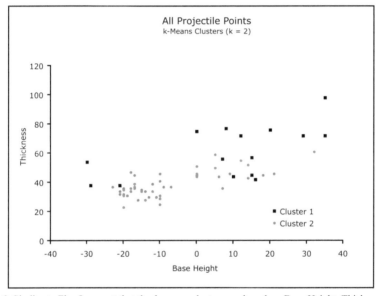

Fig. 6. Similar to Fig. 5, except that the *k*-means clusters are based on Base Height, Thickness and Length. There is extensive misclassification of the points (Reprinted from Read 2007a: Fig. 8.13 with permission of publisher).

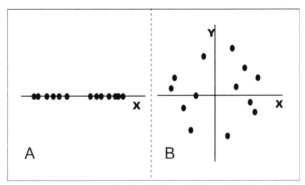

Fig. 7. (A) Two non-ambiguous clusters for the single variable, *x*. (B) Plot of the data in (A) after a second variable, *y*, with frequency distribution independent of the clusters in (A), is added to the data set (Reprinted from Read 2007a: Fig. 5.3 with permission of publisher).

same cluster when one point of the pair is moved upwards and the other point is moved downwards. Thus, the addition of the length dimension, with its frequency distribution independent of the cluster structure, causes some pairs of points, with one point in the pair from one cluster and the other point in the pair from the other cluster, to now be close to each other, while for some pairs of points from the same cluster the distance between the pair of points is now large. Consequently, the frequency distribution for the distance measures in the three dimensions is no longer determined by the cluster structure in two dimensions. Read (2007a: Appendix) has proven this as a general result: "the effect of the added variables is to reduce discrimination among the subgroups [i.e., clusters]. If the reduction is sufficient, multimodality [in the data structure] will be swamped and the subgroups will become indistinguishable. *We thus conclude that the converge assumption* [of converging onto the data structure by adding more variables] *is false*" (Read 2007a: 309, emphasis added).

It follows that we cannot simply take the set of variables, as measured, and use them in a clustering procedure under the presumption that the clustering procedure will converge on the data structure as the number of variables increases, even when the variables are selected to measure, in a representative manner, the properties upon which clustering is to be based—morphological shape in the case of the projectile points. Of course, when clusters are found that are internally cohesive and externally isolated, they are representative of the data structure. The problem lies with everything else: failure to find clusters, clusters that do not satisfy the clustering criterion and the possibility of additional structure within clusters that are found but not identified by the clustering procedure. The later problem can be seen in Fig. 5 in which there are two distinct clusters but with no evidence that the lower left cluster may be subdivided into two more clusters based on a bimodal distribution for the width of the convex-base points.

Variable Reduction Procedures Presume Relevant Data Structure is Already Known

The objection might be raised that this is precisely why all variables are included in the analysis and methods such as principal component analysis (or factor analysis, multidimensional scaling for ordinal variables or correspondence analysis for nominal variables) can reduce the original set of variables to the underlying dimensions measured jointly by these variables. This, however, is not a solution. To see why, we need to return to the aggregation problem. Over what aggregate do we do the variable reduction? While principal component analysis will find the axes of a joint-normal distribution, it may not be evident what the principal components measure, though, when the data are a mixture of several joint-normal (or approximately joint-normal) distributions, as happens in Fig. 5. The archaeological rationale for employing principal component analysis lies in the analytical goal of determining the dimensions along which the artisans structured the artifacts, such as the dimensions shown in Fig. 5. There is no assurance, though, that the principal components determined from the set of eight variables will represent those dimensions. Further, the projectile points do not, as a whole, have a single joint-normal distribution; otherwise we would not have the unmistakable clusters shown in Fig. 5. Figure 5 also implies that when the data points, as a whole, consist of at least two unimodal, approximately joint-normal distributions, each corresponding to clusters that are part of the data structure, a principal component analysis for the suite of variables used to measure the projectile points will find dimensions that are a composite of the internal structure of the two clusters, the relative location of those clusters in the measurement space, and the relative frequency of these two clusters in the data set. Thus, there is no *a priori* reason to assume that the dimensions found by a principal component analysis of the set of 8 morphological variables will be the two-dimensional subspace of the 8-dimensional measurement space in which the two clusters shown in Fig. 5 are located.

For the projectile points, by happenstance a principal component analysis of these eight variables does find an axis along which the groups are distinguishable (Christenson and Read 1977), namely the dimension going from the lower left to the upper right in Fig. 5, along which the two clusters also happen to be separated. Yet the same principal component analysis fails to find the dimension along which the concave-base projectile points are structured into narrow concave-base and wide concave-base points. In addition, simulation experiments (Read 1992) show that, in general, with a data set having structure based on a few dimensions, several dimensions added to the data set whose frequency distribution are independent of that structure, and all variables having equal variance, then the data structure is not recovered from a principal component analysis of the full complement of variables.

Double-Bind Problem Revisited

This brings us back to the double-bind problem: (1) to solve the aggregation problem, we need to subdivide the data set in accordance with its inherent data structure, but the means for so doing, typically some form of cluster analysis, depends on having already identified the aggregate specific dimension(s) along which the data are structured and (2) to solve the problem of reducing the dimensionality of a set of variables, the means for so doing typically assume all data points are from one and the same distribution, yet the data structure may be differentiated into structurally distinct clusters (similar problems arise with multidimensional scaling and correspondence analysis for ordinal and nominal variables, respectively). Because of what archaeological data represent, the variables selected initially for analysis will most likely be an unknown mixture of culturally salient modes, artisan specific attributes, and other dimensions. In addition, the data as initially organized for research purposes are not likely to be aggregated in the manner needed to meet the assumption of homogeneous data that is ubiquitous to statistical model fitting when the goal is valid abductive reasoning. That is, for a model $M = M(X_1, ..., X_m; c_1, ..., c_n; \theta_1, ..., \theta_o)$ that expresses the relationships among m variables $X_1, ..., X_m$ using n constants $c_1, ..., c_n$ and o parameters $\theta_1, ..., \theta_o$, with the latter to be estimated from sample data (or computed over population data), model fitting leading to parameter values that will be part of abductive argumentation depends on all data points being the outcome of the same model with the same parameters and parameter values. If some data points are the consequence of, say, model M_1 and other data points in the same data set are the consequence of, say, model M_2, then parameter values determined when fitting, say, model M to the data set may utilize parameters unrelated to the parameters of M_1 or M_2, or may have parameter values that are an unknown combination of the parameter values for M_1 and M_2, hence using those parameters and their estimated values as part of abductive reasoning will lead to invalid explanatory arguments.

Thus, a typology to be used as part of abductive arguments needs to be made up of types that correspond to the distinctions acted upon by the makers and users of the artifacts, otherwise a type distinguished by the archaeologist will simply be an unknown mixture of artifacts that relate to different contexts, behaviors and ideas from the perspective of the makers and users of the artifacts, hence without meaningful interpretation when used as part of abductive reasoning. As a consequence, the suggestion (e.g., Dunnell 1971, O'Brien and Lyman 2001, 2003, Lyman and O'Brien 2002) that typologies should consist of all possible types formed by a series of measurement ranges (referred to as *paradigmatic classification*; see Chapter 11, this volume), imposed as subdivisions when using a continuous variable, or as combinations of such measurement ranges for several variables, is a misguided and misleading attempt to construct objective units (see discussion in Read 2007a: Chapter 6). A simple example of the problems that arise from a classification based on imposed subdivisions can be seen with the (objective) definition of a *blade* as a *flake* that is at least three times longer than it is wide. Depending on the archaeological context, the length and width variables for flakes may have a single joint normal distribution, in which case the definition leads to an unwarranted and arbitrary division of the flakes into two types. Further, there is no reason to expect that the ratio corresponds to any aspect of the joint-normal distribution such as its axes. Even more, the joint distribution may not be unimodal and instead may be composed of two or more distinct clusters, none of which correspond to the 3:1 ratio for length:width, in which case the imposed types fail to recover the inherent data structure, and so on.

A Recursive Solution to the Double-Bind Problem

Solution of the double-bind problem requires that variable choice and definition of aggregates be resolved simultaneously, hence an exploratory, rather than a model fitting, approach is needed. For Fig. 3, the transformation is from not well-identified data and not well-identified variables (lower right corner) directly to well-identified data and well-identified variables (upper left corner). The goal of the exploratory data analysis is to divide the initial aggregate over which the research problem is defined into new aggregates and sets of variables that specify the culturally salient dimensions and normative values that guided the production of artifacts by the artisans. The means for so doing will necessarily involve recursive subdivision of aggregates and sets of variables aimed at reducing heterogeneity in aggregates since each subdivision into (sub) aggregates that reduces the heterogeneity of an aggregate from a previous stage of the transformation procedure may make evident further divisions that should be made.

More formally, let P be the initial aggregate and let $X_1, \ldots, X_n$ be the variables that have been selected for measurement. We will denote the aggregate and its associated variable set by $< P; X_1, \ldots, X_n >$. We want to restructure $< P; X_1, \ldots, X_n >$ as $< P_1, X_{11}, \ldots, X_{1n_1} > + < P_2; X_{21}, \ldots, X_{2n_2} > + \ldots + < P_m; X_{m1}, \ldots, X_{mn_m} >$, where each X_{ij} is a mode (as discussed by Rouse), each X_{ij} is (for simplicity of notation) one of the variables $X_1, \ldots, X_n$ and, for each i, $1 \leq i \leq m$, the joint distribution of variables $X_{i1}, \ldots, X_{in_i}$ over the aggregate P_i corresponds to the expected joint distribution for these variables when all artisans are making a single type of artifact; that is, where the expected value $E[X_{ij}] = \mu_{ij}$ regardless of the artisan, thus μ_{ij} is the normative value shared by all artisans for variable X_{ij} and aggregate P_i. As mentioned previously, by way of the Central Limit Theorem we expect the variables $X_{i1}, \ldots, X_{in_i}$ to have a jointly unimodal, if not joint-normal, distribution over their associated aggregate P_i. Consequently, for a variable X_i that is associated with, say, both aggregate P_j and aggregate P_k (that is, $X_{jg} = X_i = X_{kh}$ for some g with $1 \le g \le n_j$ and some h with $1 \le h \le n_k$), the variable X_i will have a bimodal distribution with one mode centered at $E[X_{jg}]$ for aggregate P_j and the other mode centered at $E[X_{kh}]$ for aggregate P_k. This implies that the criterion for the division of an aggregate into sub aggregates is a dimension along which the data form a bimodal distribution, as we observed, for example, with the maximum width variable for the concave-base projectile points. Hence the procedure for solving the double-bind problem should begin by searching for bimodality in one (or more) of the variables $X_1, ..., X_n$ over the aggregate P.

Whether bimodality for variable X_i due to the initially unknown aggregates P_i and P_j can be observed over the aggregate P depends on the degree of heterogeneity of P as a compilation of the aggregates P_i that we are attempting to discover. The extent to which P is a heterogeneous compilation of aggregates, each satisfying the well-identified data and well-identified variables criterion, is related to the extent P is an aggregate encompassing multiple cultural contexts (either between or within relevant social units). Thus (following the procedures discussed in Read 2007a), we initially need to subdivide P according to what we understand, as archaeologists, regarding different cultural contexts, e.g., regional data may need to be divided into site-specific data, site-specific data divided into structurally differentiated portions of a site, and so on. At a finer level, we may subdivide an aggregate according to qualitative differences such as topological differences in pottery forms (e.g., jars with a spout versus jars without a spout), geometric differences within topologically similar objects (e.g., bowls versus jars without handles or spouts), and qualitative differences emically imposed on a quantitative dimension (e.g., when pottery has sherd temper or sand temper but not mixed temper, as is generally the case for Paleo Indian pottery in the southwestern parts of the US), and so on (see Read 2007a for more details).[6]

We then use an iterative procedure to search for quantitative dimensions with multi-modality as follows:

1) examine each variable separately for a multi-modal distribution over a currently distinguished aggregate since a single variable may measure a dimension that is a mode (in Rouse's use of the term);

---

[6] Some commentators on lithic classifications incorrectly contrast intuitive, morphological types with quantitatively determined types and prefer the former as being intuitively satisfactory (e.g., Odell 2004). This contrast, however, makes the invalid assumption that quantitative typologies are opposed to, rather than complementary with, qualitative typologies. Instead, artifact classifications should begin with qualitative distinctions and then incorporate quantitative dimensions in order to find subdivisions within qualitative divisions (Read 2007a) that "are not readily visible to intuitive typologists" (Odell 2001: 46). However, Odell does not seem to find any value, as an archaeologist, in so doing, thereby turning the archaeologist's intuitions into the arbiter of what is important regarding the traces found by the archaeologist of the distinctions, decisions and actions of the maker and users of artifacts, thereby assuming that somehow the archaeologist's intuitions are a good arbiter of what is a valid taxonomy even though the archaeologist's intuitions are a consequence of his or her experiences (Adams and Adams 1991), hence what is intuitively obvious to one archaeologist is not intuitively obvious to another. Odell (2001, 2004) rejects an objectively derived utilized flake typology (Read and Russell 1996) on the grounds that it is "not very satisfying intuitively" (Odell 2004: 104), despite the fact that the groupings determined objectively correspond precisely to how a flake would be held and how this relates, in a functionally expected manner, to the attributes of the utilized edge and so they implement the intuitions of the archaeologist (see Decker 1976) regarding the way utilized flakes would have been held and how that would affect which edge is utilized.

2) subdivide the data set into sub data sets by the mode(s) of the multi-modal distribution, if any;

3) repeat, beginning with Step (1), with each sub data set determined in (2), until no further subdivisions are possible;

4) examine pairs of variables (thereby allowing for a relevant dimension to be defined by two variables) for a multi-modal distribution for each of the currently determined (sub) data sets;

5) subdivide a data set by the modes of the multi-modal distribution, if any;

6) repeat, beginning with Step (1) on each subset determined in (5) until no further subdivisions are possible;

7) continue in this manner with other combinations of variables and with dimensions obtained through variable reduction procedures over the currently distinguished aggregates.[7]

Figures 8 (A and B) show the result of applying this iterative procedure (see Read 2007a for details) to the projectile points from 4 VEN 39 and to a pottery assemblage from the Niederwil Site, a Late Neolithic site in Switzerland.[8] The pottery typology illustrates the double-bind problem especially well. There is no reason to expect, *a priori*, that the two ratio measures, Rim Diameter/Belly Diameter and Total Height/Belly Height, are also dimensions that are modes as defined by Rouse (see right side of Fig. 8B), nor was the data structure shown in Fig. 8(B) recovered by clustering procedures that begin with Niederwil pots as the aggregate and with all measurements made on the pots used as the variables for the cluster analysis (see discussion by Whallon 1982).[9]

---

[7] Mucha (2013) has proposed using an iterative procedure for variable selection that only includes Steps (1), (4) and (7) of the procedure outlined here. In so doing, a well-identified data set and not well-identified variables are implicitly assumed; hence the proposed procedure corresponds to the arrow going from the lower left to upper left in Fig. 3 and does not solve the double-bind problem. For a well-identified data set, his procedure provides a useful alternative to the dimensionality reduction procedures shown in Fig. 3.

[8] The site was excavated by H.T. Waterbolk and J.D. Van der Waals and measurements were made by J.P. de Roever (Waterbolk and van Zeist 1967, 1978).

[9] All taxonomies can be converted into paradigmatic classifications and vice-versa. The Niederwil pottery taxonomy becomes a paradigmatic classification with 540 classes (see Read 2007a for details), of which only 21 (incorrectly stated as 17 in Read 2007a) have nonzero values, namely the classes shown in Fig. 8B. However, recasting the taxonomy as a paradigmatic classification with 519 empty classes begs the question of what is to be gained by doing so, contra the assumption by some archaeologists of the primacy of paradigmatic classifications (e.g., Odell 2001, O'Brien and Lyman 2001, 2003, Lyman and O'Brien 2002; Chapter 11, this volume). Even more critical, those advocating paradigmatic classifications typically utilize an *a priori* division for each dimension, hence divisions that are arbitrary, due to the double-blind problem, with regard to the distinctions of concern to the makers and users of the artifacts, thereby constructing classes that will crosscut a taxonomic structure such as the one shown in Fig. 8B and would likely exclude derived dimensions such as the two ratio measures. It is unclear how a crosscutting classification would help form abductive arguments leading to valid explanatory hypotheses. The classes shown in Fig. 8B reflect the decisions the potters made about pot shapes, so crosscutting classes cannot be interpreted in this manner and would lack any obvious, valid interpretation regarding the production and use of artifacts. The paradigmatic classifications advocated by O'Brien and Lyman are neither "the only ... " nor even a "... systematic means of tracking variation at various scales" (O'Brien and Lyman 2009: 231), especially when the goal is to make evident the cultural and other constraints acting on artifact production and use.

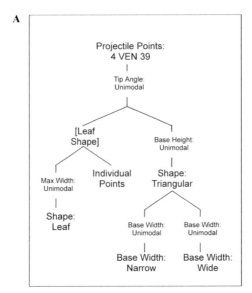

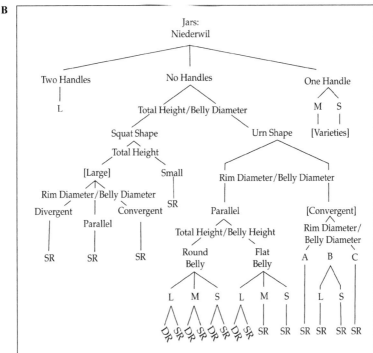

Fig. 8. (A) Iteratively generated typology for projectile points at 4 VEN 39. (B) Iteratively generated typology for pots at the Niederwil site. Text written across two or more lines identifies a measure for which a multi-modal distribution was found. Text written across a single line indicates a variable that is a mode (in Rouse's sense of the term) that is unimodal, hence normatively constrained. Class names are at the end of lines. Names in square brackets indicate that contrasting classes such as Parallel/Convergent are further subdivided by different variables (Reprinted from Read 2007a: Fig. 8.12, 8.23 with permission of publisher).

Conclusion

Ignoring the double bind problem by applying procedures based on homogeneous data to heterogeneous data may both mask actual structure and create invalid structure. This can lead to incorrect explanatory arguments. All too frequently, archaeological accounts justify the application of statistical models to archaeological data through surface level analogies based on similarity in the vocabulary used by archaeologists and statisticians to describe model structure and data pattern (Read 1989a). Rather than surface analogies, though, connection between model and data should integrate statistical reasoning with archaeological theorizing, as has been discussed here, and thereby develop concordance between formal relations generating model form and structuring processes hypothesized to underlie data structure.

References Cited

Adams, W.Y. and E.W. Adams. 1991. Archaeological Typology and Practical Reality: A Dialectical Approach to Artifact Classification and Sorting. Cambridge University Press, Cambridge, Massachusetts.

Bennardo, G. and D. Read. 2007. Cognition, algebra, and culture in the Tongan kinship terminology. J. of Cognit. Sci. 7: 49–88.

Carr, C. (ed.). 1989. For Concordance in Archaeological Analysis. Waveland Press, Prospect Heights, Illinois.

Christenson, A. and D. Read. 1977. Numerical taxonomy, r-mode factor analysis and archaeological classification. Amer. Antiq. 42: 163–179.

Decker, D. 1976. A typology for the Chevelon flaked lithic implements. pp. 92–106. *In*: F. Plog, J. Hill and D. Read (eds.). Chevelon Archaeological Research Project. Monograph II. Department of Anthropology, University of California, Los Angeles.

Douven, I. (ed.). 2011. The Stanford Encyclopedia of Philosophy. Stanford University Press, Stanford. http://plato.stanford.edu/archives/spr2011/entries/abduction/Accessed 7 November 2013.

Dunnell, R.C. 1971. Systematics in Prehistory. The Free Press, New York.

Fisher, R. 1954. Statistical Methods for Research Workers. Oliver & Boyd, London.

Gandon, E., R.J. Bootsma, J.A. Endler and L. Grosman. 2013. How can ten fingers shape a pot? Evidence for equivalent function in culturally distinct motor skills. PLoS ONE 8(11): e81614. doi:10.1371/journal.pone.0081614.

Gregorius, H.R. 2006. The isolation principle of clustering: structural characteristics and implementation. Acta Biotheor. 54(3): 219–233.

Kaufman, L. and P.J. Rousseeuw. 1990. Finding Groups in Data. An Introduction to Cluster analysis. John Wiley & Sons, New York.

Lane, D., R.M. Maxfield, D. Read and S. van der Leeuw. 2009. From population to organization thinking. pp. 11–42. *In*: D. Lane, D. Pumain, S. van der Leeuw and G. West (eds.). Complexity Perspectives on Innovation and Social Change. Springer Verlan, Berlin.

Leaf, M. and D. Read. 2012. The Conceptual Foundation of Human Society and Thought: Anthropology on a New Plane. Lexington Books, Lanham, Massachusetts.

Lyman, R.I. and M.J. O'Brien. 2002. Classification pp. 69–88 *In*: I.P. Hart and I.E. Terrell (eds.). Darwin and Archaeology: A Handbook of Key Concepts. Bergin and Garvey, Westport, Connecticut.

Mucha, H.-J. 2013. Variable selection in cluster analysis using resampling techniques: A proposal. Paper give at the AGDANK/BCS Meeting, 8–9 November 2013, University College London, London, UK.

O'Brien, M.J. and R.L. Lyman. 2001. Applying Evolutionary Archaeology: A Systematic Approach. Kluwer Academic Press, New York.

O'Brien, M.J. and R.L. Lyman. 2003. Cladistics and Archaeology. The University of Utah Press, Salt Lake City.

O'Brien, M.J. and R.L. Lyman. 2009. Darwinism and historical archaeology. pp. 227–252. *In*: T. Majewski and D. Gaimster (eds.). International Handbook of Historical Archaeology, Springer, New York.

Odell, G.H. 2001. Stone tool research at the end of the millennium: Classification, function, and behavior. J. Archaeol. Res. 9(1): 45–100.

Odell, G.H. 2004. Lithic Analysis. Kluwer Academic, Plenum Publishers, New York.

Peirce, C.S. 1935. *In*: C. Hartshorne, P. Weiss and A. Burks (eds.). Collected Papers of Charles Sanders Peirce, vol. 5, Harvard University Press, Cambridge, Massachusetts.

Read, D. 1982. Toward a theory of archaeological classification. pp. 56–92. *In*: R. Whallon and J.A. Brown (eds.). Essays on Archaeological Typology. Center for American Archaeology Press, Evanston, Illinois.

Read, D. 1984. An algebraic account of the American kinship terminology. Curr. Anthropol. 25(4): 417–449.

Read, D. 1987. Archaeological theory and statistical methods: Discordance, resolution and new directions. pp. 151–184. *In*: M. Aldenderfer (ed.). Quantitative Research in Archaeology: Progress and Prospects. Sage Publications, Walnut Creek, California.

Read, D. 1989a. Statistical methods and reasoning in archaeological research: A review of praxis and promise. J. Quant. Anthropol. 1: 5–78.

Read, D. 1989b. The substance of archaeological analysis and the mold of statistical method: Enlightenment out of discordance? pp. 45–86. *In*: C. Carr (ed.). For Concordance in Archaeological Analysis: Bridging Data Structure, Quantitative Technique and Theory. Waveland Press, Prospect Heights, Illinois.

Read, D. 1992. The convergence assumption: A fatal flaw in clustering algorithms. UCLA Statistics Series 105: 1–25.

Read, D. 2007a. Artifact classification: A conceptual and methodological approach. Left Coast Press, Walnut Creek, California.

Read, D. 2007b. Kinship theory: A paradigm shift. Ethnology 46(4): 329–364.

Read, D. 2008. A formal explanation of formal explanation. Struct. Dynam. 2, http://repositories.cdlib.org/imbs/socdyn/sdeas/vol3/iss2/art4/.

Read, D. 2010. The generative logic of Dravidian language terminologies. Math. Anthrop. Cult. Th. 3(7), http://mathematicalanthropology.org/Pdf/Read.0810.pdf.

Read, D. 2011. Mathematical representation of cultural constructs. pp. 229–253. *In*: D. Kronenfeld, G. Bennardo, V.C. de Munck and M.D. Fischer (eds.). A Companion to Cognitive Anthropology. Wiley-Blackwell, Oxford.

Read, D. 2012. How Culture Makes Us Human: Primate Evolution and the Formation of Human Societies. Left Coast Press, Walnut Creek, California.

Read, D. and G. Russell. 1996. A Method for taxonomic typology construction and an example: Utilized flakes. Amer. Antiq. 61(4): 663–684.

Rouse, I. 1939. Prehistory in Haiti: A Study in Method. Yale University Press, New Haven, Connecticut.

Schiffer, M. 1987. Formation Processes of the Archaeological Record. University of New Mexico Press, Albuquerque.

Sokal, R.R. and P.H.A. Sneath. 1963. Principles of Numerical Taxonomy. W.H. Freeman & Co., San Francisco.

Waterbolk, H.T. and W. van Zeist. 1967. Preliminary report on the Neolithic bog settlement of Niederwil. Palaeohistoria 12: 559–580.

Waterbolk, H.T. and W. van Zeist. 1978. Niederwil, eine Siedlung der Pfyner Kultur. Vol. Band I: De Grabungen. Band II: Beilagen. Paul Haupt, Bern and Stuttgart.

Whallon, R. 1982. Variables and dimensions: The critical step in quantitative typology. pp. 127–161. *In*: R. Whallon and J.A. Brown (eds.). Essays on Archaeological Typology. Evanston: Center for American Archaeology Press, Evanston, Illinois.

Wiessner, P. 1983. Kalahari san projectile points. Amer. Antiq. 48(2): 253–276.

5

Social Network Analysis for Sharing and Understanding Archaeology

Luciana Bordoni

Introduction

In the age of Facebook and Twitter, social networks are a mainstay of the human experience that, in the simplest of terms consist of a set of actors—individuals, communities, or even organizations—and the connections among them. "Connections" can represent any number of relationships between pairs of actors: familial ties, friendships, acquaintances, frequent interactions, exchange partnerships, or political alliances, among others. Based in graph theory, social network analysis (SNA) is a developing field that most often evaluates these kinds of connections in today's world, as a means of systematically exploring interaction. The graph theory has been widely used in SNA due to its representational capacity and simplicity. Essentially, the graph consists of nodes and of connections which connect the nodes. In social networks the representation by graphs is also called sociogram, where the nodes are the actors or events and the lines of connection establish the set of relationships in a two dimensional drawing. Graphs enable many interesting analyses to be made and have visual appeal which helps us to understand the network under study. However, for networks with many actors and connections, this becomes impossible. Similarly, some important information, such as the frequency of occurrence and of specific values, are difficult to apply in a graph. To resolve this problem, we use the matrices developed by sociometrics which produce what we call sociomatrices. Thus, sociometrics and its sociomatrices complement graph theory, establishing a mathematical basis for the analysis of social networks.

ENEA/UTICT, Lungotevere Thaon di Revel 76, Roma (Italy).
Email: luciana.bordoni@enea.it

The use of graphs and sociomatrices is necessary in order to create models, or simplified representation systems of networks of relationship. Over the past decade the term network has become a "buzzword" in many disciplines across the humanities and sciences. Social networking data comes today in many forms: blogs (Blogger, LiveJournal), micro-blogs (Twitter, FMyLife), social networking (Facebook, LinkedIn), wiki sites (Wikipedia, Wetpaint), social bookmarking (Delicious, CiteULike), social news (Digg, Mixx), reviews (ePinions, Yelp), and multimedia sharing (Flickr, Youtube). Online social networking represents a fundamental shift of how information is being produced, transferred and consumed. Contrary to other disciplines like sociology, in which ideas from graph theory were rapidly adopted and where SNA has developed into a major paradigm (Freeman 2004, Wasserman and Faust 1994), network analysis has only recently become popular as a method for archaeological research, whose applications share a number of issues (Brughmans 2010). One interesting attempt consisted in applying SNA to relationships in the distant past (http://www.archaeologysouthwest. org). Although these analyses conducted by Southwest Social Networks (SWSN) project team members draw on somewhat different techniques and evidence, they all focus on related questions. How did patterns of interaction and exchange change through time at local and regional scales? How might the structure and organization of networks of interaction among settlements have influenced the long-term success or failure of settlements or regions? Researchers in archaeology and history, in particular, are increasingly exploring network-based theory and methodologies drawn from complex network models as a means of understanding dynamic social relationships in the past, as well as technical relationships in their data.

The chapter is organised as follows. After this introduction, Section 2 provides background and related work. Section 3 explores how network analysis can be applied successfully to research issues in archaeology. Section 4 discusses how digital technologies and social networks can be integrated to support the documentation and sharing of intangible heritage knowledge and practices. Conclusions and recommendations are made in Section 5.

Background: Networks and Analysis of Networks

The study of networks, in the form of mathematical graph theory, is one of the fundamental pillars of discrete mathematics. Euler's celebrated 1735 solution of the Königsberg bridge problem is often cited as the first true proof in the theory of networks, and during the twentieth century graph theory has developed into a substantial body of knowledge. Typical network studies in sociology involve the circulation of questionnaires, asking respondents to detail their interactions with others. One can then use the responses to reconstruct a network in which vertices represent individuals and edges the interactions between them. Typical social network studies address issues of centrality (which individuals are best connected to others or have most influence) and connectivity (whether and how individuals are connected to one another through the network). A set of vertices joined by edges is only the simplest type of network; there are many ways in which networks may be more complex than this. For instance, there may be more than one different type of vertex in a network, or more than one different type of edge. Vertices or edges may have a variety of properties, numerical or otherwise, associated with them. Taking the example of a social network of people, the vertices may represent men or women, people of different nationalities, locations, ages, incomes, or many other things. Edges may represent friendship, but they could also represent animosity, or professional acquaintance, or geographical proximity. They can carry weights, representing, for example, how well two people know each other.

Empirical studies of the structure of networks include: social networks, information networks, technological networks and biological networks. A social network is a set of people or groups of people with some pattern of contacts or interactions between them (Scott 2000, Wasserman and Faust 1994). The patterns of friendships between individuals (Moreno 1934, Rapoport and Horvath 1961), business relationships between companies (Mariolis 1975, Mizruchi 1982), and intermarriages between families (Padgett and Ansell 1993) are all examples of networks that have been studied. The classic example of an information networks is the network of citations between academic papers (Egghe and Rousseau 1990) but a very important example of an information network (also called "knowledge networks") is the World Wide Web. Technological networks are man-made networks designed typically for distribution of some commodity or resource, such as electricity or information. A very wide example is the Internet, i.e., the network of physical connections between computers. Since there is a large and ever changing number of computers on the Internet, the structure of the network is usually examined at a coarse-grained level, either the level of routers, special-purpose computers on the network that control the movement of data, or "autonomous systems", which are groups of computers within which networking is handled locally, but between which data flows over the public Internet. An important class of biological network is the genetic regulatory network. Neural networks are another class of biological networks of considerable importance. Measuring the topology of real neural networks is extremely difficult, but has been done successfully in a few cases.

To study networks of various relationships in an objective way, models need to be created to represent them. There are three *notations* currently in use in the social network analysis:

- *Graph Theory*—the most common model for visual representation, it is graph based;
- *Sociometrics*—proposes matrices representation, also called *sociomatrices;*
- *Algebraic*—proposes algebraic notations for specific cases, especially for multiple relationships (Wasserman and Faust 1994).

Each notation scheme has different applications and different developments and analyses. The combination of *graphs* and *sociomatrices* has significantly helped the evolution of social network analysis. For making the analysis of networks with many actors and connections possible, the matrices developed by sociometrics, *sociomatrices,* are being used. Thus, sociometrics and its sociomatrices complement the graph theory, establishing a mathematical basis for analyses of social networks. Formal network analysis and modelling hold potential especially for archaeological studies of past communication. Modern archaeology offers datasets of great complexity, which may preserve patterns of such networks and which calls for formal analysis.

The following section presents how network analysis can be applied successfully to research issues in archaeology.

Network Analysis and Archaeology

The application of network analysis and modelling has begun to make an impact on archaeological studies in recent years (Brughmans 2010, 2012a, 2012b). The most definitive results have been obtained so far through modelling based on geographical parameters, as pioneered by Hage and Harary (1991, 1996) and Broodbank (2000), and more recently explored by Knappett et al. (2008, 2011) and Knappett (2013). It is characteristic that all major studies concern the island networks, in which the nodes are defined by natural geography and the space can be modelled as the isotropic surface of the sea. Attempts have also been made to use network analysis to quantify statistical properties in archaeological data relating to interaction, in particular to define measures of centrality and resilience (Johansen et al. 2004, Sindbæk 2007, Mizoguchi 2009, Brughmans 2010). These and other pilot studies demonstrate that network properties can indeed be distilled from the archaeological record and for that reason other archaeologists have based their explorations into network analysis.

Recent studies have sought ways to characterize social and cultural boundaries and community structure by applying network analysis to material evidence (Sindbæk 2010, Terrell 2010). Terrell demonstrates the power of network analysis to facilitate structural comparison by contrasting the affiliations in language and material culture within the set of island communities in Papua New Guinea. The study of Sindbæk, based on archaeological records, integrates the distribution of a group of common artefacts type in contemporary excavated sites in the southern Baltic Sea region in the early Middle Ages, in order to characterize the combined patterns of cultural affinity. Isaksen's (2008) analysis of transport networks in Roman Baetica is another archaeological application of network analysis. Based on Antonine Itineraries and the Ravenna Cosmography, networks were created in which towns formed the nodes and transport routes were represented by the connections between nodes. The Antonine Itineraries, a collection of routes within the Empire in which the itineraries for Iberia, Gaul, Italy, and Britain are considered, formed the basis for another archaeological application of network analysis. Graham (2006) used a social network analysis method to examine how these itineraries presented geographical space to the reader or traveller, in order to reflect on the Roman perception of space. The itineraries were transformed into a network, consisting of routes of travel between places throughout the empire, and the structure of parts of the empire-wide network was explored. The shortest, longest and average path lengths between places in regions were calculated, suggesting a stronger homogeneity within regions than for the empire as a whole. However, both authors interpret their results similarly, focusing on the implications of structural features for communication and transportation. Graham (2009) in his work on Roman networks, intends to prove the potential of a network science for understanding the Roman world. He affirms that "if we have put dots on the map, the science of networks might be able to tell us something of what happened in the space between those dots and what those dots represent". In this view, cities are nodes of social relations in time and space. Graham presents a study of the individuals active in the Roman brick industry in central Italy. By combining information on brick-producing centres, derived from an analysis of clay sources, a social network of people is constructed and analysed. This example illustrates how a social network approach can provide an innovative view on old data, and allows archaeologists to study the relationships, of whatever nature, between individuals in the past directly.

Through a case study of Roman table wares in the eastern Mediterranean, Brughmans (2010) provides a number of issues with network analysis as a method for archaeology. He states (Brughmans 2010: 284) that the following issues must be addressed if we want to continue applying network analysis in the archaeological discipline:

- The role of archaeological data in network;
- The diversity of network structures, their consequences and their interpretation;
- The critical use of quantitative tools;
- The influence of other discipline, especially sociology.

This confirms that network analysis is not a fixed method with a clearly defined set of analytical tools, and that this inherent flexibility allows for diverse applications of network analysis in the archaeological discipline. Although a growing number of software packages can be used to perform SNA techniques (Huisman and van Duijn 2005), most archaeologists use either Pajek (de Nooy et al. 2005) or UCINET (Borgatti et al. 2002), arguably the two most popular programs in SNA that are frequently expanded with new SNA techniques.

Social networks present high dynamics and a continuous transformation by adding new nodes and edges. This behaviour causes changes in the nature of the social interaction and the structure of the network. For various domains it would be a great benefit to be able to understand and therefore control the mechanism of evolution of social networks. By combining network analysis and agent-based modelling, Graham (2006) touched upon the issue of evolving networks. There are few examples of the examination of evolving archaeological networks (Knappett et al. 2011, Bentley and Shennan 2003) nevertheless both static and dynamic network approaches are potentially informative for the archaeological discipline.

Digital Technologies for Communicating Archaeology

Hypertext is typically conceptualized as a network of linked documents, where the links between documents are referred to as hyperlinks, providing a means of navigating/browsing through the network by selecting a path of links through the documents. Social networks are analogous aggregations of interconnected people. Formally, networks of people and networks of documents can both be described by graphs, with nodes and edges/links as common abstract representations of each type of network. The web is a form of social hypertext (Erickson 1996) where the hyperlinks in the hypertext can indicate a social relationship between web pages. Blogs are an excellent example of social hypertext because the comments from blogs are hyperlinks to other sources on the web and to the commenter's blog. Since the comments are associated with a particular blog's post and are in chronological order, blogs can facilitate members' social interactions (Nardi et al. 2004, Blanchard 2004) and provide conversation as studied by Herring et al. (2005).

Blogs (Weblogs) represent special kinds of dynamic documents that contain periodic, reverse chronologically ordered posts on a common webpage. The format varies from simple bullet-lists of hyperlinks to article summaries with user-provided comments and ratings. Typically, blogs are written by a single author and are closely identified with that person. Thus, a blog functions as an amalgam of document and person, and blogs link hypertext networks with social networks. The patterns of interconnections between large numbers of blogs thus forms a social hypertext, and this social hypertext not only conveys hyperlinked information, but also may create forms of virtual community as people establish online presences and locales, and communicate with each other, through webs of interconnected blogs.

A common theme that has emerged from past research is that the concepts of sense of community and virtual settlement are prerequisites to finding virtual community (McMillan and Chavis 1986, Jones 1997). These prerequisites can then be supplemented with links from blogs and clustering algorithms to indicate the shape or structures of potentially overlapping communities. Community is an ambiguous term with over 120 definitions noted by Poplin (1979). Virtual communities have been described by Rheingold (1993) as "social aggregations that emerge from the Net when enough people carry on those public discussions long enough, with sufficient human feeling, to form webs of personal relationships in cyberspace."

Jones (1997) argues that researchers need to differentiate between the technology on which the virtual group exists and the actual virtual community. Jones (1997) proposes, and others concur (Liu 1999, Nocera 2002), that we should first consider the *virtual settlement* within which virtual communities exist. Jones defines virtual settlements as the virtual place in which people interact. He uses the analogy of archaeology to develop his model: archaeologists understand a village by understanding the cultural artefacts (e.g., arrowheads, pots, etc.) that they find. Similarly, Jones argues that we can understand virtual communities by understanding the artefacts of its virtual settlement: its postings, structure and content. Jones defines virtual settlements as the virtual place in which people interact.

Jones (2003) developed the idea of a cyber-archaeology to suggest a new way of understanding virtual communities through the study of their cultural artefacts. He was influenced in particular by Fletcher's (1995) work on settlement growth, and the ways in which theories about settlement growth from archaeology might be applied to virtual worlds. His approach is to attempt to understand aspects of a virtual community through the study of its virtual material culture. A number of researchers have proposed using network analysis as a method for identifying community. However, there is no universal consensus as to which measures may be most closely related to different aspects of community.

Virtual communities have been defined and classified along different dimensions (Hagel and Amstrong 1997, Kozinets 1999, Blanchard and Horan 1998). The definitions of a virtual community in computer-mediated communities concern people that form a social network by sharing information and knowledge, achieving socialization, or by making transactions. Some of its characteristics include using common language and ease of communication; public space; common interests, values, and goals; persistence of common meaning; use of information technology for interaction, not physical space; overcoming time and space barriers; and using digitized identities as a substitute for physical being (Stanoevska-Slabevam and Schmid 2000, Lechner and Schmid 2000). The interesting dimensions include who, how, why, and the potential results of interactions. Through the interaction it is possible to communicate of, new insights and understandings about the past to the widest possible audience. As mentioned above

Jones (1997) argued that the prerequisite for an online community is the presence of a "virtual settlement" that meets four conditions:

- interactivity,
- more than two communicators,
- common-public-place where members can meet and interact, and
- sustained membership over time.

A strength of Jones's notation is that it is grounded in the combination of computer-mediated communication, cyber-archaeology and virtual communities. To apply Jones's notation, researchers need to examine the "artefacts" that community members create (in the case of Twitter, artefacts are source-follow relationships and their posting) and, based on these artefacts, quantify each of the four conditions described above (e.g., Blanchard 2004). Efimova and Hendrick (2005) use Jones theory of a virtual settlement and archaeological metaphor to address research challenges of locating weblog communities. They suggest an iterative approach that includes refinement of research methods based on assumptions about community norms, practices and artefacts, and propose which artefacts could serve as indicators of a community presence. This pilot study indicates the potential existence of a virtual settlement of a weblog community.

In the following some examples and investigations for online communities to become actively engaged in the publishing process, contribute their knowledge and partake in a dynamic creation and conceptualisation of the cultural resources, have been illustrated.

Case studies from zooarcheology present how and why social media are playing an increasing role in professional communications. Zooarchaeology has a highly active research community and has a 700-member international organization, the International Council for Archaeozoology (ICAZ), whose members meet and communicate regularly. The materials (animal remains) and methodologies are global in scope; scholars working in Europe, for example, may struggle with similar questions and methodological challenges as their colleagues in South America. Finally, zooarchaeology overlaps with other disciplines such as ecology, geology, paleontology, and biodiversity, so it benefits from tools that facilitate scholarly exchange and discovery of research content beyond its disciplinary boundaries. ZOOARCH (https://www.jiscmail.ac.uk/cgi-bin/webadmin?A0=zooarch) is an email list dedicated to zooarchaeology-related discussion that currently reaches over 1,000 subscribers worldwide. Questions, which come to the list almost daily, involve a wide diversity of needs amongst the zooarchaeological community. These can range from requests for publications to identification of "mystery bones", to people searching for lost email addresses. A researcher can pose a question to a whole community of people most likely to have the answer, and with just a few responses, has made some helpful steps forward in his or her research.

The Web plays a limited direct role with the ZOOARCH list, which remains rooted within the world of email communications. Nevertheless, the ZOOARCH list helps to illustrate how multiple communication channels and technologies can complement one another.

BoneCommons (http://alexandriaarchive.org/bonecommons/) was developed in 2006 by the Alexandria Archive Institute as an open access Web based system to complement the ZOOARCH list and help advance communication within the global zooarchaeological community. It was conceived as a community hub, where people would gather and relevant content could find them, instead of expecting them to go out and find it. Soon after its inception, the site became associated with the ZOOARCH email list in order to provide a place where people who wanted to post materials to ZOOARCH could upload attachments, which are not allowed on ZOOARCH. This relationship proved extremely helpful to the zooarchaeological community by facilitating discussions around an image. In late 2009 James Morris launched the zooarchaeology social network (http://www.academia.edu/226215/Zoobook_A_Zooarchaeology_Social_Network). It provides a "private" space for communications and offers enough new tools that address some of the outstanding needs of the user community. The main goal of these projects is to use the inherent capabilities of Web 2.0 technologies and platforms to make archaeology more collaborative and more transparent.

Conclusions

This chapter, highlighted that there exists a need for collaborative practice in archaeology, suggesting that community involvement in an archaeological investigation gives access to the considerable amount of knowledge within it. As stated, archaeological data can be presented as a network and the relationships between archaeological data can be examined directly. In many works above cited (Graham 2006, Knappett et al. 2008, Sindbæk 2007) networks are considered a medium for social interaction, an idea central to social networks analysis from which it was undoubtedly adopted in the archaeological discipline (Brughmans 2010, 2012a). In this sense a deconstruction of past archaeology methods through the potential of network analysis was illustrated. Network approaches are becoming increasingly popular among archaeologists and historians. They provide a broad range of models and methods that inspire scholars in both disciplines to original analyses of various past networks and present datasets. However, more and more questions arise regarding their possibilities and limitations. Brughmans affirms that as yet the potential of formal network-based models and techniques in archaeology have been insufficiently explored and in order to move towards richer archaeological applications of formal network methods archaeological network analysts should become better networked both within and outside their discipline. However, digital technologies hold great promise for improving and integrating the processes of the cultural heritage and in particular of archaeology. A major phenomenon that has emerged with the advent of Internet is virtual communities, field of study and research for SNA. Virtual communities and cyber-archaeology, proposed by Jones, were presented for showing as an archaeological metaphor to help to identify a weblog community in terms of an archaeological excavation community. This demonstrates the need to examine the nature of the relationship between the virtual spaces typically used for shared public online-interactions, their technological platforms, and the behaviours such systems contain. All this includes a mutual interest in artefacts, for the archaeologist items like pottery and arrowheads, for HCI researchers items like listserv postings, web site structures, Usenet content, user logs, etc.

Some cases for example, concerning online archaeological communities in which the potential of community approach is highlighted, were presented. Carl Knappett argued that "for new network approaches to be successful in archaeology they have to be as profoundly transdisciplinary as we can possibly make them" (Knappett et al. 2011: 37). The combination of SNA techniques and complex network modelling (e.g.,

Coward 2010, Graham 2006), which is considered to have great potential according to John Scott (as well as other social network analysts, e.g., Borgatti et al. 2009), is particularly promising for archaeology as it allows for a top-down as well as a bottom-up perspective to explore the multi-scalar nature of network thinking (Knappett et al. 2011). The adoption or development of network methods should, however, always be motivated by specific archaeological research questions.

As said, network analysis constitutes a strong prospective as a method for archaeology and this contribution wishes to emphasise the great value of the interaction between two disciplines such as archeology and mathematics.

References Cited

http://www.academia.edu/226215/Zoobook_A_Zooarchaeology_Social_Network

http://alexandriaarchive.org/bonecommons

http://www.archaeologysouthwest.org

https://www.jiscmail.ac.uk/cgi-bin/webadmin?A0=zooarch

Bentley, R.A. and S.J. Shennan. 2003. Cultural transmission and stochastic network growth. American Antiquity 68(3): 459–485.

Blanchard, A. and T. Horan. 1998. Virtual communities and social capital. Social Science Computer Review 16(3): 207–293.

Blanchard, A. 2004. Blogs as virtual communities: Identifying a sense of community in the Julie/Julia project. *In*: L. Gurak, S. Antonijevic, L. Johnson, C. Ratliff and J. Reyman (eds.). Into the Blogosphere; Rhetoric, Community and Culture of Weblogs. University of Minnesota.

Borgatti, S.P., M.G. Everett and L.C. Freeman. 2002. Ucinet 6 for Windows: software for social network analysis, analytic technologies. Harvard University.

Borgatti, S.P., A. Mehra, D.J. Brass and G. Labianca. 2009. Network analysis in the social sciences. Science 323(4): 892–896.

Broodbank, C. 2000. An Island Archaeology of the Early Cyclades. University Press, Cambridge.

Brughmans, T. 2010. Connecting the dots: towards archaeological network analysis. Oxford Journal of Archaeology 29(3): 277–303.

Brughmans, T. 2012a. Thinking through networks: a review of formal network methods in archaeology. Journal of Archaeological Method and Theory. Springer.

Brughmans, T. 2012b. Facebooking the past: a critical social network analysis approach for archaeology. *In*: A. Chrysanthi, P. Flores and C. Papadopoulos (eds.). Thinking Beyond the Tool: Archaeological Computing and the Interpretative Process. British Archaeological Reports International Series. Archaeopress, Oxford.

Coward, F. 2010. Small worlds, material culture and ancient Near Eastern social networks. Proceedings of the British Academy 158: 453–484.

de Nooy, W., A. Mrvar and V. Batagelj. 2005. Exploratory social network analysis with Pajek. Cambridge: Cambridge University Press.

Egghe, L. and R. Rousseau. 1990. Introduction to Informetrics, Elsevier, Amsterdam.

Efimova, L. and S. Hendrick. 2005. In Search for a Virtual Settlement: An Exploration of Weblog Community Boundaries (https://doc.novay.nl/dsweb/Get/Document-46041/).

Erickson, T. 1996. WWW as Social Hypertext. Communications of the ACM 39(1): 15–17.

Fletcher, R. 1995. The limits of settlement growth: A theoretical outline. Cambridge University Press.

Freeman, L.C. 2004. The Development of Social Network Analysis. Empirical Press, Vancouver, BC Canada.

Graham, S. 2006. Networks, agent-based models and the Antonine itineraries: implications for Roman archaeology. Journal of Mediterranean Archaeology 19(1): 45–64.

Graham, S. 2009. The space between: the geography of social networks in the Tiber valley. *In*: F. Coarelli and H. Patterson (eds.). Mercator Placidissimus: the Tiber Valley in Antiquity. New research in the upper and middle river valley. Rome: Edizioni Quasar.

Hage, P. and F. Harary. 1991. Exchange in Oceania: a Graph Theoretic Analysis. Oxford University Press.

Hage, P. and F. Harary. 1996. Island Networks: Communication, Kinship and Classification Structures in Oceania. Cambridge University Press.

Hagel, J. and A.G. Amstrong. 1997. Net Gain: Expanding Markets Through Virtual Communities. McKinsey & Company, Boston.

Herring, S.C., I. Kouper, J.C. Paolillo, L.A. Scheidt, M. Tyworth, P. Welsch, E. Wright and N. Yu. 2005. Conversations in the blogosphere: An analysis from the bottom-up. In Proceedings of the 38th Hawaii International Conference on System Sciences (HICSS'05), Los Alamitos: IEEE Press.

Huisman, M. and M.A.J. van Duijn. 2005. Software for social network analysis. pp. 270–316. *In*: P.J. Carrington, J. Scott and S. Wasserman (eds.). Models and Methods in Social Network Analysis. Cambridge: Cambridge University Press.

Isaksen, L. 2008. The application of network analysis to ancient transport geography: a case study of Roman Baetica. Digital Medievalist 4.

Johansen, K.L., S.T. Laursen and M.K. Holst. 2004. Spatial patterns of social organization in the Early Bronze Age of South Scandinavia. Journal of Anthropological Archaeology 23(1): 33–65.

Jones, Q. 1997. Virtual-communities, virtual settlements and cyber-archaeology: A theoretical outline. Journal of Computer Supported Cooperative Work 3(3).

Jones, Q. 2003. Applying Cyber-Archaeology. *In*: K. Kuutti, E.H. Karsten, G. Fitzpatrick, P. Dourish and K. Schmidt (eds.). ECSCW 2003: Proceedings of the Eighth European Conference of Computer Supported Cooperative Work 14–18 September 2003, Helsinki Finland; Netherlands: Kluwer Academic Publishers.

Knappett, C., T. Evans and R. Rivers. 2008. Modelling maritime interaction in the Aegean Bronze Age. Antiquity 82(318): 1009–1024.

Knappett, C., T. Evans and R. Rivers. 2011. The Theran eruption and Minoan palatial collapse: new interpretations gained from modelling the maritime network. Antiquity 85(329): 1008–1023.

Knappett, C. 2013. Network Analysis in Archaeology. Oxford University Press.

Kozinets, R.V. 1999. E-tribalized marketing? The strategic implications of virtual communities of consumption. European Management Journal 17(3): 252–264.

Lechner, U. and B.F. Schmid. 2000. Communities and media—Towards a reconstruction of communities on media. In Proc. of the 33rd Hawaii Int. Conf. on System Sciences.

Liu, G.Z. 1999. Virtual community presence in Internet relay chatting. Journal of Computer-Mediated Communication [online], 5(1).

Mariolis, P. 1975. Interlocking directorates and control of corporations: The theory of bank control. Social Science Quarterly 56: 425–439.

McMillan, D.W. and D.M. Chavis. 1986. Sense of community: A definition and theory. Journal of Community Psychology 14(1): 6–23.

Mizoguchi, K. 2009. Nodes and edges: a network approach to hierarchisation and state formation in Japan. Journal of Anthropological Archaeology 28(1): 14–26.

Mizruchi, M.S. 1982. The American Corporate Network, 1904–1974, Sage, Beverley Hills.

Moreno, J.L. 1934. Who Shall Survive? Beacon House, Beacon, NY.

Nardi, B.A., D.J. Schiano, M. Gumbrecht and L. Swartz. 2004. Why we blog. Communications of the ACM 47(12): 41–46.

Nocera, J.L.A. 2002. Ethnography and hermeneutics in cybercultural research accessing IRC virtual communities. Journal of Computer Mediated Communication 7(2).

Padgett, J.F. and C.K. Ansell. 1993. Robust action and the rise of the Medici, 1400–1434, Am. Journal Sociol. 98: 1259–1319.

Poplin, D.E. 1979. Communities: A Survey of Theories and Methods of Research, 2nd ed., MacMillan Publishing Co., Inc. New York.

Rapoport, A. and W.J. Horvath. 1961. A study of a large sociogram. Behavioral Science 6: 279–291.

Rheingold, H. 1993. The Virtual Community: Homesteading on the Electronic Frontier. Addison-Wesley, Toronto, Canada.

Scott, J. 2000. Social Network Analysis: A Handbook. Sage Publications, London, 2nd ed.

Sindbæk, S.M. 2007. The small world of the Vikings. Networks in Early Medieval communication and exchange Norwegian Archaeological Review 40: 59–74.

Sindbæk, S.M. 2010. Re-assembling regions. The social occasions of technological exchange in Viking Age Scandinavia. *In*: Randi Barndon, Ingvild, Øye and Asbjørn Engevik (eds.). The Archaeology of Regional Technologies: Case Studies from the Palaeolithic to the Age of the Vikings. Edwin Mellen Press.

Stanoevska-Slabevam, K. and B.F. Schmid. 2000. A generic architecture of community supporting platforms based on the concept of media. In Proc. of the 33rd Hawaii International Conference on System Sciences.

Terrell, J.E. 2010. Language and material culture on the Sepik Coast of Papua New Guinea: using social network analysis to simulate, graph, identify, and analyze social and cultural boundaries between communities. The Journal of Island and Coastal Archaeology 5(1): 3–32.

Wasserman, S. and K. Faust. 1994. Social Network Analysis: Methods and Application, University Press. Cambridge.

The Mathematics of
Ancient Artifacts

6

Shape Analysis and Geometric Modelling

*Martin Kampel** and *Sebastian Zambanini*

Introduction

In order to quantitatively describe the relationship between 2D image structures and their corresponding real world structures, methods for extracting 3D information have to be investigated. According to (Marr 1982) we see 3D vision as a 3D object reconstruction task to describe 3D shape in a co-ordinate system independent of the viewer. The shape of a 3D object is represented by a 3D model. Two main classes of 3D models are identified: volumetric models represent the "inside" of a 3D object explicitly, while surface models use only object surfaces. Unlike 3D models, depth maps or range images describe relative distances from the viewer of surfaces detected in the scene.

Archaeology is at a point where it can benefit greatly from the application of 3D Vision methods, and in turn provides a large number of new, challenging and interesting conceptual problems and data for 3D Vision. In particular, a major obstacle to the wider use of 3D object reconstruction and modelling, e.g., for classification, is the extent of manual intervention needed. The purpose of classification is to get a systematic view of the material found, to recognize types, and to add labels for additional information as a measure of quantity. In order to standardize the classification, which is based on an object's structure, it can be divided into two main parts, shape features and properties. The classification of shape defines the process by which archaeologists distinguish between various features such as a profile or the dimensions of the object like diameter and type of surface.

The topics in this chapter can be subdivided by the sensed data they rely on (see Fig. 1 for an illustration): 3D images represent 3D objects, which allow us to consider a 3D volume as part of the entire 3D world. A 3D representation is a transition of an object in the real 3D world to an object-centered co-ordinate system, allowing the object

Computer Vision Lab, Vienna University of Technology, Favoritenstr. 9-11 A-1040 Vienna, Austria.
  Emails: kampel@caa.tuwien.ac.at; zamba@caa.tuwien.ac.at
* Corresponding author

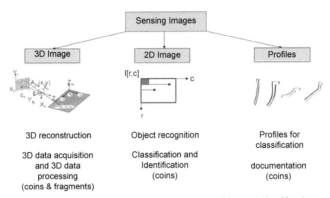

Fig. 1. Sensing images for reconstruction, recognition and classification.

descriptions to be viewer-independent (Sonka et al. 2007). A digital image is defined as a 2D image I[r, c] represented by a discrete 2D array of intensities (Shapiro and Stockman 2001). A profile of an archaeological fragment describes the vessel unambiguously.

In this chapter we describe the necessary methodology for describing shapes from archaeological objects like ceramic fragments or ancient coins. We present experiments and results on real data of archaeological ceramics and ancient coins.

Segmentation of Archaeological Profiles for Classification

A large number of ceramic fragments, called sherds, are found at excavations. These fragments are documented by being photographed, measured, and drawn; then they are classified and stored in boxes and containers. The purpose of classification is to get a systematic view on the excavation finds. As the conventional method for documentation is often unsatisfactory (Orton and Hughes 2013), an automated archival system with respect to archaeological requirements is investigated that tries to combine the traditional archaeological classification with new techniques in order to get an objective classification scheme. A graphic documentation by hand additionally raises the possibility of errors. This leads to a lack of objectivity in the documentation of the material found.

By classifying the parts of the profile, the vessel is classified; missing parts may be reconstructed with the expert knowledge of the archaeologist (Shoukry and Amin 1983). Segmentation of the profile is done for three reasons: to complete the archive drawing, to classify the vessel and to reconstruct missing profile parts. Following this manual strategy, the profile should first be segmented into its parts, the so-called *primitives*, automatically. Our approach is a hierarchical segmentation of the profile into rim, wall, and base by creating segmentation rules based on expert knowledge of the archaeologists and the curvature of the profile. The segments of the curve are divided by so-called segmentation points. If there is a corner point, that means a point where the curvature changes significantly, the segmentation point is obvious. If there is no corner point, the segmentation point has to be determined mathematically.

The curve is characterized by several points. Figure 2 shows the segmentation scheme of an S-shaped vessel as an example. A set of points is defined like,

inflexion point (IP): a point, where the curvature changes its sign;
*local maximum (*MA*)*: point of vertical tangency;
*local minimum (*MI*)*: point of vertical tangency;
*orifice point (*OP*)*: outermost point, where the profile line touches the orifice plane;
*base point (*BP*)*: outermost point, where the profile line touches the base plane;
*point of the axis of rotation (*RP*)*: point, where the profile line touches the axis of rotation.

By means of these curve points several main segments of a vessel are distinguished: rim, upper part, lower part, neck, shoulder, belly and bottom. On the basis of the number and characteristics of these segments different kinds of vessels can be classified.

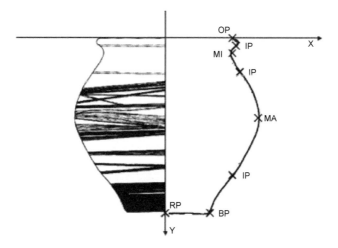

Fig. 2. S-shaped vessel: profile segmentation scheme.

Automated Segmentation

The profile sections are achieved automatically by a 3D measurement system based on structured light and a two laser-technique (Sablatnig et al. 2003). The profile determined has to be converted into a parameterized curve and the curvature has to be computed (Bennett and Donald 1975). Local changes in curvature (Rosenfeld and Nakamura 1997) are the basis for rules required for segmenting the profile. Our formalized approach uses mathematical curves to describe the shapes of the vessels and their parts. The profile is thus converted into one or more mathematical curves. We apply four methods for interpolation and four methods for approximation by B-splines on the reconstruction of the vessel profiles (i.e., the profiles are projected into the plane). The following definitions were adopted from DeVore and Lorentz (1993). We suppose that the planar closed curve *r* to be fitted (interpolated or approximated) will be represented by parametric equations

$$r(t) = [x(t), y(t)] \tag{1}$$

in an interval in the Cartesian coordinates $\mathcal{R}^2$ and has continouos second derivates. The curve is given by a set of points $P_i = [x(t), y(t)]$ together with the non decreasing sequence of knots $\{t_i, i = 1, \dots, n + 1\}$ of parameter t. Constructing a curve $S(t)$, which approximates the function given by the points can be done by a cubic spline with an adequate parameterization and external conditions. The curve must be initially divided into sub-intervals, where functional approximation and interpolation methods can be applied.

The support of a cubic spline is 5 intervals. Denote by B_i^4 a $k-th$ order spline ($k \le 3$) whose support is $[t_i, t_{i+4}]$. Then, it is possible to normalize these splines so that for any $x \in [a, b]$

$$\sum_{i=-3}^{n+3} B_i^4(x) = 1 \qquad (2)$$

Any cubic spline $S_n(x)$ with knots $t_0, \dots, t_n$ and coefficients $a_{-3}, a_{-2}, \dots, a_n$ can be written in the form

$$S_n(x) = \sum_{i=-3}^{n} a_i B_i^4(x) \qquad (3)$$

There are $n + 3$ coefficients a_i in representation (3) showing that the vector space of cubic splines has dimension $n + 3$, so that the $n + 1$ functional values will not determine $S_n(x)$ uniquely—two additional constraints must be supplied. Cardinality of the basis has been sacrificed for small support in the basis. Consequently, in evaluating $S(x)$ for any $x \in [a, b]$, only four terms at most in the sum (3) will be non-zero.

The basis cubic splines can be constructed by the following recurrent relationship:

$$B_i^n(x) = \frac{x - t_i}{t_{i+n-1} - t_i} B_i^{n-1}(x) + \frac{t_{i+n} - x}{t_{i+n} - t_{i+1}} B_{i+1}^{n-1}(x) \qquad (4)$$

with $i = -3, \dots, n - 1$ and $n = 1, 2, 3, 4$. A useful convention is to define the first-order splines as *right-continuous* so that

$$B_i^1(x) = \delta_i \text{ for } x \in [t_i, t_{i+1}), i = -3, -2, \dots, n + 3. \qquad (5)$$

The method is of local character: The change of the position of one control vertex influences only 4 segments of the curve. The resulting curve is in particular coordinates a polynomial of $3 - rd$ degree for $t \in (t_j, t_{j+1})$ and has continuous all derivatives in these coordinates.

Since $B_i^n(x)$ is nonzero only in the interval $[t_i, t_{i+4}]$, the linear system for the B-spline coefficients of the spline to be determined, by interpolation or least-squares approximation, is banded, making the solving of that linear system particularly easy.

$$S^4(x_j) = \sum_{i=0}^{n} B_i^4(x_j) a_i = y_j, \ j = 0, \dots, n \qquad (6)$$

for the unknown B-spline coefficients a_j in which each equation has at most 4 nonzero entries.

We selected four interpolation methods:

a) Cubic spline interpolation with Lagrange end-conditions (*cs*1) (i.e., it matches end slopes to the slope of the cubic that matches the first four data at the respective end);

b) Cubic spline interpolation with not-a-knot end-condition (*cs*2);

c) Spline interpolation with an acceptable knot sequence (*cs*3);

d) Spline interpolation with an optimal knot distribution (*cs*4). As 'optimal' knot sequence the optimal recovery theory of (Winograd et al. 1976) is used for interpolation at data points $\tau(1)$, ..., $\tau(n)$ by splines of order k.

All the discussed interpolation methods satisfy the Schoenberg-Whithey conditions, i.e., the achieved representation is unique for the method, the given data and knot sequences. These methods were applied from the point of view of their approximation error (least mean square of the differences of the input value and the spline value) on the given data.

We observed that spline interpolation with an acceptable knot sequence in all intervals of all profiles approximated the data with a smaller error than spline interpolation with optimal knot distribution.

We select an 'optimal' method according to the following criteria: the first criterion for selection of the most appropriate interpolation method is the minimal approximation error on the data in the corresponding interval. To exclude ambiguity, the second third criterion is applied: the priority of the interpolation method based on the statistical observations. The priority of the methods was achieved experimentally on profiles and their particular intervals and expresses a 'statistical' ordering according to the smallest approximation error over all intervals of the tested profiles.

After the most appropriate interpolation and approximation methods are computed and selected for each of the intervals of the curve, the method with a smaller error (in case of ambiguity, the interpolation method is preferred) is selected for the interval. The approximation error of the representation over the whole curve is computed. This representation is unique and optimal with respect to the above-mentioned criteria. The method was tested on computed profiles like the ones shown in Fig. 3.

All interpolation and approximation methods are applied for every sub-interval of the curve after each run of the program. While the curve is generated gradually for each sub-interval of the curve, the overall approximation error is computed. As a result the profile is constructed from the selected methods and is compared to the data set.

Table 1 displays the approximation errors for all methods in all intervals of the leftmost profile in Fig. 3, including the selected interpolation and approximation methods for the corresponding interval and the selected overall method for the whole profile. The whole data sets contained approximately 350 data points and the length of the whole curve was approximately 400 points.

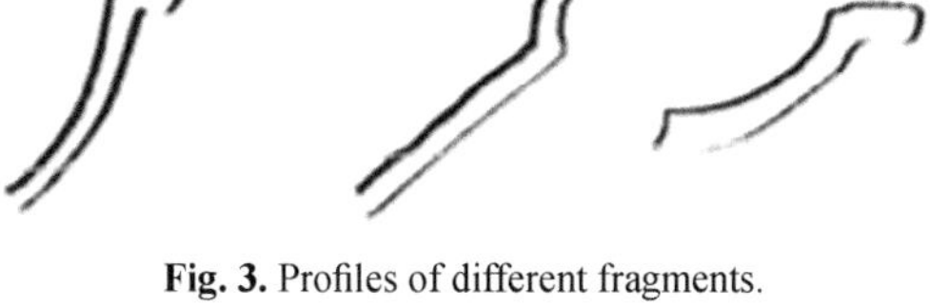

Fig. 3. Profiles of different fragments.

Table 1. Approximation errors for all methods in 8 intervals.

method/interv.	1	2	3	4	5	6	7	8
cs1	0,2163	0	0,6047	0,0781	1,1685	2,2497	1,1424	0,0884
cs2	0,2163	0	0,5994	0,0782	1,1686	2,2514	0,1433	0,0884
cs3	0,2163	0	0,5994	0,0782	1,1686	2,2514	0,1430	0,0883
cs4	0,2163	0,6169	2,1080	0,0877	1,4510	2,3485	0,1615	0,0991
cs5 (tol = 5)	0,2163	2,3114	0,5994	1,1816	2,9430	2,2514	2,2073	0,0884
cs6 (p = 1)	0,1350	0	0,6229	0,0781	1,1687	2,2496	0,1646	0,0884
cs7	0,2163	5,9470	5,5298	0,5015	6,9127	6,2323	0,8617	1,0675
cs8	0,2163	0,0032	0,6014	0,1308	1,1850	3,8347	0,1430	0,2551
select. intp.	1	1	2	1	1	1	1	1
select. appr.	6	6	5	6	6	6	8	6
overall	6	1	2	1	1	6	1	6

The most frequently selected interpolation method was cs1 and the most frequently selected approximation method was cs6 in our experiments. An interpolation method was preferred in the intervals where a sufficient number of data with respect to the length of the interval was given. An approximation method was preferred in the intervals where there was a lack of data.

The right half of Fig. 4 shows one example of an automatically segmented pot with the characteristic points detected. The appropriate manual segmentation is shown on the left of Fig. 4.

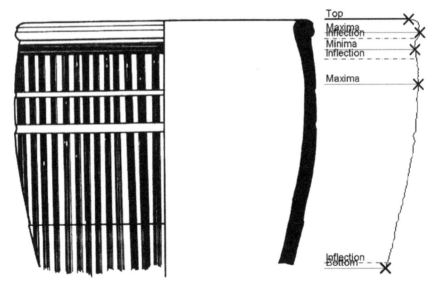

Fig. 4. Manual drawing (left) and detected characteristic points (right).

Fusion of Volume and Surface Data

An algorithm for the automatic construction of a 3D model of archaeological vessels using two different 3D algorithms is presented. In archaeology the determination of the exact volume of arbitrary vessels is of importance since this provides information about the manufacturer and the usage of the vessel. To acquire the 3D shape of objects with handles is complicated, since occlusions of the object's surface are introduced by the handle and can only be resolved by taking multiple views. Therefore, the 3D reconstruction is based on a sequence of images of the object taken from different viewpoints with different algorithms; shape from silhouette and shape from structured light. The output of both algorithms is then used to construct a single 3D model.

Images for both algorithms are acquired by rotating the object on a turntable in front of a stationary camera. Then an octree representation of the object is built incrementally, by performing limited processing of all input images for each level of the octree. Beginning from the root node at the level 0, a rough model of the object is obtained quickly and is refined as the processed level of the octree increases. Results of the algorithm developed are presented for both synthetic and real input images.

The combination of the Shape from Silhouette method with the Shape from Structured Light method aims to provide an objective and automated method for classification and reconstruction of archaeological pottery. The final goal is to provide a tool which helps archaeologists in their classification process.

Pottery was made in a very wide range of forms and shapes. The purpose of classification is to get a systematic view of the material found, to recognize types, and to add labels for additional information as a measure of quantity (Orton and Hughes 2013). In this context, decoration of pottery is of great interest. Decoration is difficult to illustrate since it is a perspective projection of an originally spherical surface. This fact induces distortions that can be minimized by 'unwrapping' the surface. In order to be able to unwrap the surface it is necessary to have a 3D representation of the original surface. Furthermore, the exact volume of the vessel is of great interest to archaeologists too, since the volume estimation allows also a more precise classification (Orton and Hughes 2013). Since pottery is manufactured on a turntable we use a turntable based method for the 3D-reconstruction of the original. To acquire images from multiple views we put the archaeological vessel on a turntable which rotates in front of a stationary camera.

Shape from Silhouette is a method of automatic construction of a 3D model of an object based on a sequence of images of the object taken from multiple views, in which the object's silhouette represents the only interesting feature of the image (Szeliski 1993). The object's silhouette in each input image corresponds to a conic volume in the object real-world space. A 3D model of the object can be built by intersecting the conic volumes from all views, which is also called Space Carving (Kutulakos and Seitz 2000),

Shape from Silhouette is a computationally simple algorithm (it employs only basic matrix operations for all transformations). It can be applied on objects of arbitrary shapes, including objects with certain concavities (like a handle of a cup), as long as the concavities are visible from at least one input view. This condition is very hard to hold since most of the archaeological vessels do have concavities (like a cup for instance) that have to be modelled. Therefore a second, active shape determination method has to be used to discover all concavities. The acquisition method used for estimating the

3D-shape of objects is shape from structured light, based on active triangulation (Besl 1988). Both methods require only a camera and illumination devices as equipment, so they can be used to obtain a quick initial model of an object which can then be refined stepwise.

Model Representation

There are many different model representations that are used in computer vision and computer graphics. Here we will mention only the most important ones. Surface-based representations describe the surface of an object as a set of simple approximating patches, like planar or quadric patches. Generalized cylinder representation (Shirai 1987) defines a volume by a curved axis and a cross-section function at each point of the axis. Overlapping sphere representation (O'Rourke and Badler 1979) describes a volume as a set of arbitrarily located and sized spheres. Approaches such as these are efficient in representing a specific set of shapes but they are not flexible enough to describe arbitrary solid objects.

Two of the most commonly used representations for solid volumes are boundary representation (B-Rep) and constructive solid geometry (CSG) (Shirai 1987). The B-Rep method describes an object as a volume enclosed by a set of surface elements, typically sections of planes and quadratic surfaces such as spheres, cylinders and cones. The CSG method uses volume elements rather than surface elements to describe an object. Typical volume elements are blocks, spheres, cylinders, cones and prisms. These elements are combined by set operations into the modelled object. The B-Rep and CSG method suffer from quadratic growth of elemental operations as the complexity of the modelled object increases.

An octree (Chen and Huang 1988) is a tree-formed data structure used to represent 3-dimensional objects. Each node of an octree represents a cube subset of a 3-dimensional volume. A node of an octree which represents a 3D object is said to be:

- black, if the corresponding cube lies completely within the object;
- white, if the corresponding cube lies completely within the background, i.e., has no intersection with the object;
- gray, if the corresponding cube is a boundary cube, i.e., belongs partly to the object and partly to the background. In this case the node is divided into 8 child nodes (octants) representing 8 equally sized sub-cubes of the original cube;

All leaf nodes are either black or white and all intermediate nodes are gray. An example of a simple 3D object and the corresponding octree is shown in Fig. 5. This

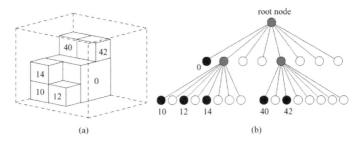

Fig. 5. A simple object (a) and the corresponding octree (b).

octree contains binary information in the leaf nodes and therefore it is called a binary octree and it is suitable for representation of 3D objects where the shape of the object is the only object property that needs to be modelled by the octree. Non-binary octrees can contain other information in the leaf nodes, e.g., the cube colour in RGB-space. For the Shape from Silhouette algorithm presented, a binary octree model is sufficient to represent 3D objects.

Construction of an Octree

The algorithm builds up a 3D model of an object in the following way: first, all input images are transformed into binary images where a "black" pixel belongs to the object observed and a "white" one to the background (Fig. 6a). In the implementation, black means background and white means object, but it is more intuitive to describe an object pixel as "black" and a background pixel as "white". Then, the initial octree is created with a single root node (Fig. 6b) representing the whole object space, which will be "carved out" corresponding to the shape of the object observed.

Then, the octree is processed in a level-by-level manner: starting from level 0 (with root node as the only node), all octree nodes of the current level marked as "black", i.e., belonging to the object, are projected into the first input image (Fig. 6c) and tested for intersection with the object's image silhouette. Depending on the result of the intersection test, a node can remain to be "black", it can be marked as "white" (belonging to the background) or in case it belongs partly to the object and partly to the background, it is marked as "gray" and divided into 8 black child nodes of the next higher level (Fig. 6d). The remainder of the black nodes of the current level are then projected into the next input image where the procedure of intersection testing with the object's silhouette is repeated. Once all input images have been processed for the current octree level, the current level is incremented by one and the whole procedure (starting from the projection of the black nodes of the current level into the first input image) is repeated, until the maximal octree level has been reached. The remaining octree after the processing of the last level is the final 3D model of the object (Fig. 6e).

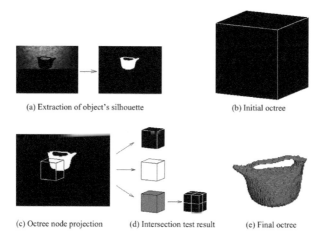

(a) Extraction of object's silhouette   (b) Initial octree

(c) Octree node projection   (d) Intersection test result   (e) Final octree

Fig. 6. Algorithm overview.

Determination of Octree Nodes

An octree node is projected into the image plane in the following way: as a pre-processing step, the translation and rotation matrices and are multiplied for all possible view angles α, and the resulting matrices of these multiplications are stored in a lookup table. This is done before any processing of octree nodes starts. Once it starts, all vertices of the current node are projected into the image plane by multiplying their octree coordinates with the matrix from the lookup table corresponding to the current view angle and then multiplying the result with the appropriate scaling matrix. The results of the projection of an octree node into the image plane are image coordinates of all of the vertices of the node's corresponding cube. In the general case, the projection of a node looks like a hexagon, as depicted in Fig. 7(a). To find the hexagon corresponding to the eight projected vertices is a costly task, because it requires to determine which points are inside and which outside the hexagon, and there can be hundreds of thousands of octree nodes that need to be processed. It is much simpler (and therefore faster) to compare the bounding box of the eight points. Figure 7 shows a projected octree node and the corresponding bounding box.

The bounding box is tested for intersection with the object's silhouette in the current input (binary) image. All image pixels within the bounding box are checked for their colour, whether they are black or white. The output of the intersection testing procedure is percentage of the black pixels of the bounding box, i.e., the percentage of pixels belonging to the object. If this percentage is equal or higher than a user definable threshold for black nodes, the node is marked as black. If the percentage is smaller than or equal with a user definable threshold for white nodes, the node is marked as white. Otherwise, the node is marked as gray and it is divided into eight child nodes representing eight sub-cubes of finer resolution.

The calculated image coordinates of the cube's vertices can lay between two image pixels and a pixel is the smallest testable unit for intersection testing. Which pixels are considered to be "within" the bounding box? Figure 8 illustrates our answer to this question. We decided to test only pixels that lie completely within the bounding box (Fig. 8a), because that way the number of pixels that need to be tested is smaller than testing all pixels that are at least partly covered by the bounding box. The pixels at the border of the bounding box are excluded, because most of them do not lie within the hexagon approximated by the bounding box. In the special case if there are no pixels that lie completely within the bounding box (Fig. 8b) the pixel closest to the centre of the bounding box is checked for the colour.

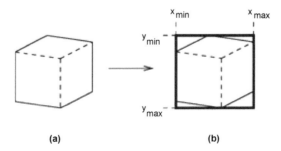

(a) **(b)**

Fig. 7. Projection of a node (a) and its bounding box (b).

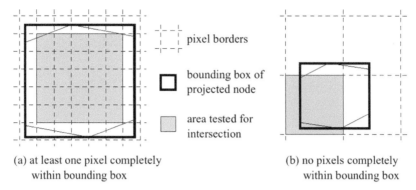

pixel borders

bounding box of projected node

area tested for intersection

(a) at least one pixel completely within bounding box

(b) no pixels completely within bounding box

Fig. 8. Selection of pixels for the intersection test.

The octree representation has several advantages (Chen and Huang 1988): for a typical solid object it is an efficient representation, because of a large degree of coherence between neighbouring volume elements (voxels), which means that a large piece of an object can be represented by a single octree node. Another advantage is the ease of performing geometrical transformations on a node, because they only need to be performed on the node's vertices. The disadvantage of octree models is that they digitize the space by representing it through cubes whose resolution depend on the maximal octree depth and therefore cannot have smooth surfaces.

Combination of Algorithms

An input image for Shape from Silhouette defines a conic volume in space which contains the object to be modelled (Fig. 9a). Another input image taken from a different view defines another conic volume containing the object (Fig. 9b). Intersection of the two conic volumes narrows down the space the object can possibly occupy (Fig. 9c). With an increasing number of views the intersection of all conic volumes approximates the actual volume occupied by the object better and better, converging to the 3D visual hull of the object. Therefore by its nature Shape from Silhouette defines a volumetric model of an object.

An input image for Shape from Structured Light using laser light defines solely the points on the surface of the object which intersect the laser plane (Fig. 10a). Using multiple views provides a cloud of points belonging to the object surface (Fig. 10b), i.e., with the surface model of the object.

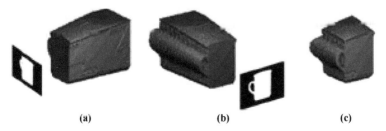

(a)   (b)   (c)

Fig. 9. Two conic volumes and their intersection.

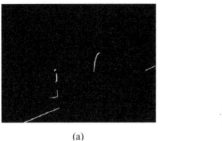

(a)   (b)

Fig. 10. Laser projection and cloud of points.

The main problem that needs to be addressed in an attempt to combine these two methods is how to adapt the two representations to one another, i.e., how to build a common 3D model representation. This can be done in several ways:

- Build the *Shape from Silhouette*'s volumetric model and the *Shape from Structured Light*'s surface model independently from one another. Then, either convert the volumetric model to a surface model and use the intersection of the two surface models as the final representation or convert the surface model to a volumetric model and use the intersection of the two volumetric models as the final representation.
- Use a common 3D model representation from the ground up, avoiding any model conversions. That means either design a volume based Shape from Structured Light algorithm or a surface based Shape from Silhouette algorithm.

With the former method, both underlying algorithms would build their "native" model of the object. However, conversion and intersection of the models would not be a simple task. While conversion of the Shape from Silhouette's volumetric model to a surface model is straightforward—one only has to find 3D points of the volume belonging to the surface—an intersection of two surface models can be rather complex. One could start from the points obtained by Shape from Structured Light (because they really lie on the object's surface, whereas points on the surface of the volume obtained by Shape from Silhouette only lie somewhere on the object's visual hull) and fill up the missing surface points with points from the Shape from Silhouette model.

There are several problems with this approach. There could be many "jumps" on the object surface, because the points taken from the Shape from Silhouette model might be relatively far away from the actual surface. The approach would also not be very efficient, because we would need to build a complete volumetric model through Shape from Silhouette, then intersect it with every laser plane used for Shape from Structured Light in order to create a surface model, and then, if we also want to compute the volume of the object, we would have to convert the final surface model back to the volumetric model.

Another possibility would be converting the surface model obtained by Shape from Structured Light to a volumetric model and intersect it with the Shape from Silhouette's model. In this case the intersection is the easier part—for each voxel of the space observed one would only have to look up whether both models "agree" that the voxel belongs to the object—only such voxels would be kept in the final model and all others defined as background. Also the volume computation is simple in this case—it is a multiplication of the number of voxels in the final model with the volume of a single voxel. But the problem with this approach is the conversion of the Shape from Structured Light's surface model to a volumetric model—in most cases, the surface model obtained using laser plane is very incomplete, so one would have to decide how to handle the missing parts of the surface. And generally, the conversion of a surface model to a volumetric model is a complex task, because if the surface is not completely closed, it is hard to say whether a certain voxel lies inside or outside the object. With closed surfaces one could follow a line in 3D space starting from the voxel observed and going in any direction and count how many times the line intersects the surface. For an odd number of intersections one can say that the voxel belongs to the object. But even in this case there would be many special cases to handle, e.g., when the chosen line is tangential to the object's surface.

This reasoning leads us to the following conclusions:

- Building a separate Shape from Structured Light surface model and a Shape from Silhouette volumetric model followed by converting one model to the other and intersecting them is mathematically complex and computationally costly.
- If we want to estimate the volume of an object using our model, any intermediate surface models should be avoided because of the problems of conversion to a volumetric model.

For tests with real objects we use 7 objects shown in Fig. 11: a metal cuboid, a wooden cone, a coffee mug, two archaeological vessels and two archaeological sherds.

The cuboid and the cone have known dimensions so we can calculate their volumes analytically and compare them with the volumes of their reconstructed models. Using these two objects we can also measure the impact of ignoring camera lens distortion on the accuracy of the models. The other objects have unknown volume, so we will just show the models constructed. All models shown in this section are built using 360 views, with constant angle of 1° between two neighbouring views.

The exact volumes of these objects are unknown and therefore the accuracy of the volume calculated through reconstruction cannot be estimated. However, we can measure the bounding cuboids and compare it with the dimensions of the model.

Fig. 11. Real objects used for tests.

Table 2 summarizes the results. The resulting models, shown from three views, are depicted in Fig. 12. All models are built with an octree resolution of 256^3 and using 360 views.

Table 2. Reconstruction of two vessels, two sherds and a cup.

Object	Voxel size	Measured dimensions (mm)
Vessel #1	0.74 mm	141.2 x 84.8 x 93.7
Vessel #2	0.53 mm	114.2 x 114.6 x 87.4
Sherd #1	0.84 mm	51.8 x 67.0 x 82.2
Sherd #2	0.76 mm	76.0 x 107.3 x 88.5
Cup	0.66 mm	113.3 x 80.0 x 98.9
Object	**Volume (mm³)**	**Calculated dimensions (mm)**
Vessel #1	336131	139.2 x 83.2 x 91.4
Vessel #2	263696	113.0 x 111.9 x 86.4
Sherd #1	35911	51.0 x 66.0 x 79.4
Sherd #2	38586	74.9 x 103.9 x 86.2
Cup	276440	111.6 x 79.0 x 98.3

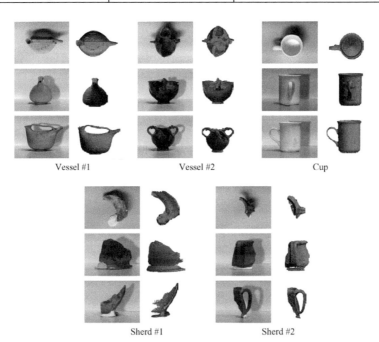

Vessel #1   Vessel #2   Cup

Sherd #1   Sherd #2

Fig. 12. 3D models of two vessels, two sherds and a cup.

Coins

Ancient coins, like the Roman Republican coins shown in Fig. 13, are fundamental objects of Numismatic research and thus demand a detailed analysis for a wide variety of research tasks like coin classification, specimen identification, die studies in coin hoards

Fig. 13. Ancient Roman Republican coins from the Museum of Fine Arts Vienna.

processing, re-assembly of coin fragments, etc. (Zambanini et al. 2008). Acquiring, modeling and processing their 3D shape is of high importance for an automated or semi-automated analysis as on ancient coins the motives, symbols and legends do not appear as colour variations but as relief-like structures. As a consequence, traditional 2D imaging suffers from a high degree of information loss and is vulnerable to the illumination conditions during image acquisition (Zambanini and Kampel 2008).

Dependent on the needed application and research question to tackle, there are basically three ways of acquiring and modelling coin shapes: (a) the 2D outline formed by the coin border, (b) the surface normal field captured by Polynomial Texture Mapping, and (c) the fully registered 3D surface captured by a high-accuracy 3D scanner. In the given order, the three options provide an increasing degree of information content and number of potential investigations, but at the price of higher acquisition effort and complexity.

2D Coin Border

In contrast to their present-day counterparts, ancient coins have not been manufactured industrially and have altered by abrasion and fouling over the centuries. Therefore, every ancient coin can be seen as a unique specimen that exhibits individual identification features which can be extracted from 2D coin images. It has been shown by (Huber-Mörk et al. 2011) that the shape of the coin border is a very discriminative feature for coin identification.

Before modelling this shape in order to compute a similarity metric, it has to be extracted from the 2D images first. This process is called coin segmentation and can be characterized as the separation of the image pixels into binary sets "coin" and "background". This separation is commonly accomplished by thresholding the image against pre-processed image intensity values, whereas the optimal threshold/ segmentation is determined by optimizing an objective function describing the resulting shape. As the shape of the coin border is nearly circular, a measure of compactness (Huber-Mörk et al. 2011) or roundness (Zambanini and Kampel 2009) can be used. Figure 14 shows an example of this operation: the input image (14a) is transformed to the segmented image (14b), where the white pixels identify the coin region and the black pixels identify the background region.

In order to derive a mathematical description of the 2D coin border from the binary segmentation image, the deviation from a circle is computed at N discrete sample points

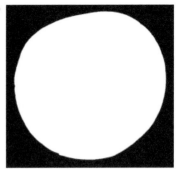

Fig. 14. Coin segmentation.

$s_i = (x_i, y_i)$ along the border pixels (Huber-Mörk et al. 2011) . The radius r of the circle is determined by

$$r = \frac{1}{N} \sum_{i=1}^{N} \|s_i - c\|_2$$

where c denotes the centre of gravity of the coin region. The descriptor of the 2D coin border is then given by the vector $F = (f_1, ..., f_N)$ where

$$f_i = \frac{\|s_i - c\|_2 - r}{r}$$

Please note that the descriptor is invariant to scale changes due to the division by r. In order to detect similar coin borders from this descriptor for coin identification, a metric of dissimilarity is needed. Huber-Mörk et al. (2011) propose to use a weighted sum of a Fourier-based and a geometry-based dissimilarity. The Fourier-based dissimilarity $d_F(F, F')$ is computed as

$$d_F(F, F') = \sum_{i=1}^{N} \frac{\|\hat{f}_i' - \hat{f}_i\|_2}{N}$$

where $\hat{f}_i'$ and $\hat{f}_i$ denote the magnitudes of the i-th coefficient of the Fourier spectrums of the descriptors F and F', respectively. The geometry-based dissimilarity d_G is computed as the minimum mean squared error between corresponding sample points

$$d_G(F, F') = \min_{u=0, ..., N-1} \frac{1}{N} \sum_{i=1}^{N} (f_i - f_{i+u}')^2$$

where $i + u$ is taken modulo N. This measure has to be computed for all shifts u to account for different orders of the border point samples due to rotation differences between the coins in the two images.

The identification power of 2D coin border shape matching was tested on 2400 images of 240 ancient coin specimens of the same class. The proposed method achieved an identification rate of 97.04% which could be further improved to 98.54% by combining it with local feature matching of the interior coin region (Zaharieva et al. 2007).

Surface Normal Field

Instead of capturing the true 3D information of a coin in the form of registered surfaces, one can derive the surface normals only as well. These normals can be used to obtain realistic renderings using computer graphic algorithms (Hughes et al. 2013). As coins are relatively flat objects, interactively changing the virtual light position gives a realistic impression of the coin's surface from a frontal view. The rendering style can also be changed to non-photorealistic images in order to accentuate certain features of the coins.

A technique for capturing surface normals as well as additional properties needed for realistic renderings like inter-reflections and self-shadowing is called Polynomial Texture Mapping (PTM) (Malzbender et al. 2001). Acquisition for PTM is accomplished by taking multiple images of an object with a static camera under varying lighting conditions. Therefore, instead of an expensive 3D scanning device only a framework that allows to fix the camera and light source is needed. From the acquired image series, the object colour variation of each pixel can be approximated with a biquadratic polynomial which provides a compact representation that can be exploited for real-time rendering. Consequently, the polynomial texture map is able to reconstruct the appearance of changing light source directions on the object.

PTM has been first applied to historical coins by Mudge et al. (2005). In their setup the coins were placed on a flat light position template that indicated the positions of the light source which was placed on a stand. The idea of using this technique for historical coins has been adopted by others, and their results can be interactively viewed online.[1] Compared to traditional coin digitalization and documentation, PTM provides a reasonable compromise between the level of information about the geometric structure of a coin and the effort to acquire this structure. It is also mentioned by Mudge et al. (2005) that certain features which are hard to recognize with the naked eye can be made more visible by special enhancement techniques of the viewer software.

Fully Registered 3D Surface

Full 3D coin models in form of polygon meshes offer the most accurate and richest information about the 3D geometry of a coin. However, compared to simpler documentation techniques that accept a certain degree of information loss, acquiring 3D coin models is more expensive and laborious. This is mainly due to the challenging nature of historical coins: their relatively small size and the highly specular reflectance of their metallic surfaces desires high-end 3D scanning devices that provide the necessary level of accuracy.

For 3D coin acquisition, synchronous acquisition of 3D data and texture/colour information is needed. This leads to a realistic representation of a coin and allows the subsequent combination of 2D (colour, gray value) and 3D (geometry) analysis. Another requested feature of the scanning device is portability since coin acquisition preferably takes place where the coins are kept.

First results of acquiring fully registered 3D coin models have been presented by Zambanini et al. (2009). The authors used the active stereo scanner *Breuckmann*

---

[1] For an example visit http://www.wessexarch.co.uk/computing/ptm/examples (accessed 14 October 2013).

stereoSCAN 3D (see Fig. 15). The main components of this system are a projector and two cameras serving as a stereo camera pair. These three components allow for the computation of the depth of object points through the principle of triangulation (Hartley and Sturm 1997): corresponding points are detected in the two camera images by means of the known light pattern of the projector. As the geometry of the two cameras is known, the depth of the corresponding 3D point can be estimated based on the disparity between the 2D points in the two camera images.

For full 3D models several scans from different sides need to be acquired and registered in subsequent post processing steps. In order to ease the registration later on, the coins were placed on a rotation/tilt table in front of the active stereo system and were scanned from eight viewing positions. The difference of the viewing positions is known from the controlled rotation on the table. Therefore, the scanned eight 3D point clouds could be roughly aligned and an automatic iterative closest point algorithm (Besl and McKay 1992) could be applied for the accurate registration. Finally, all eight point clouds were merged into a polygon mesh.

In total, 25 historical coins have been scanned and documented in Zambanini et al. (2009)—16 ancient coins from the Roman era and 9 tornese coins from the medieval age. According to the theoretical accuracy of the provided 3D scanner, the final models have a resolution of 20 μm in x- and y-direction and 1 μm in z-direction, respectively. Examples of coin models visualized with a professional rendering tool can be seen in Fig. 16.

Please notice the small real world size of these objects, as the given coins have a diameter range of 17–33 mm. The accuracy of the scanned 3D models was evaluated by comparing actual measurements on the real coins with measurements on their virtual counterparts (Hödlmoser et al. 2010). These experiments revealed that the average variation coefficient of coin diameter and volume measurements was about 1.26% and 0.26%, respectively, which emphasizes the usefulness of 3D coin models for numismatic studies and documentation.

Fig. 15. The Breuckmann stereoSCAN 3D.

Fig. 16. Renderings of 3D coin models.

Conclusion

As the variety of different archaeological objects is huge in terms of size and surface condition, the number of methods to acquire their 3D structure is versatile as well. The choice of a proper method is thus always application-dependent: for coin identification the extraction of the 2D coin border is sufficient, whereas for more detailed analyses like coin die studies accurate 3D models are needed. We have shown that the same principle also applies to other archaeological objects like vessels and sherds.

The application area is also the main motivation for choosing the right data structure for processing the 3D data. 2D profiles like vessel outlines and coin borders can be represented by splines or by deviations from a circle. 3D structures can be defined by volumetric or surface representations and we have discussed how these representations can be combined to a joint 3D model. As an intermediate representation between 2D and 3D representations, surface normal fields provide a convenient description of the surface conditions of objects without explicit 3D data.

References Cited

http://www.wessexarch.co.uk/computing/ptm/examples (accessed 14 October 2013).

Bennett, J.R. and J.S. Mac Donald. 1975. On the Measurement of Curvature in a Quantized Environment. IEEE Transactions on Computers 24(8): 803–820.

Besl, P.J. 1988. Active, optical range imaging sensors. Machine Vision and Applications 1(2): 127–152.

Besl, P.J. and N.D. McKay. 1992. A Method for Registration of 3-D shapes. Pattern Analysis and Machine Intelligence 14(2): 239–256.

Chen, H.H. and T.S. Huang. 1988. A survey of construction and manipulation of octrees. Computer Vision, Graphics, and Image Processing 43(3): 409–431.

DeVore, R.A. and G.G. Lorentz. 1993. Constructive approximation, Springer.

Hartley, R.I. and P. Sturm. 1997. Triangulation. Computer Vision and Image Understanding 68(2): 146–157.

Hödlmoser, M., S. Zambanini, M. Kampel and M. Schlapke. Evaluation of Historical 3D Coin Models. in Proc. of the 38th Conf. on Computer Applications and Quantitative Methods in Archaeology (CAA'10), Granada/Spain, Apr. 2010

Huber-Mörk R., S. Zambanini, M. Zaharieva, M. Kampel. Identification of Ancient Coins Based on Fusion of Shape and Local Features", Machine Vision and Applications, pp. 1–12, 2010.

Hughes, J.F., Andries van Dam, Morgan McGuire, David F. Sklar, James D. Foley, Steven K. Feiner and Kurt Akeley. 2013. Computer graphics: principles and practice, Addison-Wesley Professional.

Kutulakos, K.N. and S.M. Seitz. 2000. A Theory of Shape by Space Carving. International Journal of Computer Vision 38(3): 199–218.

Malzbender, T., D. Gelb and H. Wolters. 2001. Polynomial texture maps. In Proceedings of the 28th Annual Conference on Computer Graphics and Interactive Techniques, pp. 519 528.

Marr, D. Vision - A Computiational Investigation into the Human Representation and Processing of Visual Information. Freeman, San Francisco, 1982.

Mudge, M., Jean-Pierre Voutaz, Carla Schroer and Marlin Lum. 2005. Reflection transformation imaging and virtual representations of coins from the hospice of the grand st. bernard. In Proceedings of the 6th International conference on Virtual Reality, Archaeology and Intelligent Cultural Heritage, pp. 29–39.

O'Rourke, J. and N. Badler. 1979. Decomposition of three-dimensional objects into spheres. IEEE transactions on pattern analysis and machine intelligence 1(3): 295–305.

Orton, C. and M. Hughes. 2013. Pottery in archaeology, Cambridge University Press.

Rosenfeld, A. and A. Nakamura. 1997. Local deformations of digital curves. Pattern Recognition Letters.

Sablatnig, R., S. Tosovic and M. Kampel. 2003. Next view planning for a combination of passive and active acquisition techniques. Fourth International Conference on 3-D Digital Imaging and Modeling, 2003. 3DIM 2003. Proceedings, pp. 62–69.

Shapiro, L. and G. Stockman. 2001. Computer Vision. 2001, Prentice Hall.

Shirai, Y. 1987. Three-Dimensional Computer Vision, Berlin, Heidelberg: Springer Berlin Heidelberg.

Shoukry, A. and A. Amin. 1983. Topological and statistical analysis of line drawings. Pattern Recognition Letters 1: 365–374.

Sonka, M., V. Hlavac and R. Boyle. 2007. Image Processing, Analysis, and Machine Vision, Thomson-Engineering.

Szeliski, R. 1993. Rapid Octree Construction from Image Sequences. CVGIP: Image Understanding 58(1): 23–32.

Winograd, S., T.J. Rivlin and C.A. Micchelli. 1976. The optimal recovery of smooth functions. Numerische Mathematik 26: 191–200.

Zaharieva, M., M. Kampel and S. Zambanini. 2007. Image Based Recognition of Ancient Coins. Proceedings of the 12th international conference on Computer Analysis of Images and Patterns, pp. 547–554.

Zambanini, S. and M. Kampel. 2008. Coin Data Acquisition for Image Recognition. In Proceedings of the 36th Annual Conference on Computer Applications and Quantitative Methods in Archaeology (CAA'08).

Zambanini, S. and M. Kampel. 2009. Robust Automatic Segmentation of Ancient Coins. In 4th International Conference on Computer Vision Theory and Applications (VISAPP'09), pp. 273–276.

Zambanini, S., M. Kampel and M. Schlapke. 2008. On the Use of Computer Vision for Numismatic Research M. Ashley et al. (eds.). 9th International Symposium on Virtual Reality, Archaeology and Cultural Heritage 45(7): 17–24.

Zambanini, S., M. Schlapke and M. Kampel. 2009. Historical Coins in 3D: Aquisition and Numismatic Applications. In 10th International Symposium on Virtual Reality, Archaeology and Cultural Heritage (VAST'09), pp. 49–52.

7

Curvature-Based Method for the Morphometric Analysis of Archaeological Shapes

Michael L. Merrill

Introduction

This contribution will look at a particularly effective method of curvature-based analysis of two-dimensional outlines. See Kampel and Zambanini (Chapter 5, this volume), who introduce the general idea of curvature-based analysis. I apply this method to examine global and localized differences in the endocranial outline of a small sample of fossil hominins and modern humans. A previous morphometric study (Bookstein et al. 1999), which used procrustes analysis of semi-landmarks, compared the endocranial outlines of 21 fossil hominins and modern humans provided a surprising result. Specifically, that the shape of the mid-sagittal endocranial frontal bone profile of the fossil and modern humans was very similar. The fossil hominin sample used in the Bookstein et al. study consisted of three *Homo heidelbergensis* (Bodo, Kabwe, and Petralona), the Spanish Atapuerca SH5 cranium (possibly proto-Neanderthal), and a "classic" *Homo neanderthalensis* skull, Guattari I.

Further Bookstein et al. use an earlier although subjective observation (Schultz 1931) to support their results. Schultz observed that a median-sagittal endocast of Kabwe yielded a curve similar to that of modern humans. I will use my method to suggest this is not the case for the mid-sagittal section of the endocranial outline adjacent to the parietal region of the brain, contrary to Shultz's subjective observation, by examining global shape characteristics.

Also, as I will discuss later, the results of an analysis of maximal curvature, suggest there are significant differences in the locations of maximal curvature between the fossil and modern human samples in the interval of the midsagittal endocranial outline adjacent to the parietal region.

School of Human Evolution and Social Change, Arizona State University.
  Email: Michael.L.Merrill@asu.edu

In this chapter, no biological interpretations are made, but I use the endocranial outline case study to demonstrate the utility of my curvature-based method for identifying global and localized shape characteristics of two-dimensional curves.

Finally, Bookstein's procrustes and semi-landmark based method compares shapes by asking how one shape can be distorted into another shape, and how this affects the respective grid systems based on landmarks and semi-landmarks (which Bookstein had to invent because of the paucity of real, biological landmarks), whereas the curvature method I introduce in this chapter (which can only be used to study curves and not surfaces) enables the detailed characterization of a curve, which is not the case with Bookstein's method.

Elliptical Fourier Function Analysis

A closed curve in two-dimensional space, such as an endocranial outline, can be parameterized in terms of two periodic functions. In this chapter, hominin endocranial outlines are fit to what is known as an elliptical Fourier function, or simply an EFF, using the Kuhl and Giardina method (Kuhl and Giardina 1982) to arrive at the following parametric solution in terms of x-y coordinates as functions of a third variable t.

$$x(t) = A_0 + \sum_{n=1}^{N} a_n \cos(nt) + \sum_{n=1}^{N} b_n \sin(nt)$$

$$y(t) = C_0 + \sum_{n=1}^{N} c_n \cos(nt) + \sum_{n=1}^{N} d_n \sin(nt)$$

In the above formula A_0, C_0, a_n, b_n, c_n and d_n are constant real numbers called harmonic coefficients, n is the harmonic number, N the maximum harmonic number, and t is a real number multiple of π radians in the open interval $[0 \leq t \leq 2\pi]$, which constitutes a period or in other words a single trace of the outline. Therefore $x(0) = x(2\pi)$ and $y(0) = y(2\pi)$.

EFFs provide global shape estimates of smooth and complex two-dimensional outlines but are not capable of identifying localized shape information (Lestrel et al. 2005, Lestrel 1997, Diaz et al. 1997). A recently introduced method that uses EFFs to generate the raw data (predicted point coordinates of a two-dimensional outline from an EFF) for computing continuous wavelet transforms (Lestrel et al. 2005) is also able to identify localized features of the outline (specifically localized and gradual changes in the curvature along the outline). The drawbacks to Lestrel et al. method are: (1) its complexity, (2) it only provides indirect and scaled measures of curvature, and (3) has a number of major constraints, which they discuss in detail on page 611 of their paper. The method I introduce in this chapter is much easier to apply than the Lestrel et al. method, and (in a similar manner to the Lestrel et al. approach) uses EFFs to provide the raw data (which for my method are EFF harmonics), which are used to directly compute curvature at any position along the outline and at any level of resolution.

Previous and Present Use of Curvature

The development of methods for the accurate and reproducible measurement of the curvature of two and three-dimensional images has been a popular endeavor in the computer vision and analysis community (Duncan et al. 1990, Teh and Chin 1989). These methods have seen a number of applications in the biological sciences, especially medical imaging. This is primarily due to the following reasons:

1) Curvature provides a convenient measure of shape in objects that are either fixed or whose shape is changing over time.
2) Curvature is amenable to measurement at both local (fine) and regional (coarse) scales.
3) Curvature can be measured independent of reference systems, indexing (landmarks), and coordinate systems (Mancini et al. 1985, 1987).

In the present work curvature will be examined at a coarse scale to study global shape patterns and at a fine scale to detect localized maxima in the curvature of two-dimensional hominin endocranial outlines. The acquisition of the raw data used in my study was very time consuming, because outlines were traced by hand and then input manually into a computer file using a digitizing board. Currently much more efficient and accurate methods are available that obtain points from digital pictures using computer software (e.g., MLmetrics, available from Pete Lestrel and Charles Wolfe @ http://www.plestrel.com/).

Calculation of Curvature

In the real Euclidean plane, curvature is defined as the rate of change of slope as a function of arc length. For the curve $y = f(x)$, this can be expressed in terms of derivatives as

$$k = \frac{\left|\frac{d^2y}{dx^2}\right|}{\left[1 + \left(\frac{dy}{dx}\right)^2\right]^{3/2}}$$

In the present chapter only the magnitude of curvature is being evaluated and not inflections of the outlines from concave to convex. As will be seen later, in each of the samples the segments of the endocranial outline with significant local curvature are all concave. To discriminate between concave and convex intervals the absolute value in the numerator of the preceding equation is removed, so that convex segments have positive curvature and concave segments negative curvature, or vice versa (if the numerator is multiplied by –1).

To parameterize the preceding curvature equation in terms of a third variable t we start with an arbitrary curve in the real Euclidean plane defined by

$$C = \{x(t), y(t)\}$$

Define

$$x'(t) = \frac{dx}{dt}, \qquad y'(t) = \frac{dy}{dt}, \qquad x''(t) = \frac{d^2 x}{dt^2}, \qquad y''(t) = \frac{d^2 y}{dt^2}$$

Observe that

$$\frac{dy}{dt} = \frac{dy}{dx}\frac{dx}{dt}$$

Remembering that $y(t)$ is a function of $x(t)$ it follows that

$$\frac{d^2 y}{dt^2} = \frac{d^2 y}{dx^2}\frac{dx}{dt}\frac{dx}{dt} + \frac{dy}{dx}\frac{d^2 x}{dt^2} = \frac{d^2 y}{dx^2}\left(\frac{dx}{dt}\right)^2 + \frac{dy}{dx}\frac{d^2 x}{dt^2}$$

Observe that

$$\frac{dy}{dx} = \frac{\dfrac{dy}{dt}}{\dfrac{dx}{dt}}$$

Substituting the preceding equation into the equation of the second derivative of y with respect to t we get the following

$$\frac{d^2 y}{dt^2} = \frac{\dfrac{d^2 y}{dx^2}\left(\dfrac{dx}{dt}\right)^3 + \dfrac{dy}{dt}\dfrac{d^2 x}{dt^2}}{\dfrac{dx}{dt}}$$

Solving the preceding equation in terms of the second derivative of y with respect to x we get

$$\frac{d^2 y}{dx^2} = \frac{\dfrac{dx}{dt}\dfrac{d^2 y}{dt^2} - \dfrac{dy}{dt}\dfrac{d^2 x}{dt^2}}{\left(\dfrac{dx}{dt}\right)^3}$$

Substituting the above equation and the parameterized form of the first derivative of y with respect to x into the un-parameterized equation of curvature we have

$$k = \frac{\left|\dfrac{\dfrac{dx}{dt}\dfrac{d^2 y}{dt^2} - \dfrac{dy}{dt}\dfrac{d^2 x}{dt^2}}{\left(\dfrac{dx}{dt}\right)^3}\right|}{\left[1 + \left(\dfrac{\dfrac{dy}{dt}}{\dfrac{dx}{dt}}\right)^2\right]^{\frac{3}{2}}} = \frac{\left|\dfrac{\dfrac{dx}{dt}\dfrac{d^2 y}{dt^2} - \dfrac{dy}{dt}\dfrac{d^2 x}{dt^2}}{\left(\dfrac{dx}{dt}\right)^3}\right|}{\dfrac{\left[\left(\dfrac{dx}{dt}\right)^2 + \left(\dfrac{dy}{dt}\right)^2\right]^{\frac{3}{2}}}{\left(\dfrac{dx}{dt}\right)^3}} = \frac{\left|\dfrac{dx}{dt}\dfrac{d^2 y}{dt^2} - \dfrac{dy}{dt}\dfrac{d^2 x}{dt^2}\right|}{\left[\left(\dfrac{dx}{dt}\right)^2 + \left(\dfrac{dy}{dt}\right)^2\right]^{\frac{3}{2}}}$$

The above equation can be written more compactly using the previously defined notation for the first and second derivatives of x and y with respect to t. With this notation

$$k = |x'(t)y''(t) - y'(t)x''(t)| \Big/ [(x'(t))^2 + (y'(t))^2]^{3/2}$$

Another important contribution of the new method is that it shows unequivocally that the claim by Pavlidis, which applies to the preceding differential geometric formula of curvature, is incorrect.

Unfortunately, a direct measurement of curvature is not always feasible because of noise. The formula given in calculus texts requires taking a second derivative and thus cannot be used in any practical situation (Pavlidis 1982: 160).

In terms of the Kuhl and Giardina EFF formula, the first and second derivatives are as follows.

$$x'(t) = n\left[-\sum_{n-1}^{N} a_n \sin(nt) + \sum_{n=1}^{N} b_n \cos(nt)\right]$$

$$x''(t) = -n^2\left[\sum_{n=1}^{N} a_n \cos(nt) + \sum_{n=1}^{N} b_n \sin(nt)\right]$$

$$y'(t) = n\left[-\sum_{n=1}^{N} c_n \sin(nt) + \sum_{n=1}^{N} d_n \cos(nt)\right]$$

$$y''(t) = -n^2\left[\sum_{n=1}^{N} c_n \cos(nt) + \sum_{n=1}^{N} d_n \sin(nt)\right]$$

To operationalize the above equation of curvature with the preceding first and second derivatives, at any desired (though constant) incremental value, I wrote a c++ program, which uses harmonic coefficients (a_n, b_n, c_n, d_n) as raw data and the maximum harmonic number N and incremental size of t as interactive input.

Ostelogical Material

The cranial radiographs of one hundred adult modern humans of European and African descent were obtained by Loïsa deFelice from the Robert J. Terry Anatomical Collection at the Smithsonian Natural Museum of Natural History (Washington, DC) for her master's research.

Eight specimens from deFelice's sample were chosen, which represent both typical, and extremes (Table 1) in variation of the correlation between the shape of the exocranium and endocranium. The modern sample also consists of black and white female and male pairs of different ages. For example the pair, BF241500 and BF421198 are black females, 24 and 42 years old, respectively. The fossil endocranial radiographs were obtained from the sources in Table 2.

Table 1. Extremes in the modern endocranial sample. BF = black female, BM = black male, WF = white female, and WM = white male. The first two numbers are the age of the specimen.

Specimen	Pearson correlation coefficient	Reason for selection
WF34745R	0.99	Cluster at the first step for both the exocranium and the endocranium, which implies that they are similar in both internal and external cranial morphology (deFelice 2004: 99).
WF411612	0.977	Lowest correlation coefficient in the white female group.
BM38826	0.98	Clustered at the extremes in the endocranial analysis but cluster at step five in the exocranial analysis (deFelice 2004: 98). More similar exocranially than endocranially.
WM431089	0.972	Lowest correlation coefficient in the white male group.

Table 2. Fossil endocranial sample.

Fossil	Origin	Age (kya)	Species	X-ray Source
Peking	China	400–780	*H. erectus*	Weidenreich 1943
Kabwe	Zambia	125–300	*H. heidelbergensis*	Skinner and Sperber 1992
Petralona	Greece	150–300	*H. heidelbergensis*	Murrill 1981
Bodo 1	Ethiopia	600	*H. heidelbergensis*	Conroy et al. 2000
Monte Circeo 1	Italy	70–80	*H. neanderthalensis*	Silipo et al. 1991

Data Collection

Leisa deFelice prepared the endocranial radiographs for elliptical Fourier function analysis as follows.

The radiographs for both the modern and fossil samples were initially scanned in grayscale at 300 dots per inch using a CanonScan D1250U2F scanner and saved to disk. Each of the grayscale images was digitally enhanced using Microsoft PictureIt! 7.0, resulting in a 200 percent increase in sharpness and a 100 percent increase in contrast.

Next each radiograph was traced on a sheet of graphical acetate using a 0.3 mm mechanical pencil. The nasion was chosen as the homologous endpoint. The homologous start point is a tangent that was drawn through the nasion and porion so that it intersects the base of the skull.

Then, an eight-degree angle was drawn from this tangent to move this point high enough on the occiput so that angular values could be accurately and consistently computed for all specimens from this new start point (hereafter referred to as A1).

Lines were drawn through A1 and B1 that were perpendicular to the cranial suture within +/– five degrees. These lines were then extended until they intersected; this point is called the vector center (VC). Lines were next drawn between A1 and B1. The intersection between the perpendicular and A1B1 forms a triangle that has the same three angles in all specimens (Fig. 1).

Finally, using a protractor, the angle representing A1-VC-B1 was determined to be 110 degrees. Therefore the area of study encompasses an angular distance of 250 degrees.

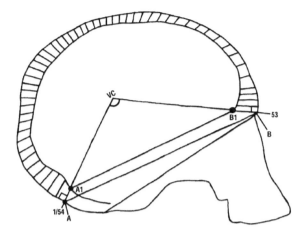

Fig. 1. Digitization of points on the endocranial outline (deFelice 2004).

The EFF23 software (Wolfe et al. 1999) used in de Felice's study (as well as this one) requires that the total number of points used for raw data be divisible by six. de Felice decided to place points at five-degree intervals and add three points to the supraorbital region, for a total of 54 points. Also, points one and 54 are equivalent since the points are on a closed contour. The 54 points were drawn on the endocranial outline tracing using a protractor centered at the VC (Fig. 1).

Next the radiographic tracings were digitized using a Calcomp 300 digitizer pad accurate to 0.01 inch. The VC was chosen as the origin (point (0, 0)) in the two-dimensional x-y coordinate system. The line A1B1 was selected as the x-axis (Fig. 1).

The raw data resulting from digitizing points on the endocranial tracings are ASCII files consisting of 54 x-y coordinates. These files are the raw data, entered into the EFF23 software for computing oriented EFF harmonic coefficients. In turn, the EFF harmonic coefficients are the raw data for a c++ program that I wrote to compute curvature.

Noise Filtering

Discarding higher order harmonics in a Fourier analysis, functions as a low-pass digital filter (Bow 1984, Rohlf 1986). In the following analyses eleven out of twenty-seven EFF harmonics are retained. EFF amplitude (defined by amplitude $= 0.5 * \sqrt{a_n^2 + b_n^2 + c_n^2 + d_n^2}$, where n is the harmonic number) plots (which are also proportional to variance) were used to decide on a cut-off for the number of harmonics to retain (Fig. 2). By the eleventh harmonic nearly all of the variance is accounted for in each of the samples (Fig. 2).

To further reduce noise in the endocranial curvature plots (due also to trying to fit what is nearly a straight line along the base of the endocranium with trigonometric functions), a moving average filter was used to smooth the endocranial curvature plots (Fig. 3). Note that the 1.2 to 1.8 radians/π interval of the endocranial outline corresponds to the base of the endocranium, which was not measured by deFelice, but is created in the predicted endocranial outlines using the EFF harmonic coefficients. In this case,

A

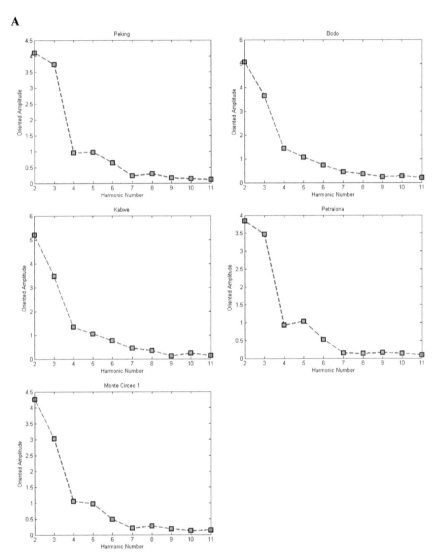

Fig. 2. (A) Amplitudes of the second to the eleventh oriented harmonics of the endocranial outlines in *Homo erectus* (Peking), *Homo heidelbergensis* (Bodo, Petralona, and Kabwe), and *Homo neanderthalensis* (Monte Circeo 1). (B) Amplitudes of the second to the eleventh oriented harmonics of the endocranial outlines of the modern human sample.

Fig. 2. contd.

Fig. 2. contd.

B

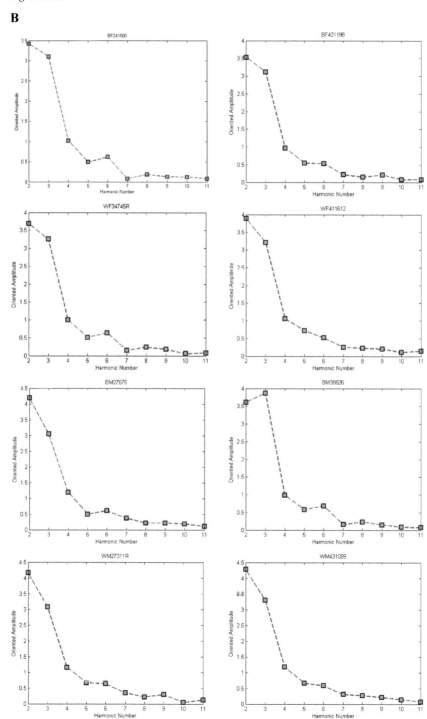

A

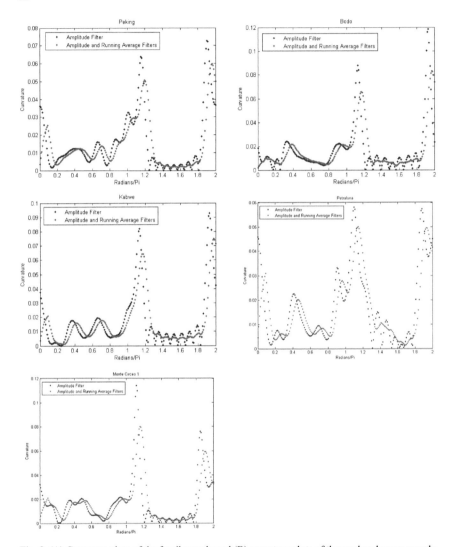

Fig. 3. (A) Curvature plots of the fossil sample and (B) curvature plots of the modern human sample.

Fig. 3. contd....

Fig. 3. contd.

B

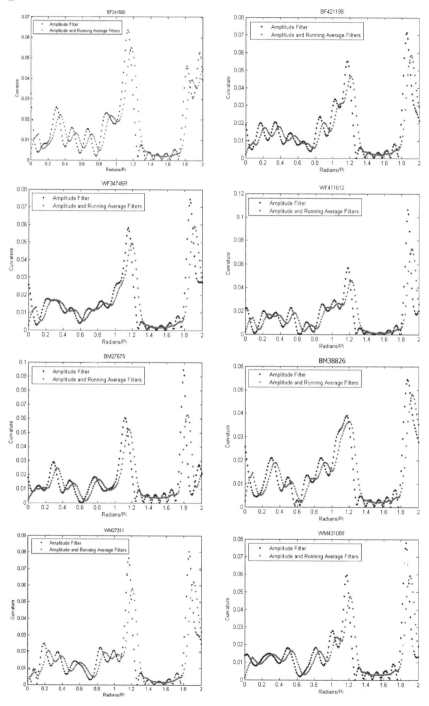

the trigonometric functions sine and cosine are being used to fit a straight line, which (combined with noise) results in the oscillations in the amplitude of only filtered curvature plots in the 1.2 to 1.8 radians/π interval (Fig. 3). Adding the running average filter significantly flattens this section of the curvature plots (Fig. 3), and reduces noise in the remaining intervals as well. After some experimentation, I determined that an interval (step size) of ten in the moving average procedure was close to optimal for the data in the current study. This is because with a step size of ten the "noise" resulting from the EFFs along the base of the endocranium nearly vanishes (as is evident in the endocranial curvature plots), and at the same time the size and shape of the endocranial outline is nearly unaffected.

Global Shape Patterns

My curvature-based method can be used to evaluate global shape characteristics of two-dimensional curves. For example, by examining the distribution of curvature on the mid-sagittal endocranial outline using 31 regularly spaced (spaced at 0.01 radians/π, although any choice of scale is possible with this method) points in the 0.2 to 0.5 interval, provides a large sample that can be used to compare global shape characteristics of the mid-sagittal endocranial outline in the fossil hominin and modern human samples. To illustrate one way such a comparison can be done, I first consider the *H. heidelbergensis* and *H. neanderthalensis* fossils.

Consider the null hypothesis in which the global shape (measured by the set of 31 regularly spaced curvature points) of the 0.2 to 0.5 interval of the mid-sagittal endocranial outline is assumed to be similar in the *H. heidelbergensis* and *H. neanderthalensis* samples. I test this hypothesis using a Smirnov two sample test with Monte Carlo analysis (Baglivo 2005: 151–152) for a pair-wise comparison of the empirical cumulative distributions (or empirical CDFs) of the curvature sets for each combination of fossils (Fig. 4, Table 4).

The null hypothesis is rejected (at $\alpha = 0.05$) for the comparisons of Kabwe with the two other *H. heidelbergensis* (Bodo and Petralona), and the *H. neanderthalensis* (MC1) fossils (Table 3). In the comparison of Kabwe with the other two *H. heidelbergensis*

Table 3. Summary of results of the Smirnov two sample test.

	Smirnov Statistic (Observed Maximum Difference between ECDF plots)	Maximum Difference Between ECDF plots (Random Partition)	P-value (Number ≥ Observed)/2000
Bodo vs. Kabwe	0.419355 at 0.00492438	0.193548 at 0.00545864	11/2000 = 0.0055
Bodo vs. Petralona	0.290323 at 0.00853275	0.225806 at 0.0126688	301/2000 = 0.1505
Bodo vs. MC1	0.258065 at 0.0179934	0.129032 at 0.0175495	496/2000 = 0.248
Kabwe vs. Petralona	0.387097 at 0.00364456	0.16129 at 0.0119016	44/2000 = 0.022
Kabwe vs. MC1	0.419355 at 0.0160669	0.193548 at 0.00480185	18/2000 = 0.009
Petralona vs. MC1	0.290323 at 0.00622502	0.225806 at 0.0175167	290/2000 = 0.145

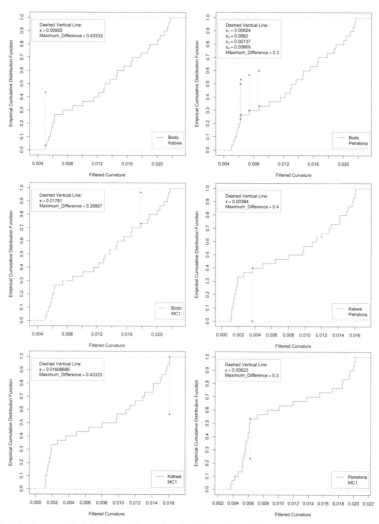

Fig. 4. Pairwise empirical CDF plots for each pair of the *H. heidelbergensis* and *H. neanderthalensis fossils*. Each empirical CDF plot results from the curvature values of 31 regularly spaced points on the endocranial outline in the 0.2 to 0.5 interval. Note that these plots were made with R code I wrote and with the R ecdf function. This resulted in minor differences in the observed maximum difference and x values in the plots of this figure compared to the corresponding values in the second column of Table 3, which were computed in Mathematica with code provided by Baglivo 2005.

fossils it is clear that the maximum difference between the empirical CDFs is in the lower end of the curvature distribution (Fig. 4). Specifically, in the 0.2 to 0.5 interval, the Kabwe outline is more flattened, and has a significantly greater proportion of curvature values in the low end of the distribution. Similarly, the comparison of Kabwe with the *H. neanderthalensis* fossil shows that the Kabwe outline has less overall curvature in the 0.2 to 0.5 interval, and (in contrast to the previous comparisons) has a significantly lower proportion of curvature values in the high end of the distribution (Fig. 4).

Recalling Schultz's suggestion that the median-sagittal endocranial outline of Kabwe is similar to that of modern humans, we can objectively evaluate his interpretation, using the Smirnov two sample test. We can also extend it to include the Bodo, Petralona, and MC1 fossils, and restrict our attention to the 0.2 to 0.5 interval.

Our second null hypothesis is that the global shape (measured using the set of 31 regularly spaced curvature values) of the 0.2 to 0.5 interval of the mid-sagittal endocranial outline is similar in the *H. heidelbergensis* and *H. neanderthalensis* samples, to the global shape of the outline in this interval in the anatomically modern human sample. Figure 5 shows the proportion of the modern human sample in which the null hypothesis is not rejected using the Smirnov two sample test. The null hypothesis is completely rejected for Kabwe and Petralona, and is mostly rejected for MC1 (Fig. 5). In the case of Bodo, the null hypothesis is rejected for slightly less than half of the modern human sample (Fig. 5). Clearly this analysis does not support Schultz's suggestion, and (more generally), suggests there may be substantial variability in the global shape of the segment of the mid-sagittal endocranial outline adjacent to the parietal region in anatomically modern humans, as well as in *H. heidelbergensis* and *H. neanderthalensis*.

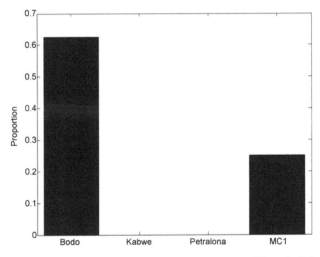

Fig. 5. Proportion of the modern human sample for which the second null hypothesis is not rejected.

Localized Maximal Curvature

My method is also capable of identifying localized shape characteristics. For example, in the shape patterning of a biological structure, such as the curvature patterns of an endocranial outline, it is plausible that localized areas of maximal curvature may correlate with specific biological processes.

To illustrate how my method can be used to identify localized patterns of maximal curvature, I chose to focus on the 0.2 to 0.5 radians/π interval of the endocranial outline. This interval (as discussed earlier) is adjacent to the parietal region of the human brain (Fig. 6), which is important to language (Gazzaniga et al. 2008, Sakai 2005), tool use (Peeters et al. 2009) and other cognitive functions that are uniquely human (Price 2000). This interval also has one or two peaks in curvature (Fig. 7) in each of the samples. I used the Matlab max function to identify the points of maximal curvature in the 0.2 to

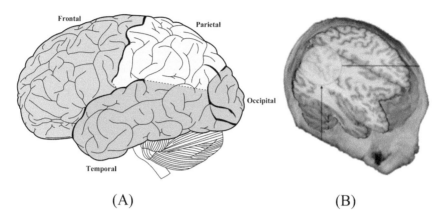

(A)   (B)

Fig. 6. (A) Human brain showing the four lobes and displayed in the left hemisphere. (B) Human brain showing both hemispheres (with a side view of the right hemisphere). The circle encloses the upper parietal region of the right hemisphere (Bruner 2010). Note the mid-sagittal line that divides the two hemispheres.

0.5 interval of the endocranial outline. Also, note that the right hemisphere of the brain is the orientation used for digitizing points on the endocranial outline (Fig. 1) and in the resulting predicted endocranial plots (Fig. 7), whereas the left hemisphere of the brain is used in (A) of Fig. 6.

The left side of A, B, and C in Fig. 7 shows the points of maximal curvature on a curvature plot of the 0.2 to 0.5 interval of the endocranial outline. The right side of A, B, and C in Fig. 7 is the predicted endocranial outline (using eleven EFF harmonics), which is plotted in 0.01 increments in the clockwise direction. With the 0.01 increments there are 200 points total. The x and y coordinates of the predicted endocranial plots were computed with a c++ program I wrote for this purpose. This program uses EFF harmonics (which were computed in EFF23) as raw data. The X in the endocranial plots identifies the first x-y coordinate produced by the c++ program. The predicted endocranial plots are used only to show where maximal curvature points are on the 0.2 to 0.5 interval of the endocranial outline.

Using this approach, we observe that the maximum curvature in the 0.2 to 0.5 radians/π interval is a lot less for the *Homo erectus* fossil compared to the other fossil and modern human samples (Fig. 8). Interestingly, the difference between the fossils and modern humans in the location and magnitude of maximal curvature on the 0.2 and 0.5 radians/π interval of the endocranial outline is statistically significant in the one-way MANOVA (Wilks' Lambda = 0.2737, p-value = 0.001535), using the method of Nath and Pavur (1985). Much of the difference results from the position of maximal endocranial curvature (in the 0.2 to 0.5 radians/π interval) of four of the five fossils interval being anterior to that of all of the modern humans (Fig. 8).

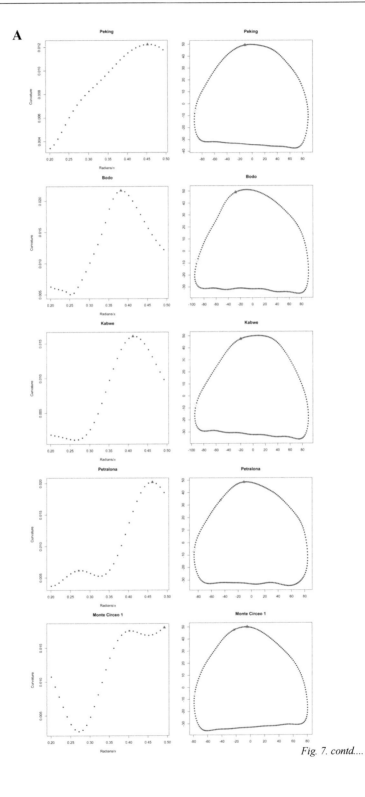

Fig. 7. contd....

Fig. 7. contd.

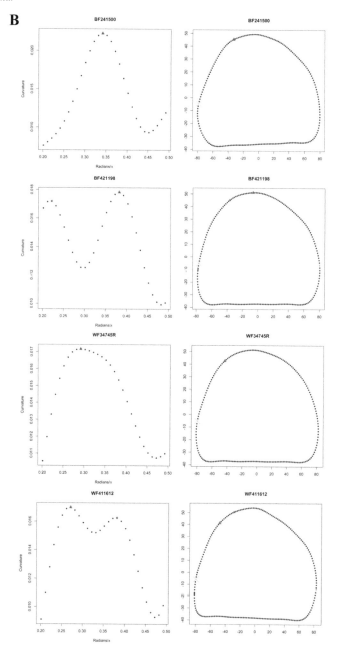

Fig. 7. (A) Curvature on the 0.2 to 0.5 radians/π interval of the endocranial outline (left) and predicted endocranial outline (right) for the fossil sample. (B) and (C) Curvature on the 0.2 to 0.5 radians/π interval of the endocranial outline (left) and predicted endocranial outline (right) for the modern human sample. In (A), (B), and (C) the point of maximal curvature is indicated by a triangle, and a square indicates the point of second maximal curvature.

Fig. 7. contd....

C

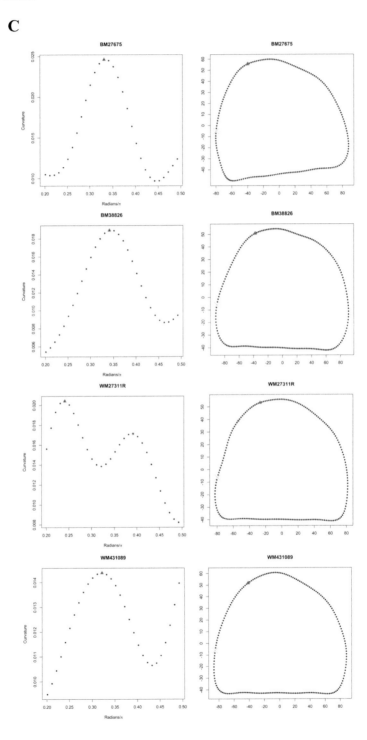

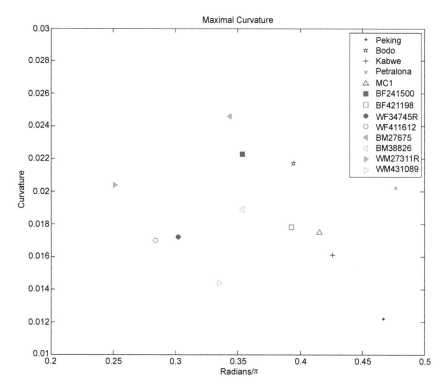

Fig. 8. Plot of maximal curvature on the 0.2 to 0.5 radians/π interval for the fossil and modern human samples.

Conclusion

The curvature-based method introduced in this chapter provides a powerful analytical tool for the study of two-dimensional curves. It is useful for studying global shape characteristics of two-dimensional curves, but unlike other morphometric techniques, it provides a geometrically direct way to study local shape patterns in two-dimensional curves, with the only constraint being that the curve is smooth. For example, my method detects and quantifies fine-scaled patterns in shape that could be used to identify and compare growth pathways in biological systems that may be of interest to archaeologists and paleoanthropologists, in accordance with the following comment by Read.

> Though selection is the ultimate arbitrator of form, selection does not act upon random variation in form but on variation in those forms that can be realized through growth [emphasis added]. Comparison of forms, in this view, becomes comparison of forms seen as the consequence of alternate growth pathways rather than the comparison of the products of those growth pathways considered in isolation (Read 1997: 70).

In the case study used to illustrate the method, localized patterns of curvature were identified, which revealed a statistically significant difference in the positioning of the point of maximal curvature on the mid-sagittal endocranial outline adjacent to the parietal region of the brain, between a small sample of modern humans and fossil hominins. Biological and structural explanations for this interesting pattern may be revealed in future paleoanthropological studies.

Finally, the curvature-based method introduced in this chapter offers a number of significant advantages and provides information on two-dimensional curves that cannot be obtained with other elliptical Fourier function (EFF) techniques or any other morphometric methods, that include: (1) it is able to directly detect highly localized changes in the curvature of two-dimensional boundaries of shapes, which provides a highly detailed and direct representation of the form, and (2) it is based on an internally defined reference system, which in some instances may useful for understanding structure modifying processes, such as growth. In fact, Read makes a strong argument for using an internally defined reference system to study growth and form (which is not the case for any morphometric method based on coordinates, such as those that use homologous points).

> Such a representation must ultimately be linked to genetically encoded information. The genetic linkage argues against using a representation or comparison based on an externally defined reference system, such as a coordinate system, as it is difficult to imagine how an externally defined coordinate system could be genetically encoded (Read 1997: 66).

Other archaeological applications of this curvature-based method include the shape analysis and classification of artifacts with smooth outlines, such as groundstone and shell preciosities. For example, the curvature of points along two-dimensional outlines of artifacts could be used as a multivariate raw data set for a classification analysis based on shape, which I will present in a future paper.

References Cited

Baglivo, J.A. 2005. Mathematica Laboratories for Mathematical Statistics: Emphasizing Simulation and Computer Intensive Methods. Society for Industrial and Applied Mathematics, Philadelphia, Pennsylvania.

Bookstein, F., K. Schafer, H. Prossinger, H. Seidler, M. Fieder, C. Stringer, G.W. Weber, J.-L. Arsuaga, D.E. Slice, F.J. Rohlf, W. Recheis, A.J. Mariam and L.F. Marcus. 1999. Comparing frontal cranial profiles in archaic and modern homo by morphometric analysis. The Anatomical Record 257: 217–224.

Bow, S.T. 1984. Pattern Recognition. Marcel Dekker, New York, New York.

Conroy, G.C., G.W. Weber, H. Seidler, W. Recheis, D. Zur Nedden and J.H. Mariam. 2000. Endocranial capacity of the Bodo cranium determined from three-dimensional computed tomography. American Journal of Physical Anthropology 113: 111–118.

deFelice, L. 2004. A radiographic analysis of middle pleistocene hominin cranial morphology: implications for taxonomy and methodology in human evolution. M.A. Thesis, University of California, Los Angeles, Los Angeles, California.

Diaz, G., C. Cappai, M., D. Setzu, S. Sirigu and A. Diana. 1997. Elliptic Fourier descriptors of cell and nuclear shapes. pp. 307–321. *In*: P.E. Lestrel (ed.). Fourier Descriptors and their Applications in Biology. Cambridge University Press, New York, New York.

Duncan, J., R. Owen, P. Anandan, L. Staib, T. McCauley, A. Salazar and F. Lee. 1990, September. Shape-based tracking of left ventricular wall motion. In: Computers in Cardiology 1990, Proceedings. (pp. 41-44). IEEE.

Gazzaniga, M., R.B. Ivry and G.R. Mangun. 2008. Cognitive Neuroscience: The Biology of the Mind, 3rd Edition. W.W. Norton & Company, Inc., New York, New York.

Kuhl, F.P. and C.R. Giardina. 1982. Elliptic Fourier features of a closed contour. Computer Graphics and Image Processing 18: 236–258.

Lestrel, P.E. 1997. Fourier Descriptors and their Applications in Biology, Cambridge University Press, New York, New York.

Lestrel, P.E., R.M. Cesar Jr., O. Takahashi and E. Kanazawa. 2005. Sexual dimorphism in the Japanese cranial base: a Fourier-wavelet representation. American Journal of Physical Anthropology 128: 608–622.

Mancini, G.B., S.F. DeBoe, E.A. Anselmo, S.B. Simon, M.T. LeFree and R.A. Vogel. 1987. Quantitative regional curvature analysis: an application of shape determination for the assessment of segmental left ventricular function in man. American Heart Journal: 326–334.

Mancini, G.B., M.T. LeFree and R.A. Vogel. 1985. Curvature analysis of normal ventriculograms: fundamental framework for the assessment of shape changes in man. Computers in Cardiology: 141–144.

Murrill, R.I. 1981. Petralona Man: A Descriptive and Comparative Study, with New Important Information on Rhodesian Man. Charles C. Thomas, Publisher, Springfield, Illinois.

Nath, R. and R. Pavur. 1985. A new statistic in the one way multivariate analysis of variance. Computational Statistics and Data Analysis 2: 297–315.

Pavlidis, T. 1982. Algorithms for Graphics and Image Processing. Computer Science Press, Rockville, Maryland.

Peeters, R.S., K. Nelissen, M. Fabbri-Destro, W. Vanduffel, G. Rizzolatti and G.A. Orban. 2009. The representation of tool use in humans and monkeys: common and uniquely human features. Journal of Neuroscience 29: 11523–11539.

Price, C.J. 2000. The anatomy of language: contributions from neuroimaging. Journal of Anatomy 197: 335–359.

Read, D.W. 1997. Growth and form revisited. pp. 45–73. *In*: P.E. Lestrel (ed.). Fourier Descriptors and their Applications in Biology. Cambridge University Press, New York, New York.

Rohlf, F.J. 1986. Relationships among eigenshape analysis, Fourier analysis and analysis of coordinates. Mathematical Geology 18: 845–854.

Sakai, K.L. 2005. Language acquisition and brain development. Science 310: 815–819.

Schultz, B.K. 1931. Der innenraum des schaedels in stammesgeshichtlicher betrachtung mit besonderer beruecksichtigung des rhodesiafundes. Verhandlungen der Gesellschaft fuer Physische Anthropologie 5: 30–38.

Silipo, P., M. Dazzi, M. Feliciani, G. Guglielmi, G. Guidetti, S. Martini, D. Massani, S. Mori and G. Tanfani. 1991. Computerized tomographic analysis of the Neanderthal cranium of the Guattari cave. pp. 513–538. *In*: M. Piperno and G. Sachilone (eds.). The Circeo 1 Neanderthal Skull, Studies and Documentation. Libreria dello Stato, Rome.

Skinner, M.F. and G.H. Sperber. 1992. Atlas of the Radiographs of Early Man. Liss, New York, New York.

Teh, C.-H. and R.T. Chin. 1989. On the detection of dominant points on digital curves. IEEE Transactions on Pattern Analysis and Machine Intelligence 11: 859–872.

Weidenreich, F. 1943. The skull of *Sinanthropus pekinensis*; a comparative anatomy on a primitive hominid skull. Lancaster Press, Inc., Lancaster, Pennsylvania.

Wolfe, C.A., P.E. Lestrel and D.W. Read. 1999. EFF23 2-D and 3-D elliptical Fourier functions. Getting Started, Van Nuys, California.

8

Logratio Analysis in Archeometry: Principles and Methods

Josep A. Martín-Fernández,[1,*] *Jaume Buxeda i Garrigós*[2] *and Vera Pawlowsky-Glahn*[3]

Introduction

Compositional data (CoDa) quantitatively describe the components of some whole and these components are usually termed *parts*. According to the experimental field, CoDa appear as vectors of percentages, parts per unit, parts per million, or other non-closed units such as molar concentrations or absolute frequencies. The units used are irrelevant, because the total sum of the vector is not informative, i.e., the information is relative rather than absolute, and lies in the ratios of the parts. For example, in Archeometry the total weight of a material sample is irrelevant when the aim is to analyze the chemical composition of an artifact. Formally, CoDa are considered to be equivalence classes. They thus embrace scale invariance and the sample space of CoDa, the simplex, is $S^D = \{ \mathbf{x} \in \mathbb{R}_+^D : \sum_{j=1}^D x_j = k \}$, where the value of the constant k is irrelevant—a popular choice is $k = 1$—, and D is the number of parts.

Standard data analysis techniques applied to CoDa carry technical and conceptual problems and may result in misleading conclusions. Over recent decades, numerous new

[1] Dept. Computer Science, Applied Mathematics, and Statistics, Campus, Montilivi, Edif. P-IV, Universitat de Girona (UdG), Girona (E-17071) Spain.
  Email: josepantoni.martin@udg.edu
[2] Cultura Material i Arqueometria UB (ARQ|UB, GRACPE), Dept. de Prehistòria, Història Antiga i Arqueologia (Despatx 1029), Facultat de Geografia i Història, Universitat de Barcelona, C/de Montalegre, 6, 08001 Barcelona (Catalonia, Spain).
  Email: jbuxeda@ub.edu
[3] Dept. Computer Science, Applied Mathematics, and Statistics, Campus Montilivi, Edif. P-IV, Universitat de Girona (UdG), Girona (E-17071) Spain.
  Email: vera.pawlowsky@udg.edu
* Corresponding author

ideas and strategies for analyzing CoDa were presented at the five CoDaWork meetings (e.g., Hron et al. 2013) and collected in special publications, such as the recent review of the state-of-the-art, which provides a detailed account of these issues, edited by Pawlowsky-Glahn and Buccianti (2011). The classic monograph by Aitchison (1986) introduced the first consistent methodological proposal to deal with CoDa. The focal point of this methodology is the statement of the principles of CoDa analysis, the core of the statistical analysis of logratios. According to Mateu-Figueras et al. (2011), logratio analysis can be reduced to three steps, termed principle of working in coordinates. Step 1: represent CoDa in logratio type coordinates; (2) apply usual statistical analysis to the coordinates as real random variables; and (3) interpret results in coordinates and/ or in terms of the original units. In the following section, some anomalous behavior of the standard statistical methods when applied to raw archeometric CoDa is illustrated. Next, the basic principles of CoDa analysis are introduced, and geometric settings are presented. Afterwards, descriptive elements and techniques of logratio analysis are defined and applied to a typical dataset. Final remarks are given in the last section.

Standard Statistical Methods Applied to Raw Archeometric CoDa

According to Baxter and Freestone (2006), the most common archeometric CoDa set is the chemical concentration of D parts (oxides/elements) determined from n artifacts. The set of D parts corresponding to the measured chemical concentrations will be referred to from here on as full-composition or as sub-composition if it is a normalized subset of them. In the following, the customary statistical methods, applied to some such datasets, are given. Figure 1 shows the ternary diagram for a full-compositional dataset (Baxter

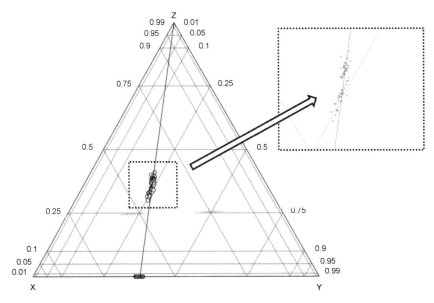

Fig. 1. CoDa set (Baxter and Freestone 2006, Table 1) in the ternary diagram. Empty circles are the [X, Y, Z] full-composition. Filled circles on the edge of parts X and Y are the $[s_X, s_Y]$ sub-composition. The continuous line is the sub-compositional projection line. The dotted square shows a zoom of the dataset.

and Freestone 2006, Table 1) involving three parts [X, Y, Z]. According to standard criteria, the large variability appears in the Z part. The continuous line represents the projection line from the simplex S^3 of the full-composition [X, Y, Z] to the simplex S^2 of the sub-composition $[s_X, s_Y]$, which shows little variability. Formally the sub-composition $[s_X, s_Y]$ is obtained by applying the closure operation $\mathbb{C}$ to the raw parts [X, Y], i.e. $[s_X, s_Y] = \mathbb{C}([X, Y]) = [X/(X+Y), Y/(X+Y)]$. Akin to any projection in any Euclidean space, the sub-composition operation generally produces a loss of information and this effect, for example, could be crucial in statistical analysis, as existing clusters in the full-composition may collapse into a sub-composition.

When only the typical scatter plot of the raw parts [X, Y] is observed (Fig. 2), a large variability of these parts may be inferred, which is a typical example of a misleading interpretation. The small dots in the ternary diagram are the full-composition. The vertical dotted lines represent its projection to the plane of the raw parts [X, Y]. The sub-composition $[s_X, s_Y]$ is represented on the edge of parts X and Y. Note that, due to the fact that X+Y+Z= 1, when the raw parts [X, Y] are plotted, the information of Z is actually included in the relationship as well.

One can assume that concentrations X and Y are, respectively, obtained from the closure of quantities W_X and W_Y against a total T, i.e. $X = W_X/T$ and $Y = W_Y/T$. Consequently, part Z is a residual part, $Z = 1-X-Y$, and $W_Z = T-W_X-W_Y$, where T= $W_X+W_Y+W_Z$. From this point of view, one can see that the raw parts [X, Y] = [W_X/T, W_Y/T] include a spurious relationship between X and Y, a misleading effect which was first mentioned by Pearson (1897). In other words, the typical correlation coefficient r_{XY}= 0.83 includes the effect of the residual part Z. On the other hand, the sub-composition

$$[s_X, s_Y]= \left[\frac{X}{X+Y}, \frac{Y}{X+Y}\right] = \left[\frac{W_X/T}{W_X/T+W_Y/T}, \frac{W_Y/T}{W_X/T+W_Y/T}\right] = \left[\frac{W_X}{W_X+W_Y}, \frac{W_Y}{W_X+W_Y}\right]$$

is completely free of the spurious influence of the residual part Z. However, again because of the relationship induced by closure, it makes no sense to calculate the usual correlation coefficient between both parts ($r_{s_X s_Y}$).

Table 1. Variation array of Romano–British pottery dataset (Tubb et al. 1980). Lower triangle: logratio means; Upper triangle: logratio variances. The bottom row contains the center of the dataset; the right column the clr-variances. Largest and smallest values in bold.

	Al_2O_3	Fe_2O_3	MgO	CaO	Na_2O	K_2O	TiO_2	MnO	BaO	clrVar
Al_2O_3		0.50	0.76	2.03	0.82	0.19	0.26	2.75	**0.05**	0.36
Fe_2O_3	−1.12		0.27	0.93	0.29	0.27	0.78	1.09	0.40	**0.05**
MgO	−2.04	−0.92		1.70	0.65	0.30	0.97	1.04	0.55	0.24
CaO	−4.05	−2.94	−2.01		0.93	1.74	2.50	1.31	1.97	1.00
Na_2O	−4.45	−3.33	−2.41	**−0.40**		0.63	1.15	1.22	0.71	0.25
K_2O	−1.62	−0.51	0.42	2.43	2.83		0.44	1.91	0.14	0.17
TiO_2	−2.96	−1.84	−0.92	1.09	1.49	−1.34		**3.08**	0.29	0.60
MnO	−5.90	−4.79	−3.86	−1.85	−1.45	−4.28	−2.94		2.34	**1.18**
BaO	**−6.84**	−5.72	−4.80	−2.79	−2.39	−5.22	−3.88	−0.93		0.26
Cen(**X**) (%)	**57.54**	18.80	7.47	1.00	0.67	11.33	2.98	0.16	**0.06**	

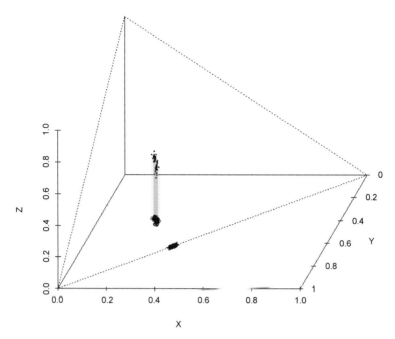

Fig. 2. CoDa set (Baxter and Freestone 2006, Table 1) in the ternary diagram (dots). Vertical lines represent a typical projection on the plane of the raw parts [X, Y] (circles). Points on the edge of parts X and Y correspond to the sub-composition $[s_X, s_Y]$.

To the contrary of the differences X–Y and s_X–s_Y, the ratios $X/Y = s_X/s_Y = W_X/W_Y$, provide the same information in both the full and sub-composition. Note that, the closure $\mathbb{C}$ is not a required operation when analyzing a ratio. This property suggests a way of analyzing CoDa and avoiding spurious correlation. Indeed, as the total sum of a vector is irrelevant, the information carried in any D-composition $\mathbf{x} = [x_1, x_2, ..., x_D]$ is the same as in $c \cdot \mathbf{x} = [c \cdot x_1, c \cdot x_2, ..., c \cdot x_D]$ $\forall c > 0$. In particular, the vectors of D–1 ratios $[x_1/x_D, x_2/x_D, ..., x_{(D-1)}/x_D, 1]$ or $[x_1/x_D, x_2/x_D, ..., x_{(D-1)}/x_D]$ completely determine a composition $\mathbf{X}$ (Aitchison 1986). For example, the full-composition [X, Y, Z] is completely determined by the vector of ratios [X/Z, Y/Z] and the ratio X/Y determines the sub-composition $[s_X, s_Y]$. Figure 3 shows the scatter plot of ratios [X/Z, Y/Z]. Observe that the cloud is very similar to the cloud of the full-composition in the ternary diagram (Fig. 1). In the margins, the boxplots of the ratios are also plotted. Note that the slight skewness would suggest using logarithms to improve the symmetry of the ratios.

The Main Principles and Geometric Settings of CoDa Analysis

Based on the difficulties of typical statistical techniques, Aitchison (1986) introduced two main principles for analyzing CoDa: *scale invariance* and *sub-compositional coherence*. Scale invariance means that vectors with proportional positive components represent the same composition. In other words, a D-composition $\mathbf{x}$ is an equivalence class whose unitary ($k = 1$) representation is $\mathbb{C}(\mathbf{x}) = \left[\dfrac{x_1}{\sum x_j}, \dfrac{x_2}{\sum x_j}, ..., \dfrac{x_D}{\sum x_j} \right]$.

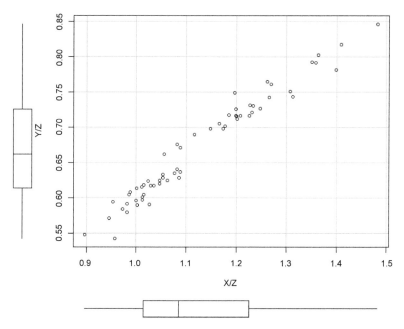

Fig. 3. Scatter plot of vector of ratios [X/Z, Y/Z].

According to Aitchison (1986), scale invariance is the equivalent of stating that "any meaningful function of a composition can be expressed in terms of a set of component ratios". From a statistical point of view, this property is crucial when defining a distance function or a probability density function on the simplex. Sub-compositional coherence means that interpretations and results provided by any analysis of a subset of parts must not depend on the rest of parts or lead to results contradicting an analysis of the full composition. Taking a sub-composition in the simplex is the analogue to taking a marginal in real space. In real space, geometrically speaking, one projects the dataset onto a sub-space and the distances become smaller. Typical statistical analysis based on Euclidean distance or multivariate normal distribution violates these principles when applied to raw CoDa (e.g., Chacón et al. 2011, Palarea-Albaladejo et al. 2012). One explanation for this is that typical statistical techniques are based on measuring absolute differences $\mathbf{x}-\mathbf{y}$. On the other hand, working with differences based on ratios $\mathbf{x}/\mathbf{y}$ automatically involves scale invariance and sub-compositional coherence. Consequently, the natural movement from composition $\mathbf{x}$ to composition $\mathbf{y}$ should be based on a component wise product. This operation is termed perturbation and was defined (Aitchison 1986) as $\mathbf{x} \oplus \mathbf{y} = \mathbb{C}\left[x_1 \cdot y_1, \ldots, x_D \cdot y_D\right]$; the perturbation difference is $\mathbf{x} \ominus \mathbf{y} = \mathbb{C}\left[x_1/y_1, \ldots, x_D/y_D\right]$, and the neutral element is the composition $\mathbf{n} = [1,1,\ldots,1]$. The second operation that forms the Aitchison geometry of the simplex (Pawlowsky-Glahn and Egozcue 2001) is the powering of a composition $\mathbf{x}$ by a real number α defined $\alpha \odot \mathbf{x} = \mathbb{C}\left[x_1^{\alpha}, \ldots, x_D^{\alpha}\right]$. Perturbation and powering defined in the simplex satisfy the principles of compositional analysis and define a vector space structure of dimension $D-1$ (Pawlowsky-Glahn and Egozcue 2001). Pawlowsky-Glahn and Buccianti (2011) illustrate the role played by both operations in the most common statistical models along with its usefulness in many applications in different fields. In Archeometry, the alteration of ceramics is a problem that can be modeled by a perturbation process (e.g., Buxeda 1999). For example, $[X_1, Y_1, Z_1] = [44.52, 25.43, 30.05]$ and $[X_{61}, Y_{61}, Z_{61}] = [36.70, 22.41, 40.90]$ are, respectively, the compositions with the minimum and maximum concentration in the residual part Z of the dataset analyzed in the previous section (Baxter and Freestone 2006, Table 1). In a simple exercise, consider the preceding compositions as the result of a continuous alteration of the initial composition. In this case the non normalized/closed perturbation difference vector is equal to [0.824, 0.881, 1.36], whose interpretation is, respectively, a 17.6% and 11.9% decrease of X and Y, and a 36% increment of Z, and whose unitary representation is [26.88, 28.74, 44.38].

Once a ratio x/y between two concentrations is computed, the result is within the interval $[0, +\infty)$. The interval $(0, 1)$ corresponds to the "$x < y$" case and the interval $(1, +\infty)$ to the opposite "$x > y$". This asymmetry recommends using logarithms. In addition, dealing with ratios is more complex than with logratios because "ratio" is a multiplicative way of thinking. On the other hand, dealing with "logratios" consists of the typical additive way of computation: $\ln(x/y) = \ln(x) - \ln(y)$. Given a D-composition $\mathbf{x} = [x_1, x_2, \ldots, x_D]$, the general logratio is defined as

$$\ln \left(\prod_{j=1}^{D} x_j^{\alpha_j} \right) = \sum_{j=1}^{D} \alpha_j \cdot \ln(x_j), \tag{1}$$

where only when $\Sigma \alpha_j = 0$ the logratio is a scale invariant function. In this case, expression (1) is known as a log-contrast (Aitchison 1986). Probably, the most frequently applied log-contrast in the literature is the ith centered logratio (clr_i) whose expression applied to a composition $\mathbf{x}$ is

$$\text{clr}_i(\mathbf{x}) = -\frac{\ln(x_1)}{D} - \cdots - \frac{\ln(x_{i-1})}{D} - \frac{(1-D) \cdot \ln(x_i)}{D} - \frac{\ln(x_{i+1})}{D} - \cdots - \frac{\ln(x_D)}{D} = \ln \left(\frac{x_i}{g_m(\mathbf{x})} \right) \tag{2}$$

where $g_m(\mathbf{x})$ is the geometric mean of $\mathbf{x}$. When clr_i log-contrast is applied to all parts, the vector of clr-coefficients (Aitchison 1986) is obtained: $\text{clr}(\mathbf{x}) = (\text{clr}_1(\mathbf{x}), \ldots, \text{clr}_D(\mathbf{x}))$, whose inverse transformation $\mathcal{C}(\exp(\text{clr}(\mathbf{x})))$ gives the unitary representation of $\mathbf{x}$. The vector space structure defined by perturbation and powering operations can be easily extended to produce a metric vector space. This extension is based on the Aitchison distance, d_a, defined as $d_a(\mathbf{x},\mathbf{y}) = d(\text{clr}(\mathbf{x}),\text{clr}(\mathbf{y}))$, where $d(\cdot,\cdot)$ denotes the Euclidean distance in R^D. The norm and inner product, consistent with the Aitchison distance, are respectively defined by $\|\mathbf{x}\|_a = d_a(\mathbf{x},\mathbf{n})$ and $<\mathbf{x}, \mathbf{y}>_a = \Sigma_{j=1}^{D} \text{clr}_j(\mathbf{x}) \cdot \text{clr}_j(\mathbf{y})$, leading to a Euclidean space structure (Pawlowsky-Glahn and Egozcue 2001). All these metric elements satisfy the principles of CoDa analysis, and suggest exploiting the well-known properties of Euclidean spaces: orthonormal basis, orthogonal projections, angles, ellipses, etc. These properties are the basis of multivariate statistical techniques such as cluster analysis, principal component analysis, linear regression or discriminant analysis. In Pawlowsky-Glahn and Buccianti (2011), several contributions present the state-of-the-art of these techniques, including basic elements of simplicial linear algebra and geometry, differential calculus and statistical modeling.

An important step in using these statistical techniques is to build orthonormal bases in the simplex and express any composition $\mathbf{x}$ in the corresponding coordinates. Let $\mathbf{e}_1, \mathbf{e}_2, \ldots, \mathbf{e}_{D-1}$ be an orthonormal basis of S^D. The orthonormal coordinates of a composition $\mathbf{x}$ are obtained using the *isometric logratio* function $\text{ilr}(\mathbf{x}) = [<\mathbf{x}, \mathbf{e}_1>_a, \ldots, <\mathbf{x}, \mathbf{e}_{D-1}>_a]$. It can be proven (Egozcue et al. 2003) that these coordinates are log-contrasts

(Eq. 1) and isometric, $d_a(\mathbf{x}, \mathbf{y}) = d(ilr(\mathbf{x}), ilr(\mathbf{y}))$. As these orthonormal coordinates are indeed log-contrasts, it is advisable to build log-contrasts which are capable of adequately interpreting the problem studied. A Sequential Binary Partition (SBP) of the parts of a composition consists of $D-1$ steps, where an orthonormal coordinate, now called balance, is built into each step of the partition. A practical implementation, and a representation in a dendrogram-like structure, can be found in Thió-Henestrosa et al. (2008) or Pawlowsky-Glahn and Egozcue (2011). In the first step, an SBP splits the parts of the composition $\mathbf{x}$ into two groups, which are indicated by +1 and −1. In consecutive steps, each previously created group of parts is split again into two groups. The partition process ends when the groups are made up of a unique part. Each SBP step generates one log-contrast of an orthonormal basis, i.e., one orthonormal coordinate. These coordinates are called balances because they have a very peculiar log-contrast expression. In the jth step of an SBP, denoting as $\mathbf{x}+$ the group of r parts marked with a +1 and as $\mathbf{x}-$ the group of s parts marked with a −1, the corresponding balance, b_j, is

$$b_j = \sqrt{\frac{r \cdot s}{r+s}} \ln\left(\frac{g_m(\mathbf{x}+)}{g_m(\mathbf{x}-)}\right), \tag{3}$$

where $g_m(\cdot)$ is the geometric mean of the involved parts of $\mathbf{x}$. Balances have an easy interpretation because they are the logratios of the geometric means of the groups of parts. Table 2 shows the SBP used in the analysis of the dataset from Tubb et al. (1980). This dataset consists of the concentrations in a sub-composition of 9 chemical elements for 48 samples of Romano–British pottery. The first step in the SBP consists of separating the chemical elements Al_2O_3, K_2O, TiO_2 and BaO (+1) from the rest (−1). In other words, the balance b_1 is

$$b_1 = \sqrt{\frac{20}{9}} \ln\left(\frac{\sqrt[4]{Al_2O_3 \cdot K_2O \cdot TiO_2 \cdot BaO}}{\sqrt[5]{Fe_2O_3 \cdot MgO \cdot CaO \cdot Na_2O \cdot MnO}}\right). \tag{4}$$

In the fifth step, the parts Fe_2O_3, MgO, CaO, Na_2O, and MnO that are in the denominator of balance b_1, are split into the groups $\{Fe_2O_3, CaO, Na_2O\}$ and $\{MgO, MnO\}$ to define balance b_5

$$b_5 = \sqrt{\frac{6}{5}} \ln\left(\frac{\sqrt[3]{Fe_2O_3 \cdot CaO \cdot Na_2O}}{\sqrt[2]{MgO \cdot MnO}}\right).$$

The analyst is completely free to define the SBP based on their expertise and/or on a previous exploratory analysis of the dataset.

Table 2. SBP of CoDa set from Tubb et al. (1980), represented in Fig. 5 as a CoDa-dendrogram.

Balance	Al_2O_3	Fe_2O_3	MgO	CaO	Na_2O	K_2O	TiO_2	MnO	BaO
b_1	+1	−1	−1	−1	−1	+1	+1	−1	+1
b_2	+1	0	0	0	0	−1	+1	0	−1
b_3	+1	0	0	0	0	0	−1	0	0
b_4	0	0	0	0	0	+1	0	0	−1
b_5	0	+1	−1	+1	+1	0	0	−1	0
b_6	0	−1	0	+1	−1	0	0	0	0
b_7	0	−1	0	0	+1	0	0	0	0
b_8	0	0	−1	0	0	0	0	+1	0

Basic Elements for an Exploratory Analysis

In CoDa analysis, to define the basic elements mean and variability, the standard approach is modified to take into account the Aitchison geometry. Let $\mathbf{X}$ be a random composition in S^D. According to Pawlowsky-Glahn and Egozcue (2001), the variability of a random composition $\mathbf{X}$, with respect to a composition $\mathbf{x}$, is $\text{Var}(\mathbf{X}, \mathbf{x}) = \text{E}(d_a^2(\mathbf{X}, \mathbf{x}))$, where E is the usual expectation in real space. Furthermore, the expectation or center of $\mathbf{X}$ is $\text{Cen}(\mathbf{X}) = \min_{\mathbf{x} \in S^D} d_a^2(\mathbf{X}, \mathbf{x})$ and the total variance of $\mathbf{X}$ is $\text{totVar}(\mathbf{X}) = \text{E}(d_a^2(\mathbf{X}, \text{Cen}(\mathbf{X})))$. In practical terms, center and total variance are calculated using the expressions

$$\text{Cen}(\mathbf{X}) = \text{ilr}^{-1}\left(\text{E}(\text{ilr}(\mathbf{X}))\right) = \mathbb{C}[\exp(\text{E}(\ln X_1)),\ldots,\exp(\text{E}(\ln X_D))],$$

$$\text{totVar}(\mathbf{X}) = \frac{1}{D}\sum_{i=1}^{D-1}\sum_{j=i+1}^{D} \text{Var}\left(\ln\frac{X_i}{X_j}\right) = \sum_{j=1}^{D-1} \text{Var}\left(\text{ilr}_j(X)\right) = \sum_{j=1}^{D} \text{Var}\left(\text{clr}_j(X)\right),$$

where an estimate of $\exp(\text{E}(\ln(\cdot))$ is the geometric mean of a part $g(\cdot)$. Therefore, $\text{Cen}(\mathbf{X})$ is the unitary representation of the geometric mean of $\mathbf{X}$. The variation array (Aitchison 1986) is a very informative way to present logratio expectations and variances. The lower triangle of the array contains the values of the sample means of the logratios of the corresponding two parts (numerator by row, denominator by column). The upper triangle of the array contains the sample variances of the same logratios. This array can be easily extended by adding a column to the right, which collects the values of the variances of the clr-coefficients, and a row at the bottom for the Cen($\mathbf{X}$). Table 1 shows the variation array for the Romano–British pottery dataset (Tubb et al. 1980). The sum of the right column equals totVar and is 4.11. Note that MnO and CaO are the elements showing the largest relative variability, while Fe_2O_3 and K_2O have little variability. With the logratios, $\text{Var}(\ln(Al_2O_3/BaO)) = 0.05$ is the smallest, suggesting that these chemical elements are approximately proportional in the sample (Aitchison 1986). On the other hand, MnO and TiO_2 have the largest logratio variance, suggesting slight or no association between them. Observe that the center in the bottom row suggests that Al_2O_3 depicts the largest percentages, whereas the concentrations of BaO are very small. As these elements are approximately proportional and $\text{E}(\ln(Al_2O_3/BaO))$ = −6.84, one can assume that on average $Al_2O_3 \approx e^{-6.84} \cdot BaO$. Observe that the sign of $\text{E}(\ln(\cdot/\cdot))$ indicates which element shows higher concentrations, while a value close to zero suggests that both elements are similarly present in the artifacts. For example, $\text{E}(\ln(K_2O/MgO)) = 0.42$ indicates that, on average, K_2O has a slightly higher presence than MgO and because the corresponding logratio variance equals 0.30, one can expect that many artifacts contain more K_2O content than MgO. Finally, note that the major elements (Al_2O_3, Fe_2O_3, and K_2O) have a comparable logratio variance to some trace elements (BaO, Na_2O). The trace elements MnO and CaO show the largest logratio variances, suggesting an important role in the analysis.

The variation array provides information about relationships between pairs of chemical elements; however, more complex associations may exist in a random composition. According to Aitchison and Greenacre (2002), the clr-biplot of a CoDa set is an appropriate graphical tool for analyzing these associations. The clr-biplot represents a projection (usually bidimensional) of the clr-logratio coefficients of samples in the same plot as the projection of the centered clr variables. The coordinates of samples and variables in the plot are calculated using elements provided by a Singular Value Decomposition (SVD) of the clr-coefficients data matrix. The principal elements are the clr transformation of the vectors of an orthonormal basis of the simplex. SVD associates one singular value to each of these vectors, where the squares of the singular values add up to the total variance. As the order of the vectors in the basis decreases according to the singular value, by taking the two first vectors one can calculate the proportion of variance retained by the clr-biplot. The coordinates of the samples represented in the clr-biplot are the coordinates of the centered data, with respect to the first two vectors of the orthonormal basis. Consequently, the Euclidean distance between two samples in the clr-biplot is an approximation of the Aitchison distance between the corresponding compositions in the simplex.

In addition to the typical properties of a biplot, Aitchison and Greenacre (2002) introduce particular features for CoDa sets. To illustrate these particularities, the clr-biplot for the Romano–British pottery dataset (Tubb et al. 1980) is shown (Fig. 4). Note that the proportion of variability retained by the first two axis of the clr-biplot is 85.03%. The squared length of a ray is proportional to the clr-variance of the corresponding chemical element. Observe that the shortest ray is $clr(Fe_2O_3)$ and the longest is $clr(MnO)$; which is in agreement with the values collected in Table 1. The squared length of the link between two vertices of rays is approximately equal to the variance of the logratio of the parts corresponding to the numerator of the clr-rays. In this sense, the closer the vertices are, the more proportional the concentrations of the chemical elements are. Observe (Table 1, Fig. 4) that the closest vertices are those of Al_2O_3, BaO, K_2O, and TiO_2. The cosine of the angle between two links approaches the correlation coefficient

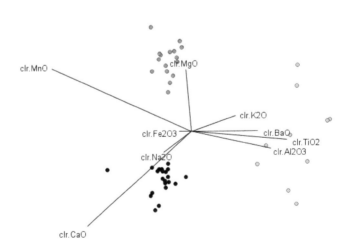

$$Z_{clr} = U(V\Gamma)^T$$

Fig. 4. Clr-biplot (1st and 2nd axis) of Romano–British pottery dataset (Tubb et al. 1980). Red lines are the rays of the clr-variables. Colors (black, dark gray, light gray) represent three groups provided by the hierarchical Ward method with squared Aitchison distance.

between the corresponding simple logratios. Consequently, the orthogonality of the links suggests that there is no correlation of the logratios. For example, the linear correlation coefficient between $\ln(Al_2O_3/Fe_2O_3)$ and $\ln(MgO/Na_2O)$ equals 0.06, accordingly the link between the vertices of the rays $clr(Al_2O_3)$ and $clr(Fe_2O_3)$ is approximately orthogonal to the link of $clr(MgO)$ and $clr(Na_2O)$ vertices.

The clr-biplot suggests the existence of clusters which are also detected using the hierarchical Ward method with squared Aitchison distance. Observe that groups 1 and 2 are separated by the first axis, whereas both are separated from group 3 by the second axis. In other words, group 3 is on the positive part of the first axis, and samples from groups 1 and 2 have negative first coordinates. Group 2 is on the positive part of the second axis, and samples from group 1 have negative second coordinates.

The log-contrast of the first axis is

$$\ln\left(\frac{Al_2O_3^{0.35} \cdot K_2O^{0.19} \cdot TiO_2^{0.42} \cdot BaO^{0.29}}{Fe_2O_3^{0.05} \cdot MgO^{0.02} \cdot CaO^{0.45} \cdot Na_2O^{0.12} \cdot MnO^{0.61}}\right),$$

which, in terms of the original chemical elements, has a difficult and vague interpretation. Indeed, to improve the interpretation of the orthonormal log-contrast provided by the SVD, a specific SBP has been selected (Table 2).

A CoDa-dendrogram is a plot that summarizes the structure of the SBP, the ilr decomposition of the total variance, and the mean and dispersion of each balance. Figure 5 shows the CoDa-dendrogram for the Romano–British pottery dataset (Tubb et al. 1980), following the SBP shown in Table 2. The lengths of the vertical bars, which connect two groups of parts, are proportional to the variance of the balance. Observe that, in this case, the first balance has the largest variance. This first balance is the interpretation of a logratio of two geometric means of groups of parts (Eq. 4). Anchored on each vertical bar there is a horizontal bar. The contact of the vertical bar with the horizontal bar is the mean balance (coordinate of the sample centre). For the first balance, the mean equals 0.98 and the contact point of the vertical bar of the first balance is close to the middle of the horizontal bar. On the other hand, the mean of

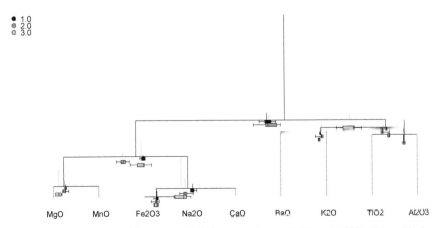

Fig. 5. CoDa-dendrogram of the Romano–British pottery dataset (Tubb et al. 1980). Colors (black, dark gray, light gray) represent three groups provided by the hierarchical Ward clustering method with squared Aitchison distance.

the second balance ($b_2 = \ln\left(\frac{\sqrt[2]{Al_2O_3 \cdot TiO_2}}{\sqrt[2]{K_2O \cdot BaO}}\right)$) equals 2.75 and the contact point is on the right. On each horizontal bar a boxplot of quantiles (0.05, 0.25, 0.50, 0.75, 0.95) of the corresponding balance for each group is represented to visualize sample dispersion. For instance, the location of the boxplots of the first balance indicates that samples from group 3 have positive coordinates, but samples from the other groups take negative values. The definition of the log-contrast of the first balance (Eq. 4) suggests that samples from group 1 have, on average, a higher concentration of elements Al_2O_3, K_2O, TiO_2, and BaO than of the rest of the chemical elements.

Conclusion

CoDa frequently appear in many archeometric studies. Particular characteristics of CoDa require a coherent statistical analysis in order to avoid misleading results and conclusions. Aitchison, in the 1980s, established the principles of CoDa analysis, i.e., the basis of a methodology free of spurious correlation which relies on the study of logratios. In this sense, compositions (e.g., concentrations in percentages) are represented by logratio orthonormal coordinates that can be studied with the usual statistical methods. The use of coordinates requires interpreting log-contrasts to obtain conclusions. These interpretations are made easier using an appropriate SBP based on the expertise of the analyst and/or on a previous exploratory analysis.

As in many other data analysis techniques, logratio requires complete datasets. When measuring chemical concentrations, often some elements are not present in sufficient concentrations and measuring instruments report them as values below detection limits; usually labeled as "<DL". In these scenarios, data processing software will not compute logratio coordinates, thus calling for imputation strategies for the "<DL" values. In the literature this issue is also known as the rounded zero problem. The imputation strategies complete a data matrix, replacing non-detected values by reasonable estimates, allowing the computation of any log-contrast and any multivariate data analysis. The interested reader can refer to Palarea-Albaladejo and Martín-Fernández (2013), who encompass the recent advances in this area.

The programming of the data analyses discussed in this work was carried out using own R routines (R development core team 2013) and the open-source package CoDaPack (Comas-Cufí and Thió-Henestrosa 2011). Other useful R packages for performing CoDa analysis are "compositions" and "robCompositions". Computer routines implementing the methods, as well as other related compositional techniques, can be obtained from the website http://www.compositionaldata.com.

Acknowledgments

This research has been supported by the Spanish Ministry of Economy and Competitiveness under the project "METRICS" Ref. MTM2012–33236; and by the Agència de Gestió d'Ajuts Universitaris i de Recerca of the Generalitat de Catalunya under the project "COSDA" Ref. 2014SGR551.

References Cited

http://www.compositionaldata.com.

Aitchison, J. 1986. The Statistical Analysis of Compositional Data. Monographs on Statistics and Applied Probability. Chapman and Hall Ltd (reprinted 2003 with additional material by The Blackburn Press), London (UK).

Aitchison, J. and M. Greenacre. 2002. Biplots for compositional data. Applied Statistics 51(4): 375–392.

Baxter, M.J. and I.C. Freestone. 2006. Log–ratio compositional data analysis in archaeometry. Archaeometry 48(3): 511–531.

Buxeda i Garrigós, J. 1999. Alteration and contamination of archaeological ceramics–the perturbation problem. Journal of Archaeological Science 26: 295–313.

Chacón, J.E., G. Mateu–Figueras and J.A. Martín–Fernández. 2011. Gaussian kernels for density estimation with compositional data. Computer & Geosciences 37: 702–711.

Comas-Cufí, M. and S. Thió-Henestrosa. 2011. CoDaPack 2.0: a stand-alone, multi-platform compositional software. *In*: J.J. Egozcue, R. Tolosana-Delgado and M.I. Ortego (eds.). CoDaWork'11: 4th International Workshop on Compositional Data Analysis. Sant Feliu de Guíxols.

Egozcue, J.J., V. Pawlowsky–Glahn, G. Mateu–Figueras and C. Barceló–Vidal. 2003. Isometric logratio transformations for compositional data analysis. Mathematical Geology 35(3): 279–300.

Hron, K., P. Filzmoser and M. Templ (eds.). 2013. Proceedings of the 5th International Workshop on Compositional Data Analysis, Codawork'13 June 3–7, Vorau, Austria. ISBN: 978-3-200-03103-6.

Mateu-Figueras, G., V. Pawlowsky-Glahn and J.J. Egozcue. 2011. The principle of working on coordinates and compositional data analysis. *In*: V. and A. Compositional Data Analysis: Theory and Applications. Wiley & Sons (UK).

Palarea-Albaladejo, J. and J.A. Martín-Fernández. 2013. Values below detection limit in compositional chemical data. Analytica Chimica Acta 764: 32–43.

Palarea–Albaladejo, J., J.A. Martín–Fernández and J.A. Soto. 2012. Dealing with Distances and Transformations for Fuzzy C–Means Clustering of Compositional Data. Journal of Classification 29(2): 144–169.

Pawlowsky-Glahn, V. and A. Buccianti (eds.). 2011. Compositional Data Analysis: Theory and Applications. John Wiley & Sons, Chichester (UK).

Pawlowsky–Glahn, V. and J.J. Egozcue. 2001. Geometric approach to statistical analysis on the simplex. Stochastic Environmental Research and Risk Assessment (SERRA) 15(5): 384–398.

Pawlowsky-Glahn, V. and J.J. Egozcue. 2011. Exploring Compositional Data with the Coda-Dendrogram. Austrian Journal of Statistics (1-2): 103–113.

Pearson, K. 1897. Mathematical contributions to the theory of evolution. On a form of spurious correlation which may arise when indices are used in the measurement of organs. Proceedings of the Royal Society of London LX, 489–502.

R development core team 2013. R: A language and environment for statistical computing: Vienna, http://www.r-project.org.

Thió–Henestrosa, S., J.J. Egozcue, V. Pawlowsky–Glahn, L.O. Kovács and G. Kovács. 2008. Balance–dendrogram a new routine of CoDaPack. Computer and Geosciences 34(12): 1682–1696.

Tubb, A., A.J. Parker and G. Nickless. 1980. The analysis of Romano–British pottery by atomic–absorption spectrophotometry. Archaeometry 22: 153–171.

9

An Introduction to Clustering with Applications to Archaeometry

Hans-Joachim Mucha,[1,*] *Hans-Georg Bartel,*[2] *Jens Dolata*[3] and
Carlos Morales-Merino[4]

Introduction

The aim of cluster analysis is to find sub-populations such as proveniences of pottery or of other artifacts (Kaufman and Rousseeuw 1990). Here the variability as the key property of mathematical statistics plays an important role. More precisely, the within-cluster variability should be as small as possible. It is, therefore, important to know at first the details about the internal variability (homogeneity) of the archaeological objects under consideration. In this chapter, coarse ceramics such as tiles and bricks will be thoroughly investigated. First, to get an idea about intrinsic inhomogeneity, different areas of the tile depicted in Fig. 1 have to be analyzed. For details, see Dolata and Werr (1998/99). Such an investigation provides us estimates of the variability of each chemical element of the archaeological object. This variability can be considered for the clustering step later.

[1] Weierstrass Institute for Applied Analysis and Stochastics (WIAS), Mohrenstraße 39, 10117 Berlin, Germany.
Email: mucha@wias-berlin.de
[2] Department of Chemistry at Humboldt University Berlin, Brook-Taylor-Straße 2, 12489 Berlin, Germany.
Email: hg.bartel@yahoo.de
[3] Head Office for Cultural Heritage Rhineland-Palatinate (GDKE), Office of State Archaeology, Große Langgasse 29, 55116 Mainz, Germany.
Email: dolata@ziegelforschung.de
[4] Rathgen-Forschungslabor, Schlossstraße 1A, 14059 Berlin, Germany.
Email: c.morales-merino@t-online.de
* Corresponding author

Fig. 1. *Later* of the (gionis) XXII P(rimigeniae] P(iae) ---> (io) XXII P(rimigenia] P(ia).

Specifically, this chapter will be concerned about the following main topics:

- the archaeological problem,
- cluster analysis in a nutshell,
- applications to archeometry, and
- validation of clustering results.

In the last point, the assessment of stability in cluster analysis is strongly related to the main difficult problem of determining the number of clusters present in the data. This has been the subject of many investigations and papers considering different resampling techniques as practical tools.

Archaeological Material and Questions

About 1,500 Roman stamped bricks and tiles from the Upper Rhine area have been investigated by chemical analysis. The combination of archaeological and especial epigraphic information, and the results of mineralogy and chemistry allowed archaeologists to develop a complex model of brick and tile making in Germany in Roman times. Coarse ceramics are important findings in archaeological excavations, particularly if they are marked with stamps of the manufacturing workshop. These are references concerning their chronology and function. The evaluation of stratigraphic-documented excavations is possible by systematic and comparative investigations of brickstamps.

In south-west Germany, in parts of the former Roman province *Germania Superior*, only a few large brickyards existed, whose operating authority was the Roman army. These produced all kind of building materials such as roof tiles, bricks for the underfloor heating system called hypocaust, and tubes for water supply in all public buildings. These architecture elements are very common in the Transalpine area because of the existing rougher climates when compared to the Mediterranean. Not only the commander's residence, but all functional buildings and the barracks in the major military camps in Mainz and Straßburg were tiled, and even in the winter half-year hot tube-baths, heated by hypocaust, were in full action. The period of running the brickyards was from the middle of the first century A.D. until the end of the fourth century. The locations were Rheinzabern, Straßburg-Königshofen, Windisch and Worms on the banks of the River Rhine, Frankfurt-Nied and Groß-Krotzenburg at the River Main. These brickworks belonged to different legions and were run in different periods. Several of them still have an unknown origin.

One of the most important and most investigated areas for Roman brick-production is located at the lower Main riverbank at Frankfurt-Nied (Dolata et al. 2007). The running of this site has its coherence with the construction of buildings of the Roman border zone in Upper Germany, the so-called "Obergermanischer Limes" (*Limes Germaniae Superioris*). The beginning of the production of brick in Frankfurt-Nied is to be set in the years 83/85 A.D., and the main output is that of the *legio XXII Primigenia* at the beginning of the second century A.D. More than 500 different brickstamps of this military unit used in Frankfurt-Nied can be distinguished. Archaeologists are most interested in the chronology of the production marks, which are found on the building materials.

Once the proveniences (places of manufacture) are detected by clustering, they can be characterized by their chemical fingerprint, and another important archaeological task may occur: the identification of new findings without a stamp or with a known or yet unknown stamp. However, this is not the topic of the present discussion. For detail concerning statistical identification, see Bartel et al. (2008) and Mucha et al. (2008b).

An Appetizer: from Univariate Sorting to High-dimensional Clustering

It is well known that a set of numbers is easier for a human to interpret when they are sorted. In the univariate case, cluster analysis is nothing else than reordering the set of objects based on a single variable followed by dividing the total order of objects into homogeneous regions. Especially hierarchical clustering appears self-explanatory: the corresponding dendrogram (binary tree) presents all the clusters (either disjointed or including one into the other) that are established during the agglomeration or the division process. It presents a complete hierarchy of partitions as it is shown in Fig. 2. Those algorithms are agglomerative or divisive. Agglomerative hierarchical clusterings proceed from an initial partition, in which each cluster contains a single observation, by successive merging of pairs of clusters. In contrast, divisive hierarchical clusterings start with the whole set of observations as a cluster and proceed the other way around by successive bi-partitions. Those techniques are rarely used (Späth 1980). Therefore, without loss of generality, we consider here only agglomerative hierarchical clustering.

Figure 2 shows the univariate clustering of $I = 18$ archaeological objects. In this case the hierarchical minimum variance method of Ward is used (see below and Ward 1963). This method is based on the squared Euclidean distance. These objects are tiles that were produced in two different Roman military units, namely *Cohors I Asturum* (samples H269–H274: their measurement values were published in Mucha and Ritter 2009: 122, see Fig. 3) and *Legio XXI Rapax* (samples G006, G035, G038, G160–G163, G165, G166, G167, H253, H254, see Mucha and Ritter 2009: 117–120, 122, and see also Fig. 13 below). In Fig. 2, the objects are represented by points (black or light gray

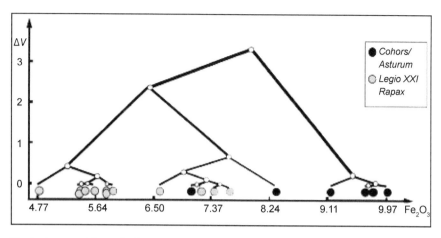

Fig. 2. A non-equidistant dendrogram of 18 measurements (in weight-%) of Fe_2O_3 located on the real line (axis of abscissae).

Fig. 3. H273: *Tegula* fragment with stamp of *COH(ors) I AS(turum)* (Archäologisches Museum Frankfurt α 23719, find spot: Frankfurt-Heddernheim, chemical analysis by Schneider (FU Berlin) based on RFA-WD, provenance: Frankfurt-Nied.

filled circles). The univariate information (variable) is the content of Fe_2O_3 that is located on the real line (axis of abscissae), and so the approximate measurement values can be taken from the picture. Each point is a terminal node (trivial cluster) in the tree. It is marked by a circle of different gray level according to its military unit. The ordinate represents the increment of within-clusters variance ΔV when merging two clusters (see below and formulae (10) and (12)). It must be pointed out here that the ΔV values have been transformed into $\ln(\Delta V + 1)$ values (logarithmic scale of the ordinate). Each non-trivial cluster (non-terminal node) is marked by a white filled circle.

In the univariate case, there is a total order, see Fig. 2. The dream of statisticians is a univariate setting (i.e., the simplest model) that explains the archaeological model. But, that is often a dream only. When moving to two and more variables the total order is usually lost (Mucha et al. 2005b). In addition, a scaling problem occurs when the variables are measured in different scales such as weight percent and ppm. Figure 4 shows the scatterplot of the two variables Fe_2O_3 (in %) and Zr (in ppm). Here the two groups of tiles look better separated than it is the case in the univariate considerations. Additionally, two lines are inserted in the plot. The solid gray arrow represents the first principal component based on the correlation. Both the dashed line and *Ward*'s bivariate clustering separates the two archaeological groups perfectly (Fig. 5). By the way, multivariate projection techniques such as principal component analysis and linear discriminant analysis allow a visualization of high-dimensional data in a low-dimensional space such as R^2 or R^3, see, for instance, Ripley (1996).

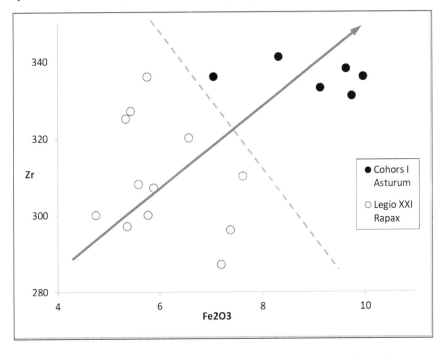

Fig. 4. Scatterplot of the two groups of bricks which are distinguished by the stamp of their military unit.

In general, the hope is that the separation of clusters can be further improved by including additional variables. Concerning more details about some investigations/problems of bottom-up variable selection the chapter of Read in this book is recommended. However, both the scatterplot (Fig. 4) and the plot-dendrogram in substitute for: Fig. 5 show considerable amount of within-group variability. This is mainly because these archaeological objects are coarse ceramics. Dolata and Werr (1998/99) investigated the variability within one tile by taking ten different samples on it. Figure 1 shows this Roman tile under investigation. The aim is to assess the importance of individual variables (chemical contents) with regard to the classification task. As a specific result, the recommendation was that the variable P_2O_5 should be descarded.

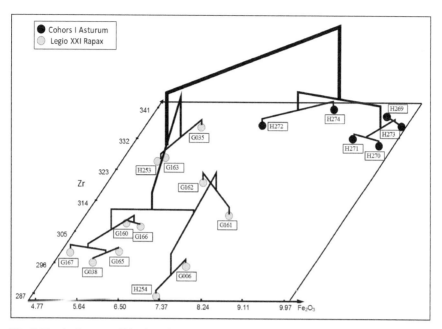

Fig. 5. Plot dendrogram of bivariate clustering.

Without any doubt, this small example shows that better results can be obtained by adding further variables. Usually, multivariate statistical methods such as cluster analysis can consider several variables simultaneously. For pairwise data clustering (see Sec. Additive Models: Minimum Variance Clustering, and Fig. 6), the good news is that the size of the distance matrix remains the same whatever the number of variables is: it is a $I \times I$ matrix as it is in the univariate case. As already mentioned, the (theoretical) hope is, that the more information is used the better are the statistical results and conclusions. In applications, however, our experience is that the model has to be chosen as simple as possible, but not simpler. That is a wide field of investigation, simply too extensive for this introduction to clustering. It is an ongoing research topic (see, for instance, Mucha 2013).

Cluster Analysis in a Nutshell

Let us introduce the problem of finding clusters of observations. Clustering the variables can be often done in a similar way, for instance, in the case of binary data or contingency tables (Greenacre 1988, Mucha 2014). Without loss of generality we focus on additive models based on pairwise distances such as minimum variance methods.

Some Notations for (Model-based) Clustering

Let a sample of I independent observations (objects) be given in R^J and denoted by $X = (x_{ij})$ the corresponding data matrix consisting of I rows and J columns (variables), where the element x_{ij} provides a value for the jth variable describing the ith object.

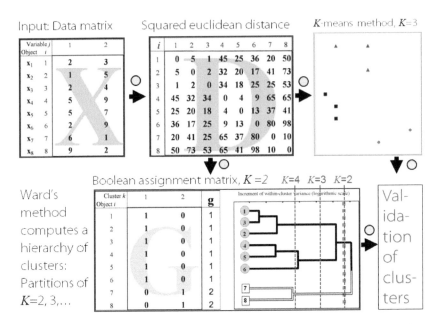

Fig. 6. Toy data example: From a data matrix **X** to a distance matrix **D** with squared Euclidean distances (6) as elements, and further on to a hierarchy or to a partition.

Objects can be tiles, bricks, plants, animals, individuals, countries, enterprises, and so on. Further, let $\mathcal{C} = \{\mathbf{x}_1, ..., \mathbf{x}_i, ..., \mathbf{x}_I\}$ denote the finite set of the I objects (see on the outer left hand side at the top of Fig. 6). Alternatively, let us write shortly $\mathcal{C} = \{1,..., i,..., I\}$.

Generally, the clusters that we are looking for should be as homogeneous as possible in some sense. Figure 6 illustrates schematically the typical steps of cluster analysis starting either with a data matrix or a distance matrix. At the top, the figure presents two dimensional toy data and introduces the data matrix $\mathbf{X} = (x_{ij})$ and the corresponding distance matrix $\mathbf{D} = (d_{ih})$ with the element $d_{ih} = (x_{i1} - x_{h1})^2 + (x_{i2} - x_{h2})^2$. In some applications the distance matrix **D** (or a proximity matrix) may arise directly. Therefore, and because of its more general meaning, a distance matrix will be our preferable starting point for practical cluster analysis. **D** is symmetric with $d_{ih} = d_{hi}$.

But now to our basic problem: finding clusters. In what follows, we focus on model-based Gaussian clustering of observations in its simplest setting. It results in the sum of squares and logarithmic sum of squares methods. These are additive models. It should be mentioned that both methods can be extended to adaptive techniques (Mucha 1992: 141). These simple methods would become slightly flexible by weighting objects and/or variables, and thus they get more practical relevance. The general model-based Gaussian clustering approach is described in several publications such as in Banfield and Raftery (1993). Because we will make use only of the simple model-based Gaussian clustering based on pairwise distances in this chapter, we briefly introduce first some general underlying facts and notations.

Let us formalize the simplest (elementary) solution to the clustering problem with a fixed number of clusters K: the Boolean assignment matrix $\mathbf{G} \in \{0,1\}^{IxK}$ with the restriction of uniqueness and exhaustive assignment (completeness) $\sum_{k=1}^{K} g_{ik} = 1$ for every object i. Formally, the mapping G is

$$G : \mathscr{C} \times \{1,2,...,K\} \rightarrow \{0,1\}$$

with
$$g_{ik} = \begin{cases} 1 & \text{if } i \text{ comes from the cluster (subset) } k \\ 0 & \text{otherwise.} \end{cases}$$

Indeed, the cluster mapping G induces a partition $\{\mathscr{C}_1, ..., \mathscr{C}_K\}$ of C. Hereby

$$\bigcup_{k=1}^{K} \mathscr{C}_k = \mathscr{C} \quad \text{and} \quad \mathscr{C}_k \cap \mathscr{C}_l = \varnothing$$

for every pair of clusters $\mathscr{C}_k$ and $\mathscr{C}_l$, $k, l = 1, 2, \ldots, K$, $k \neq l$. This cluster mapping yields exactly K clusters (subsets), where the numbering of the clusters is arbitrary because it usually depends on the applied clustering algorithm. Alternatively, let $\mathbf{g} = (g_1, ..., g_I)^T$ denote the identifying labels for the clustering and thus for the cluster mapping $\mathbf{G}$, where $g_i = k$ if the ith object $\mathbf{x}_i$ comes from the kth cluster. One can understand $\mathbf{g}$ as a categorical variable or partition variable with K different nominal states $\{1, 2, \ldots, K\}$. Formally, $\mathbf{g} = \mathbf{Ge}$, where the vector $\mathbf{e} = (1, 2, 3, \ldots, K)^T$ has K entities.

Figure 6 illustrates (at the bottom) an example of the Boolean assignment matrix $\mathbf{G} = (g_{ik})$ that maps all I objects into $K = 2$ clusters and the corresponding partition $\mathbf{g}$ of the data matrix $\mathbf{X}$ that is given at the top. Here $\mathbf{G}$ is the clustering result that can be obtained by cutting the dendrogram (at the bottom on the right hand side) at a certain level of cluster distance (as it is indicated by the expanded dashed vertical line). In addition, two dashed vertical lines indicate partitions in $K = 3$ clusters and $K = 4$ clusters, respectively. Here the hierarchical cluster analysis method finds the same optimum partition in $K = 2$ clusters as the partitional K-means method does (see below for details). Usually, the partitional K-means method and the hierarchical *Ward*'s method find only sub-optimum solutions. The latter presents usually a unique solution. The iterative solution of the K-means method depends strongly on the initial (start) partition (Mucha 2009). In Fig. 6, the dendrogram is the result of hierarchical clustering by *Ward*'s incremental sum of squares method based on the corresponding distance matrix $\mathbf{D}$. By cutting a dendrogram at several different levels of cluster distances one gets a set of partitions.

Banfield and Raftery (1993) developed a model-based framework for clustering by parameterizing the covariance matrix in terms of its eigenvalue decomposition. The most general model-based Gaussian clustering is given when the covariance matrix $\mathbf{W}_k$ of each cluster k is allowed to vary completely. Then the log-likelihood (Mucha 1992: 28) is maximized whenever the cluster mapping $\mathbf{G}$ of I observations in K clusters minimizes

$$W_K(\mathbf{G}) = \sum_{k=1}^{K} g_k \log \frac{|\mathbf{W}_k|}{g_k}. \tag{1}$$

Here

$$\mathbf{W}_k = \sum_{i=1}^{I} g_{ik} (\mathbf{x}_i - \overline{\mathbf{x}}_k)(\mathbf{x}_i - \overline{\mathbf{x}}_k)^T \tag{2}$$

is the sample cross-product matrix for the kth cluster C_k, and

$$\overline{\mathbf{x}}_k = \frac{1}{g_k} \sum_{i=1}^{I} g_{ik} \mathbf{x}_i \tag{3}$$

is the usual maximum likelihood estimate of expected values in cluster $\mathscr{C}_k$. Further, g_k is the cardinality of cluster $\mathscr{C}_k$, that is, $g_k = \sum_i g_{ik}$. In this general Gaussian clustering model, the shapes, volumes and directions of the clusters can vary completely. To our knowledge and experience, the parameters of such general models can only be estimated in a stable manner if the number of observations is suitably large in relation to the number of variables. For further details and real world applications, see Mucha et al. 2002.

Additive Models: Minimum Variance Clustering

In this subsection the focus lies on special assumptions about the covariance structure of the clusters. The aim is to allow applications to archaeometry, even though the number of variables is large in relation to the number of archaeological objects under investigation. When the covariance matrix is constrained to be diagonal and uniform across all K assumed clusters, the sum of within-clusters sum of squares criterion (shortly: sum of squares = SS)

$$W_K(\mathbf{G}) = \mathrm{tr}(\sum_{k=1}^{K} \mathbf{W}_k) \tag{4}$$

has to be minimized with respect to $\mathbf{G}$ for a fixed K, where $\mathbf{W}_k$ is the sample cross product matrix (2) for the kth cluster $\mathscr{C}_k$. This is the simplest additive Gaussian model. The SS is fundamental for the inferential statistics and the descriptive statistics. In (4), no pairwise distances occur directly in the case of a Gaussian distribution, but indirectly they are introduced via the corresponding density function (Mucha 1992: 30). It is well known that criterion (4) can be written in the following equivalent form without the explicit specification of cluster centers (centroids) $\overline{\mathbf{x}}_k$

$$W_K(\mathbf{G}) = \sum_{k=1}^{K} \frac{1}{2g_k} \sum_{i=1}^{I} \sum_{h=1}^{I} g_{ik} g_{hk} d_{ih} \tag{5}$$

and

$$d_{ih} = d(\mathbf{x}_i, \mathbf{x}_h) = (\mathbf{x}_i - \mathbf{x}_h)^T (\mathbf{x}_i - \mathbf{x}_h) = \| \mathbf{x}_i - \mathbf{x}_h \|^2 \tag{6}$$

is the squared Euclidean distance between two objects i and h. It is also well known that this criterion is dependent on the scales of the variables. Different scales can be formalized by introducing weights of variables $q_j > 0$, $j = 1, 2, \dots, J$. Behind this special use, the variables can be weighted generally by giving more important variables more weight (i.e., to gain in importance). Taking into account the weights of the variables the squared weighted Euclidean distance

$$d_{ih} = d_Q(\mathbf{x}_i, \mathbf{x}_h) = (\mathbf{x}_i - \mathbf{x}_h)^T \mathbf{Q}(\mathbf{x}_i - \mathbf{x}_h) \tag{7}$$

generalizes formulae (5), where the $J \times J$ matrix $\mathbf{Q}$ is restricted to be diagonal. With $\mathbf{Q}$ = diag($q_1, q_2, \ldots, q_J$), where q_j ($=q_{jj}$) denotes the weight of the jth variable, we can write simply

$$d_{ih} = \sum_{j=1}^{J} q_j (x_{ij} - x_{hj})^2 \tag{8}$$

Because the distance matrix $\mathbf{D} = (d_{ih})$ is additive De Carvalho et al. (2012) and many other authors alternatively work with J distance matrices $\mathbf{D}^{(j)} = (d^{(j)})$, $j = 1, 2, \ldots, J$. Then, according to (8), we have

$$\mathbf{D} = \sum_{j=1}^{J} \mathbf{D}^{(j)}.$$

By doing this, at least the scaling problems can be handled fashionably without any data preprocessing step such as the standardization of variables. For example, using the weights q_j equals to the inverse total variance of the corresponding variable j in (8) means nothing else than standardization of the variables to variance equals 1.

Moreover adaptive weights of variables can be used that are estimated during the iteration process of clustering (For details, also in the frame of principal components analysis (PCA) and in terms of the sample cross-product matrices (2), see Mucha 1995). Of course, the statistical distance (7) with a positive definite matrix $\mathbf{Q}$ can be generalized to cluster specific statistical distances, where instead of $\mathbf{Q}$ the K inverse within-cluster covariance matrices $\mathbf{Q}_k$ are used (Späth 1985). This corresponds to the general criterion (1).

Now we are able to forget (2) and thus the corresponding estimates (3). Keeping in mind, pairwise distances $\mathbf{D}$ are more general as a starting point for (exploratory) cluster analysis and data analysis than a data matrix $\mathbf{X}$. However, the criterion (5) presents practical problems of storage and computation time for increasing I because of the quadratic increase of memory capacity of $\mathbf{D}$, as Späth (1985) pointed out. Meanwhile, a new generation of computers can deal easily with both problems and also for I approximately 10,000.

In order to carry out clustering of a practically unlimited number of objects based on criterion (5), let us generalize by introducing positive weights of objects $u_i > 0, i = 1, 2, \ldots, I$ that will be also referred to as masses. Instead of dealing with millions of objects directly in (5), their appropriate representatives are classified. These representatives are the result of an (usually fast) preprocessing step of data aggregation. Such an aggregation is like smoothing and it has a stabilizing effect. Especially the influence of outliers can be handled in this way to some degree. Obviously, the estimates (2) and (3) are affected by masses, but the distances (6) are independent of masses. That means, from the computational point of view, that distances are most suitable for simulation studies because they need to be figured out only once (see below in Sec. Validation of Clustering Results). The criterion (5) becomes the generalized form

$$W_K^*(\mathbf{G}) = \sum_{k=1}^{K} \frac{1}{2U_k} \sum_{i=1}^{I} u_i \sum_{h=1}^{I} g_{ik} g_{hk} u_h d_{ih} \tag{9}$$

that has to be minimized by incorporating positive masses (weights of objects) and distances (6), where $U_k = \sum_i g_{ik} u_i$ and u_i denote the mass of cluster $\mathscr{C}_k$ and the mass of object i, respectively (Bartel et al. 2003). In the case of weighted observations, the sample cross-product matrix for the kth cluster $\mathscr{C}_k$ (2) becomes

$$\mathbf{W}_k^* = \sum_{i=1}^{I} g_{ik} u_i (\mathbf{x}_i - \bar{\mathbf{x}}_k^*)(\mathbf{x}_i - \bar{\mathbf{x}}_k^*)^T$$

with

$$\bar{\mathbf{x}}_k^* = \frac{1}{U_k} \sum_{i=1}^{I} g_{ik} u_i \mathbf{x}_i.$$

As already mentioned above, the principle of weighting the observations is a key idea for handling cores (i.e., micro-clusters, sets of already aggregated observations) and outliers. In the case of outliers one has to downweight them in some way in order to reduce their influence. In the case of representatives of cores, one has to weight them, for example, proportionally to the cardinality of the cores. Often, centroids or most typical objects are chosen as representatives of micro-clusters.

This simplest model-based clustering criterion (9) becomes flexible if special weights of rows u_i and special weights of columns q_j in (7) are used. Thereby, for example, the decomposition of the chi-squared statistic of a contingency table can be obtained, for details, see Mucha (2014). For further reading concerning correspondence analysis and related methods, see Greenacre (1988, 1989). In archaeology, other successful applications of correspondence analysis are seriation (Van de Velden et al. 2009) and classification (Mucha et al. 2014: dual scaling classification based on the pioneering work of Nishisato (1980)).

It is worth noting that by moving from pair wise squared Euclidean distances d_{ih} to within-cluster sum of squares $w\{i, h\}$ of the two objects i and h, the following formula generally holds

$$w\{i, h\} = \frac{u_i u_h}{u_i + u_h} d_{ih} \tag{10}$$

Obviously, in the case of unit masses $u_i = u_h = 1$, it is

$w\{i, h\} = d_{ih}/2$.

This way is correct if the simplest Gaussian model is assumed. The first step of the hierarchical *Ward*'s method is the agglomeration of the two observations (i.e., terminal clusters) to one cluster which criterion (10) has the minimum value. At the same moment, this is the first ΔV, see the ordinate in Fig. 2. In general, the increment of within-clusters variance ΔV is given by

$$\Delta V_K = W_{K-1}^* - W_K^*, \tag{11}$$

if we move from K clusters to $K - 1$ cluster(s). In particular, in hierarchical *Ward's* method, the increment is simply

$$\Delta V_K = W_{\mathscr{C}_k \cap \mathscr{C}_l} - W_{\mathscr{C}_k} - W_{\mathscr{C}_l}, \tag{12}$$

i.e., if the clusters $\mathscr{C}_k$ and $\mathscr{C}_l$ of the partition in K clusters are merged together. The advantage of distances is that they are always fixed independently of weighting the corresponding observations. In contrast, the sample cross-product matrices (2) and the cluster centers (3) are affected by changing the weights of objects. Therefore, "soft bootstrapping" by random weighting the objects or subsampling can be performed with an unchanged distance matrix **D**, for details see next Sec.and Mucha and Bartel (2014 a, b). For example, thinking of doubling the sample as using double weights of the observations, i.e., $u_i = 2, i = 1,2,\ldots,I$. In this special case, where the pairwise distances become the sum of squares of the corresponding sets consisting of pairs of objects, the result of clustering should be the same.

There are at least two well-known clustering techniques for minimizing the sum of squares criterion based on pairwise distances: the partitional K-means method minimizes the criterion (5) for a single partition **G** by exchanging objects between clusters (Steinhaus 1956, MacQueen 1967), and the hierarchical *Ward's* method minimizing (5) stepwise by agglomerative grouping (Ward 1963). The well-known K-means method becomes a special case of partitional pairwise clustering if it is based on squared Euclidean distances without using any centroids (Späth 1985, Bartel et al. 2003). Also, one usual definition between many possible definitions of the most typical object (MTO) of a cluster is: A MTO is the most similar object to the centroid. In pairwise clustering, a (general) MTO is the one that minimizes the sum of the (pairwise) distances to the other members of the cluster. Of course, in the case of Gaussian normals, this general most typical object is usually located near the expected value, i.e., the centroid, of the cluster.

If the covariance matrix of each cluster $\mathscr{C}_k$ ($k = 1,2,\ldots,K$) is constrained to be diagonal, but otherwise allowed to vary between groups, the logarithmic sum-of-squares criterion

$$V_K(\mathbf{G}) = \sum_{k=1}^{K} g_{.k} \log \text{tr}\left(\frac{\mathbf{W}_k}{g_{.k}}\right) \tag{13}$$

Once again the following equivalent formulation can be derived

$$V_K(\mathbf{G}) = \sum_{k=1}^{K} g_{.k} \log \left(\sum_{i=1}^{I}\sum_{h=1}^{I} \frac{g_{ik} g_{hk}}{2 g_{.k}^2} d_{ih}\right) \tag{14}$$

Considering formulae (14) (and (13) in the case of formulation with sample cross-product matrices, respectively) and weights of observations u_i, the logarithmic sum-of-squares criterion can be generalized to

$$V_K^*(\mathbf{G}) = \sum_{k=1}^{K} U_k \log \left(\sum_{i=1}^{I}\sum_{h=1}^{I} \frac{u_i u_h}{2 U_k^2} g_{ik} g_{hk} d_{ih}\right) \tag{15}$$

According to this logarithmic sum-of-squares criterion, the partitional *K-means*-like clustering algorithm is also called *Log-K-means*. Respectively, the hierarchical *Ward*-like agglomerative method is called *LogWard*. Concerning the hierarchical algorithms, special numerical treatments of observations having small weights are necessary. Such special tricks are essential because the original *Ward's* hierarchical agglomerative clustering is based on a minimum incremental of sum of squares, therefore all observations with zero (or quasi-zero) weight would be merged together into one cluster, whatever the level of distance values may be. It might also be observed that *K-means* and *Log-K-means* based on pairwise distances (7) are also more general because they never require an $(I \times J)$-data matrix **X**.

Application to Archaeometry

As introduced in the beginning, the archaeological objects under investigation are Roman bricks and tiles from the German Rhine area. These objects are described by the following 19 chemical variables Fe_2O_3, MnO, SiO_2, CaO, TiO_2, MgO, Al_2O_3, Na_2O, K_2O, Cr, Sr, Zr, Zn, Y, Ni, Nb, Rb, V, and Ba. There are nine oxides and ten trace elements.

This application to archaeometry is based on the set of 613 Roman bricks and tiles. These objects were the starting point of our statistical investigations in 1999, see Mucha et al. 2002. This data set was completely published as part of the report of Mucha and Ritter (2009: 114–125).

Due to the scaling problem, adaptive weights of variables such as values proportional to the inverse pooled within-cluster variances are recommended (Mucha et al. 2002).Those variances can be estimated in the adaptive *K*-means method in an iterative way (Mucha 1995). We found out that these adaptive weights perform similar to the weight proportional to the inverse square root of the mean of the variable *j*:

$$q_j = \frac{1}{\overline{\mathbf{x}}_j} \tag{16}$$

Later, many other different weights were investigated (Bartel et al. 2002, Mucha et al. 2005a, Mucha 2006). For example, the logarithmic transformation is applied in order to handle both the scaling problem and the outliers (Morales-Merino et al. 2012). We also have investigated the log-ratio transformation for compositional data (Mucha et al. 2008a). However, in our case we have to do with sub-compositional data. Concerning its application to archaeometry the previous chapter (authors: Martín-Fernández and collaborators) is recommended. Moreover, we studied the transformation into ranks, and made comparisons of different transformations (Mucha et al. 2008a). From the statistical point of view, the use of rank data leads to robust versions of multivariate statistical methods.

Without loss of generality, here we prefer the scaling (16) in (8). Furthermore, the same holds for principal components analysis (which also depends on the scale).

Figure 7 shows a projection of the 613 archaeological objects (marked as points) by principal components analysis (PCA). In addition, a continuous view is given by cutting the corresponding nonparametric density at several levels. Without any doubt, this result suggests that there are clusters. With respect to the number of variables, the quality of the projection is high. Especially the first component is very important. It counts for 61% of the total variance.

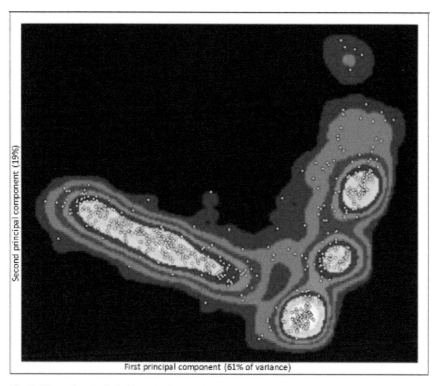

Fig. 7. The archaeological objects (points) and several cuts of the bivariate density in the plane of the first two principal components.

Color image of this figure appears in the color plate section at the end of the book.

The hierarchical *Ward*'s method is applied to this data set using the scaling (16) in (8). From the statistical point of view, at least, eight clusters are stable (Mucha et al. 2002) because they can be reproduced to a high degree by simulations based on resampling techniques. Visually, regions of high density in Fig. 7 coincide with the clusters found by the *Ward*'s method.

Now we can use the results of cluster analysis and PCA to make an informative fingerprint of the data matrix **X** (Fig. 8) and a heatplot of the symmetric distance matrix **D** (Fig. 9), respectively. Here the observations are reordered according to the clusters and to the first principal component. The fingerprint represents the measurements by their corresponding gray level. The darker the pixel is, the higher is the value. In addition, the clusters (locations of military brickyards searched or found, respectively) are indicated. Such a presentation is ready to an easy interpretation by the archaeologists. Obviously, there is a structure in the data and the contributions of the variables to this clustering look different.

Figure 9 shows clear cluster wise patterns (blocks) of distance levels. Such fingerprints and heatplots are helpful tools for the interpretation of the cluster analysis results. As already mentioned, the size of such a heatplot of **D** is independent from the number of variables *J*. In any case, it shows the full distance information. This is different from projection methods such as the PCA.

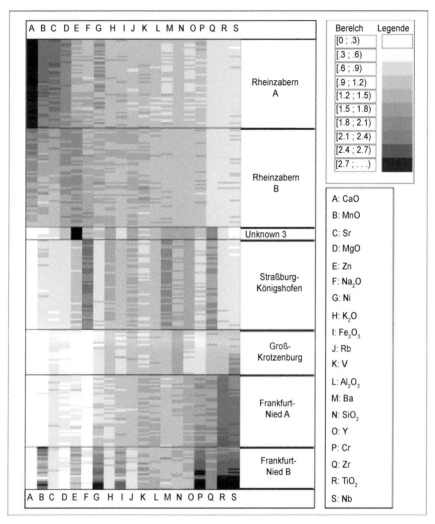

Fig. 8. Fingerprint of ordered data, where the observations and the variables are rearranged.

Validation of Clustering Results

As already shown above, clustering techniques can find homogeneous groups of bricks and tiles. However, cluster analysis usually presents always clusters, even in the case of no structure in the data. Moreover, hierarchical clustering presents nice dendrograms in any case containing all the clusters that are established during the process of agglomeration (see Fig. 2). Obviously, in every agglomeration step an increasing amount of information is lost. The main question is: How many clusters are there? Or, in other words, when should the agglomeration process be stopped? Or, in terms of the density estimation in Fig. 7, what is the right cut-off density level for fixing clusters?

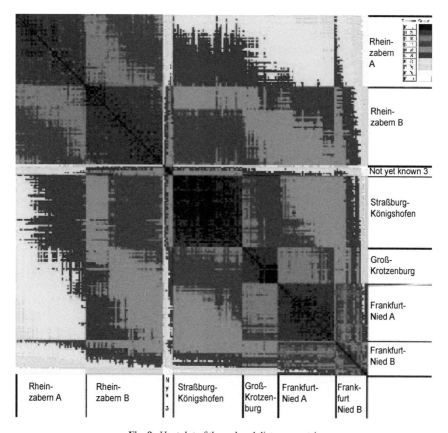

Fig. 9. Heatplot of the ordered distance matrix.

Color image of this figure appears in the color plate section at the end of the book.

Several different clustering algorithms already exist and new ones appear daily in the literature. Often they do their job and, usually, they give a solution in almost all cases. Let us suppose that they do a good (accurate) job, i.e., they are able to reflect an appropriate model of the data (because otherwise the validation can give the right answer to the wrong question). The main question then will be: Is there really a cluster structure in the data? And if it is the case, how many clusters are there?

Therefore, a validation of clustering results based on resampling techniques is recommended. This validation can be considered as a three level assessment of stability:

- The first and most general level is to decide on the appropriate number of clusters. This decision is based on such well-known measures of correspondence between partitions like the *Rand*'s index (Rand 1971), the *adjusted Rand*'s index of Hubert and Arabie (1985), and the index of Fowlkes and Mallows (1983). These are pair-counting-measures.

- Second, the stability of each individual cluster $\mathscr{C}_k$ is assessed based on measures of similarity between sets (Hennig 2004, Mucha 2009), e.g., the asymmetric measure of cluster agreement η_k and the symmetric measures of *Jaccard* γ_k and *Dice* τ_k. From many applications we know that it makes sense to investigate the (often quite different) specific stability of clusters of the same clustering on the same data. One can often observe that the clusters have a quite different stability. Some of them are very stable. Thus, they can be reproduced and confirmed to a high degree, for instance, by bootstrap simulations. They are homogeneous inside and well separated from each other. Moreover, sometimes they are located far away from the main body of the data like outliers. On the other side, hidden and tight neighboring clusters are more difficult to detect and they cannot be reproduced to a high degree. In order to assess the stability of a partition into K clusters, these individual stability values can be aggregated in some way to a total measure η_K, γ_K, and τ_K.
- In the third and most detailed level of validation, the reliability of the cluster membership of each individual object can be assessed.

We call this assessment of stability at three levels a built-in validation because it is an integrated part of the outcome of clustering by using our statistical software *ClusCorr98* (Mucha 2009). In any case, it is recommended that the stability of the obtained clusters has to be assessed by using validation techniques (Jain and Dubes 1988, Mucha 1992, Hennig 2007). To be concrete, the built-in validation of clustering results based on resampling techniques is recommended. There are different resampling techniques.

For example, bootstrapping is resampling with replacement. It generates multiple observations. In clustering, this is often a disadvantage because they can be seen as mini-clusters themselves. In average, more than a third of the original observations are absent in a bootstrap sample. The case of small sample size can lead to another disadvantage. Therefore, Mucha and Bartel (2014a) recommended soft bootstrapping that allocates a small weight $u_i \ll 1$ to observation i that are not part of the original bootstrap sample. Besides bootstrapping an alternative is the subsampling (resampling taken without replacement from the original data). In addition, a parameter $L < I$ is needed in this case: the cardinality of the drawn sample. Another alternative is jittering: adding random noise to every measurement of each single observation, i.e., disturbing the data by randomly generated errors. These techniques or a combination of them are often applied in order to investigate the stability of clustering results. For further details, see Mucha and Bartel (2014a, 2014b).

We do not consider here special properties like compactness and isolation as it is done by Jain and Dubes (1988). Our built-in validation is a general purpose technology for validation that works very well especially in high dimensional settings. In low dimensional cases and in cases where projection methods result in good approximations into R^2 or R^3, graphical methods are often the better and more efficient choice for validation.

How many clusters? About the number of clusters, Hubert and Arabie (1985) recommended the adjusted *Rand* index R. Usually it is a pair-counting-measure. But, it can be also figured out based on the contingency table $\mathbf{N} = (n_{mk})$ that is obtained by crossing two partitions into M and K clusters, namely the vectors $\mathbf{f}$ and $\mathbf{g}$, respectivelly (see Fig. 10). Well-known measures of correspondence between such categorical

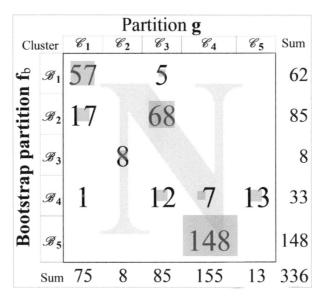

Fig. 10. Example of a confusion matrix (or contingency table) **N** that results from crossing two partitions of the same data set. The partition **g** is the original unique result of *Ward*'s method (see Fig. 12). Here, the other one is the result of Ward's method applied to a soft bootstrap sample. The size of a square is proportional to the count in the corresponding cell of the confusion matrix.

variables are based on such a contingency table, which is also called the pivot table or confusion matrix. Alternatively, such a contingency table **N** can be formulated by simple matrix notation

$$\mathbf{N} = \mathbf{F}^T\,\mathbf{G}$$

based on the corresponding two Boolean assignment matrices **F** and **G**. Figure 10 shows the contingency table that results from crossing a partition of the Troia data set of 336 objects with the cluster analysis result of a bootstrapped sample (Mucha et al. 2013a). In this case, the objects are clay sediments collected within a radius of about 5 km around the archaeological site of Troia in Northwest Turkey. The clay deposits in the plain of Troia originate predominantly from alluvial sediments of two rivers. These sediments probably provided the resources for the production of ancient pottery. The concentrations of the following 26 elements were determined by neutron activation analysis: Na, K, As, Sb, Ba, La, Sm, Yb, Lu, U, Sc, Cr, Fe, Co, Ni, Zn, Rb, Zr, Cs, Ce, Nd, Eu, Tb, Hf, Ta, and Th.

The adjusted *Rand* index R is appropriate to decide on the number of clusters K. The reason is that it takes the value 0 when the *Rand* index (Rand 1971) is equal to its expected value for each K. Figure 11 shows the simulation results. The median of the *adjusted Rand*'s values (the scale is on the right hand side of the plot) reaches its maximum value for two clusters. Moreover, the standard deviations of the 250 *adjusted Rand*'s values for each partition into $K = 2, 3, \ldots$ cluster has its minimum value also for $K = 2$ clusters. They correspond to the bars and scale on the left hand side. They support the decision for the solution into two clusters. More than two clusters are less likely because the clusters cannot be confirmed to a high degree and their stability

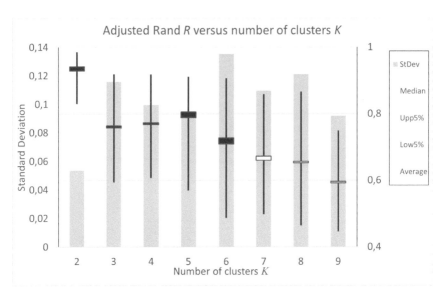

Fig. 11. Statistics of the adjusted Rand index *R* versus the number of clusters. For interpretation and further details see Mucha (2004).

decreases rapidly. For further details concerning this application to archaeometry, see Morales-Merino et al. (2010) and Mucha et al. (2013a).

What are stable clusters from a general statistical point of view? These clusters can be confirmed and reproduced to a high degree. To define stability with respect to the individual clusters, measures of correspondence between two sets such as γ, τ, and η are in use (Hennig 2007, Mucha 2009). They attain their minimum 0 only for disjoint sets. Their maximum is 1.

Figure 12 shows the schematic dendrogram of *Ward*'s clustering of the Troia data set of clay sediments (Mucha et al. 2013a). It shows only a small part of the hierarchy for up to 5 clusters. Each node (cluster) of the binary tree is denoted by the corresponding number of objects (symbol #) and the Jaccard index γ_k in %. The two-cluster solution is emphasized in bold type. While looking for stable clusters and for an outstanding number of clusters one should keep in mind that a hierarchy is a set of nested partitions. Therefore it is recommended to walk step by step through the binary tree (dendrogram) from the right hand side (that corresponds to the root of the tree) to the left. At each step $K-1$ clusters remain unchanged and one cluster is divided only into two parts. Usually, the larger the number of clusters K becomes during the trip through the dendrogram the smaller the amount of changes of the total measures of stability can be expected. Such measures summarized over all individual clusters can also be used for a decision about the number of clusters. To our experience, the total Jaccard value behaves similar to the adjusted *Rand* index *R*.

Some of the clusters remain unchanged during many steps such as the cluster of 168 observations at the bottom of Fig. 12. However, the value of the stability of this cluster decreases from 96% for the partition into two clusters to 92% for the partition into four clusters because of the altering clusters in its neighborhood.

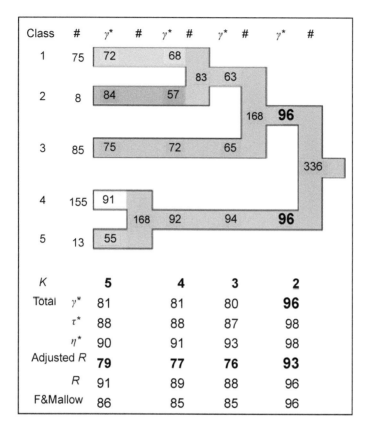

K	5	4	3	2
Total γ^*	81	81	80	**96**
τ^*	88	88	87	98
η^*	90	91	93	98
Adjusted R	**79**	**77**	**76**	**93**
R	91	89	88	96
F&Mallow	86	85	85	96

Fig. 12. Informative dendrogram: a binary tree with results of validation of hierarchical clustering.

Color image of this figure appears in the color plate section at the end of the book.

In the third and most detailed level of validation, the reliability of the cluster membership of each individual observation will be assessed. It is a decomposition of the stability of a cluster according to a measure such as Jaccard into the stability of its members. So, it is simply a byproduct of assessing the stability of individual clusters. Let us look at a special application by returning to clustering of Roman bricks and tiles and the corresponding validation. Later on we will investigate locally the clusters Frankfurt-Nied A and Frankfurt-Nied B (see Figs. 8 and 9) because from the archaeological point of view, bricks and tiles of Frankfurt-Nied come from the same brickyard, see also Dolata et al. (2007). In this case the question arises: Are there really at least two clusters? Concretely, the set of 137 objects is investigated by partitional cluster analysis (K-means method based on pairwise distances) including validation based on the contents of 9 oxides and 10 chemical trace elements. The simulation concerning bootstrapping the adjusted *Rand*'s measure finds out that the two cluster solution is the most likely one with 106 objects in the very compact cluster 1 and only 31 objects in the widespread cluster 2, respectively (Fig. 13). Concretely, the partition into two clusters is the most stable one with the highest median of the 250 adjusted *Rand* indexes (= 0.96). Moreover, the 250 adjusted Rand values have the smallest standard deviation.

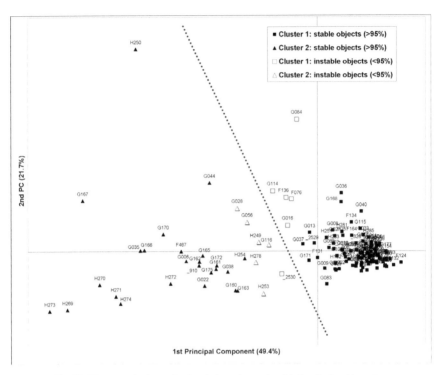

Fig. 13. Cluster analysis results: local clustering and validation by bootstrapping.

Figure 13 shows the cluster membership of each object. Additionally, instable observations are marked by light symbols. Instable means that the observation is not assigned with high probability to the same cluster by repeating the cluster analysis based on bootstrap samples. The most instable objects are G114 and G116 with a rate of reliability of 0.53 and 0.63, respectively. That means that object G114 was classified with nearly the same frequency into cluster 1 or into cluster 2. As a result of these validations, the archaeologists can now modify and consolidate their assumptions. In this way archaeologists obtain additional information about the objects. Thus, the principal aim of cluster analysis is to give assistance to the experts. The corresponding graphical and numerical outputs are helpful to them.

Archaeological Results and Historical Interpretation

Based on the investigation of different types of brickstamps from various military units, it is possible to interpret the results obtained by statistical examination. The objects assigned to Frankfurt-Nied can be separated into two clusters. In the small cluster, 6 samples are situated with brickstamps of the *cohors I Asturum*, and 12 samples are contained with brickstamps of the *legio XXI Rapax*. This result is archaeologically very important. Brickstamps of the auxiliary-unit *cohors I Asturum* are very rare. The analyzed examples have been found in the brickyard of Frankfurt-Nied and in the civil settlement of Frankfurt-Heddernheim. Archaeologists still do not know what date soldiers of this unit were sent by the Roman governor to the brickyard for making bricks and what the historical context was. Brickstamps of a second military-unit are found together in the statistical cluster: stamps of the *legio XXI Rapax*. The chronological assignment of bricks and tiles from this *legio* is very precise. The *legio XXI Rapax* was transferred by the Emperor *Domitianus* in 83 A.D. from *Bonna* in *Germania Inferior* to *Mogontiacum* in *Germania Superior*. Already in 89/90 A.D. the unit had been withdrawn from *Germania Superior* as a consequence of the rebellion of *Saturninus*. It is known that one was a stamp-type used in the brickyard of Rheinzabern and only two were stamp-types used in the brickyard of Frankfurt-Nied. Because of the shifting of brick-production from the Rhine to Main in 83/85 A.D. it is possible to date the 12 samples separated in the cluster in a very short period of time: between 83/85 and 89/90 A.D. We consider that the stamps of *cohors I Asturum* were pressed in bricks and tiles of similar clay, perhaps at the same time. In the future this hypothesis has to be proofed by investigation of additional objects. In this context, other open questions on the archaeological model of Roman brick and tile production in the Upper Rhine area could be answered. Our way of investigation and consolidation of archaeological ideas of a special field of Roman military-history can also be used in other provinces of the Roman Empire.

Conclusion

In this study, we have considered chemical measurement data only. Usually, archaeologists record additional information such as survey coordinates, epigraphy, or type of brick (*later*, *tegula*, etc.). With reference to an application of spatial statistical analysis to archaeometry, see Dolata et al. (2010), and, for the application of classification /clustering based on mixed data (metric and non-metric information) to archaeometry, see Mucha et al. (2014). Both papers are also related to correspondence analysis which allows appropriate multivariate projections.

Acknowledgements

We are grateful to Dwight Read, the Referee, and especially Juan A. Barcelo for their interest in our work as well as for the constructive and valuable suggestions which greatly improved the manuscript.

References Cited

Aitchison, J. 1986. The Statistical Analysis of Compositional Data. Chapman and Hall, London.

Banfield, J.D. and A.E. Raftery. 1993. Model-Based Gaussian and non Gaussian Clustering. Biometrics 49: 803–821.

Bartel, H.-G., H.-J. Mucha and J. Dolata. 2002. Automatische Klassifikation in der Archäometrie: Berliner und Mainzer Arbeiten zu oberrheinischen Ziegeleien in römischer Zeit. Berliner Beiträge zur Archäometrie 19: 31–62.

Bartel, H.-G., H.-J. Mucha and J. Dolata. 2003. Über eine Modifikation eines graphentheoretisch basierten partitionierenden Verfahrens der Clusteranalyse. Match 48: 209–223.

Bartel, H.-G., H.-J. Mucha and J. Dolata. 2008. Über Identifikations-methoden, dargestellt am Beispiel römischer Baukeramik aus Obergermanien. Berliner Beiträge zur Archäometrie 21: 115–132.

De Carvalho, F., Y. Lechevallier and F.M. De Melo. 2012. Partitioning Hard Clustering Algorithms Based on Multiple Dissimilarity Matrices. Pattern Recognition 45: 447–464.

Dolata, J. and U. Werr. 1998/1999. Wie gleich ist derselbe?—Homogenität eines römischen Ziegels und Aussagegrenzen geochemischer Analytik aufgrund von Meßtechnik und Materialvarietät. Mainzer Archäologische Zeitschrift 5/6: 129–147.

Dolata, J., H.-J. Mucha and H.-G. Bartel. 2007. Uncovering the Internal Structure of the Roman Brick and Tile Making in Frankfurt-Nied by Cluster Validation. pp. 663–670. *In*: R. Decker and H.-J. Lenz (eds.). Advances in Data Analysis, Springer, Berlin.

Dolata, J., H.-J. Mucha and H.-G. Bartel. 2010. Mapping Findspots of Roman Military Brickstamps in Mogontiacum (Mainz) and Archaeometrical Analysis. pp. 595–603. *In*: A. Fink, B. Lausen, W. Seidel and A. Ultsch (eds.). Advances in Data Analysis, Data Handling and Business Intelligence, Springer, Berlin.

Fowlkes, E.B. and C.L. Mallows. 1983. A Method for Comparing two Hierarchical Clusterings. JASA 78: 553–569.

Greenacre, M.J. 1988. Clustering the Rows and Columns of a Contingency Table, Journal of Classification 5: 39–52.

Greenacre, M.J. 1989. Theory and Applications of Correspondence Analysis. 3rd printing, Academic Press, London.

Hennig, C. 2007. Clusterwise Assessment of Cluster Stability. Computational Statistics and Data Analysis 52: 258–271.

Hubert, L.J. and P. Arabie. 1985. Comparing Partitions. Journal of Classification 2: 193–218.

Jain, A.K. and R.C. Dubes. 1988. Algorithms for Clustering Data. Prentice Hall, New Jersey.

Kaufman, L. and P.J. Rosseeuw. 1990. Finding Groups in Data. Wiley, New York.

MacQueen, J. 1967. Some Methods for Classification and Analysis of Multivariate Observations. pp. 281–297. *In*: L. Lecam and J. Neyman (eds.). Proc. 5th Berkeley Symp. Math. Statist. Prob., Vol. 1. Univ. California Press, Berkeley.

Mardia, K.V., J.T. Kent and J.M. Bibby. 1979. Multivariate Analysis. Academic Press, London.

Morales-Merino, C., H.-J. Mucha and H.-G. Bartel. 2012. Multivariate Statistical Analysis of Clay and Ceramic-data for Provenance of Bronze Age Pottery from Troia. Metalla Sonderheft 5, Deutsches Bergbau-Museum Bochum, Bochum, 174–176.

Morales-Merino, C., H.-J. Mucha, H.-G. Bartel and E. Pernicka. 2010. Clay Sediments Analysis in the Troad and its Segmentation. Metalla Son-derheft 3, Deutsches Bergbau-Museum Bochum, Bochum, 122–124.

Mucha, H.-J. 1992. Clusteranalyse mit Mikrocomputern. Akademie Verlag, Berlin.

Mucha, H.-J. 1995. XClust: Clustering in an Interactive Way. pp. 141–168. *In*: W. Härdle, S. Klinke and B.A. Turlach (eds.). XploRe: An Interactive Statistical Computing Environment. Springer, New York.

Mucha, H.-J. 2004. Automatic Validation of Hierarchical Clustering. pp. 1535–1542. *In*: J. An-toch (ed.). COMPSTAT 2004. Physica-Verlag, Heidelberg.

Mucha, H.-J. 2006. Finding Meaningful and Stable Clusters Using Local Cluster Analysis. pp. 101–108. *In*: V. Batagelj, H.-H. Bock, A. Ferligoj and A. Ziberna (eds.). Data Science and Classification, Springer, Berlin.

Mucha, H.-J. 2007. On Validation of Hierarchical Clustering. pp. 115–122. *In*: R. Decker and H.-J. Lenz (eds.). Advances in Data Analysis. Springer, Berlin.

Mucha, H.-J. 2009. ClusCorr98 for Excel 2007: clustering, multivariate visualization, and validation. pp. 14–40. *In*: H.-J. Mucha and G. Ritter (eds.). Classification and Clustering: Models, Software and Applications, Report No. 26, WIAS, Berlin.

Mucha, H.-J. 2013. Variable Selection in Cluster Analysis Using Resampling Techniques: A Proposal. AG DANK/British Classification Society Meeting 2013 at University College London. http://www.homepages.ucl.ac.uk/ũcakche/agdank/agdank2013presentations/mucha.pdf, London.

Mucha, H.-J. 2014. Pairwise data clustering accompanied by validation and visualisation. pp. 47–57. *In*: W. Gaul, A. Geyer-Schulz, Y. Baba and A. Okada (eds.). German-Japanese Interchange of Data Analysis Results, Springer, Cham.

Mucha, H.-J. and H.-G. Bartel. 2014a. Soft Bootstrapping in Cluster Analysis and Its Comparison with Other Resampling Methods. pp. 97–104. *In*: M. Spiliopoulou, L. Schmidt-Thieme and R. Janning (eds.). Data Analysis, Machine Learning and Knowledge Discovery, Springer, Berlin.

Mucha, H.-J. and H.-G. Bartel. 2014b. Resampling Techniques in Cluster Analysis: Is Subsampling Better Than Bootstrapping? *In*: S. Krolak-Schwerdt, B. Lausen and M. Böhmer (eds.). European Conference on Data Analysis. Springer, Berlin, forthcoming.

Mucha, H.-J., H.-G. Bartel and J. Dolata. 2002. Exploring Roman Brick and Tile by Cluster Analysis with Validation of Results. pp. 471–478. *In*: W. Gaul and G. Rit-ter (eds.). Classification, Automation, and New Media, Springer, Heidelberg.

Mucha, H.-J., H.-G. Bartel and J. Dolata. 2005a. Model-based Cluster Analysis of Roman Bricks and Tiles from Worms and Rheinzabern. pp. 317–324. *In*: C. Weihs and W. Gaul (eds.). Classification—the Ubiquitous Challenge, Springer, Berlin.

Mucha, H.-J., H.-G. Bartel and J. Dolata. 2005b. Techniques of Rearrangements in Binary Trees (Dendrograms) and Applications. Match 54: 561–582.

Mucha, H.-J., H.-G. Bartel and J. Dolata. 2008a. Effects of Data Transformation on Cluster Analysis of Archaeological Data. pp. 681–688. *In*: C. Preisach, H. Burkhardt, L. Schmidt-Thieme and R. Decker (eds.). Data Analysis, Machine Learning and Applications, Springer, Berlin.

Mucha, H.-J., H.-G. Bartel and J. Dolata. 2014. Dual Scaling Classification and Its Application to Archaeometry. pp. 105–113. *In*: M. Spiliopoulou, L. Schmidt-Thieme and R. Janning (eds.). Data Analysis, Machine Learning and Knowledge Discovery, Springer, Cham.

Mucha, H.-J., H.-G. Bartel and C. Morales-Merino. 2013a. Visualisation of Cluster Analysis Results. pp. 261–270. *In*: A. Giusti, G. Ritter and M. Vichi (eds.). Classification and Data Mining, Springer, Berlin.

Mucha, H.-J. and G. Ritter. 2009. Classification and Clustering: Models, Software and Applications. Weierstrass Institute for Applied Analysis and Stochastic (WIAS) Report No. 26: http://www.wias-berlin.de/report/26/wias_reports_26.pdf, Berlin.

Mucha, H.-J., U. Simon and R. Bruggemann. 2002. Model-based Cluster Analysis Applied to Flow Cytometry Data of Phytoplankton. Weierstrass Institute for Applied Analysis and Stochastic (WIAS) Technical Report No. 5: http://www.wias-berlin.de/techreport/5/wias_technicalreports_5.pdf, Berlin.

Nishisato, S. 1980. Analysis of Categorical Data: Dual Scaling and its Applications. University of Toronto Press, Toronto.

Rand, W. M. 1971. Objective Criteria for the Evaluation of Clustering Methods. Journal of the American Statistical Association 66: 846–850.

Ripley, B.D. 1996. Pattern Recognition and Neural Networks. Cambridge University Press, Cambridge.

Späth, H. 1980. Cluster Analysis Algorithms for Data Reduction and Classification of Objects. Ellis Horwood, Chichester.

Späth, H. 1985. Cluster Dissection and Analysis. Ellis Horwood, Chichester.

Steinhaus, H. 1956. Sur la division des corps matériels en parties. Bulletin de l'Académie Polonaise des Sciences, Vol. IV, 12: 801–804.

Van De Velden, M., P.J.F. Groenen and J. Poblome. 2009. Seriation by Constrained Correspondence Analysis: A Simulation Study. Computational Statistics & Data Analysis 53(8): 3129–3138.

Ward, J.H. 1963. Hierarchical Grouping to Optimize an Objective Function. Journal of the American Statistical Association 58: 236–244.

10

Archaeological Discriminant Applications of the Lubischew Test

José Antonio Esquivel,[1] *Alexia Serrano*[1,*] and
Juan Manuel Jiménez-Arenas[2]

Introduction

Classification has played an important role in Archaeology. In fact, one of the main aims of this discipline is to determine and establish the laws that order the diversity of material culture. Archaeologists have made use of qualitative values and, since the 1960s, of quantitative techniques in the so-called Numerical Taxonomy, with analytic methods mainly of classification, and discriminant analyses playing a major role. These applications, firstly developed by R.A. Fisher in 1936, have been of great importance because they search for lineal combinations of variables to characterize or distinguish observation in two or more groups. Also, this is because their use can establish the probability that one observation will belong to a previously defined group. Therefore, two of the main characteristics of discriminant analyses are: 1) their ability to maximize the differences between groups, and 2) their predictive character, i.e., the ability to assign individuals to groups (predictive discriminant analyses).

The combination of the above two characteristics has enabled past professionals in science to establish the belonging of unknown or unclassified archaeological items to previously predetermined groups, and therefore we can make use of them in the

---

[1] Department of Prehistory and Archaeology, University of Granada, 3D Archaeological Laboratory Modelling, University of Granada.
Emails: esquivel@ugr.es; alisera@correo.ugr.es
[2] Department of Prehistory and Archaeology, University of Granada, Institut of Peace and Conflicts, University of Granada (Spain).
Email: jumajia@ugr.es
* Corresponding author

context of "actualistic comparison". In the case of palaeontological, archaeozoological, palaeobotanical, or archaeobotanical records, this is possible through the comparison of current species and/or varieties whose systemic (including both taxonomy and behaviour) is known. Also, in the case of material culture, comparison with experimental elements, based on the archaeological material, allows us to generate hypotheses about the technologies used in the past.

Among all of the available methods of numerical classification, discriminant function analysis (DFA) is the most widely used. Several reasons can be argued to explain such preponderance. One is the easy interpretation of the results compared to other techniques, such as the artificial neural networks (ANN). Thus, DFA presents lineal functions and is based mostly on logistic or hyperbolic tangents (Du Jardin et al. 2009). Another reason is that most computing packages usually provide this as the default method. For example, DFA has successfully been used to establish the sex of individuals corresponding to archaeological human populations of known sex, fundamentally from the hipbone, and moreover from bones or complex bones that do not reflect sexual dimorphism so clearly (e.g., the femur, Murphy 2005). Other contexts have been: to establish whether the bone remains of certain canids belong to domestic dogs or wolves (Benecke 1987), and with suids too (Evin et al. 2013); to determine whether there is a gender bias among mountain gazelles hunted *in situ* (Munro et al. 2011); to study the changes in the lithic industries during the transit of hunter-gatherer societies towards the agricultural producers in the British islands (Pitts and Jacobi 1979); to analyse the presence of fitolites and their relationship with the routes and distribution rates of this cereal along America (Hart and Martson 2009); or to trace the marble used by Michelangelo to sculpt *David* (Attanasio et al. 2005).

On the contrary, DFA has important use limitations, especially in archaeology. The first limitation is clearly shown by the sensitivity of this technique regarding the composition of the sample. This can prove problematic in archaeological samples composed of a low observation number, which makes them especially sensitive to outliers. On the other hand, it is well known that the higher the number of variables, the more accurate the prediction. However, another problem archaeologists must face is the fragmentary character of the archaeological artefacts, which can often be small parts of the original items. Finally, other limitation derives from the data structure, because several times we do not know the original individual data but only few parameters (i.e., average, number of observations and standard deviation).

For the above reasons, it is important for archaeology to be equipped with classification tools that allow us to work with a low or moderate number of observations while enabling the use of the fewest possible number of variables and/or parameters.

Among the techniques of numerical classification, the Lubischew test (1962) has remained forgotten for decades. Curiously, a search in the main bibliographic databases (Web of Science, Google Scholar), using the keyword "Lubischew test" locates mostly studies that use data from the genus *Chaetocnema* used by the Russian biologist but not the discriminant function developed by him. However, in recent years there has been a revitalization of this since it is particularly useful to assess the overlap/discrimination from simple variables, provided there are two groups of affiliation.

Thus the Lubischew test has been used in Palaeontology and Forensic Archaeology. Specifically, in the first case, it has been employed to evaluate whether the variability of the genus *Megantereon* has two species, or whether it is a taxon with a high sexual dimorphism (Palmqvist et al. 2007); to establish whether a variable of the shape of the cranium is valid, from species of current hominoids, to evaluate the taxonomy of the genus *Homo* (Jiménez-Arenas et al. 2011); and to evaluate the similarity of the endocraneal volume between Neanderthals and anatomically modern humans (Serrano Ramos et al. 2013). In the second case, in Forensic Archaeology it has been successfully applied to evaluate sexual dimorphism and identify sex from the femur (Jiménez-Arenas 2009), humerus (Jiménez-Arenas 2010) and the crania of a medieval archaeological population from the southern Iberian Peninsula (Jiménez-Arenas and Esquivel 2013).

Based on all the above, the aim of the present study seeks to characterize the use of the Lubischew test on archaeological samples to determine the conditions under which this method is most effective, the type of data that can provide the best results and, therefore, contribute to the resolution of archaeological problems where the objective is to characterize dichotomous samples from single variables in many fields of Archaeology, specifically when we only know the number of observations, average and standard deviation.

Materials and Methods

Statistical methods were applied from the origins of the discipline to analyse a great variety of data. In this chapter, we apply the Lubischew test to four datasets to evaluate different contexts: a) typologies of classic amphorae (20 from Dressel 10 type, and 7 from Dressel 12 type) belonging to Roman sites of the East Andalusian area; b) the study of trace elements in archaeometric pottery analysis of *lantathum* (19 from Copper Age, and 10 from Bronze Age), and *chromium* (21 clays and 78 archaeological ceramics samples) both belonging to *Ronda La Vieja* archaeological site (Málaga, Spain); c) experimental lithic knapping using two different types of hammers; and d) chemical composition of Bronze fibulae in the Iberian Peninsula.

The Lubischew Test

An important problem in taxonomic research is the identification of species, mainly when there are no characters whose specific ranges do not overlap. Also, many species cannot be identified using a single metric trait, and the use of qualitative variables in an independent form is sometimes an insoluble problem. Using a unique dichotomous variable, two principal criteria for the selection of characters are the relation of interspecific and intraspecific variability and the comparison of intraspecific and interspecific correlation of characters (Lubischew 1962). The interspecific relationship is the interaction between organisms of different species in some ecosystem, and the intraspecific relation is the interaction between organisms belonging to the same species.

One of most widely used statistical methods used to discriminate is a variation of Student's t-test based on an unique metric trait developed by A.A. Lubischew (1962). The null hypothesis is that there are no statistically significant differences in mean between two populations (e.g., testing the differences between sex in a species). In this way, this method has been used to analyse both the interspecific and intraspecific variation.

On the basis of two samples of sizes n_1 and n_2, the suitability of a single character x for discriminating between two species is tested by the t statistical parameter

$$t = \frac{\overline{x}_1 - \overline{x}_2}{S_{x_1 x_2}} \Rightarrow F = t^2 = \frac{n_1 n_2 (\overline{x}_1 - \overline{x}_2)^2}{(n_1 + n_2) S_{x_1 x_2}^2}$$

being $\overline{x}_1$ and $\overline{x}_2$ the sample means in each group. The parameter is the pooled estimated of within-species variance assumed to be common, a biased estimation of variance of population:

$$S_{x_1 x_2} = \sqrt{\frac{n_1 S_{x_1}^2 + n_2 S_{x_2}^2}{n_1 + n_2}}$$

The particular case simplifies the expression, giving:

$$F = \frac{n(\overline{x}_1 - \overline{x}_2)^2}{2 S_{x_1 x_2}^2} = \frac{n(\overline{x}_1 - \overline{x}_2)^2}{2v}$$

and the elimination of n originates the so-called "estimated coefficient of discrimination" (Lubischew 1962):

$$K = \frac{(\overline{x}_1 - \overline{x}_2)^2}{2 S_{x_1 x_2}^2} = \frac{(\overline{x}_1 - \overline{x}_2)^2}{2v} \Rightarrow K = \frac{F}{n} \quad if \quad n_1 = n_2$$

The significance of the difference is given by F, and any standard error is quoted for K. If n is not too small, assuming x to be normally distributed within species, the K value can be used to estimate the probability of misclassification when the character x is used alone (Lubischew 1962). Previously it was necessary to use a normality test such as the Shapiro-Wilk test, which is one of the most powerful tests for normality, especially for small samples ($n < 30$) (Shapiro 1965). Figure 1 gives an example having the same distribution for both sexes, assuming that the mean for males is greater than for the females (Josephson et al. 1996).

The estimate K can be used to estimate the probability p of misclassification when the character x is used alone. This probabilistic value is adjusted very well using a random normal deviate; that is, this value is approximately equal to the probability

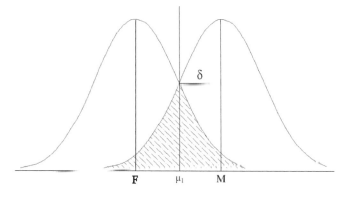

Fig. 1. Graphics of two normal univariate distributions with the same parameters.

that a random normal deviate exceeds the value. If the size of n_1 and n_2 are large, the approximation is very good. The probability for the more usual range of coefficient of discrimination K and the probability of misclassification p (Lubischew 1962) appear in Table 1.

In general research, it is usual to calculate K for scores of characters and selecting those with the largest K. Nevertheless, when we use two or more characters, the intraspecific and interspecific correlations, there are two possibilities of error: 1) ignoring the correlation that induces an overestimation in the degree of discrimination, and 2) choosing a set of additional characters that produces an insignificant increase in information. This case needs be to be studied using the correlations and the discriminant functions to try to avoid these failures (Lubischew 1962).

Let us assume two numeric characters having equal variances: 1) an increment of x implies a decrease of y (the interspecific correlation of the characters is negative), and 2) an increment of x implies an increment of y (the interspecific correlation of the characters is positive). In the absence of intraspecific correlation, the two characters overlap in the shaded area and disappear when the characters show a marked positive intraspecific correlation. The lines connecting the two means are almost perpendicular to the long axes of the correlation ellipses, very similar to the situation in which the two means are almost parallel to the long axes. The overlap remains practically at the former level, and the increase in information is almost negligible (Lubischew 1962).

The most important selection of characters was analysed by Fisher and the solution presents a different problem (resolved by Fisher) that proposes to determine the linear function that maximizes the ratio of the difference between the specific means to the within-species standard deviation of the function. When the characters are to be chosen, the problem is to select the minimal number of variables which can be compounded to give a maximal discrimination between species.

Table 1. Some value of the coefficient of discrimination and the probability of misclassification.

K	p	K	p	K	p	K	p
0	0.5	4	0.0793	11	0.0096	25	0.000208
0.5	0.309	5	0.0570	12	0.0072	30	0.0000544
1	0.239	6	0.0418	14	0.0041	40	0.0000041
2	0.159	8	0.0228	16	0.0026	45	0.00000107
3	0.111	10	0.0132	18	0.0013	50	0.000000030

Comparing the Lubischew Test with other Discriminant Analysis

Two of us have recently published a paper wherein we compare the Lubischew test with other discriminant analysis, and more specifically with the discriminant function analysis (DFA) (Jiménez-Arenas and Esquivel 2013). The main difference between the Lubischew test and the most used quantitative discriminant technique, i.e., DFA is related with the different formulation of the two procedures. While DFA uses both within-group and between-group variation, the K coefficient of discrimination incorporates only between-group variability. Thus, DFA calculates the average variability with reference to the whole mean, while Lubischew's test provides the difference between the means of each group. Moreover, Lubischew's test is specifically operational for small samples, a usual scenario in Archaeology, because the factor 'number of observations' is omitted in calculations.

An additional positive aspect of the Lubischew test lies in that we must not know the original individual data in order to calculate the overlapping/discrimination percentages. This feature is a milestone because it permits us to use, as noted above, only three of the main statistical parameters: sample size, average and standard deviation. Therefore this characteristic supposes an advantage regarding the possibility of assessing discrimination analysis based on data from literature because in most of the publications, only such statistical parameters are given.

Applications of the Lubischew Test

1. The Roman Amphorae

Roman amphorae have various typologies with usually clearly distinguishable characteristics. These shapes of vessels are usually explained by the different uses they can have, such as conservation and transport of different products (e.g., grapes/ wine, olives/oils, fish/salted fish, cereals), etc. Because of the great number of these types of amphorae, their long evolution, and the extensive space of their use during ancient times, these artefacts became an important element for the relative dating of the archaeological ensembles. In this example, we applied the Lubischew test to amphorae belonging Dressel 10 (N=20) which are salted-fish containers made in the southern Iberian Peninsula and Roman amphorae (N=7) Dressel 12 typologies (Moscati 1989) (Fig. 2).

For this example we have chosen as discriminatory variable the neck length (Table 2).

Table 2. Main statistical parameters used in the first archaeological application: The Roman amphorae neck length. Legend. n: number of observations; x: average; σ: standart deviation; K: coefficient of discrimination; R: square root of ($K/2$).

Dressel 10	$n_1 = 20$	$\overline{x_1} = 11.10$	$\sigma_1 = 0.95$	$K = 93.49$	$R = 6.84$
Dressel 12	$n_2 = 7$	$\overline{x_2} = 24.04$	$\sigma_2 = 0.92$		

10

12

Fig. 2. Roman amphorae of 10 y 12 Dressel typologies.

The results show a discrimination of 100%, thus having a probability of misclassification p = 0, meaning that these two types of amphorae can be correctly classified by the length of their necks. This agrees with the presumption that the Roman amphorae have different typologies with clearly distinguishable characteristics, at least between these two types.

2. Elements in clay composition

An important study feature of the ceramics is ceramic production, from gathering the raw material (including the evaluation of ceramic microstructures), pottery production (distinguishing production methods), forms and functionalities, and decorative techniques, in order to help establish a relative chronology. Moreover, the analyses of the raw material (clay, but also the so called "opening materials") helps define the geographic origin of the pottery and can provide information about the size of the cultural areas and the trade routes during a given period. Thus, the study of trace elements has been key in archaeometric research in ceramics.

The archaeological pre- and protohistorical pottery of Ronda la Vieja (Ronda, Málaga, Spain) has been studied in order to establish the procedence of the pottery and the level of specialization reached over two thousand years in this settlement (Padial 1999), by performing mineralogical and chemical analyses. In the present work, we took the data of the *lanthanum* content, taking into account the chronocultural period to which the two groups of pottery belong, i.e., the Bronze Age group and the Copper Age (Table 3).

Table 3. Main statistical parameters used in the second archaeological application: Elements in clay composition (lanthanum). Legend. to see Table 2.

Lanthanum					
Bronze Age	$n_1 = 19$	$\overline{x_1} = 42.03$	$\sigma_1 = 9.04$	$K = 1.42$	$R = 0.84$
Copper Age	$n_2 = 10$	$\overline{x_2} = 44.82$	$\sigma_2 = 9.58$		

The percentage of archaeological misclassified pottery was only 79.49% ($p = 0.2051$), showing that the lanthanum is a good discriminatory factor to identify the production period. This result indicates a break between the two periods with respect to the content of lanthanum, perhaps due to shortage of this element or/and cultural changes.

On the other hand, the analysis of contents in chromium to distinguish between both, extant (experimental) and archaeological clays used for making pottery show a different result (Table 4).

Table 4. Main statistical parameters used in the second archaeological application: Elements in clay composition (chromium). Legend. to see Table 2.

Chromium					
Clay	$n_1 = 21$	$\overline{x_1} = 111.61$	$\sigma_1 = 10.78$	$K = 0.01$	$R = 0.01$
Ceramics	$n_2 = 78$	$\overline{x_2} = 108.40$	$\sigma_2 = 26.80$		

The percentage of misclassification is 47.37% ($p = 0.4737$). This constitutes a great overlap, signifying that chromium is not a good factor to discriminate the use of clays in the present (experimental archaeology) and in the past, which could represent a good reproduction of the process of the pottery manufacture (same clay, same temperature, etc.).

3. Experimental flint-knapping

Experimental archaeology constitutes useful matters to apply discriminant methods. The lithic material came from a flint-knapping experience by the students of the Master's degree in Archaeology in 2010–2011 (University of Granada, Spain) in the *Algaba de Ronda* (a private research centre specialized in Experimental Archaeology located at Ronda, Spain). The students, with little or no experience, knapped flint both with a hard hammer (stone) and a soft one (deer's antler). From this knapping, 56 products were gathered, differentiating by the kind of hammer used. Afterwards at the lab, the blades were photographed and measurements were taken by Autocad. Several variables were introduced into a database and statistical analyses were made in order to analyze the different knapping marks left by the two kinds of hammers because experimentation applied to lithic knapping in the experimental products allowed us to recognize a group of characteristics and technical markers that can be extrapolated to the archaeological record and therefore enabled us to recognize the knapping techniques used.

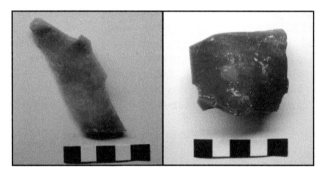

Fig. 3. Blades resulting of the flint knapping experience with soft hammer (left) and hard one (right).

Color image of this figure appears in the color plate section at the end of the book.

In this example, we used the main metric variables—maximum length, maximum breadth and maximum thickness, and some indexes derivative from these length indices (max. length/max. breadth) and roundness index (max. length/max. thickness)—in order to establish a quantitative value for the difference between the lithic material produced by the two types of hammers, i.e., to distinguish the kind of hammer used in each case (Table 5).

Table 5. Main statistical parameters used in the third archaeological application: Experimental flint knapping. Legend. to see Table 2.

		Max. Length	Max. Breadth	Max. Thickness	Length Index	Roundness Index
	N	20	20	20	20	20
Hammer	$\bar{x}$	8.43	9.25	4.43	0.96	2.24
1	σ	3.15	3.89	2.86	0.26	0.68
	N	36	36	36	36	36
Hammer	$\bar{x}$	4.26	3.55	1.3	1.27	4.24
2	σ	1.71	1.17	1.03	0.53	2.48
H_1 vs. H_2	K	1.60	2.58	1.35	0.23	0.48
	% Discrim.	81.33	87.08	79.39	63.31	68.79
	% Overlap.	18.67	12.92	20.61	36.69	31.21

The hard hammer (stone) produced bigger size blades, tending to be wider rather than longer, and therefore, they present a lower lengthening and roundness index. Meanwhile, the blades produced by the soft hammer (antler) tended to be smaller, more homogeneous (indicating a more controlled production) and trended to be longer than wide: however, this pattern was interrupted by distal fractures, produced both by knapping accidents and tectonic fractures in the flint's cores. The resulting blades were slender and presented higher lengthening and roundness indices.

The Lubischew test indicated that the maximum length of the blades, as well as the maximum breadth, are variables that can be well used to distinguish the type of hammer used with a good value of correct classification.

4. Chemical composition of Bronze fibulae from the Iberian Peninsula

Bronze is considered the most important metallic alloy used in prehistoric societies for producing weapons, tools, jewellery, sculptures, coins, and brooches.

The generalization of bronze technology spans a long period of experimentation and metallurgic development to overcome the arsenical bronze. Bronze is composed of about 88% of copper and 12% of tin. In nature, mines containing both minerals are scarce, so that the spread of this alloy is linked to the trade networks of tin. The bronze chemical composition can vary according to the skills of the metal workers, the availability of the raw material, the object to be made, and even cultural patterns can vary.

Data were taken from Carrasco Rus et al. (2005) and a statistical analysis was made on the chemical composition of Huelva-type elbow fibulae of from the Iberian Peninsula. In this example, we apply the Lubischew test in order to study the metallic components of the Later Bronze fibulae coming from two areas of Spain: Granada and Huelva (Zone B) (Table 6).

The results indicate that the best discriminant elements are tin, copper, nickel, and silver, ranging between 79.39% and 88.88% (Table 7), the amount of tin being the main discriminant. However, there were other elements with high degrees of overlap, such as arsenic, lead, iron, and antimony.

Table 6. Size sample (N), mean $\bar{x}$ and standard deviation (σ) for several chemical elements in each area: copper (Cu), arsenic (As), tin (Sn), lead (Pb), silver (Ag), nickel (Ni), iron (Fe), and antimony (Sb).

		Cu	As	Sn	Pb	Ag	Ni	Fe	Sb
	N	12	12	12	12	9	12	10	10
Granada	$\bar{x}$	93.09	0.15	6.27	0.17	0.04	0.07	0.06	0.08
	σ	2.91	0.9	2.89	0.18	0.03	0.13	0.09	0.11
	N	8	8	8	8	8	8	8	8
Huelva	$\bar{x}$	86.93	-	12.8	0.10	0.004	0.33	0.10	0.11
	σ	2.38	-	2.31	0.10	0.002	0.17	0.10	0.13

Table 7. Results of the Lubischew test for the content of different elements present in fibulae from Granada and Huelva.

	Cu	As	Sn	Pb	Ag	Ni	Fe	Sb
R	1.14	0.11	1.22	0.23	0.82	0.88	0.21	0.18
% DISCRIM	87.08	53.98	88.88	58.71	79.39	81.06	58.32	57.14
% OVERLAP	12.92	46.02	11.12	41.29	20.61	18.94	41.68	42.86

Conclusions

In order to assess the Lubischew test as a useful tool for discrimination and classification in Archaeology, we have passed through four different archaeological examples. All of them come from data contained in bibliographic sources. The main conclusion we can draw is that the Lubischew test is a reliable way to evaluate discrimination and classification, from a quantitative point of view. Our numerical results are consistent with previous approaches. The other conclusion is that the Lubischew test works efficiently with different kinds of samples that have different data composition. Thus the case of the Roman Amphorae exemplifies the cases in which the variable under study is completely discriminatory. The Lubischew test shows, as in the case of the Roman amphorae, that it is possible to objectively classify these amphorae using only the length of the neck. That is, the Lubischew test detects the cases in which the discrimination is perfect using a single variable. From an archaeological standpoint, pieces are often incomplete, and therefore, to be able to use only one variable for discrimination gives the test added value. Trace elements are key to discriminate provenance for different materials. The analysis of the composition of Copper and Bronze Age ceramics shows lanthanum to be a good discriminant factor for both periods. These results confirm the importance of so called rare lands to Identity the possible provenience of ceramics.

Focusing on Experimental Archaeology results, we have characterized two groups of blades by the kind of hammer used (stone vs. antler). Different metric variables and presence of several production marks were used to extrapolate these characteristics to the archaeological record. Our results show that the length and the width of the blades serve as good discriminant factors to recognize the kind of hammer used. In the fourth example, we have analyzed elements present in two sets of fibulae. This is the most complex example and has allowed distinguishing the Huelva fibulae from the Granada ones on the basis of the percentage of strontium and coopering.

Finally, the Lubischew test is a statistical technique that enables us to know and evaluate in a numerical way the overlap between two samples from the same population, and it is used in an isolated way or combined with other statistical methods, such as discriminant analysis. Also, it is useful to evaluate samples easily and can be applied to samples with a low number of observations.

References Cited

Attanasio, D., R. Platania and P. Rocchi. 2005. The marble of the David of Michelangelo: A multi-method analysis of provenance, Journal of Archaeological Science 32(9):1369–1377.

Benecke, N. 1987. Studies On Early Dog Remains From Northern Europe. J. Archaeol. Sci. 14(1): 31–49.

Carrasco, J., J.A. Pachón and J.A. Esquivel. 2005. Nuevos datos para el estudio metalúrgico de la fíbula de codo tipo Huelva. *In*: A.L. Cortés, M.L. López-Guadalupe and F. Sánchez-Montes (eds.). Estudios en Homenaje al Profesor José Szmolka. Editorial Universidad de Granada, Granada. 21–39.

du Jardin, P., J. Ponsaillé, V. Alunni-Perret and G. Quatrehomme. 2009. A comparison between neural network and other metric methods to determine sex from the upper femur in a modern French population. Forensic Sci. Int. 192(1-3): 127.e1–6.

Evin, A., T. Cucchia, A. Cardini, U.S. Vidarsdottir, G. Larson and K. Dobney. 2013. The long and winding road: identifying pig domestication through molar size and shape. Journal of Archaeological Science 40(1): 735–743.

Fisher, R.F. 1936. The use of multiple measurements in taxonomic problems. Annals of Eugenics 7: 179–188.

Hart, J.P. and R.G. Matson. 2009. The use of multiple discriminant analysis in classifying Prehistoric phytolith Assemblages Recovered from Cooking residues. Journal of Archaeological Science 36:74–83.

Jimenez-Arenas, J.M. 2009. Discriminación de sexo en una población medieval del sur de la Península Ibérica. Cuadernos de Prehistoria y Arqueología de la Universidad de Granada 19: 463–477.

Jiménez-Arenas, J.M. 2010. Sex Discrimination in a Middle Age Population of the Southern Iberian Peninsula by the Use of Simple Variables. International Journal of Morphology 28(3): 667–672.

Jiménez-Arenas, J.M., P. Palmqvist and J.A. Pérez-Claros. 2011. A probabilistic approach to the craniometric variability of the genus Homo and inferences on the taxonomic affinities of the first human population dispersing out of Africa. Quaternary International 243(1): 219–230.

Jiménez-Arenas, J.M. and J.A. Esquivel. 2013. Comparing two methods of univariate discriminant analysis for sex discrimination in an Iberian population. Forensic Science International 228(1-3): 175.e1–175.e4.

Josephson, S.C., K.E. Juell and A.R. Rogers. 1996. Estimating Sexual Dimorphism by Method-of-Moments, American Journal of Physical Anthropology 100: 191–206.

Lubischew, A.A. 1959. On the use of biometry in systematics. (In Russian). Vestnik Leningradsk. Universit. 9: 128–36.

Lubischew, A.A. 1962. On the use of discriminant functions in taxonomy, Biometrics, Biometrics 18: 455–477.

Moscati, P. 1989. Archeologia e Calcolatori, Giunti, Firenze.

Munro, N.D., G. Bar-Oz and A.C. Hill. 2011. An exploration of character traits and linear measurements for sexing mountain gazelle (Gazella gazella) skeletons. Journal of Archaeological Science 38(6): 1253–1265.

Murphy, A.C.M. 2005. The femoral head: sex assessment of prehistoric New Zealand Polynesian skeletal remains Forensic Science International 154(2–3): 210–213.

Padial, B.R. 1999. La producción alfarera pre y protohistórica del asentamiento de Ronda la Vieja (Málaga). Aspectos tecnológicos y sociales, Editorial Universidad de Granada, ISNB 987-84-9028-072-0.

Palmqvist, P., V. Torregrossa, J.A. Pérez-Claros, B. Martínez-Navarro and A. Turner. 2007. A re-evaluation of the diversity of Megantereon (Mammalia, Carnivora, Machairodontinae) and the problem of species identification in extinct carnivores, Journal of Vertebrate Paleontology 27(1): 160–175.

Pitts, M.W. and R.M. Jacobi. 1979. Some aspects of change in flaked stone industries of the mesolithic and neolithic in Southern Britain. Journal of Archaeological Science 6(2): 163–177.

Serrano-Ramos, A., J.M. Jimenez-Arenas and J.A. Esquivel. 2013. A quantitative approach for late Pleistocene hominins' brain size, American Journal of Physical Anthropology 150, S56: 250–251.

Shapiro, S.S. 1965. An analysis of variance test for normality (complete samples). Biometrika 52 (3-4): 591–611. doi:10.1093/biomet/52.3–4.591.

11

Phylogenetic Systematics

Michael J. O'Brien,[1,*] *Matthew T. Boulanger,*[1] *R. Lee Lyman*[1] and
Briggs Buchanan[2]

Introduction

Archaeologists have long used changes in artifact form—in evolutionary terms, changes from one character state to another—to measure the passage of time (Lyman and O'Brien 2006). If evolved character states are ordered correctly, a historical sequence of forms is created, although independent evidence is needed to root the sequence—that is, to determine which end of the sequence is older. Such evidence could come, for example, from chronological dating (e.g., stratigraphy) or historical sources. Over the past several years, archaeologists have grown to appreciate that the methods biologists have developed to reconstruct the evolutionary, or phylogenetic, relationships of species can help them create not only historical sequences of artifact forms—what came before or after what—but sequences based on heritable continuity—what produced what.

One such method is cladistics, the extension of which into the cultural realm is based on the recognition that artifacts—pottery vessels, stone projectile points, and the like—comprise any number of parts that act in concert to produce a functional unit. The kinds of changes that occur over generations of, say, spear-point or ceramic-vessel manufacture are constrained in that new structures and functions usually arise through modification of existing structures and functions—descent with modification—as opposed to arising *de novo*. As with DNA, the history of cultural changes is recorded in the similarities and differences in characters (attributes of phenomena) as they are modified over time by subsequent additions, losses, and transformations (Brown and Lomolino 1998).

[1] Department of Anthropology, University of Missouri, Columbia, Missouri, 65211, USA.
[2] Department of Anthropology, University of Tulsa, Tulsa, Oklahoma, 74104, USA.
* Corresponding author: obrienm@missouri.edu

Here we describe the basic cladistic method, focusing first on distinguishing between homologous and analogous characters and, in the case of the former, distinguishing between derived and ancestral characters. We then turn attention to how trees are constructed, dividing the process into four steps: (1) generating a character-state matrix; (2) establishing the direction of evolutionary change in character states; (3) constructing branching diagrams of taxa; and (4) generating an ensemble tree. We then introduce an example of a cladistic analysis, the phylogenetic history of early Paleoindian-period (ca. 13,300–11,900 calendar years before the present [calBP]) projectile points from the southeastern United States. We stress that our discussion of cladistics is not intended to replace standard texts on the subject (for readable accounts see Brooks and McLennan 1991, Kitching et al. 1998, O'Brien and Lyman 2003, Williams and Knapp 2010); rather, it is a brief introduction to the logic behind, and key methodological elements of, cladistics.

Cladistics

Cladistics defines phylogenetic relationships in terms of relative recency of common ancestry: Two taxa are deemed to be more closely related to one another than either is to a third taxon if they share a common ancestor that is not also shared by the third taxon. The evidence for exclusive common ancestry is evolutionarily novel, or derived, character states. Two taxa are inferred to share a common ancestor to the exclusion of a third taxon if they exhibit derived character states that are not also exhibited by the third taxon.

For example, Fig. 1 is a cladogram, or phylogenetic tree (we use the terms here interchangeably), that classifies four taxa. It tells us that based on a certain character distribution, taxa C and D are more similar to one another than either is to any other taxon. It also says that taxa B, C, and D are more similar to one another than any of the three is to Taxon A. We know that taxa A–D evolved from one or more ancestral taxa, although at this point we know little or nothing about those ancestors except that with respect to certain characteristics taxa C and D look more like their immediate common ancestor (x) than they do the one (y) that unites them with Taxon B. Likewise taxa B, C, and D look more like their common ancestor (y) than they do the one (z)

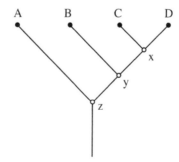

Fig. 1. A phylogenetic tree showing the historical relationship of four taxa (A–D) and three ancestors (x–z). Based on a certain character-state distribution (not shown), taxa C and D are more similar to one another than either is to any other taxon. Also, taxa B, C, and D are more similar to each other than any of the three is to Taxon A. Related taxa and their ancestors form ever-more-inclusive groups, or clades: C + D + x is one clade; B + C + D + y is a second; and A + B + C + D + z is a third.

that unites them with Taxon A. In cladistics, convention is to place *nodes* at the points where branches meet and to refer to the nodes as ancestors that produced the *terminal taxa* (those at the branch tips). In our tree, taxa C + D, together with their hypothetical common ancestor (node x), form a *monophyletic group*, or *clade*. Taxa D + C + B, together with their common ancestor (node y), form another, more inclusive clade, and taxa D + C + B + A, together with their common ancestor (node z), form yet another, and the most inclusive, clade.

One common misconception is that the interior nodes—"ancestors"—are somehow "real". They are not, and in fact, ancestors play no analytical role in cladistics because we can never be sure exactly what produced what. We know that taxa do not die when they produce offspring—Taxon z in Fig. 1 did not die when it produced Taxon A and Taxon y—so we show the ancestor as a sister taxon. It is simply a matter of convention to circumvent the illogical problem of having parents die when offspring are born (Sober 1988).

Characters and Character States

The key to cladistics lies in the kind of characters and character states that it employs. Two broad kinds of characters and character states occur in the natural world—analogs and homologs. Analogs are functionally similar characters (or character states) that evolve separately in two or more lineages after those lineages diverge. Thus they are of no utility in reconstructing lineages. In contrast, homologs *are* useful for tracking heritable continuity because they are holdovers from a previous time when two lineages were a single lineage. Darwinian theory provides the explanation for homology—descent with modification—but it does not tell us how to identify it. Although "similarity is the factor that compels us to postulate homology" (Cracraft 1981: 25), simple similarity in form is not a particularly useful criterion for homology. The reason for this is clear: Similarity can result from convergence. Thus whereas similarity is factual, homology must remain a hypothesis (Patterson 1988). But if it is a hypothesis, then it is testable (e.g., Brady 1985, Lyman 2001, McKitrick 1994).

Constructing Phylogenetic Trees

In its simplest form, cladistic analysis proceeds via four steps, the end process being the construction of phylogenetic trees that are useful in understanding not only the evolutionary relationships among taxa but also the evolutionary changes in characters that the taxa exhibit.

Step 1: Generating a Character-state Matrix

The data set used in any cladistic analysis is a matrix that lists the taxa and the various states that their characters exhibit. How do we choose appropriate characters, with appropriateness meaning how well a character performs in allowing us to separate taxa phylogenetically? In other words, how do we know a priori that a particular character will produce a phylogenetic signal? The bottom line is we don't. In reality, character choice is a classic case of trial and error, with a good measure of inductive reasoning thrown in. Archaeologists, like biologists, do not go into phylogenetic reconstruction with *no* prior knowledge of which characters might be useful. They know from chronological tracking—by means of, for example, radiometric dating, seriation, or superposition—or they strongly suspect, how certain characters change states over time. Other character polarity might not be so obvious in terms of which states are ancestral and which are derived, but even here prior knowledge offers at least a reasonable means of selecting useful characters.

Step 2: Establishing the Direction of Evolutionary Change in Character States

Several methods have been developed to facilitate establishing the direction of evolutionary change in character states, one of which is outgroup analysis (Maddison et al. 1984). Basic to the method is identifying a close relative of the taxa in the study group. The logic is this: When a character occurs in two states among taxa in the study group, but only one of the states is found in the outgroup, the principle of parsimony is invoked, and the state found only in the study group is deemed to be evolutionarily novel with respect to the outgroup state. It is important to make clear that analytical use of the term "parsimony" has nothing to do with whether evolution itself is parsimonious. Rather, it has to do with logical argumentation: It is more parsimonious to make as few ad hoc phylogenetic hypotheses as possible (Sober 1983).

Step 3: Constructing Branching Diagrams of Taxa

After the probable direction of change for the character states has been determined, usually through the use of computer programs designed for that purpose (see below), the third step is to construct a branching diagram that shows phylogenetic relationships of the taxa. This is done by joining the two most derived taxa by two intersecting lines and then successively connecting each of the other taxa according to how they are derived. Again, this is usually done with the assistance of computer programs. Various methods have been used for phylogenetic inference, each based on different models and each having its own strengths and weaknesses (Archibald et al. 2003, Goloboff and Pol 2005, Pol and Siddall 2001, Sober 2004). One, maximum parsimony, is based on a model that seeks to identify the least number of evolutionary steps required to arrange the taxonomic units under study. Parsimony trees are evaluated on the basis of the minimum number of character-state changes required to create them, without assuming a priori a specific distribution of trait changes. Two other commonly used methods, maximum likelihood and Bayesian inference, are probabilistically based, where the criterion for constructing trees is calculated with reference to an explicit evolutionary model from which the data are assumed to be distributed identically (Kolaczkowski and Thornton 2004). Cultural phylogenies that are based on language evolution have relied largely on probabilistic methods (e.g., Currie and Mace 2011, Gray et al. 2009). Those not based on language evolution—archaeological phylogenies, for example, which are more prospective—tend to rely on parsimony (e.g., Buchanan and Collard 2007, 2008, García Rivero and O'Brien 2014, O'Brien et al. 2001, 2012, Tehrani and Collard 2002).

Each group of taxa defined by a set of intersecting lines corresponds to a clade. Ideally, the distribution of character states among the taxa will be such that all the character-state relationships are congruent, but we have never witnessed such a happy event. Far more likely, a tree will contain multiple character states that show up in lines not related directly through one common ancestor. These are referred to as *homoplasies*. One kind of homoplasy results from character-state reversals—meaning, for example, that character state A changed to state A' and then at some later point in the lineage reverted to state A. We view this kind of homoplasy more as a classification problem, meaning that rarely if ever will precisely the same character state reemerge after it disappears. More likely, the classification system being used makes it *appear* as if the new character state is a homoplasy. Another kind of homoplasy results from parallelism or convergence—organisms, perhaps because of anatomical and/or environmental constraints (the first the result of common history, the second because of similar environments), independently evolve the same character state.

Step 4: Generating an Ensemble Tree

The fourth step is to generate an ensemble tree that is consistent with the largest number of characters and therefore requires the smallest number of homoplasies to account for the distribution of character states among the taxa. There are several ways of generating such a "consensus" tree, one of which is to construct a majority-rule consensus tree, which places taxa in their most common positions across the sample of trees (Swofford 1991). The percentage of trees in which the taxa must occur in the same positions can be varied between 50 percent and 100 percent. An example of a 50-percent majority-rule consensus tree is shown in Fig. 2. Notice that the G + H + I clade has the same arrangement in two out of the three trees; thus that arrangement is the one shown in the consensus tree. The same is true for the E + F clade. It also holds true for the other four taxa, although it is not as readily apparent. Note also that the middle tree just happens to have the same arrangement of taxa as the consensus tree.

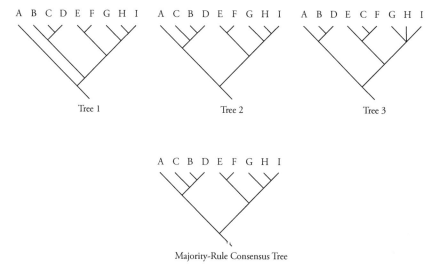

Fig. 2. Fifty-percent majority-rule consensus tree based on three trees of equal length.

Computers and Cladistics

As the number of taxa and/or the number of characters and character states increases, constructing trees and resolving them becomes impossible to do by hand. There are numerous programs available to perform all kinds of calculations and tree building. By far the best resource for perusing the various programs and finding out what they can do is on the Web at http://evolution.genetics.washington.edu/phylip/software.html. The site is maintained by Joseph Felsenstein of the Department of Genome Sciences at the University of Washington. At the time of this writing, June 2014, Felsenstein had listed almost 400 phylogeny packages and 54 free Web servers.

Perhaps the most widely used series of programs, and certainly one of the more user-friendly, is PAUP* (Phylogenetic Analysis Using Parsimony [and Other Methods]), written by David Swofford (1998) of the Department of Biological Science and the School of Computational Science and Information Technology at Florida State University. It currently is in version 4.0 (beta) and is a cross-platform program, running on Macintosh, Windows, and UNIX machines. Because the Macintosh version offers pulldown menus for all commands and settings, it is by far the simplest to use. However, the current GUI-based version is compatible only with MacOS 9 and earlier. Users with access only to MacOS 10 and later machines are restricted to the command-line UNIX version of PAUP. Originally, PAUP implemented only parsimony, but starting with version 4.0 (when the program became known as PAUP*) it also supports distance matrix and likelihood methods.

PAUP* and other tree-building programs can use several search methods to generate the shortest possible trees. There are two main kinds of searches: exact methods and heuristic methods. Which method should one use? That depends in large part on the amount of time available for analysis. Tree building is not a rapid process, especially with a large number of taxa. Lipscomb (1998: 40) summed up the bottom line beautifully: "Balancing the need for precision in finding the shortest tree against a reasonable amount of computation time is one of the most difficult computational problems for systematists."

Exact methods are guaranteed to find all shortest trees—if one has the time and computational equipment. The exhaustive search sorts through every possible tree until it finds the shortest one(s). The branch-and-bound search works by checking only those trees that are likely to be shorter than the shortest tree already found. It first creates a tree—any tree—and begins creating other trees to compare against it. As soon as it finds a tree of shorter length, that tree becomes the one against which to compare new trees. If a certain partial arrangement of taxa looks as if the trees it will produce are going to be longer than the comparative tree, the program doesn't waste time continuing to build trees in that direction. It abandons that direction and takes off in another one. Once it finds a partial solution that looks promising, it continues building in that direction until it finds a shorter solution or decides it's moving toward a longer tree and abandons that search vector.

If the data set is small, exact searches might be feasible, but for large data sets we might have to turn to heuristic methods. In heuristic searches there are no guarantees that even one shortest possible tree will be found, but we might get close. One heuristic method is branch swapping, of which there are two kinds. In local swapping, adjacent branches of a tree are systematically swapped until a shorter length is found. The routine continues swapping branches until no shorter trees are found (or until the operator terminates the search). In global swapping, the program slices the trees into "subtrees" and then rearranges the various "subtrees" into new trees and calculates their length.

An Archaeological Example

The earliest well-documented human occupation of North America is marked by the occurrence of bifacially chipped and fluted projectile points (Fig. 3) that date roughly 13,300–11,900 calBP (Faught 2008, Hamilton and Buchanan 2007)—a time period referred to as the Early Paleoindian period. Despite the fact that it is marked by the presence of fluted points, the Early Paleoindian period encompassed a range of spatial and temporal variation in such things as settlement pattern, diet, and technology (Haynes 2002, Meltzer 1993). The picture that has emerged for the first several hundred years of the Early Paleoindian period is one of hunters who targeted a wide range of large game animals, including mammoth, mastodon, bison, and, in the eastern woodlands, caribou (Cannon and Meltzer 2004, Robinson et al. 2009, Surovell and Waguespack 2009).

The most widely accepted hypothesis for the origin of Paleoindian peoples is that hunter-gatherer groups migrated by way of Beringia, the landmass between Siberia and North America that was exposed by sea-level reduction during glacial intervals (Haynes 2005). Once in eastern Beringia, the groups gained entry to the interior of the continent, specifically the Great Plains, by way of an ice-free corridor between two ice sheets that is hypothesized to have opened around 14,000 calBP (Catto 1996). Thereafter, Early Paleoindians spread rapidly throughout North and South America, reaching the Patagonian Plateau within just a few centuries (Fiedel 2000).

Much of our work over the last 15 years has focused on flaked-stone weaponry from the Early Paleoindian period (Buchanan and Collard 2007, 2008, Buchanan and

Fig. 3. Examples of Paleoindian fluted projectile points from North America: (a) Clovis (Logan Co., Kentucky); (b) Cumberland (Colbert Co., Alabama); (c) Crowfield (Addison Co., Vermont); (d) Dalton (Lyon Co., Kentucky); (e) Gainey/Bull Brook (Essex Co., Massachusetts); (f) Suwannee (Santa Fe River, Florida).

Color image of this figure appears in the color plate section at the end of the book.

Hamilton 2009, Buchanan, Kilby, Huckell et al. 2012, Buchanan, O'Brien, Kilby et al. 2012, Buchanan et al. 2011, 2014, Collard et al. 2010, Eren et al. 2013, Hamilton and Buchanan 2007, 2009, O'Brien, Boulanger, Buchanan et al. 2014, O'Brien, Boulanger, Collard et al. 2014, O'Brien et al. 2001, 2002, 2012). At the center of analysis is the Clovis point—an elongated symmetrical form that exhibits a concave base and a series of flake-removal scars on one or both faces that extend distally (toward the pointed end) (Fig. 3a). Despite the fact that a single type name, Clovis, is applied to many of these projectile points, considerable regional variation is evident across North America (Buchanan et al. 2014). For example, there are significant differences between points from the East and the West and among points from some subregions, one conclusion being that Clovis people modified their points to suit the characteristics of local prey and/or the habitats in which they hunted. Sometimes other projectile-point type names are given to regional variants (Fig. 3b–f).

One set of analyses was geared to understanding the evolution of points in the southeastern United States (O'Brien et al. 2001, 2002). The first problem encountered was with the taxa that could be used. Although types are commonly used to classify points from eastern North America, there are two major problems with this approach. One is a lack of redundancy in the characters used to create types. In the case of projectile points, one point type may be defined primarily by blade length and curvature, whereas another point type may be defined by basal shape and curvature. The other problem is that types are extensionally defined (Dunnell 1986, Lyman and O'Brien 2000, 2002), meaning that definitions are derived from the sorting of specimens into groups based on overall similarity and then describing the average properties of each group of specimens. Extensionally defined types are often fuzzy amalgams of character states because such units (types) conflate the taxa and the specimens in them (Lyman and O'Brien 2002, O'Brien and Lyman 2002).

To circumvent the problems of using established types, O'Brien and colleagues (2001, 2002) turned to paradigmatic classification (Dunnell 1971) in order to create classes (taxa).[1] Each class comprises eight unweighted characters, each of which has a variable number of character states (Table 1). The characters are defined as follows; locations of the characters are shown in Fig. 4:

I. Height of maximum blade width—the quarter section of a specimen in which the widest point of the blade occurs.

II. Overall base shape—qualitative assessment of the shape of the basal indentation.

III. Basal indentation ratio—the ratio between the medial length of a specimen and its maximum length; the smaller the ratio, the deeper the indentation.

IV. Constriction ratio—the ratio between the minimum blade width (proximal to the point of maximum blade width) and the maximum blade width; the smaller the ratio, the higher the amount of constriction.

V. Outer tang angle—the degree of tang expansion from the long axis of a specimen; the lower the angle, the greater the expansion.

VI. Tang-tip shape—the shape of the tip ends of tangs.

VII. Fluting—the removal of one or more large flakes (> 1 cm long) from the base of a specimen and parallel to its long axis; subsequent flake removal may obliterate earlier flake scars.

VIII. Length/width ratio—the maximum length of a specimen divided by its maximum width.

Table 1. System Used to Classify Projectile Points from the Southeast

Character Character State	Character Character State
I. Location of Maximum Blade Width	V. Outer Tang Angle
1. proximal quarter 2. secondmost proximal quarter 3. secondmost distal quarter 4. distal quarter	1. 93°–115° 2. 88°–92° 3. 81°–87° 4. 66°–80° 5. 51°–65°
II. Base Shape	6. ≤ 50°
1. arc-shaped 2. normal curve 3. triangular 4. Folsomoid	VI. Tang-Tip Shape 1. pointed 2. round 3. blunt
III. Basal-Indentation Ratio[a]	VII. Fluting
1. no basal indentation 2. 0.90–0.99 (shallow) 3. 0.80–0.89 (deep)	1. absent 2. present
IV. Constriction Ratio[b]	VIII. Length/Width Ratio
1. 1.00 2. 0.90–0.99 3. 0.80–0.89 4. 0.70–0.79 5. 0.60–0.69 6. 0.50–0.59	1. 1.00–1.99 2. 2.00–2.99 3. 3.00–3.99 4. 4.00–4.99 5. 5.00–5.99 6. ≥ 6.00

[a] The ratio between the medial length of a specimen and its total length; the smaller the ratio, the deeper the indentation.

[b] The ratio between the minimum blade width (proximal to the point of maximum blade width) and the maximum blade width as a measure of "waistedness"; the smaller the ratio, the higher the amount of constriction.

Seventeen classes had a minimum of four specimens each (83 specimens total), and they were the ones used in the analysis. The resulting tree—and, interestingly, there was only one most parsimonious tree—is shown in Fig. 5. It has a length of 22, a retention index (RI) of 0.70, and a CI of 0.59.[2] The RI and CI values are high enough to offer encouragement that the tree is fairly representative of the true phylogeny, but there are still some problematic features. Notice that the tree contains several *polytomies*, or points at which the program cannot make a simple dichotomous split. For example, there is a polytomy in the form of a trichotomous branching that produces KC, CU, and the ancestor of the clade comprising BQDU + DUCold + DCQURSuw + DAQS + QC + QUDG + QDG. Phylogenetic analysis often assumes that diversification occurs only by a series of bifurcations, but this assumption is unnecessary and may obscure reality (Hoelzer and Melnick 1994). In fact, cultural transmission may result more often in polytomies than in simple bifurcation. Alternatively, in reality there may be a bifurcation, but the phylogenetic signal is too weak for the program to resolve the pattern. Thus it treats it as a polytomy.

Character-state changes—there are 22 of them, hence the tree length of 22—are represented by the small boxes in Fig. 5. Each box has two numbers associated with

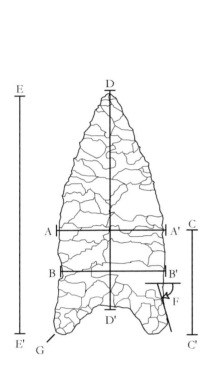

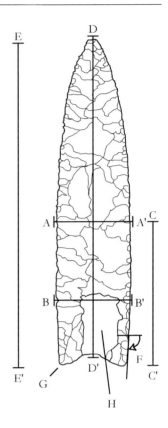

Landmark Characters

A–A' = maximum blade width

B–B' = minimum blade width

C–C' = height of maximum blade width

D–D' = medial length

E–E' = maximum length

F = outer tang angle

G = tang tip

H = flute

Base Shapes

arc-shaped

normal curve

triangular

Folsomoid

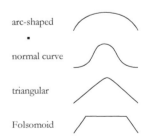

Fig. 4. Locations of characters used in the analysis of projectile points (see Table 1 for character states). Character states for base shape are shown at the lower right.

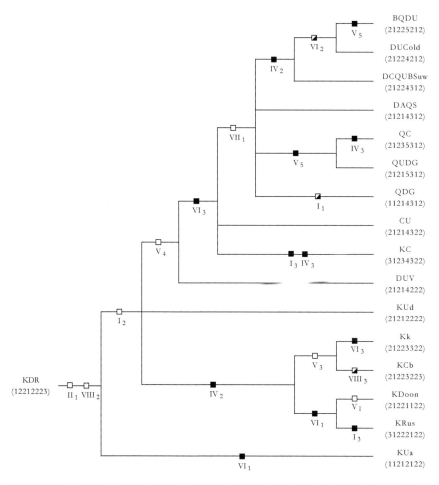

Fig. 5. Phylogenetic tree of 17 projectile-point classes. The tree has a length of 22 and a consistency index of 0.59. For simplicity, KDR is shown as an ancestor as opposed to a terminal taxon. Changes in character state are denoted by boxes, Roman numerals denote characters, and subscript numbers denote character states. For example, the boxes at the far left indicate that Class KDR underwent changes in characters II (to state 1) and VIII (to state 2) to produce the ancestor of the other 16 classes.

it: The Roman numeral refers to the character (Table 1 and Fig. 4), and the subscript Arabic number indicates the evolved state of that character (moving from left to right across the tree). Open boxes indicate nonhomoplasious changes. For example, KDR produced a descendant in which character II changed states from 2 to 1. In fact, all descendants of KDR exhibit character-state 1. Once that character state appeared, it never disappeared and reappeared. Similarly, KDR's direct descendant changed states in character VIII from 3 to 2. Later taxa might exhibit a different state of character VIII, but state 2 arose only once. Shaded boxes indicate parallelism or convergence (homoplasy)—that is, a change to a particular character state occurs more than once within the entire set. For example, character IV changes to state 3 in both the line leading to QC and the line leading to KC. Finally, half-shaded boxes indicate characters that reverted to an ancestral state. For example, character VI began in KDR, the outgroup, as state 2; it later changed to state 3 and then changed back to state 2 in the line that produced BQDU + DUCold.

Conclusion

As we have noted previously (O'Brien et al. 2012), the growing interest in cultural phylogenetics evident over the last two decades marks a return to the questions on which the founding of much of anthropology rests—a return that is important to the growth and continued health of anthropology. Why? Because, as Linder and Rieseberg (2004), point out, a reconstructed phylogeny helps guide interpretation of the evolution of traits by providing hypotheses about the lineages in which those traits arose and under what circumstances. Thus, it plays a vital role in studies of adaptation and evolutionary constraints.

Archaeology is a historical science, and its sole claim to unique status within the human sciences is its access to portions of past phenotypes—something that ethnographers, sociologists, psychologists, historians, and others who study humans do not have. Only archaeologists have access to the entire time span of culture, however it is defined. Historical questions are the most obvious ones archaeologists can ask, although this is hardly a strong warrant for asking them (Eldredge 1989). However, we believe archaeologists should ask historical questions not only because they have access to data that provide a direct test of historical hypotheses but also because answers to historical questions are critical to gaining a complete understanding of cultural manifestations occupying particular positions in time and space. To understand the operation of historical processes means that we not only have the units upon which the processes act in correct historical sequence but that we have them in a sequence that is correct from the standpoint of heredity. Cladistics is a method that allows us to do that.

Acknowledgments

We thank Juan Barceló and Igor Bogdanovic for their kind invitation to contribute to this volume. We also thank Barceló and several of the authors whose work appears in this volume for their suggestions for improving the paper.

Note

1. We have also employed other methods of capturing shape, including creating a series of landmarks and then computing interlandmark distances for each specimen (Buchanan and Hamilton 2009; Hamilton and Buchanan 2009) and using geometric morphometric methods, in which shape is defined as the geometric properties of an object that are invariant to location, scale, and orientation (e.g., Buchanan et al. 2011, 2013). All have their strengths and weaknesses, but where our interests are in phylogenetic relationships and plotting spatially various character states, we have found paradigmatic classification to produce superior results (e.g., O'Brien, Boulanger, Buchanan et al. 2014).

2. How robust are the trees we generate? That is, how well do they approximate the one true phylogeny we assume exists? One calculation is the consistency index (CI), which measures the amount of homoplasy in a data set (Farris 1989b, Goloboff 1991). The index ranges from 0 (complete homoplasy) to 1.0 (no homoplasy) and is calculated by dividing the number of characters in the data matrix by the number of characters on the tree. There are several potential drawbacks to the consistency index, one of which is that the CI value is not independent of the number of taxa. When the number of taxa increases, the CI value will decrease. To overcome some of the problems, Farris (1989a, 1989b) developed the rescaled consistency index and the retention index (RI), the latter of which measures the fit of characters to a cladogram (the ratio of apparent synapomorphy to actual synapomorphy). The RI is calculated as

$$\frac{\text{Max. steps in matrix} - \text{No. of characters on tree}}{\text{Max. steps in matrix} - \text{No. of characters in matrix}}$$

References Cited

Archibald, J.K., M.E. Mort and D.J. Crawford. 2003. Bayesian inference of phylogeny: A non-technical primer. Taxon 52: 187–191.

Brady, R.H. 1985. On the independence of systematics. Cladistics 1: 113–126.

Brooks, D.R. and D.A. McLennan. 1991. Phylogeny, Ecology, and Behavior: A Research Program in Comparative Biology. University of Chicago Press, Chicago.

Brown, J.H. and M.V. Lomolino. 1998. Biogeography (2nd ed.). Sinauer, Sunderland, MA.

Buchanan, B. and M. Collard. 2007. Investigating the peopling of North America through cladistic analyses of early Paleoindian projectile points. J. Anthropol. Archaeol. 26: 366–393.

Buchanan, B. and M. Collard. 2008. Phenetics, cladistics, and the search for the Alaskan ancestors of the Paleoindians: A reassessment of relationships among the Clovis, Nenana, and Denali archaeological complexes. J. Archaeol. Sci. 35: 1683–1694.

Buchanan, B., M. Collard, M.J. Hamilton and M.J. O'Brien. 2011. Points and prey: A quantitative test of the hypothesis that prey size influences early Paleoindian projectile point form. J. Archaeol. Sci. 38: 852–864.

Buchanan, B. and M.J. Hamilton. 2009. A formal test of the origin of variation in North American Early Paleoindian projectile points. Am. Antiquity 74: 279–298.

Buchanan, B., J.D. Kilby, B.B. Huckell, M.J. O'Brien and M. Collard. 2012. A morphometric assessment of the intended function of cached Clovis points. PLoS ONE 7(2): e30530.

Buchanan, B., M.J. O'Brien and M. Collard. 2014. Continent-wide or region-specific? A geometric morphometrics-based assessment of variation in Clovis point shape. Archaeol. Anthropol. Sci. 6: 145–162.

Buchanan, B., M.J. O'Brien, J.D. Kilby, B.B. Huckell and M. Collard. 2012. An assessment of the impact of hafting on Paleoindian projectile-point variability. PLoS ONE 7(5): e36364.

Cannon, M.D. and D.J. Meltzer. 2004. Early Paleoindian foraging: Examining the faunal evidence for large mammal specialization and regional variability in prey choice. Quaternary Sci. Rev. 23. 1955–1987

Catto, N.R. 1996. Richardson Mountains, Yukon–Northwest Territories: The northern portal of the postulated "ice-free corridor". Quaternary Int. 32: 21–32.

Collard, M., B. Buchanan, M.J. Hamilton and M.J. O'Brien. 2010. Spatiotemporal dynamics of the Clovis–Folsom transition. J. Archaeol. Sci. 37: 2513–2519.

Cracraft, J. 1981. The use of functional and adaptive criteria in phylogenetic systematics. Am. Zool. 21: 21–36.

Currie, T.E. and R. Mace. 2011. Mode and tempo in the evolution of socio-political organization: reconciling 'Darwinian' and 'Spencerian' approaches in anthropology. Phil. Trans. R. Soc. B 366: 1108–1117.

Dunnell, R.C. 1971. Systematics in Prehistory. Free Press, New York.

Dunnell, R.C. 1986. Methodological issues in Americanist artifact classification. Advances Archaeol. Method Theory 9: 149–207.

Eldredge, N. 1989. Punctuated equilibria, rates of change, and large-scale entities in evolutionary systems. J. Biol. Soc. Struc. 12: 173–184.

Eren, M.I., R.J. Patten, M.J. O'Brien and D.J. Meltzer. 2013. Refuting the technological cornerstone of the North Atlantic Ice-Edge Hypothesis. J. Archaeol. Sci. 40: 2934–2941.

Farris, J.S. 1989a. The retention index and homoplasy excess. Syst. Zool. 38: 406–407.

Farris, J.S. 1989b. The retention index and the rescaled consistency index. Cladistics 5: 417–419.

Faught, M.K. 2008. Archaeological roots of human diversity in the New World: A compilation of accurate and precise radiocarbon ages from earliest sites. Am. Antiquity 73: 670–698.

Fiedel, S.J. 2000. The peopling of the New World: Present evidence, new theories, and future directions. J. Archaeol. Res. 39–103.

García Rivero, D. and M.J. O'Brien. 2014. Phylogenetic analysis shows that Neolithic slate plaques from the southwestern Iberian Peninsula are not genealogical recording systems. PLoS One 9(2): e88296.

Goloboff, P.A. 1991. Homoplasy and the choice among cladograms. Cladistics 7: 215–232.

Goloboff, P. and D. Pol. 2005. Parsimony and Bayesian phylogenetics. pp. 148–159. In: V.A. Albert (ed.). Parsimony, Phylogeny, and Genomics. Oxford University Press, New York.

Gray, R.D., A.J. Drummond and S.J. Greenhill. 2009. Language phylogenies reveal expansion pulses and pauses in Pacific settlement. Science 323: 479–483.

Hamilton, M.J. and B. Buchanan. 2007. Spatial gradients in Clovis-age radiocarbon dates across North America suggest rapid colonization from the north. Proc. Nat. Acad. Sci. 104: 15625–15630.

Hamilton, M.J. and B. Buchanan. 2009. The accumulation of stochastic copying errors causes drift in culturally transmitted technologies: Quantifying Clovis evolutionary dynamics. J. Anthropol. Archaeol. 28: 55–69.

Haynes, C.V. Jr. 2005. Clovis, pre-Clovis, climate change, and extinction. pp. 113–132. In: R. Bonnichsen, B.T. Lepper, D. Stanford and M.R. Waters (eds.). Paleoamerican Origins: Beyond Clovis. Texas A&M Press, College Station.

Haynes, G. 2002. The Early Settlement of North America. Cambridge University Press, Cambridge.

Hoelzer, G.A. and D.J. Melnick. 1994. Patterns of speciation and limits to phylogenetic resolution. Trends Ecol. Evol. 9: 104–107.

Kitching, I.J., P.L. Forey, C.J. Humphries and D.M. Williams. 1998. Cladistics: The Theory and Practice of Parsimony Analysis. Oxford University Press, Oxford.

Kolaczkowski, B. and J.W. Thornton. 2004. Performance of maximum parsimony and likelihood phylogenetics when evolution is heterogeneous. Nature 431: 980–984.

Linder, C.R. and L.H. Rieseberg. 2004. Reconstructing patterns of reticulate evolution in plants. Am. J. Bot. 91: 1700–1708.

Lipscomb, D. 1998. Basics of Cladistic Analysis. http://www.gwu.edu/~clade/faculty/lipscomb/Cladistics.pdf.

Lyman, R.L. 2001. Culture historical and biological approaches to identifying homologous traits. pp. 69–89. In: T.D. Hurt and G.F.M. Rakita (eds.). Style and Function: Conceptual Issues in Evolutionary Archaeology. Bergin and Garvey, Westport, CT.

Lyman, R.L. and M.J. O'Brien. 2000. Measuring and explaining change in artifact variation with clade-diversity diagrams. J. Anthropol. Archaeol. 19: 39–74.

Lyman, R.L. and M.J. O'Brien. 2002. Classification. pp. 69–88. In: J. Hart and J.E. Terrell (eds.). Darwin and Archaeology: A Handbook of Key Concepts. Bergin and Garvey, Westport, CT.

Lyman, R.L. and M.J. O'Brien. 2006. Measuring Time with Artifacts: A History of Methods in American Archaeology. University of Nebraska Press, Lincoln.

Maddison, W.P., M.J. Donoghue and D.R. Maddison. 1984. Outgroup analysis and parsimony. Syst. Zool. 33: 83–103.

McKitrick, M.C. 1994. On homology and the ontological relationship of parts. Syst. Biol. 43: 1–10.

Meltzer, D.J. 1993. Is there a Clovis adaptation? pp. 293–310. In: O. Soffer and N.D. Praslov (eds.). From Kostenki to Clovis: Upper Paleolithic-Paleo Indian adaptations. Plenum, New York.

O'Brien, M.J., M.T. Boulanger, B. Buchanan, M. Collard, R.L. Lyman and J. Darwent. 2014. Innovation and cultural transmission in the American Paleolithic: Phylogenetic analysis of eastern Paleoindian projectile-point classes. J. Anthropol. Archaeol. 34: 100–119.

O'Brien, M.J., M.T. Boulanger, M. Collard, B. Buchanan, L. Tarle, L.G. Straus and M.I. Eren. 2014. On thin ice: Problems with Stanford and Bradley's proposed Solutrean colonization of North America. Antiquity 88: 606–624.

O'Brien, M.J., B. Buchanan, M. Collard and M.T. Boulanger. 2012. Cultural cladistics and the early prehistory of North America. pp. 23–42. *In*: P. Pontarotti (ed.). Evolutionary Biology: Mechanisms and Trends. Springer, New York.

O'Brien, M.J., J. Darwent and R.L. Lyman. 2001. Cladistics is useful for reconstructing archaeological phylogenies: Palaeoindian points from the southeastern United States. J. Archaeol. Sci. 28: 1115–1136.

O'Brien, M.J. and R.L. Lyman. 2002. The epistemological nature of archaeological units. Anthropol. Theory 2: 37–57.

O'Brien, M.J. and R.L. Lyman. 2003. Cladistics and Archaeology. University of Utah Press, Salt Lake City.

O'Brien, M.J., R.L. Lyman, Y. Saab, E. Saab, J. Darwent and D.S. Glover. 2002. Two issues in archaeological phylogenetics: Taxon construction and outgroup selection. J. Theor. Biol. 215: 133–150.

Patterson, C. 1988. Homology in classical and molecular biology. Mol. Biol. Evol. 5: 603–625.

Pol, D. and M.E. Siddall. 2001. Biases in maximum likelihood and parsimony: A simulation approach to a ten-taxon case. Cladistics 17: 266–281.

Robinson, B.S., J.C. Ort, W.A. Eldridge, A.L. Burke and B.G. Pelletier. 2009. Paleoindian aggregation and social context at Bull Brook. Am. Antiquity 74: 423–447.

Sober, E. 1983. Parsimony methods in systematics. pp. 37–47. *In*: N.I. Platnick and V.A. Funk (eds.). Advances in Cladistics (Vol. 2). Columbia University Press, New York.

Sober, E. 1988. Reconstructing the Past: Parsimony, Evolution, and Inference. MIT Press, Cambridge, MA.

Sober, E. 2004. The contest between parsimony and likelihood. Syst. Biol. 53: 644–653.

Surovell, T.A. and N.M. Waguespack. 2009. Human prey choice in the Late Pleistocene and its relation to megafaunal extinction. pp. 77–105. *In*: G. Haynes (ed.). American Megafaunal Extinctions at the End of the Pleistocene. Springer, New York.

Swofford, D.L. 1991. When are phylogeny estimates from morphological and molecular data incongruent? pp. 295–333. *In*: M.M. Miyamoto and J. Cracraft (eds.). Phylogenetic Analysis of DNA Sequences. Oxford University Press, New York.

Swofford, D. 1998. PAUP*: Phylogenetic Analysis Using Parsimony (*and Other Methods) (version 4). Sinauer, Sunderland, MA.

Tehrani, J. and M. Collard. 2002. Investigating cultural evolution through biological phylogenetic analyses of Turkmen textiles. J. Anthropol. Archaeol. 21: 443–463.

Williams, D.M. and S. Knapp (eds.). 2010. Beyond Cladistics: The Branching of a Paradigm. University of California Press, Berkeley.

12

Text Mining in Archaeology: Extracting Information from Archaeological Reports

Julian D. Richards,[1,*] *Douglas Tudhope*[2] and *Andreas Vlachidis*[3]

Introduction

Archaeologists generate large quantities of text, ranging from unpublished technical fieldwork reports (the 'grey literature') to synthetic journal articles. However, the indexing and analysis of these documents can be time consuming and lacks consistency when done by hand. It is also rarely integrated with the wider archaeological information domain, and bibliographic searches have to be undertaken independently of database queries. Text mining offers a means of extracting information from large volumes of text, providing researchers with an easy way of locating relevant texts and also of identifying patterns in the literature. In recent years, techniques of Natural Language Processing (NLP) and its subfield, Information Extraction (IE), have been adopted to allow researchers to find, compare and analyse relevant documents, and to link them to other types of data. This chapter introduces the underpinning mathematics and provides a short presentation of the algorithms used, from the point of view of artificial intelligence and computational logic. It describes the different NLP schools of thought and compares the pros and cons of rule-based vs. machine learning approaches to IE. The role of ontologies and named

[1] Department of Archaeology, University of York, The King's Manor, York, YO1 7EP, UK.
Email: julian.richards@york.ac.uk
[2] Hypermedia Research Unit, Faculty of Computing, Engineering and Science, University of South Wales, Pontypridd, CF37 1DL, Wales, UK.
Email: douglas.tudhope@southwales.ac.uk
[3] Hypermedia Research Unit, Faculty of Computing, Engineering and Science, University of South Wales, Pontypridd, CF37 1DL, Wales, UK.
Email: andreas.vlachidis@southwales.ac.uk
* Corresponding author

entity recognition is discussed and the chapter demonstrates how IE can provide the basis for semantic annotation and how it contributes to the construction of a semantic web for archaeology. The authors have worked on a number of projects that have employed techniques from NLP and IE in Archaeology, including Archaeotools, STAR and STELLAR and draw on these projects to discuss the problems and challenges, as well as the potential benefits of employing text mining in the archaeological domain.

Background

Easy access to the information locked within texts is a significant problem for the archaeological domain, in all countries. In the UK, there is an average of 6,000 interventions per annum, and there are equivalent figures for other European countries, varying only according to the extent of the legal requirement for intervention prior to development, and whether it is undertaken as a state-led operation (as in France, Germany, Greece and Italy) or commercial enterprise (as in the UK and Netherlands). In the US, in the order of 50,000 field projects a year are carried out by federal agencies under these mandates, with another 50,000 federal undertakings requiring record searches or other inquiries that do not result in fieldwork. However, there is no legal requirement to publish the outcome of all this activity, either in the USA or Europe.

On both sides of the Atlantic, therefore, this activity generates vast numbers of reports that together constitute the unpublished 'grey literature' whose inaccessibility has long been an issue of major concern. With so much work being performed and so much data being generated, it is not surprising that archaeologists working in the same region do not know of each others' work, let alone archaeologists working in different continents. Decisions about whether to preserve particular sites, how many sites of specific types to excavate, and how much more work needs to be done are frequently made in an informational vacuum. Furthermore new data is not fed into the research cycle and academic researchers may be dealing with information which is at least 10 years out-of-date. Nonetheless, the fact that such reports are not fully published should not be taken to suggest that the value of the archaeological data or interpretation is not significant enough for publication (Falkingham 2005).

In recent years the detrimental effect of inaccessibility and difficulty of discovery of the large amounts of archaeological information represented by this material has begun to be recognised by the academic community. Bradley (2006) has questioned why it is not more widely available, and several research projects have been undertaken specifically to attempt to synthesise the outcomes of development control archaeology from the grey literature (Fulford and Holbrook 2011). Digital collection and online delivery of both newly created (i.e., 'born digital') and legacy material could provide a solution to addressing these access issues. However, good access is predicated on good discovery mechanisms and these rely, amongst other things, on good data about data, or metadata.

In the UK the Archaeology Data Service (ADS) actively gathers digital versions of grey literature fieldwork reports as part of the OASIS project <http://www.oasis.ac.uk/> (Hardman and Richards 2003, Richards and Hardman 2008). The ADS grey literature library currently (as of June 2014) comprises over 26,000 reports although it is increasing at the rate of 200 per month. The Dutch e-depot for Archaeology, managed by DANS, also holds over 20,000 reports. In the UK all reports can be downloaded free of charge and there is a high level of usage. In collaboration with the British Library and Datacite each report is assigned a Digital Object Identifier, ensuring a permanent means of citation.

Each of the reports also has manually generated resource discovery metadata covering such attributes as author, publisher, and temporal and geospatial coverage, adhering to the Dublin Core metadata standard <http://dublincore.org/>. Generating metadata this way may be feasible, if time-consuming, where it is created simultaneously with the report's deposit with the ADS but it is not a feasible means of dealing with the tens of thousands of legacy reports known to exist. For any attempt to digitise these disparate and distributed sets of records to facilitate broader access, the key in terms of both cost and time would be automated metadata generation.

Within the ADS digital library there are also electronic versions of more conventional journals and reports, including a complete back run of the *Proceedings of the Society of Antiquaries of Scotland* (PSAS) going back to 1851 <http://dx.doi.org/10.5284/1000184>. Many of the same indexing issues arise with reference to digitised versions of such early or short run published material. As an increasing number of journal back-runs are digitised, and held within large online libraries such as JSTOR, or by smaller discipline-specific repositories such as the ADS, providing deeper and richer access to these resources becomes an increasing priority. Whilst the ADS repository is accessible to Google and other automated search engines these provide only free text indexing, regardless of any domain-specific controlled vocabularies, and they do not allow researchers to situate a specific term within the wider set of concepts implicit within a hierarchical thesaurus, identifying 'round barrow' and 'long barrow' as sub-types of barrow, for example, and even situating them as specific types of funerary monument. Such literal string match searches are also susceptible to large numbers of false positives, recovering 'Barrow' as a place name, or barrow as a wheelbarrow, for instance. Several research projects are now undertaking text mining on large quantities of published text in order to identify intellectual trends (Michel et al. 2011). Furthermore, in archaeology, there is the potential to provide joined-up access to published and unpublished literature within a single interface, allowing users to cross-search both types of resource. However, indexing of journal back-runs rarely goes beyond author and title. This is generally inadequate for the scholar wishing to investigate previous research on a particular site or artefact class. Furthermore, whilst modern fieldwork reports generally provide Ordnance Survey grid references for site locations, antiquarian reports use a variety of non-standard and historic place names, making it impossible to integrate this sort of information in modern geospatial interfaces. Ideally a methodology to automatically generate metadata for grey literature should be flexible enough to be applicable to this additional dataset with the minimum of reworking.

Mathematical Methods of Natural Language Processing

Statistical Information Retrieval

Information Retrieval (IR) is the activity of finding relevant information resources to satisfy specific user queries originating from generic information needs. The automatic definition of representative document abstractions (metadata or index terms) and ranking of search results is used by many different retrieval models that have been introduced in the last decades, including the Boolean model, vector space models and probabilistic models (Baeza-Yates and Ribeiro-Neto 1999). The Boolean model enables users to seek for information using precise semantics that are joined by the Boolean operators. Thus, a query is conventional Boolean expression composed of index terms linked by three operators: AND, OR, NOT, for example $[q = K_a \wedge (K_b \vee \neg K_c)]$. The model predicts that the document d_j is relevant to the query q if $sim(d_j, q) = 1$. The model supports clean formalisms but the exact match condition may lead to retrieval of too few or too many, relevant to the Boolean expression but irrelevant to user information needs, documents.

The vector model acknowledges the fact that binary weights are limited and does not provide a means for partial matching of user queries (Moens 2006). To overcome this, the vector model calculates the degree of similarity between document and query vector. The similarity between the query vector $\vec{q} = (w_{1,q}, w_{2,q}, ... w_{t,q})$ where t is the total number of index terms in the system and the vector for a document d_j, represented by $\vec{d}_j = (w_{1,j}, w_{2,j}, ... w_{t,q})$ is based on the quantification of the cosine of the angle between those two vectors. Since $w_{i,j} \geq 0$ and $w_{i,q} \geq 0$ the similarity $sim(q, d_j)$ varies from 0 to +1 and the model ranks the documents according to their degree of similarity to the query. An established threshold could dictate the retrieved matches based on the degree of similarity between query and document which might be equal to or over a given cut-off point. In this way partial match retrieval is achieved and results can be presented in a ranked order.

The vector model employs two distinct factors namely, the (tf) and the (idf) factor to provide a means of assigning weight to indexed terms. The frequency that a term ki appears within a document is known as intra-cluster similarity or tf and determines how well a term i is representative of the document contents. Inter-cluster dissimilarity or idf is a measurement of the inverse frequency of term i among the documents in collection. A term appearing in almost every document in a collection is not very useful for distinguishing a relevant document from a non-relevant one.

The probabilistic IR model is based upon the assumption that there is an 'ideal answer set' which contains exactly the relevant documents for a given user query. Knowing the description and attributes of an 'ideal answer set' will lead us to successful retrieval results. The probabilistic model defines a query q as a subset of index terms and R the set of documents known (or initially guessed) to be relevant. If $\sim R$ is the compliment of R and the set of non-relevant documents, then the probability P of the document Dj being relevant to the query q is defined as $P(R|Dj)$ (Baeza-Yates and Ribeiro-Neto 1999).

A number of variations to these IR models aim to improve and enhance their performance. The above models operate on the assumption that index terms are mutually independent and none of them acknowledge dependencies between index terms. This may result in poor retrieval performance since relevant documents not indexed by any of the query terms are not retrieved and irrelevant documents indexed with the query terms are retrieved (Smeaton 1997). The *Latent Semantic Indexing* model proposes a solution to this problem by enhancing the vector based model and matching each document and query vector to a lower dimensional space of concepts enabling concept based matching.

Information Extraction

Information Extraction (IE) is a specific NLP technique defined as a text analysis task which extracts targeted information from context (Cowie and Lehnert 1996, Gaizauskas and Wilks 1998, Moens 2006). It is a process whereby a textual input is analysed to form a textual output capable of further manipulation. Information extraction systems fall into two distinct categories; Rule-Based (hand-crafted) and Machine Learning systems (Feldman et al. 2002). During the seven Machine Understanding Conferences (MUC), the involvement of rule-based information extraction systems has been influential. The issue of information systems portability quickly gained attention and during MUC-4 the Machine Learning applications introduced a semi-automatic technique for defining information extraction patterns as a way of improving a system's portability to new domains and scenarios (Soderland et al. 1997).

Rule-based Information Extraction Systems

Rule-based systems consist of cascaded finite state traducers that process input in successive stages. Dictated by a pattern matching mechanism, such systems are targeted at building abstractions that correspond to specific IE scenarios. Hand-crafted rules make use of domain knowledge and domain-independent linguistic syntax, in order to negotiate semantics in context and to extract information for a defined problem. It is reported that rule-based systems can achieve high levels of precision of between 80–90 percent when identifying general purpose entities such as 'Person', 'Location', and 'Organisation' from financial news documents (Feldman et al. 2002, Lin 1995).

The definition of hand-crafted rules is a labour intensive task that requires domain knowledge and good understanding of the IE problem. For this reason rule-based systems have been criticised by some as being costly and inflexible, having limited portability and adaptability to new IE scenarios (Feldman et al. 2002). However, developers of rule-based systems claim that, depending on the IE task, the linguistic complexity can be bypassed and a small number of rules can be used to extract large sets of variant information (Hobbs et al. 1993). In addition, rule-based systems do not require training for delivering results in contrast to the supervised machine learning system discussed below.

Machine Learning Information Extraction Systems

The use of machine learning has been envisaged to provide a solution capable of overcoming the potential domain-dependencies of rule-based IE systems (Moens 2006, Ciravegna and Lavelli 2004). Learning in the Artificial Intelligence context describes the condition where a computer programme is able to alter its 'behaviour', that is, to alter structure, data or algorithmic behaviour in response to an input or to external information (Nilsson 2005).

Machine learning strategies can support supervised and unsupervised learning activities. The supervised learning process is based upon a training dataset annotated by human experts, which is used by the machine learning process to deliver generalisations of the extraction rules, which are then able to perform a large scale exercise over a larger corpus. It is argued that it is easier to annotate a small corpus of training documents than to create hand-crafted extraction rules, since the latter requires programming expertise and domain knowledge (Moens 2006). On the other hand, the size of the training set may depend on the range and complexity of the desired annotations and the characteristic language use in the domain.

During unsupervised learning, human intervention is not present and the output of the training data set is not characterised by any desired label. Instead a probabilistic clustering technique is employed to partition the training dataset and to describe the output result, with subsequent generalisation to a larger collection (Nilsson 2005). Unsupervised IE is very challenging and so far such systems have been unable to perform at an operational level (Uren et al. 2006, Wilks and Brewster 2009).

Information Extraction Evaluation

The evaluation of IE systems was established by the Machine Understanding Conference, MUC-2. Two primary measurements adopted by the conference, *Precision* and *Recall,* originated from the domain of Information Retrieval but were adjusted for the task of IE (template filling). According to the MUC definition, when the answer key is N_{key} and the system delivers $N_{correct}$ responses correctly and $N_{incorrect}$ incorrectly then $Recall = \dfrac{N_{correct}}{N_{key}}$ and $Precision = \dfrac{N_{correct}}{N_{correct} + N_{incorrect}}$.

The formulas examine a system's response in terms of correct or incorrect matches. This binary approach does not provide enough flexibility to address partially correct answers. A slightly scalar approach can be adopted to incorporate the partial matches. In this case, the above formulas can be defined as

$$Recall = \frac{N_{correct} + \frac{1}{2}\,Partial\ Matches}{N_{key}}, \quad Precision = \frac{N_{correct} + \frac{1}{2}\,Partial\ Matches}{N_{correct} + N_{incorrect} + Partial\ Matches}$$

Partial matches are shown weighted as 'half' matches. The value of the weight can change if partial matches seem more or less important.

The weighted average of Precision and Recall is reflected by a third metric, the F-measure score. When both Precision and Recall are deemed equally important then we can use the equation: $F_1 = 2\,\dfrac{Precision * Recall}{Precision + Recall}$. Attempts to improve Recall will usually cause Precision to drop and vice versa. High scoring of F_1 is desirable since the measure can be used to test the overall accuracy of the system (Maynard et al. 2006).

Gold Standard Measures

The Gold Standard (GS) is a test set of human annotated documents describing the desirable system outcome. An erroneous GS definition could distort the results of the evaluation and lead to false conclusions. Problematic and erroneous GS definition is addressed by enabling multiple annotations per document. The technique allows more than one person to annotate the same text in order to address discrepancies between different annotators. To calculate the agreement level between individual annotators the technique employs the Inter Annotator Agreement (IAA) metric (Maynard et al. 2006).

Manual annotation of archaeological documents is influenced by domain characteristics and embedded language ambiguities that challenge IAA scores. Such ambiguities concern the definition of domain entities; for example, the fine distinction between physical object and material entities, application of annotation boundaries and inclusion of lexical moderators. Typical IAA scores in an archaeological context range from between 60 and 80 percent (Byrne 2007, Vlachidis and Tudhope 2012, Zhang et al. 2010). In the case of a low IAA score, a final and explicit GS set is proposed by a human 'Super Annotator' who acts as a conciliator between individual annotation sets, reviewing the cases of disagreement and choosing the correct annotation (Savary et al. 2010). Normally the Super Annotator is a field expert with the experience and knowledge to reconcile individual annotation discrepancies, although it must be remembered that there may be underlying variation in language use and terminology within a domain.

Previous Work

Archaeology has excellent potential for the deployment of text mining because, despite its humanities focus, it has a relatively well-controlled vocabulary. Significant effort has been put into the development of controlled word lists or thesauri, including the UK MIDAS data standard (English Heritage 2007, Newman and Gilman 2007). However the nature of archaeological vocabulary poses some challenges in that, unlike highly specialised scientific domains with a unique vocabulary (e.g., biological or medical terms), much archaeological terminology consists of common words in an everyday sense, for example 'pit', 'well'. There is also the distinction between descriptions of the present and the archaeological past (for example, the term 'road' has much more significance if it is a 'roman road').

Within the last ten years a number of projects have attempted to deploy text mining on archaeological texts, with a specific focus on the grey literature. Amrani et al. (2008) reported on a pilot application in a relatively specialised area of archaeology; the *OpenBoek* project experimented with Memory Based Learning in extracting chronological and geographical terms from Dutch archaeological texts (Paijmans and Wubben 2008) and Byrne has also explored the application of NLP to extract event information from archaeological texts (Byrne and Klein 2010). The present authors have worked on two major projects that employed different methods of IE and these provide useful case studies of text mining in Archaeology. Archaeotools largely adopted a machine learning approach, whilst OPTIMA (which provided the basis of the STAR and STELLAR projects) adopted a rule-based approach. Both are described below.

The Archaeotools Project

In the UK, the Archaeotools project, a collaboration between the ADS and the University of Sheffield Computer Science OAK group, provided a major opportunity to deploy text mining in Archaeology (Jeffrey et al. 2009, Richards et al. 2011). The first objective of Archaeotools was to extract several types of information from a corpus of over 1000 unstructured archaeological grey literature reports, so that this corpus could be indexed and searched by a number of attributes, including subject, location, and period. These support the standard 'What', 'Where' and 'When' queries that underlie a broad range of archaeological research questions. The project employed a combination of a rule-based (KE) and an Automatic Training (AT) approach. The rule-based approach was applied to information that matched simple patterns, or occurred in regular contexts, such as national grid references and bibliographies. In order to deploy the AT approach the ADS staff, all of whom are archaeologically trained, carried out extensive annotation exercises on a subset (c.150 reports) of the grey literature corpus. The AT approach was applied to information that occurred in irregular contexts and could not be captured by simple rules, such as place names, temporal information, event dates, and subjects. In addition, both approaches were combined to identify report title, creator, publisher, publication dates and publisher contacts.

Text Mining Applied to Grey Literature

Relatively high levels of success were achieved when the above techniques were applied to the sample of 1000 semi-structured grey literature reports. By removing files which could not be converted to machine readable documents due to file formatting issues, this left a working sample of 906 reports.

The greatest problem encountered was that of distinguishing between 'actual' and 'reference' terms. As well as the 'actual' place name referring to the location of the archaeological intervention, most grey literature reports also refer to comparative information from other sites, here called 'reference' terms. The IE software returned all place names in the document, masking the place name for the actual site amongst large numbers of other names. However this was solved by adopting the simple rule that the primary place name would appear within the 'summary' section of the report. If it was not possible to identify a summary then the first ten percent of the document was used instead. Out of 1000 reports, this left 162 documents where it was not possible to identify a place name in the summary or first ten percent of the report.

Table 1. 'Actual' identifications for 1000 grey literature reports.

	No data	
What	159	17.5%
Where	162	17.9%
When	263	29.0%

However, for the documents as a whole there were only 17 where it was not possible to identify the 'What' facet, 20 with no 'Where' information, and 40 where it was impossible to identify a 'When' term.

Table 2. 'Reference' identifications for 1000 grey literature reports.

	No data	
What	17	1.9%
Where	20	2.2%
When	40	4.4%

Although these figures do not guarantee that the terms identified were meaningful, so long as users are shown why a document has been classified according to those terms they represent acceptable levels of classification.

Text Mining Applied to Historic Literature

Another strand of the Archaeotools project was to focus the NLP automated metadata extraction on the PSAS. Despite the highly unstructured nature of the text and the antiquated use of language we were surprised to find that once trained on the grey literature reports the IE software achieved comparable levels of success with the antiquarian literature. Problems were encountered with more synthetic papers, but where the primary subject of the article was a fieldwork report then it was possible to identify the key 'What', 'When' and 'Where' index terms.

After discounting prefatory papers, such as financial accounts, the PSAS corpus was reduced to 3991 papers referring to archaeological discoveries. By applying the rule that the actual 'What', 'Where' and 'When' would appear in the first ten percent of the paper it was possible to identify a subject term for all but 277 of the papers, although there was less success with a geospatial location (627 papers with no location), and least success with period terms (2056 papers with no When term).

However, these results could be improved somewhat by looking at the 'Reference' terms; although less certain to provide the primary identification of the key What, Where and 'When' for each paper these left far fewer papers unclassified:

Table 3. 'Actual' identifications for 3991 PSAS papers.

	No data	
What	277	6.9%
Where	627	15.7%
When	2056	51.5%

Table 4. 'Reference' identifications for 3991 PSAS papers.

	No data	
What	123	3.1%
Where	238	6.0%
When	1049	26.3%

Determining place names within the County-District-Parish (CDP) place name thesaurus proved a challenge, particularly given the number of historic names used in older accounts, but the geo-gazetteer web service hosted by EDINA at the University of Edinburgh was used to resolve many of the outstanding names. Extracted place names were sent directly to this service and the GeoXwalk automatically returned NGRs for the place name (centred) or in the case of some urban areas an actual polygon definition. This allowed the relevant place name from PSAS to be mapped in the Archaeotools geo-spatial interface and therefore made them as discoverable and searchable as standard monument inventory datasets.

Of the total of 3991 PSAS papers, it was initially impossible to find an Ordnance Survey grid reference for 3388 (85 percent), compared to a figure of just 185 (20 percent) for the grey literature. This reflects the fact that older reports did not tend to use precise geospatial references to refer to site or find locations. However, by using the GeoXwalk service it was possible to resolve a place name into a grid reference for all but 268 reports (6.7 percent)—for which there was no 'Where' term for 238 reports, leaving just 30 for which a place name had been identified that could not be geo-referenced by the EDINA web service. Manual checking revealed that the majority of these were instances where a county name was the most precise spatial location that had been used in the published paper.

The analysis of the PSAS also provided some tantalising glimpses into the potential of using IE tools to research the development of the use of more controlled and standardised vocabulary through time. In the process of generating the frequency counts used to identify the primary focus of each paper, the Archaeotools project produced frequency counts for each set of named entities for each article in the entire run of the PSAS available from the ADS. These represent the actual frequency of place names, period and monument types within these journals year on year, from 1851 to 1999. A superficial examination of these counts made it apparent that they detailed, metrically, what, when and where was being written about in each year and therefore what was considered significant at that time. It was clear that this could offer significant potential in the longitudinal consideration of changes in archaeological practice and thought. Bateman and Jeffrey (2011) were therefore able to give a more concrete basis to the presumed biases in subject and area believed to exist in the literature. For example, the usage of period terms varies in a non-random fashion both in the actual periods used and the number of different period terms themselves. The Roman period term was shown to dominate early articles and it is not until the 1970s that what we would recognise as the broad modern range of terms reflected in the MIDAS Heritage data standard came into use.

OPTIMA

By contrast, OPTIMA is an example of a rule-based semantic annotation system that performs the Natural Language Processing (NLP) tasks of Named Entity Recognition, Relation Extraction, Negation Detection and Word-Sense disambiguation using hand-crafted rules and terminological resources (Vlachidis 2012). Semantic Annotation refers to specific metadata which are usually generated with respect to a given ontology and are aimed to automate identification of concepts and their relationships in documents. The system associates contextual abstractions from grey literature documents with classes of the ISO Standard (ISO 21127:2006) CIDOC Conceptual Reference Model (CRM) ontology for cultural heritage and its archaeological extension, CRM-EH. The CRM entities Physical Object, Place, Time Appellation and Material are at the core of the system's semantic annotation process and form the basis of the system's acronym. In addition to the four main CIDOC-CRM entities, the system delivers a range of CRM-EH archaeology specific entities and relationships, which are expressed as annotations of 'rich' contextual phases connecting two or more individual entities. The hand-crafted rules of the system are expressed as JAPE grammars which are responsible for the delivery of the semantic annotations in context. JAPE (Java Annotation Pattern Engine) is a finite state transducer, which uses regular expressions for handling pattern-matching rules (Cunningham et al. 2000). The rules are developed and deployed within the NLP framework GATE (Cunningham and Scott 2004) and enable a cascading mechanism of matching conditions.

OPTIMA contributed to the Semantic Technologies for Archaeological Research (STAR) project (Vlachidis et al. 2010, Tudhope et al. 2011), which explored the potential of semantic technologies in cross search and integration of archaeological digital resources. STAR and the follow-on STELLAR projects were collaborations between the Hypermedia Research Unit at the University of South Wales (then the University of Glamorgan) with English Heritage and the ADS, funded by the UK Arts and Humanities Research Council (AHRC). STAR developed new methods for linking digital archive databases, vocabularies and associated unpublished on-line documents originating from OASIS (see above). The project supported the efforts of English Heritage in trying to integrate data from various archaeological projects, exploiting the potential of semantic technologies and NLP techniques to enable complex and semantically defined queries of archaeological digital resources. STAR developed a CRM-EH based search demonstrator which cross searches over five different excavation datasets, together with a subset of archaeological reports from the OASIS grey literature library (examples can be seen in Tudhope et al. 2011).

Named Entity Recognition

Named Entity Recognition (NER) is a particular IE subtask aimed at the recognition and classification of units of information within predefined categories, such as names of person, location, organisation, and expressions of time, etc. (Nadeau and Sekine 2007). The NER phase of OPTIMA employed hand-crafted rules, glossaries and thesauri to support identification of the four CRM entities (Place, Physical Object, Material and Time Appellation). Specialised vocabulary was also utilised by hand-crafted rules to support word-sense disambiguation and negation detection.

The NER phase introduced a novel approach of Semantic Expansion of the terminology-based resources contributing to the task. This invokes a controlled semantic expansion technique, which exploits synonym and hierarchical relationships of terminological resources for assigning distinct terminological and ontological definitions to the extracted results. The mechanism is capable of selective exploitation of gazetteer listings via synonyms, narrower and broader concepts relationships. Hence, the system can be configured to a range of different modes of semantic expansion depending on the aims of an IE task, i.e., being lenient and applying a generous semantic expansion or being strict and applying a limited semantic expansion.

A word-sense disambiguation module is invoked by the NER phase to resolve ambiguity between physical object and material terms by assigning appropriate terminological (SKOS) references (Isaac and Summers 2009). For example, when the term 'brick' is disambiguated as a material, a terminological reference from the *Material* thesaurus is assigned to the annotation. When the same term is resolved as a physical object, a terminological reference from the *Object Type* thesaurus is assigned instead.

The OPTIMA NER phase also implements a negation detection mechanism targeted at matching phrases which negate any of the four CRM entities (Vlachidis and Tudhope 2013). The implemented mechanism enhances the NegEx algorithm (Chapman et al. 2001) addressing known limitations and domain related issues. The primary aim of the negation module is to strengthen Precision by discarding negated matches that could reduce the validity of results (e.g., delivering a match on '*Roman settlement*' when it originates from the negated phrase '*No evidence of Roman settlement*').

Named Entity Recognition Results

The NER system's performance was conducted on summary extracts of archaeological fieldwork reports, originating from a range of archaeological contractors. The summaries present some significant advantages over other document sections as they are brief and contain rich discussion which reflects the main findings. The manual annotation task for the purposes of Gold Standard definition was conducted at the ADS by 12 staff and post-graduate students. Table 5 presents the full set of results. The Hypernym mode of semantic expansion, which exploits narrower terms of the vocabulary, delivers the best F-measure rates. However, there is a difference in the system performance between the different entity types.

Table 5. F-measure score of four CRM entities (E19.Physical Object, E49.Time Appellation, E53.Place and E57.Material) for the five modes of semantic expansion.

	E19	E49	E53	E57
Only-Glossary	0.63	0.98	0.69	0.50
Synonym	0.76	0.98	0.77	0.52
Hyponym	0.77	0.98	0.82	0.54
Hypernym	0.81	0.98	0.85	0.63
All-Available	0.73	0.98	0.83	0.57

The system performs best (98 percent) for the Time Appellation entity type (E49). The performance is the same across all 5 modes of semantic expansion because the entity is not affected by the expansion modes. This very good performance is based on the completeness of the Timeline thesaurus with its non-ambiguous terms. The Timeline thesaurus is the only terminological resource which contributes to the NER that does not have any overlapping terms with other terminological resources. The results of Physical Object (E19) and Place (E53) entities range from 63 percent to 85 percent depending on the semantic expansion mode. Places include archaeological contexts and larger groupings of contexts (locations are not the focus of the semantic annotation). The highest score for both entities is delivered by the Hypernym expansion mode reaching 81 percent and 85 percent for the Physical Object and the Place entity respectively. The system delivers the lowest F-measure score (50 percent) in the recognition of Material (E57), which can be ambiguous. For example the same concept ('iron', 'pottery', etc.) could be treated by archaeologists as a find (i.e., physical object) or as the material of an object. Although disambiguation is performed, it can still be challenging to identify. Whether the distinction is worth making might depend on the use cases for the information extraction.

Relation Extraction

Extraction of semantic relations between entities is a significant step towards the development of sophisticated NLP applications that explore the potential of natural language understanding. The OPTIMA pipeline can be configured to detect 'rich' textual phrases that connect CRM entity types in a meaningful way. The aim of the pipeline is to detect and to annotate such phrases as CRM-EH event or property entities. The pipeline uses hand-crafted rules that employ shallow parsing syntactical structures for the detection of 'rich' textual phrases. Other projects have also found shallow parsing useful for tackling the task of relation extraction (Zelenko et al. 2003).

The pair of entities that participate in an event phrase are the arguments of the event. For example the phrase '{ditch contains [pottery] of the Roman period}' delivers two CRM-EH events. One event connects 'ditch' and 'pottery' and another event connects the same 'pottery' with the 'Roman period', both events having 'pottery' as a common argument. The first event can be modelled in CRM-EH terms as a *deposition* event (EHE1004. ContextFindDepositionEvent) while the second event can be modelled as a *production* event (EHE1002.ContextFindProductionEvent). The pipeline detects contextual binary relationships, for example *"pit dates to Prehistoric period"*, *"pottery dates to Prehistoric period"*, *"ditch contains pottery"* and *"sherds of pottery"*, assigning the appropriate CRM-EH ontological annotations (Vlachidis 2012).

Conclusions and Future Work

Mathematical approaches to archaeology have generally been employed in rather esoteric research areas and have tended to lose popularity with the decline in interest in deterministic explanations and the rise of post-processual archaeology. By contrast in the last decade text-mining has increased in importance and has been employed to resolve problems of the inaccessibility of the results of day-to-day archaeological practice, previously locked up in the grey literature. It provides a good example of mathematical techniques serving the needs of the profession, to some extent bridging the gap between academic research and field practice. This has implications for the future structuring of reports in order to facilitate information extraction, for example through the provision of summaries, and the value of using controlled vocabularies. Above all it emphasises that those undertaking scanning projects of unpublished reports and back-runs of printed journals must always plan to produce machine-readable text in order to facilitate easy information extraction.

This chapter has described the underpinning algorithms and has provided an overview of the techniques employed, highlighting their strengths and weaknesses. It has highlighted two projects: Archaeotools, and OPTIMA (which underpins STAR and STELLAR).

Machine Learning and Rule based techniques are sometimes seen as competing NLP paradigms with different strengths and weaknesses. Which works best often depends on the specifics of the entities to be extracted and the language style of the text. It may also depend on the future use cases for the information extraction outputs and the applications that will consume the output.

However, the two methods can be combined, either in a complementary fashion, or sequentially in a pipeline. Archaeotools combined the two methods for different types of entities; OPTIMA employed rule-based techniques for very specific annotations involving 'rich phrases' that combined different types of entities in a meaningful way, while retaining the semantics of each entity in the ontological output. Thus the more specific OPTIMA annotations can be seen as complementary to the broader Archaeotools classifications of the main focus of the documents in question. Each can be seen as tending to serve a different use case, thus perhaps Archaeotools in classification for browsing and OPTIMA in providing more detailed annotations for semantic searching.

Looking to the future, as part of the EU-funded ARIADNE research e-infrastructure project (Niccolucci and Richards 2013) the authors are collaborating to explore the possibilities for combining rule-based approaches and machine learning sequentially as stages in a pipeline, as well as investigating the generalisation of OPTIMA rule based techniques to other European language grey literature. With the current interest in Big Data it seems that the potential of text mining to address archaeological research questions and some of the grand challenges of our domain (Kintigh et al. 2014) is only just beginning to be explored.

Acknowledgements

The STAR and STELLAR projects were supported by the Arts and Humanities Research Council [grant numbers AH/D001528/1, AH/H037357/1]. Archaeotools was funded under the AHRC/EPSRC/JISC eScience programme [grant number AH/E006175/1]. Julian Richards would like to thank Professor Fabio Ciravegna, Sam Chapman and Ziqi Zhang of the Sheffield Organisation, Information and Knowledge (OAK) group for their input to that project. Thanks are also due to Stuart Jeffrey and Lei Xia at the Archaeology Data Service, Phil Carlisle (English Heritage) for providing domain thesauri and for helpful input from Renato Souza (Visiting Fellow at University of South Wales).

References Cited

Dublin Core metadata standard <http://dublincore.org/>.
OASIS project <http://www.oasis.ac.uk/>
Proceedings of the Society of Antiquaries of Scotland (PSAS) going back to 1851
< http://dx.doi.org/10.5284/1000184>.
Amrani, A., V. Abajian, Y. Kodratoff and O. Matte-Tailliez. 2008. A chain of text-mining to extract information in Archaeology. Information and Communication Technologies: From Theory to Applications, 2008. ICTTA 2008. 3rd International Conference 1–5.
Baeza-Yates, R. and B. Ribeiro-Neto. 1999. Modern Information Retrieval. Boston, Addison-Wesley Longman Publishing Co., Inc.
Bateman, J. and S. Jeffrey. 2011. What Matters about the Monument: reconstructing historical classification. Internet Archaeology 29. http://dx.doi.org/10.11141/ia.29.6.
Bradley, R. 2006. Bridging the two cultures. Commercial archaeology and the study of prehistoric Britain. Antiquaries Journal 86: 1–13.

Byrne K. 2007. Nested named entity recognition in historical archive text. *In*: Proceedings (ICSC 2007) International Conference on Semantic Computing, Irvine, California, pp. 589–596.

Byrne, K.F. and E. Klein. 2010. Automatic Extraction of Archaeological Events from Text. pp. 48–56. *In*: B. Frischer, J.W. Crawford and D. Koller (eds.). Making History Interactive. Proceedings of the 37th Computer Application in Archaeology Conference, Williamsburg 2009. Archaeopress, Oxford.

Chapman, W.W., W. Bridewell, P. Hanbury, G.F. Cooper and B.G. Buchanan. 2001. A Simple Algorithm for Identifying Negated Findings and Diseases in Discharge Summaries. Journal of Biomedical Informatics 34(5): 301–310.

Ciravegna, F. and A. Lavelli. 2004. Learning Pinocchio: adaptive information extraction for real world applications. Natural Language Engineering 10(02): 145–165.

Cowie, J. and W. Lehnert. 1996. Information extraction. Communications ACM 39(1): 80–91.

Cunningham, H., D. Maynard and V. Tablan. 2000. JAPE a Java Annotation Patterns Engine (Second Edition). [online] Technical report CS—00—10, University of Sheffield, Department of Computer Science. Available at http://www.dcs.shef.ac.uk/intranet/research/resmes/CS0010.pdf.

Cunningham, H. and D. Scott. 2004. Software Architecture for Language Engineering. Natural Language Engineering 10(3-4): 205–209.

English Heritage 2007. MIDAS Heritage—The UK Historic Environment Data Standard (Best practice guidelines) http://www.english-heritage.org.uk/publications/midas-heritage/midasheritagepartone.pdf.

Falkingham, G. 2005. A Whiter Shade of Grey: a new approach to archaeological grey literature using the XML version of the TEI Guidelines. Internet Archaeology 17. http://dx.doi.org/10.11141/ia.17.5.

Feldman, R., Y. Aumann, M. Finkelstein-Landau, E. Hurvitz, Y. Regev and A. Yaroshevich. 2002. A Comparative Study of Information Extraction Strategies. Proceedings (CICLing-2002) Third International Conference on Intelligent Text Processing and Computational Linguistics, Mexico city, Mexico, 17–23 February.

Fulford, M. and N. Holbrook. 2011. Assessing the contribution of commercial archaeology to the study of the Roman period in England, 1990–2004. Antiquaries Journal 91: 323–345.

Gaizauskas, R. and Y. Wilks. 1998. Information extraction: beyond document retrieval. Journal of Documentation 54(1): 70–105.

Hardman, C. and J.D. Richards. 2003. OASIS: dealing with the digital revolution. pp. 325–328. *In*: M. Doerr and A. Sarris (eds.). CAA2002: The Digital Heritage of Archaeology. Computer Applications and Quantitative Methods in Archaeology 2002. Archive of Monuments and Publications Hellenic Ministry of Culture.

Hobbs, J.R., D. Appelt, J. Bear, D. Israel, M. Kameyama, M. Stickel and M. Tyson. 1993. FASTUS: A Cascaded Finite-State Transducer for Extracting Information from Natural-Language Text. In Proceedings (IJCAI 1993) 13th International Joint Conference on Artificial Intelligence, Chambery, France, 28 August–3 September.

Isaac, A. and E. Summers. 2009. SKOS Simple Knowledge Organization System Primer. [Online]. Available at: http://www.w3.org/TR/skos-primer (Accessed: 12 June 2012).

Jeffrey, S., J.D. Richards, F. Ciravegna, S. Waller, S. Chapman and Z. Zhang. 2009. The Archaeotools project: faceted classification and natural language processing in an archaeological context. *In*: P. Coveney (ed.). Crossing Boundaries: Computational Science, E-Science and Global E-Infrastructures, Special Themed Issue of the Philosophical Transactions of the Royal Society A 367: 2507–19.

Kintigh, K., J. Altschul, M. Beaudry, R. Drennan, A. Kinzig, T. Kohler, W.F. Limp, H. Maschner, W. Michener, T. Pauketat, P. Peregrine, J. Sabloff, T. Wilkinson, H. Wright and M. Zeder. 2014. Grand Challenges for Archaeology. American Antiquity 79: 5–24.

Lin, D. 1995. University of Manitoba: description of the PIE system used for MUC-6. In Proceedings (MUC 6) 6th Message Understanding Conference, Columbia, Maryland, 6–8 November.

Maynard, D., W. Peters and Y. Li. 2006. Metrics for Evaluation of Ontology-based Information Extraction, *In*. Proceedings WWW Conference 2006, Workshop on "Evaluation of Ontologies for the Web", Edinburgh, Scotland, 23–26 May.

Michel, J.-B., Y.K. Shen, A.P. Aiden, A. Veres, M.K. Gray, The Google Books Team, J.P. Pickett, D. Hoiberg, D. Clancy, P. Norvig, J. Orwant, S. Pinker, M.A. Nowak and E. Lieberman Aiden. 2011. Quantitative analysis of culture using millions of digitized books. Science 331, 176. doi:10.1126/science.1199644.

Moens, M.F. 2006. Information Extraction Algorithms and Prospects in a Retrieval Context. Dordrecht, Springer.

Nadeau, D. and S. Sekine. 2007. A survey of named entity recognition and classification. Lingvisticae Investigationes 30(1): 3–26.

Newman, M. and P. Gilman. 2007. Informing the Future of the Past: Guidelines for Historic Environment Records (2nd edition). ADS, ALGAO UK, English Heritage, Historic Scotland, RCAHMS and RCAHMW.

Niccolucci, F. and J.D. Richards. 2013. ARIADNE: Advanced Research Infrastructures for Archaeological Dataset Networking in Europe. International Journal of Humanities and Arts Computing 7(1-2): 70–88.

Nilsson, N. 2005. Introduction to Machine Learning. [Online]. Nils J Nilson publications. Available at: http://robotics.stanford.edu/people/nilsson/mlbook.html.

Paijmans, H. and S. Wubben. 2008. Preparing archaeological reports for intelligent retrieval. pp. 212–217. *In*: A. Posluschny, K. Lambers and I. Herzog (eds.). Layers of Perception. Proceedings of the 35th International Conference on Computer Applications and Quantitative Methods in Archaeology (CAA) Berlin, Germany, April 2–6, 2007. Kolloquien zur Vor- und Frühgeschichte Band 10, Bonn.

Richards, J.D. and C. Hardman. 2008. Stepping back from the trench edge. An archaeological perspective on the development of standards for recording and publication. pp. 101–112. *In*: M. Greengrass and L. Hughes (eds.). The Virtual Representation of the Past. Ashgate, London.

Richards, J.D., S. Jeffrey, S. Waller, F. Ciravegna, S. Chapman and Z. Zhang. 2011. The Archaeology Data Service and the Archaeotools project: faceted classification and natural language processing. pp. 31–56. *In*: S. Whitcher Kansa, E.C. Kansa and E. Watrall (eds.). Archaeology 2.0 and Beyond: New Tools for Collaboration and Communication. Cotsen Institute of Archaeology Press, Los Angeles.

Savary, A., J. Waszczuk and A. Przepiórkowski. 2010. Towards the Annotation of Named Entities in the National Corpus of Polish. In Proceedings (LREC'10) Fourth International Conference on Language Resources and Evaluation, Valletta, Malta, 13–17 May.

Smeaton, A.F. 1997. Using NLP and NLP resources for Information Retrieval Tasks. *In*: T. Strzalkowski (ed.). Natural Language Information Retrieval, Kluwer Academic Publishers.

Soderland, S., D. Fisher and W. Lehnert. 1997. Automatically Learned vs. Hand-crafted Text Analysis Rules, CIIR Technical Report T44, University of Massachusetts, Amherst.

Tudhope, D., K. May, C. Binding and A. Vlachidis. 2011. Connecting Archaeological Data and Grey Literature via Semantic Cross Search, Internet Archaeology 30. http://dx.doi.org/10.11141/ia.30.5.

Uren, V., P. Cimiano, J. Iria, S. Handschuh, M. Vargas-Vera, E. Motta and F. Ciravegna. 2006. Semantic annotation for knowledge management: Requirements and a survey of the state of the art. Web Semantics: Science, Services and Agents on the World Wide Web 4(1): 14–28.

Vlachidis, A., C. Binding, K. May and D. Tudhope. 2010. Excavating grey literature: a case study on the rich indexing of archaeological documents via Natural Language Processing techniques and knowledge based resources. ASLIB Proceedings 62 (4&5): 466–475.

Vlachidis, A. and D. Tudhope. 2012. A pilot investigation of information extraction in the semantic annotation of archaeological reports. International Journal of Metadata, Semantics and Ontologies 7(3): 222–235. Inderscience.

Vlachidis, A. 2012. Semantic Indexing via Knowledge Organization Systems: Applying the CIDOC-CRM to Archaeological Grey Literature. Unpublished PhD Thesis, University of South Wales (USW).

Vlachidis, A. and D. Tudhope. 2013. The Semantics of Negation Detection in Archaeological Grey Literature. pp. 188–200. *In*: E. Garoufallou and J. Greenberg (eds.). Metadata and Semantics Research. Communications in Computer and Information Science 390.

Wilks, Y. and C. Brewster. 2009. Natural Language Processing as a Foundation of the Semantic Web, Foundations and Trends in Web Science 1(3-4): 199–327.

Zhang, Z., S. Chapman and F. Ciravegna. 2010. A Methodology towards Effective and Efficient Manual Document Annotation: Addressing Annotator Discrepancy and Annotation Quality, Lecture Notes in Computer Science, 6317 pp. 301–315, Springer-Verlag: London.

Zelenko, D., C. Aone and A. Richardella. 2003. Kernel methods for relation extraction. Journal of Machine Learning Research 3: 1083–1106.

The Mathematics of Archaeological Time and Space

13

Time, Chronology and Classification

Franco Niccolucci[1,*] and *Sorin Hermon*[2,*]

Quid est ergo tempus? Si nemo ex me quaerat, scio; si quaerenti explicare velim, nescio.

St. Augustine[1]

Introduction

For archaeology, the "study of the past", time is a foundational concept. As it happens with many such concepts, they are often used in what mathematicians would call a "naïve" way, i.e., without much discussions on a concept and on the logic of its use, both tasks left to philosophers and logicians. What differentiates archaeology from history is the nature of the sources: archaeologists try to reconstruct and describe past events, people and places, basing on the study of artifacts and their context of recovery, whereas history relies mainly on the analysis of the content of written sources. Archaeology deals with material culture and, through its investigation from various perspectives, places it within spatio-temporal and cultural borders. Key elements in the archaeological research are therefore the excavation process itself and the detailed analysis of the items uncovered by this process, are grouped in categories based on various criteria and systematized in an ordered body of knowledge. Consequently, a fundamental activity in archaeology is the process of classification, which will determine the future assignment of finds in a chronology and their affiliation to a given 'culture'. Thus, the concepts of 'type' and 'time', together with 'space', are the pillars of archaeological knowledge. Despite being such, it is surprising to see that relatively little scientific effort

[1] PIN, scrl., Piazza Giovanni Ciardi, 25, 59100 Prato, Italy.
[2] The Cyprus Institute–20 Konstantinou Kavafi Street, 2121 Aglantzia, Nicosia, Cyprus.
* Corresponding authors

[1] "What is then time? If nobody asks me, I know. If I want to explain to whom is asking, I ignore". Augustinus, *Confessiones*, XI, 14, 17.

has been invested by archeologists in attempting to define and describe precisely and rigorously these basic concepts in archaeology. In most cases they are taken for granted, and classification, as a method of determining cultural/temporal/functional affinity of artifacts, is implemented in a rather straightforward, optimistic and positivistic (or anti-positivistic) approach. Here we will examine some issues arising in the assignment of 'type' and 'time', which underpin any further archaeological interpretation.

Issues with Time and Type Assignments

There has been, in fact, some debate among archaeologists about the intrinsic nature of time as a linear flow of events. Some authors attribute it to a Judaic-Christian vision of time, flowing from the creation to the end of the universe, which also influences evolutionistic theories. They compare it with other conceptions of time belonging to different cultures (Shanks et al. 1988). Others (Ramenofsky and Steffen 1997) have discussed the revolution induced in physics by the relativistic approach, which superseded the Newtonian vision of time as a continuous flux, serving as the background of events. They wonder whether a similar impact might happen for archaeology, concluding however (ibidem, page 78) that "archaeological time is closer conceptually to Newtonian time", which is "linear, continuous and non-reversible". These authors quote Renfrew's statement that "Dating is crucial to archaeology. Without a reliable chronology the past is chaotic" (Renfrew 1973: 20) to support their discussion of time units and measurement.

The scope of this chapter does not include a further discussion on the nature of time, and in particular, of archaeological time. We will place our work in the conceptual framework mentioned above, i.e., a linear, continuous and uni-directional time, as better explained in the following sections. The chapter goal is to introduce some simple logic tools to manage it properly and, above all, with a sound mathematical basis. In other words, if the activity of dating, i.e., establishing how old an artefact is, is used to place archaeological finds somewhere on a linear model of (past) time, and build thus a chronology by orderly arranging events witnessed by materials or historical references, the chapter will discuss which are the possible logical models, their features, and to which archaeological assumptions they correspond.

The most apparent feature about matching things and time is how uncertain such activity is, so all the mathematical methods that deal with uncertainty find here a very wide range of potential application. However, often random variables are difficult to manage, and a non-random value is preferred. Whenever useful, a confidence range is given, for example in Radiocarbon dating; in most of the other cases the value used is the average or the most probable value, which coincide when the variable distribution is close to the Normal (Gaussian) one as is frequently the case. Often a literal expression is used to express the uncertainty about the value, like well-known period appellations as "Bronze Age" or "Early Chalcolithic". If these expression are well understood by humans, and circumscribed to specific ambits, when using computers this may lead to misunderstandings and mistakes. In an attempt to overcome this specific issue Binding (2010) has proposed to create thesauri turning literal expressions into numeric ranges. In our opinion, this is just a patch: it does not take into account the other meanings such expressions have in different contexts, for example in diverse regions, and obliterates any vagueness the literal expression conveys.

The issue is not resolved using 'absolute dating', i.e., the dating of samples based on the measurements of various types of isotopes as the already mentioned C14 technique. When available, absolute dating is used, as detailed below, as a strong indicator of the time-span of the archaeological layer where the analyzed object was found, and extended to all the other objects found in the same layer. Absolute dating leaves ample margins of intrinsic uncertainty related to the technique used, and a coarse time granularity implied by assimilating all the objects in a given layer to the same time-span. Furthermore, it attaches a value to a layer,not a time-span. For this purpose 'relative dating' is more useful, as it places artifacts and their archaeological context 'before' or 'after' other distinct time slices. So dating often implies coarsening the time granularity to make time-span assignments feasible, or adopting approximations.

In a very similar way, assigning artifacts to predetermined types requires clustering them with some inevitable approximation and deciding if they belong, or do not belong, to a type, time-frame or culture, despite the sometimes problematic assignment of an artifact within a crisp category; namely, in other words, despite the fact that more often than not, the objects under investigation do not fit well in any category of the typological list used. Widening the cluster definition, i.e., adopting more generic type definitions, risks to void any deduction of archaeological value; keeping it narrower produces a more significant typification but makes the assignment to types more difficult and often forced.

It is clear that in both (and other) cases, the typical archaeological reasoning develops within the framework of an Aristotelian philosophical perspective, which, when transmuted into mathematical terms, becomes a Boolean approach, where artifacts do belong, or do not, to a certain type; dating has certain spans; and—but we will not deal with it in this chapter—a point in space belongs, or does not belong, to a place, for example an 'archaeological site'.

This issue affects time, classification, geographic and space definitions, and probably more. Since we have dealt with a possible solution to the classification issue in previous work, in the next section we will briefly outline those results, referring to those papers for further details. Then we will spend a bit more words on time, which has received in fact much less attention than it deserves.

Classification and Typology: The Fuzzy Logic Approach

Archaeologists recognized the problematic related to classification long time ago, and tried to remediate it, while remaining within the same Aristotelian philosophical/ logical framework, by applying various statistical techniques, such as probabilistic, or Bayesian methods. However, as it has been shown elsewhere more than ten years ago (Hermon and Niccolucci 2002), these techniques were developed for other purposes, mainly to discover future patterns or behavior of data, and not past events. In the same and also in other articles (Hermon and Niccolucci 2002, Hermon et al. 2004, Farinetti et al. 2010), it has been demonstrated that by shifting our reasoning to a different logical framework, namely the so-called fuzzy logic, we are closer to better describing archaeological reality.

Fuzzy logic is a multivalued logic. Instead of assigning two possible truth values, 'True' and 'False', to statements as in the Aristotelian logic, in fuzzy logic there is a continuum of truth values ranging from 0 (false) to 1 (true), and assuming intermediate values for statements on which the value is 'Maybe true'. In these cases the truth value corresponds to the subjective degree of belief in the truth of the statement. As

Aristotelian logic has a corresponding model in the classical set theory, fuzzy logic has a mathematical counterpart in the theory of fuzzy sets, illustrated in the following sections for time assignments. Besides being closer to the 'fuzzy' nature of archaeological types, such an approach also gives an estimate to the quality of the classification process, i.e., the scientist's level of confidence in its assignment of artifacts to particular typological classes.

In the papers quoted above, the fuzzy logic approach was extensively tested on various archaeological datasets, such as flint tools, pottery fragments, animal bone remains or use-wear types on lithic artifacts. Moreover, it has been shown how a fuzzy approach to the periodization of survey collections may give different time frames to the collections studied. Such an approach, besides reflecting closer the archaeological reality, also provides a method to determine and quantify qualitative matters of data under investigation. In the articles cited above, the authors describe a 'reliability index' which is assigned to each classification by the researcher itself. Such an index reflects the level of confidence of the research in the assignment of a particular item to a typological class. At assemblage level, the reliability index gives an indication to the 'quality' of the classification process, i.e., how well artifacts fit into selected categories.

Chronology and Dating

Dating

In his already mentioned book (Renfrew 1973), Renfrew outlines a history of chronology and archaeological dating. Actually the two concepts are different, although closely related. We will now briefly introduce the two concepts in what a mathematician would call a naïve way. In particular we will use the term 'event' in its intuitive meaning, as the happening of some phenomenon.

A **chronological relation** between events is a preorder relation on the set E of the events considered in the discourse, i.e., a binary relation $\leq_C$ which is reflexive and transitive. If E is the set of all the events, a chronology is a relation such that for any events e, e_1, e_2 and e_3 the following holds

$$e \leq_C e$$

$$\text{if } e_1 \leq_C e_2 \text{ and } e_2 \leq_C e_3 \text{ then } e_1 \leq_C e_3.$$

The above conditions are intuitively clear if $\leq_C$ is interpreted as 'does not follow in time', i.e., 'precedes or is contemporary'. This relation is clearly partial, since there may be cases in which for two given events e_1, e_2 neither $e_1 \leq_C e_2$ and $e_2 \leq_C e_1$ hold.

A **dating function** is instead a mapping d of E on the set I of intervals of real numbers, associating to an event e an interval (a, b), the time-span of e, i.e., the set of the real numbers (here, times) during which e is happening. A dating induces a chronology on E:

$$\sup (d(x)) \leq \inf(d(y)) \Rightarrow x \leq_C y \text{ for } x, y \in E$$

$$\sup (d(x)) \leq \inf(d(y)) \Rightarrow x \leq_C y \quad x, y \in E$$

The **duration** of an event is another mapping f from E to $\mathbb{R}$, which assigns a real number to an event. The duration measures the time-span of the event. If there is a dating, the duration of an event can be computed:

$$f(e) = \sup (d(e)) - \inf(d(e))$$

There may exist events that are outside of the domain of the dating function, i.e., for which no dating is available, but having a duration; and events for which neither the dating nor the duration is available. Note that in normal speech, duration may refer to the time length of an event ("a duration of four years") but also to its time-span ("the war duration was from 1939 to 1945").

Due to the fact that events extend in time, there may be overlaps which complicate the temporal sequence and the number of possible relations that may exist between two events: the first ends before the second begins; they have some overlap; one starts later and ends earlier than the other, and so on. If a dating function exists for all the events considered, i.e., if it is known, for every event, 'when' it starts and 'when' it ends, the mutual chronological relation between two events may be represented by the relations between the corresponding time-spans, the intervals assigned by this omnipotent dating function to each event. This case corresponds to the Allen algebra of time intervals as defined in (Allen 1983). Unfortunately, Allen's model that time may be represented as intervals of real numbers, is only an abstraction of reality that does not fit with archaeological use. Actually, no such thing as a universal dating function exists: or, at least, to make it practical it is necessary to deal with the granularity of time and consider for our reference that time is a discrete set instead of a continuous one.

Generally speaking, when measuring time it is impossible to consider it as continuous flow. As in watches, however precise, time advances in ticks. Two events may appear as contemporary even if they start, or end, with a difference in time so small that the clock cannot measure it, as it is easily experienced in everyday life. If one use the clock of a belfry, the timing of an event will round up to minutes, or even tens of them. With a standard watch, measuring time intervals shorter than a tenth of second is impossible, and one has to measure them using more sophisticated devices and methods. Each measuring device cannot measure time intervals shorter than a given threshold, and the related measure is consequently granular. This has little effect in real life because every domain has its own accepted granularity and time measuring devices fit for the purpose. A calendar, where the granule is the day, works for most everyday tasks. Watches with their one-second granularity are suitable for tasks more demanding in terms of precision, like catching a train or arriving at office on time. On the other hand, nuclear physics requires a much finer granularity. However, nobody would ever use an atomic clock to check train departures, or a wristwatch to compute bank interests, because the granularities don't fit: each activity and domain has its own preferred granularity and time measuring instruments devised for the purpose; and, a finer granularity does not necessarily imply better suitability. Unfortunately, this is not the case with archaeology. In this discipline, the granularity associated with dating varies very widely, from a very coarse one, with a granule of millennia, for events far in the past, to a much finer one, a century or even less, as events come closer to the present. As it will be discussed in the following section, there is an exception to this consideration, i.e., when an absolute dating is possible. So the granularity of the dating depends also on the dating methods that can be applied in different cases: each case has in fact its own time granularity.

There is, however, a fundamental question: is dating really important, beyond being instrumental to create a chronology, and allowing a language simplification in the description? A site chronology is a straightforward consequence of the stratigraphic sequence. Where chronology shows its limits is when cross-site comparisons are necessary, so reference to a common timeline makes ordering much simpler. But mixing a too coarse dating granularity with a finer one may lead to paradoxes and mistakes, or to trivial and irrelevant conclusions.

Dating in Archaeology

Building a chronology, i.e., establishing the time succession of events, is necessary to describe the story of objects, of sites and ultimately of people who lived there and used those objects. To create such a chronological order, as already mentioned, one may rely on dating, for which different methods may be used:

- Absolute dating of things, i.e., referring things to an absolute time-scale by means of a shared cross-reference to a scientific law concerning time-processes of growth/ decay, for example
  1. Isotope decay, e.g., the well-known C14
  2. Dendrochronology, i.e., the age of associated trees
     or to a well-defined chronology, for example
  3. Coinage or other datable archaeological material (e.g., inscriptions) creation
- Absolute dating of events via historical reports, linking events to other absolutely datable events (e.g., of astronomic nature), or to events belonging to a chronology or series (e.g., of kings), a calendar or other time-series (e.g., dating *ab urbe condita*, or using the Olympics), and via the latter to a standardized calendar.
- Relative dating, i.e., establishing time relations between things (i.e., chronologies, possibly distinct), especially stratigraphic relations used with the stratigraphic rule "if no external upsetting or intrusion intervened, what is in an upper layer is more recent; what is in a lower layer is older; what is in the same layer is contemporary". This expression must take into account the proviso made in section 2: in fact we are dating the strata, and not the objects inside. Also note that this kind of dating introduces a time granularity based on the time-span of each stratum, which may lack uniformity even throughout the same site.

Many examples of 'dating' things are very often instrumental to dating events. Very few things are so precious to deserve dating on their own. In general they are used as a mean to date events/periods in which they participated. This is, for example, the case with ceramics and its stylistic (or chemical, or petrographic) analysis, used for dating the phases of the site where they were found, under the assumption that co-presence implies co-existence.

Dating periods/events is (often implicitly) based on chronological rules that ultimately rely on the already mentioned Allen algebra of time intervals. Allen algebra deals with time intervals, with instants considered as time intervals of atomic granularity. It defines thirteen possible relations between time intervals, in fact seven, i.e., equality and six mutual relations with their inverses. Given two time intervals A and B, the possible relations between them are: A *equals* B; A *before* B, i.e., the end of A occurs before the beginning of B; A *meets* B, i.e., the end of A corresponds to the beginning of B; A *overlaps* B, i.e., the end of A occurs after the beginning, but before the end, of B; A *during* B, i.e., A is fully contained in B, the beginning of A being after the beginning if B, and the end of A being before the end of B; A *starts* B, i.e., the beginning if A is the same of *B*, and the end of *A* is before the end of *B*; and finally *A ends B*, i.e., the beginning of *A* is after the beginning of *B* and the two ends of *A* and *B* coincide.

The picture below summarizes the six relations that together with their inverses and equality form Allen's set.

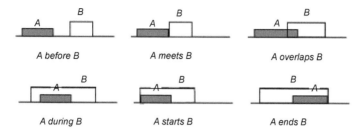

| A before B | A meets B | A overlaps B |
| A during B | A starts B | A ends B |

For example, in stratigraphy the rule is that the period *A* corresponding to a lower layer *meets* the period *B* corresponding to the layer immediately above it, and is *before* any other period *C* corresponding to layers above it, but not touching its layer.

The association between things and periods is more complicate. For example, in CIDOC-CRM, things do not have a time-span. They are however associated at least to their production, i.e., when they were created, and almost always also to their destruction, i.e., when they were destroyed or modified to give place to another thing. Transferring dating information from things to periods, e.g., the period associated to a layer, follows the rule of the *fossile directeur* in its various approaches (see Hermon and Niccolucci (2002) for more details). In general *Terminus Post Quem* (TPQ) and *Terminus Ante Quem* (TAQ) are determined using the stratigraphic sequence.

Dating imprecision

Both absolute and relative dating do not provide exact chronology. This is caused essentially by an issue of time granularity: stating that the battle of Cannae took place in the year 216 BC enables us to put it in a sequence that has the year as the finest granularity. It may be compared time-wise with other events to decide if it took place before or after only if these events happened in different years, but no time comparison may be done for events taking place in the same year unless additional information is acquired or the time granularity is made finer, e.g., stating that the battle took place on August the 2nd, unless other information is introduced: it is obvious that the battle of Cannae took place after the election of the two Roman consuls Gaius Terentius Varro and Lucius Aemilius Paulus, who led the Roman army in the defeat.

Sometimes the time granularity is so coarse that it is unsatisfactory, especially when mixed information is used, as the overall granularity must correspond to the coarsest one.

Another imprecision comes from events that do not have a well-defined beginning and end. Again, beginning and end are a matter of granularity, because besides being artificially created, the beginning of an event cannot be stated with precision. For example, the second term of Barak Obama as President of the USA started on 20 January 2013 at noon (12:00:00). Apart from these cases where a rule decides the exact starting point of a period (in the above-mentioned example, the 20th Amendment to the US Constitution), which we might call 'fiat events', all other events start gradually and at some granularity—possibly very fine—the event is still mixed with its absence. For almost all the events, at the time granularity currently used in history or archaeology, they have an initial period in which they both exist and do not exist. This complicates transitions from one status to another. Time borders between two consecutive events are intrinsically imprecise unless they are fixed by some rule, what we might call 'fiat borders'. The transition from the Roman Republic to the Empire was gradual and progressive; the transition from the French Republic to Napoleon's empire officially took place with his coronation on 2 December 1804. Contemporaneity of events is affected by similar issues.

Finally, there is a methodological aspect that must not be ignored when dealing with 'named' periods in history and archaeology. An archaeological 'named' period almost always does not simply mean giving an appellation to the corresponding time-span, for example as it happens with January, which is the name given to the time-span from the first until the thirty-first day of the year. On the contrary, an archaeological period is defined basing on a cluster of artefacts and related technologies: the expression 'Iron Age' refers for example to a period of the past in which the technology of iron production was gradually developed and eventually replaced other technologies, delivering various tools and artefacts and leading to a specific social organization. It is a result of our present apprehension of past behaviours, production processes and societal structure, in sum of what archaeologists call a 'culture'. Attaching a time-span to this construct mixes an absolute concept, the time-span, with a relative one, our mental organization of the knowledge about the past. As such, this operation is subject to critical appraisal, and in any case introduces further indeterminacy and imprecision.

Dealing with Dating Imprecision: The Rough Set Approach

Denote by ξ an event and by X the time-span during which it takes place; all times t will be henceforth assumed to be real numbers. Also denote with I the family of (time) intervals (a, b) of real numbers. X is assumed to be a convex set, i.e., an interval: if it is known that two times t_1 and t_2 belong to it, any other time between t_1 and t_2 also belongs to it. Actually, if there would be a gap in X, i.e., some time for which the event ξ does not take place, this interruption would determine two different events, one before and another one after the gap. Although X is an interval, the problem usually consists in the lack of knowledge about its extremes, for the reasons discussed above. Thus the common set algebra is insufficient to model archaeological temporal reasoning.

Possibly, even if there is no detailed information about the beginning and end of X, there may be a TPQ y_1 and a TAQ y_2 such that X is contained in the time interval Y: $Y = (y_1, y_2) \in I, X \subseteq Y$; or, in other words, no part of Ξ, the event associated with X, happens before y_1 or after y_2, i.e., outside the time span Y. In terms of Allen algebra, X is during Y. If there are many such time intervals in which X is contained for sure, the **support** of X, supp(X) is defined as the smallest interval such that $X \subseteq Y$. Thus supp(X) is the smallest time interval *with known extremes* that contains X. If X is itself an interval with known extremes, it coincides with its support.

Analogously, there may be an interval Z such that $Z \subseteq X$, i.e., such that Ξ, the event associated with X, takes place for sure at all the times $z \in Z$, and perhaps also at some other time outside Z. The union of all such intervals gives the maximal (time) interval fully contained in X, called the **core** of X, core(X).

It always holds that:

$$\text{core}(X) \subseteq X \subseteq \text{supp}(X)$$

with the equality holding if X is also an interval with known extremes.

As an archaeological example, we could use the time span X of existence and seafaring of the so-called 'Sea Peoples'. According to inscriptions, they were surely active during M, the reign of Merneptah (1213–1203 BC), so core(X) includes this time interval. Their existence took place during LBA, the Late Bronze Age (15th to 13th century BC), which is therefore part of their support. With no other information, we could say that

$$\text{core}(X) = M, \text{supp}(X) = LBA.$$

If one attempts to use classical set theory, the above definitions reduce to the trivial case, but the resulting model is of little or no use. Since this happens also in other applications, various extension of the classical set theory have been proposed to deal with uncertain cases as the above.

In an extension called *rough set theory*, ordinary sets are augmented by the introduction of so-called *rough sets*, for which no information is available except an inner approximation (what we have called the core) and an outer approximation (what we have called the support). The in-between region, called the *boundary* of X

$$\text{boun}(X) = \text{supp}(X) - \text{core}(X)$$

i.e., the set of points that belong to supp(X) but do not belong to core(X), corresponds to times for which it is undecidable if the event ξ was on-going or not. An interesting approach to rough sets use is the concept of indiscernibility. For our purpose, two elements ξ and ψ of our universe of discourse U are time-indiscernible, and are equivalent in the relation $\Re$, i.e., $\xi \Re \psi$, if it is impossible to decide if one starts or ends before the other one. Consider then the quotient set of U, usually denoted as $U/\Re$, consisting of the subsets of U formed by equivalent elements according to $\Re$:

$$A \in U/\Re \text{ if for any two } \xi, \psi \in A, \xi \Re \psi.$$

The elements of $U/\Re$ are therefore elements of the set of all subsets of U, generally denoted as 2^U.

To exemplify this rather technical definition, consider in the set U of battles the equivalence relation $\Re$ = "take place in the same year". An equivalence class is formed by all the battles taking place in the same year: within a given year it is impossible to know which element precedes in time another element of the same class. So, time-wise a battle is a rough set, because only the year is known, with core = $\varnothing$ and supp = the year in question. On the other hand, the previous construction groups together all the battles dated on the same year, and defines the same rough set.

Rough sets are usually applied to the solution of problems in information systems and to manage and classify large quantities of data. They were introduced by Pawlak in the eighties and there is an active research community working on them. An extensive bibliography is given in Polkowski et al. (2000), while the most recent advancements are presented at the annual conferences of the International Rough Set Society and published in its proceedings, the most recent being Lingras et al. (2013).

Using rough sets to manage the comparative dating of archaeological events is potentially useful when large datasets are involved. This could be considered the extension of a common practice in archaeology that uses the TAQ (*Terminus ante quem*) and TPQ (*Terminus post quem*) to assess the chronological order of events. If two events ξ_1 and ξ_2 are considered, with $TAQ(\xi_1) \leq TPQ(\xi_2)$, it is possible to deduce that ξ_1 precedes in time ξ_2. Similarly, if $TPQ(\xi_1) \geq TAQ(\xi_2)$, ξ_1 follows in time ξ_2. In these cases set theory and standard ordering help in organizing information time-wise. But when $TAQ(\xi_1) > TPQ(\xi_2)$ and $TPQ(\xi_1) < TAQ(\xi_2)$, no such chronological inference can be made. In the latter case, a mathematical theory that can deal with imprecise information may be used to manage data.

A comparison between the chronology of a few events can still be managed by hand, and using archaeological common sense to resolve any issue deriving from imprecision or different granularity—in some cases improving the fineness of the granularity, if possible. On the contrary this approach might be unfeasible when using datasets with large quantities of data, for example distinct but related pottery datasets. The various techniques and solutions used so far to manage rough sets might then reveal useful also in archaeological applications.

Dealing with Dating Imprecision: The Fuzzy Set Approach

Rough sets convey the concept of undecidability for some part of the time-span of an event, which is modelled in the boundary of the associated set. For this time extent, it is unknown whether the event was happening or not. A more sophisticated approach may give some additional information about the boundary. It may be expected that in this stage the event was starting to exist, or had some likelihood of existing. If it is possible to assign a (subjective) measure to this likelihood, fuzzy sets may be used.

In intuitive terms, a fuzzy set is a special subset of the universe for which belongingness is sure for some elements, is definitely not valid for others, and is uncertain for the remaining ones. For the latter, it is possible to assign a degree of membership, representing the likelihood of such membership according to the researcher. Fuzzy sets were introduced in by Zadeh in the sixties and there are currently a number of manuals on them, for example the extensive and authoritative Dubois and Prade (2000).

When applied to dating, a **fuzzy set** X is defined using a membership function f_X from time (modelled as the real line) to the interval [0, 1]. The value of $f_X(t)$ represents, in a scale from 0 to 1, the likelihood of a the event X taking place at time t. The extreme values correspond to the traditional truth value: when $f_X(t) = 0$, the event X does not exist; when $f_X(t) = 1$, the event X does exist. This definition incorporates the cases, called **sharp sets**, in which there is absolute certainty about the existence/non existence. The membership function may assume any shape, but for convention it is assumed to be continuous (events are not intermittent) and zero outside some—possibly large—interval (no eternity), called its **support**. Since it is a subjective measure of likelihood, simplicity is preferred so $f_X(t)$ has usually a trapezoidal shape as shown in the picture.

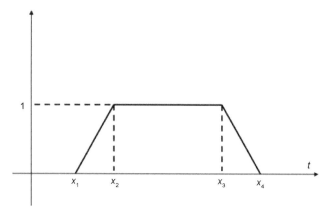

Mathematically speaking, the definition of such a function is the following:

$$f_X(t) = 0 \text{ for } t \notin (x_1, x_4); f_X(t) = 1 \text{ for } t \in (x_2, x_3)$$

$$f_X(t) = (t - x_1)/(x_2 - x_1) \text{ for } t \in (x_1, x_2); f_X(t) = (t - x_3)/(x_4 - x_3) \text{ for } t \in (x_3, x_4)$$

This means that the membership function is zero outside the interval (x_1, x_4), 1 within (x_2, x_3), and is linear and continuous in the two intervals (x_1, x_2) and (x_3, x_4). With this assumption of trapezoidal shape, $f_X(x)$ is completely described by the 4-tuple (x_1, x_2, x_3, x_4).

It is possible to describe an extension of the Allen algebra for fuzzy dating, enabling time analysis of two events. An extension usable for archaeological consideration was considered in Crescioli et al. (2002), dealing with the chronology of a necropolis. The relevant data for this case-study concerned the dating of the tombs, which was estimated by archaeologists with the usual approximate expressions "first quarter of 4th century BC", and so on. These vague expressions were turned into fuzzy sets defined as the fuzzy equivalent of any crisp time period with a slack δ at beginning and at end. So the membership function of such 'named' periods, e.g., 'first quarter of X century BC', x_2 and x_3 correspond to the literal extremes of the period while $x_1 = x_2 - \delta$ and $x_4 = x_3 + \delta$. For example, with a chosen value $\delta = 10$, the expression 'first quarter of 4th century BC' that would literally correspond to the time interval $(-399, -375)$ is interpreted as the fuzzy set $(-409, -399, -375, -365)$.

In Crescioli et al. (2002) only the constants relevant for the case study were considered, but it is simple to extend the definition to any other named time interval. In Doerr and Yiortsou (1998) this extended definition of named periods was set up again, without quoting Crescioli et al. (2002), using a slack $\delta = 25$. In Holmen and Ore (2010), some sort of fuzzy time interval is introduced without completely using fuzzy logic.

In Crescioli et al. (2002), there was additionally a hint on how to deal with the concept of fuzzy contemporariness, based on fuzzy equality, defined as follows:

Let X and Y be two fuzzy sets over the same (time) universe U, with membership functions respectively $f_X(t)$ and $f_Y(t)$, $t \in U$. The *fuzzy overlap* between X and Y is the operator $\approx$ that associates to X and Y the maximum over U of the minimum of $f_X(t)$ and $f_Y(t)$:

$$(X \approx Y) = \max \{\min(f_X(t), f_Y(t)), t \in U\}$$

If X and Y are crisp sets, i.e., $f_X(t) = 1$ for $t \in (a, b)$ and 0 otherwise, and $f_Y(t) = 1$ for $t \in (c, d)$ and 0 otherwise, fuzzy overlap reduces to the truth value of overlap: since the min is 1 for $t \in (a, b) \cap (c, d)$, and 0 otherwise, the fuzzy overlap is either 1 or 0 if the sets overlap or not.

The picture below shows how to compute the value for an example. It shows the diagram of $\min(f_X(t), f_Y(t))$: in the example X and Y overlap with a fuzzy overlap of 1, as this min function has a maximum of 1. This example corresponds to usual, non-fuzzy overlap.

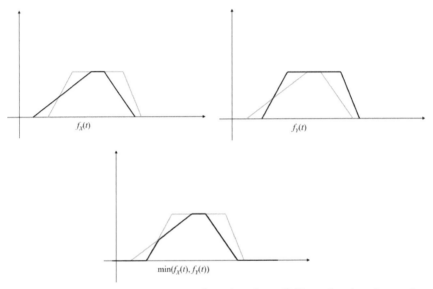

The next example shows the case where there is no 'full' overlap, i.e., there exists no time interval where both fuzzy sets have membership equal to 1.

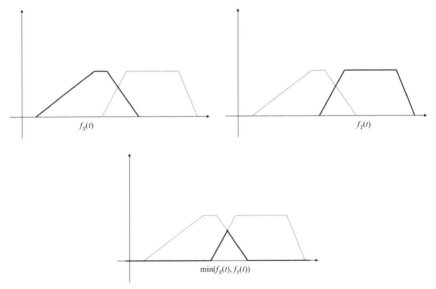

Without using fuzziness, the two crisp intervals where X and Y have membership equal to 1 do not overlap, so one would conclude that the two corresponding events have no common existence. With the fuzzy extension, instead, there is some degree of overlap, shown by the rightmost diagram. The value for the fuzzy overlap is indeed less than 1, but still (rather) larger than zero. Intuitively it corresponds to a situation similar to the evaluation of overlaps for periods such as "end of 2nd century AD" and "beginning of 3rd century AD": any archaeologist would assume some overlap, but a computer Aristotelian reasoning would not.

From this explanation it follows that the above extends of the concept of coexistence, modelled in Allen's algebra as 'overlaps', 'during', 'starts' or 'ends'. While 'overlaps' and 'during' are still meaningful in the fuzzy framework, 'starts' and 'ends' loose some of their significance because they are based on the start/end points of the time intervals, which in the fuzzy framework are indeterminate and substituted by fuzzy boundaries. However, all these Allen's relations may be easily turned in their fuzzy equivalents—keeping in mind that any such relation would turn into a fuzzified one, with multiple truth values.

Conclusions

The two approaches presented here represent two similar—but different—ways of modelling a situation in which there is uncertainty about the borders of a temporal entity. Although at first sight the fuzzy model might resemble a probabilistic approach, it is rather different, and in our opinion superior as regards modelling potential for time uncertainty in many cases. Fuzziness is about an imprecise concept; probability is about the unknown condition of a precise concept. Probability implies Boolean logic, i.e., the (so far unknown) status of a specified and verifiable condition: there is always an experiment—ideal or concrete—allowing to check this status, and if a statement based on it is false or true. The simplest case of a probabilistic event, tossing a coin, is probabilistic before tossing the coin, it becomes true or false after having tossed it, and tossing the coin is actually the verification experiment. Asking if on the 18th of January 1048 it was snowing in London is also a probabilistic situation: its truth could ideally be decided if we happen to find a written report about those weather conditions; until then, it may be modelled assigning a probability based on our present knowledge, e.g., about the average winter weather in the 11th century, the number of snow days, etc. In conclusion, a probabilistic model assumes that something is either true or false, and we do not know which is the case, for various reasons: incomplete information about the past, or because the condition concerns the future, etc. A fuzzy model concerns instead those situations in which asking if something is true or false is meaningless, because there are various degrees of being true (and false), even concerning cases which can be thoroughly inspected. Asking if my cat Agatha is grey requires only looking at her fur, but will not allow a definite answer Yes/No, since for cat skins there are (at least) fifty shades of grey. It is actually possible to attach a number to the cat's greyness, for example in the following way. Assume that $x = (r, g, b)$ are the percentages of Red-Green-Blue composing this particular grey, the one of the cat's fur, and denote with $x_0 = (r_0, g_0, b_0)$ the similar values for the 'perfect' grey according to a specified standard (e.g., a Pantone colour table). This embeds 'greyness' in a three-dimensional space with the usual Euclidean distance. For the distance *dist* of the RGB values of the cat's grey from the standard grey, since all the percentages are positive it holds that

$$(r - r_0)^2 = (r_0 - r)^2 \leq r_0^2$$

and the same holds for *g*, *b*, so:

$$0 \leq d^2 = dist(x, x_0)^2 = (r - r_0)^2 + (g - g_0)^2 + (b - b_0)^2 \leq r_0^2 + g_0^2 + b_0^2 = d_0^2$$

and consequently for

$$\mu_x = 1 - d/d_0$$

we obtain:

$$0 \leq \mu_x \leq 1$$

Therefore μ_x is a valid fuzzy membership function, which may be reasonably used to measure the degree of greyness of Agatha's fur.

On the contrary, it would make no sense asking the probability that my cat is grey, because I know her fur colour (no randomness); and verification of the statement would require being able to decide if the statement is true, which is impossible since there is no precise definition of 'grey cat', unless we stick with the 'standard' grey, which would lead to a probability of zero, as so many equiprobable shades of grey exist.

As already discussed, vagueness may be built in the concept (e.g., the Iron Age) or may concern its time-span. Fuzzy theory may be then used to model vagueness when it is possible (both conceptually and practically) to express a degree of belongingness, like in the cat fur example or for the beginning of the Iron age. There is no 'best' way to decide the value of this degree, and it can be conventionally assigned as a value, as we did in the example, or as a percentage, or whatever seems fit. In extreme cases, this value may be assigned as an expert score quantifying his or her subjective evaluation. As mentioned above, this approach was used in Hermon and Niccolucci (2002), where experts were asked to quantify a fuzzy classification and then the final membership value was computed as the mean of individual scores; and in other papers (Niccolucci and Hermon 2003, Hermon et al. 2004) to address other classification issues.

Finally, rough sets may be used in the same conditions of uncertainty but without assigning any membership degree: in this approach, the fuzzy borders are terra incognita, separating certain belongingness from certain non-belongingness. Practical usability of fuzziness and roughness in archaeology is an avenue still to be explored thoroughly. Besides the already mentioned papers and a few others, not much reconnaissance of these theoretical frameworks has taken place so far in practical examples. The reason is that for most everyday uses uncertainty is implicitly taken into account and somehow compensated when processing data in the archaeologist's mind. In the future, however, an extended accessibility of raw digital data and the integration of so far separate datasets, produced by different investigations, may require a more precise quantification of vagueness, to enable machine processing, searching and synthesizing search outcomes across different datasets or within the same dataset. As already mentioned, we have the impression that a normalizing approach like those of Binding (2010) or [Doerr et al. (2010) may turn into oversimplification. Introducing fuzzy logic may however require an extensive and cumbersome fuzzification of the data. A reasonable compromise could consist in automatically converting literal time expressions into a fuzzy thesaurus, which takes into account all the factors that may influence such a conversion, such as localization (time-spans corresponding to 'named' periods vary according to place), usage within the dataset, and so on; and creating fuzzy functionalities such as the fuzzy contemporariness relation outlined above. With a modest effort, the quality of data-based inference and interpretation might thus be greatly improved.

References Cited

Allen, James F. 1983. Maintaining Knowledge about Temporal Intervals. ACM Communications 26: 11, 832–843.

Binding, Ceri. 2010. Implementing archaeological time periods using CIDOC CRM and SKOS. pp. 273–287. *In*: L. Aroyo et al. (eds.). Proceedings 7th Extended Semantic Web Conference (ESWC 2010). Springer-Verlag.

Crescioli, Marco, Andrea D'Andrea and Franco Niccolucci. 2002. Archaeological Applications of Fuzzy Databases. pp. 107–116. *In*: Burenhult, Göran and Johan Arvidsson (eds.). Archaeological Informatics: Pushing The Envelope. Proceedings of CAA2001. Archaeopress.

Doerr, Martin and Anthi Yiortsou. 1998. Implementing a Temporal Datatype. Technical Report ICS-FORTH/TR236.

Doerr, Martin, Athena Kritsotaki and Stephen Stead. 2010. Which Period is it? A Methodology to Create Thesauri of Historical Periods. pp. 70–75. *In*: Franco Niccolucci and Sorin Hermon (eds.). Beyond the Artifact. Digital Interpretation of the Past. Proceedings CAA2004. Archaeolingua.

Dubois, Didier and Henri Prade. 2000. Fundamentals of Fuzzy Sets. Springer-Verlag.

Farinetti, Emi, Sorin Hermon and Franco Niccolucci. 2010. A Fuzzy Logic Approach to Artifact Based Data Collection Survey. pp. 123–127. *In*: Franco Niccolucci and Sermon Hermon (eds.). Beyond the Artifact—Digital Interpretation of the Past. Proceedings of CAA2004, Archaeolingua, Budapest.

Hermon, Sorin and Franco Niccolucci. 2002. Estimating subjectivity of typologists and typological classification with fuzzy logic. Archeologia e Calcolatori 13: 217–232.

Hermon, Sorin, Franco Niccolucci, Alhaique, Francesca, Iovino, Maria Rosaria and Leonini Valentina. 2004. Archaeological Typologies—an Archaeological Fuzzy Reality in AA.VV. (eds.). Enter the Past—The E-Way into the Four Dimensions of Cultural Heritage. Proceedings CAA2003. Archaeopress, 30–34.

Holmen, Jon and Christian-Emil Ore. 2010. Deducing event chronology in a cultural heritage documentation system. pp. 1–7. *In*: Frischer, Bernard, Jane Webb Crawford and David Koller (eds.). Making History Interactive. Proceedings CAA2009. Archaeopress.

Lingras, Pawan, Marcin Wolski, Chris Cornelis, Sushmita Mitra and Piotr Wasilewski (eds.). 2013. Rough Sets and Knowledge Technology. Proceedings of RSKT 2013. Springer-Verlag.

Niccolucci, Franco and Sorin Hermon. 2003. La logica fuzzy e le sue applicazioni alla ricerca archeologica, Archeologia e Calcolatori 14: 97–110.

Polkowski, Lech, Shusaku Tsumoto and Tsau Y. Lin (eds.). 2000. Rough Set Methods and Applications. Physica Verlag.

Ramenofsky, F. Ann and Anastasia Steffen. 1997. Unit Issues in Archaeology: Measuring Time, Space, and Material. University of Utah Press.

Renfrew, Colin. 1973. The Explanation of Culture Change. Duckworth.

Shanks, Michael and Christopher Tilley. 1988. Social Theory and Archaeology. University of New Mexico Press.

14

Bayesian Approaches to the Building of Archaeological Chronologies

Christopher Bronk Ramsey

Introduction

Bayesian inference has become an important tool for the construction of archaeological chronologies. This has been made possible by the development of Bayesian statistical methods (see, for example, Buck et al. 1991, 1996, Buck and Millard 2004, Nicholls and Jones 2001) and the development of both computing power and algorithms (Gelfand and Smith 1990, Gilks et al. 1996) that are able to put such methods to practical use. These have enabled the development of computer software to perform chronological analysis along with development of new Bayesian models that cover a wide range of situations (Blaauw et al. 2003, Bronk Ramsey 1995, 2008, 2009a,b, Buck et al. 1999, Christen 2003, Haslett and Parnell 2008, Jones and Nicholls 2002).

The subject of Bayesian chronology building is thus a large one and this chapter is necessarily limited in its scope. The emphasis here is on the main mathematical ideas underlying much of the chronological work, and in particular how these are relevant to the interface between mathematics and archaeology. The approach taken here is focussed on applied mathematical methods suitable for the underlying processes rather than Bayesian statistics *per se*.

One major issue relevant to all dating methods, and often overlooked, is the basis for the measure of time we are using and the numerical value we attach to our time parameters. Most chronological systems are, for historical reasons, essentially counting methods based on integer arithmetic. The numerical approaches that we use assume parameters with real-number values and if we are to have a sound basis for our analyses this needs to be properly defined. The definitions that we adopt do not quite accord with

University of Oxford, Research Laboratory for Archaeology and the History of Art, Dyson Perrins Building, South Parks Road, Oxford OX1 3QY, UK.
Email:christopher.ramsey@rlaha.ox.ac.uk

the underlying archaeological processes (for example the notion of instantaneous events) nor strictly speaking physical realities (the notion that there is a Newtonian universal timescale). We need to be aware of these limitations but in practice the definitions are usually good enough for what we are trying to do.

Probably the most important interface between archaeology and the mathematical approaches taken in chronology building is the way in which we group and relate different events to one another. The groups that we choose to define are a product of our understanding of the human activity reflected in the archaeological record. The way we relate our events also usually depends critically on our interpretation of site stratigraphy and sample taphonomy. For these reasons any Bayesian chronology building exercise is always part of the interpretation of the data rather than impartial analysis. That said, the aim of our mathematical models is to provide a degree of impartiality within that framework. The two most important classes of information that we wish to extract from any chronology are absolute date estimates and relative date estimates; both of these allow us to understand processes of change. Our models therefore aim to be unbiased in the estimates of age, and duration. What we mean by unbiased is critical in formulating our models and will be discussed in the following sections.

One of the main reasons that Bayesian methods have become so widely used is because the dating methods that we have all have uncertainty associated with them: in particular radiocarbon generated probability distribution functions which are non-normal. Despite this, because models are able to provide date and rate information for events which are not directly dated, the approaches discussed here would still be useful in cases where uncertainties are much less significant (such as is the case for dendrochronology). The outputs from our models are also in the form of probability distributions, and so mathematical methods which enable us to summarise these distributions, and compare them are also important elements of this endeavour.

Events and Processes

In archaeology what we are usually interested in is the changing nature of human behaviour, and the interaction between humans in their environment. The way we gain this understanding is to look for material evidence for past human activity as recorded in the archaeological record. Archaeology is a complex subject because the nature of the record is such that we must apply sophisticated methods, often classed as scientific, for the analysis and quantification of the physical remains, and then use a whole range of approaches to interpret that evidence in terms of human behaviour.

When it comes to chronology, our interpretation of the evidence is mostly guided by information on the speed and order of changes, rather than information on exactly 'when' things happened. However, in practice, one of the easiest ways to work out rates of change, and relative orders of events in different regions is to put events and processes onto some common timescale. In order to do this quantitatively we need mathematical methods, and we need to define how we measure our processes in a formal way. We can draw an analogy with the capture of human locomotion needed for CGI film production. The movement of a person is a very complex and subtle process, but it can often be captured in a useful way by a relatively small number of marker points on the human body. We attempt to do the same thing in chronology, by associating events, whose timing we attempt to measure with those processes of change. The simplest example of this might be a phase of activity, which we normally formalise into two defining events: the start event and the end event of that phase.

So our task in building archaeological chronologies comes down to putting events onto some common timescale. It turns out that the useful information we need to achieve this, comes from a variety of different sources: most obviously we have methods for measuring age, but equally important, methods for giving relative age to events, principally through stratigraphy, and methods for determining age differences (for example through biological processes, like tree ring growth, or through physical processes, such as sedimentation). It will be apparent already that these different kinds of information are drawn, using a complex process of inference, from the archaeological evidence, and might rightly be considered to be part of the interpretation of data. This means that chronology cannot be seen as a stand-alone plug-in module within the archaeological toolkit, but rather a key element within site interpretation. On a broader scale, the same can be said about the dating methods themselves: although they might be classes as scientific measurements, they also depend on interpretation of a complex range of environmental and archaeological information.

Bayesian Overview

Bayes theorem in its most formal terms allows us to make use of both prior and likelihood information. We will assume that we are interested in a whole set of parameters $\mathbf{t}$, for which we have a set of observed information $\mathbf{y}$. Bayes theorem can then be expressed as:

$$p(\mathbf{t}|y) = \frac{p(\mathbf{y}|t)p(\mathbf{t})}{p(\mathbf{y})} \tag{1}$$

where $\mathbf{t}$ are the set of parameters and $\mathbf{y}$ the observations or measurements made. $p(\mathbf{t})$ is the prior or the information about the parameters (mostly events) we have apart from the measurements. $p(\mathbf{y}|\mathbf{t})$ is the likelihood for the measurements given a set of parameters. $p(\mathbf{t}|\mathbf{y})$ is the posterior probability, or the probability of a particular parameter set given the measurements and the prior. Finally $p(\mathbf{y})$ is the overall probability of the observations made. In most of the cases that we will be interested in $p(\mathbf{y})$ can be treated as a normalisation constant and thus we have:

$$p(\mathbf{t}|y) \propto p(\mathbf{y}|t)p(\mathbf{t}) \tag{2}$$

In a Bayesian chronological analysis we have to express temporal information in these terms.

Most of our parameters $\mathbf{t}$ are used to express information about time. Our observations y may include measurements on isotope ratios, for example, and thus the likelihood terms are the way we relate these observations to the age parameters in the model; radiocarbon dates would thus be expressed as likelihoods in our overall scheme. Prior information is information which we know ahead of the analysis: the distinction here is slightly arbitrary, as much of this information, such as the ordering of events, is of course based on observations, and so we could consider the setting up of the prior as in itself a Bayesian exercise.

Measuring Age

One clear aim of chronology is to put events onto a single timescale so we can compare what is happening in different places in a way that allows us to understand the human activity for which we have evidence in the archaeological record. In practice all methods we have for determining ages (or dates on a time-line) have uncertainties associated with them, uncertainties which might be independent between age estimates, or might be systematic between measurements of the same kind. This section will look at different ways of measuring age and some of the issues involved in incorporating such information into chronological models.

The Timescale(s)

We have stated the aim in building chronologies as putting events onto a common timescale. We can fortunately don't have to worry about relativistic effects and can assume that time is a cartesian coordinate independent of position. There is no meaningful smallest increment of time so we consider it as real number $t \in R$. In practice, however the two most important temporal cycles for humans are the day, and (at least for non-tropical regions) the year. Because of the resolution of most archaeological research the year is the unit usually used for the measurement of time. This raises several problems: firstly the year is not an SI unit and has a number of different definitions (Julian, Siderial, Tropical, etc.), secondly the year cycle is different in the Northern and Southern hemispheres, and finally the counting of years in different calendars is notoriously complicated. For most purposes this may not have much practical consequence but it does make any formal definition of the timeline very difficult.

In practice many archaeologists simply use the integer years as a timescale. Here years AD (or CE) form one set y_a and years BC (or BCE) form another y_b. In each case the integer is definite positive (y_a, $y_b \in Z^+$). Depending on the period in question, the years may be in the Gregorian calendar, or the Julian calendar. Normally for pre-history we really just assume that the years are a count of seasonal cycles from the present. This is very unsatisfactory formalism, and astronomers have long ago (18th Century) adopted an astronomical year numbering system with a zero ($y_c \in Z$). Furthermore to add complication, an integer year notation was introduced in 1950 which is termed 'before present' where the year AD 1950 is given the value 0 ($y_d \in Z^*$).

And so:

$$y_c = y_a \tag{3}$$

$$y_c = 1 - y_b \tag{4}$$

$$y_d = 1950 - y_c \tag{5}$$

None of these really represents a reasonable basis for mathematical analysis. Even y_c in its normal definition has different year lengths for different periods, and none of the others cover the full timescale to the present. They are also all integer timescales with no ability to specify times at a finer interval without recourse to complexity of the calendar. To overcome these problems, it is much more convenient to use a real number time line with a unit of one year. Because we wish to retain synchrony with the seasons we need a year length to be a good approximation to the tropical year length and for this purpose it has been proposed (Bronk Ramsey 2009a) to use the mean Gregorian year (31,556,952s), as this can be directly related to the SI unit of the second, with the reference point being 0001-01-01 00:00:00 Universal Time under ISO-8601. Under this scheme the floating point 1.0 is the start of the year 1AD (and consequently 2001.0 is the start of the year 2001AD). Within a day or two, the calendar year (Gregorian) in its different forms can be derived as:

$$y_c = \text{floor(t)} \qquad \text{(for all } t) \qquad \qquad (6)$$

$$y_a = \text{floor(t)} \qquad \text{(for } t \geq 1) \qquad \qquad (7)$$

$$y_b = 1 - \text{floor(t)} \qquad \text{(for } t < 1) \qquad \qquad (8)$$

$$y_d = 1950 - \text{floor(t)} \qquad \text{(for } t < 1951) \qquad \qquad (9)$$

Having defined our theoretical time-line, we now have to consider how we can use information to estimate the timing of events on that scale.

Events Defined on a Calendar

Having defined the timeline in terms of current time and date standards, we can use our knowledge of different calendar systems to put any event with a defined date (and time) onto our timeline. There is a convenient tool for doing this from many of the frequently used calendar systems at http://c14.arch.ox.ac.uk/oxcalhelp/hlp_analysis_calend.html. Of course in practice we never know the time and date of an event perfectly and so usually we would define some uncertainty in the definition of that time range. This might commonly be in the form of a standard uncertainty (Normally distributed probability with a mean σ, a standard uncertainty or σ) or as a range (uniform probability in the range t_a to t_b).

$$p(t) \sim \mathcal{N}(\mu_t, \sigma_t^2) \qquad \qquad (10)$$

$$p(t) \sim \mathcal{U}(t_a, t_b) \qquad \qquad (11)$$

Other distributions, such at the Student-t distribution, might also be justified in some cases.

Scientific Dating Methods

Most scientific dating methods (Argon-Argon, Potassium-Argon, Uranium series, Luminescence methods, Amino-Acid Racemization, etc.) operate on the basis of measuring age. The uncertainties associated with the methods are often approximately Normal, and so it is possible to express the date from such measurements as a Normal probability distribution function. If the determined age is $\mu_a \pm \sigma_a$ and the sample was measured at time t_0, then the probability distribution function for the date of the event in question is:

$$p(t) \sim \mathcal{N}(t_0 - \mu_a, \sigma_a^2) \qquad \qquad (12)$$

In principle other less generic likelihood distributions could be derived. In particular for Uranium Series dating the likelihood distribution should be non-linear as you approach the limit to the technique (Ivanovich and Harmon 1992). In optically stimulated dating, there is also much more information available that in principle might be used within a Bayesian model (Rhodes et al. 2003).

Layer Counting Methods

Three layer counting techniques are often used in the generation of timescales. These are dendrochronology, varve counting and ice layer counting. In all cases the method relies on the annual cycle of the seasons to leave an annual signal in the relevant record. These methods then normally give age intervals within the archive with a precision of a year. If the archive can be extended to the present, these can be used to provide a date.

For varves and ice core layers, there are uncertainties involved in the counting process and so some uncertainty should be applied. For ice cores the normal practice is to derive a maximum counting error. This can be interpreted as a uniform distribution as in equation 11, but probably more realistically can be interpreted as a 2σ range of a normal distribution as in equation 10. Ideally, however, we should keep in mind that these uncertainties are highly correlated and we often know the relative age in such a record much better than we know the absolute age. This information can in principle be included in Bayesian models (using for example age difference priors) but a comprehensive methodology for doing this has not been developed.

Dendrochronology ideally brings in enough replication to provide a result for an extant tree ring which is precise to the year. However where there are missing rings, allowance needs to be made for this in estimating the use date (Miles 2005).

Radiocarbon Calibration

Radiocarbon dating is, mathematically, much more complex than many other dating techniques. What is measured is a radiocarbon isotope ratio, usually expressed as a fraction of the 'modern' value of reference material from 1950. This $F^{14}C$ value, f, can be converted to a radiocarbon date a using the Libby half life of 5568 a:

$$a = \frac{5568}{\ln(2)} \ln(f) \approx 8033 \ln(f) \tag{13}$$

The uncertainty in the measurement of f should be a summary of all associated analytical uncertainties and is usually assumed to be Normal in distribution:

$$f \sim \mathcal{N}(\mu_f, \sigma_f^2) \tag{14}$$

To first order this gives a normal distribution in radiocarbon date:

$$a \sim \mathcal{N}(\mu_a, \sigma_a^2) \tag{15}$$

$$\mu_a \approx 8033 \ln(\mu_f) \tag{16}$$

$$\sigma_a \approx 8033 \frac{\sigma_f}{\mu_f} \tag{17}$$

However, this approximation is only valid when $\sigma_f \ll \mu_f$, which is true for younger dates. The radiocarbon calibrations curves are generated using statistical methods that are designed to provide an estimate of the radiocarbon measurement that you would get for a sample of particular age. The curves are provided at discrete intervals but can be interpolated to give a continuous function covering the time range from 50000 calBP to 2010 (Reimer et al. 2013, Hogg et al. 2013, Hua et al. 2013). For the period after 1950, the data is circulated in terms of the parameter f but for the period before in terms of a. However, now that calibrations extends to the limit of the technique it is important that the calibration calculations are done in the parameter space of f rather than that of a given the approximation inherent in equation 17.

The curve is normally expressed as a function of time r(t) with an uncertainty which is also a function of time s(t), so that the radiocarbon value ρ is assumed to take the value:

$$p(\rho \,|\, t) \sim \mathcal{N}(r(t), s^2(t)) \tag{18}$$

You will see that from the argument above, this is more valid if ρ, r and s are expressed in terms of f, though in much of the literature they are expressed in terms of a. From this we can see that if we know the date of a particular event t_i and that the uncertainty in the measurement will be s_i it follows (Buck et al. 1992) that:

$$p(r_i \,|\, t_i, s_i) \sim \mathcal{N}(r_i - r(t_i), s^2(t_i) + s_i^2) \tag{19}$$

where r_i is the radiocarbon measurement made for that sample.

Alternatively (Bronk Ramsey 2009a) we can use the calibration curve to provide a two dimensional probability density for ρ and t. Assuming that the marginal prior for t should be uniform, this is:

$$p(\rho_i, t_i) \propto \frac{1}{s(t_i)} \exp(-(\rho_i - r(t_i))^2/(2s^2(t_i))) \tag{20}$$

And from the likelihood expression in equation 14 we have:

$$p(r_i \,|\, \rho_i, s_i) = \frac{1}{s_i\sqrt{2\pi}} \exp(-(r_i - \rho_i)^2/(2s_i^2)) \tag{21}$$

Combining these two and integrating over the parameter ρ_i we have:

$$p(r_i \,|\, \rho_i, t_i, s_i) \propto \frac{1}{s_i s(t_i)} \exp(-(r_i - \rho_i)^2/(2s_i^2))\exp(-(\rho_i - r(t_i))^2/(2s^2(t_i))) \tag{22}$$

$$p(r_i | t_i, s_i) \propto \int_{\rho_i = -\infty}^{\infty} p(y_i | \rho_i, t_i) d\rho_i \tag{23}$$

$$\propto \frac{\exp\left(-\dfrac{(r_i - r(t_i))^2}{2(s_i^2 + s^2(t_i))}\right)}{\sqrt{s_i^2 + s^2(t_i)}} \tag{24}$$

$$\sim \mathcal{N}(r_i - r(t_i), s^2(t_i) + s_i^2) \tag{25}$$

Which gives the same result as in equation 19. The value of this approach lies in understanding the probability distribution in equation 20. It this two dimensional probability distribution is marginalised over t_i it is uniform, but not if it is marginalised over ρ_i. The non-linearity of the calibration 'curve' implies that the probability density in only one of these two parameters can be uniform and clearly for dating material which might equally be any age this formulation makes most sense of the available information.

Application of Dating Methods to Single Events

Using Bayes theorem:

$$p(t_i|y_i) = \frac{p(y_i|t_i)p(t_i)}{\int p(y_i|t_i)p(t_i)dt_i} \tag{26}$$

with a uniform $p(t_i)$, and making use of the fact that $\int p(y_i|t_i)p(t_i)dt_i$ does not depend on t_i, we derive the result that:

$$p(t_i|y_i) \propto p(y_i|t_i) \tag{27}$$

What this means is that, for any dating method, the likelihood function when normalised over t_i can be used as a marginal posterior for the date of the event in question. This is why, for example, for simple calibration of single radiocarbon results (from equation 24) we can see that:

$$p(t_i|r_i, s_i) \propto \frac{\exp\left(-\dfrac{(r_i - r(t_i))^2}{2(s_i^2 + s^2(t_i))}\right)}{\sqrt{s_i^2 + s^2(t_i)}} \tag{28}$$

This is the probability distribution used for calibration of radiocarbon dates in all widely used software packages (such as Bronk Ramsey 2001, Buck et al. 1999, van der Plicht 1993).

Dealing with Outliers

In practice not every age measurement is correct. The reasons for this are manifold: there may be problems with the associated measurements, or it may be that the age measured cannot really be associated with the event envisaged. For example, a cereal grain deposited in a particular context might be residual and so the chronologist expects the measured radiocarbon date to be close to the time of deposition but the radiocarbon date reflects the growth of the grain some years earlier. There have been three main approaches taken to such outliers.

The first approach is to identify outliers, either 'by eye' or using the measures discussed in Section 6.2 below.

The second method, within the context of a Bayesian model, is to introduce a new boolean parameter for each measurement φ_i which is 1 if the sample is an outlier and 0 if it is not. We assign a prior probability q_i that the sample's measurement is an outlier for some reason. This approach has the advantage that we can look at the marginal posterior for φ_i and thereby identify probable outliers. It was first formulated by Christen (1994, 2003) and then further elaborated by Bronk Ramsey (2009b) to include a number of different types of outlier.

The final method is to modify the distribution used for the calibration from the Normal distribution in equation 28 to a student-t distribution: this is the approach taken by Christen and Pérez (2009).

Groupings, Processes and Models

The previous section dealt with how we treat individual events and the simplest approach to dealing with multiple events is just to treat them as a series of completely independent and unconnected events. Such statistical models are relevant when, for example, we are dealing with indistinguishable particles. However with events we always know something about them and, if we are studying them in relation to an archaeological problem, they do by definition have some relationship to one another.

The Null Model

At this stage we will introduce the notion of the null model. This is one under which we do assume full independence of all events. This is unrealistic but it does provide a baseline from which to consider other models. In this null model the prior can be factorized as a uniform distribution in each event parameter. This prior is thus improper in that it cannot be normalised. However under this prior the marginal posterior for each event is just directly proportional to the likelihood function (as with a single event).

The problem with such a model as a realistic description of an archaeological process comes principally from the fact that although the model has a uniform prior for each individual event (out of a total of n, where $n \geq 2$), the effective prior for the derived quantity of the span s of those events is proportional to s^{n-2}. Thus, for example 12 events would be 10^{10} times more likely to span 100 years than 10 years. This is not a reasonable assumption to make for most archaeological processes.

Constraints

Before we come on to consider our solution to the problem with the 'null model' we will introduce an important aspect of the prior and that is constraints. One of the pieces of information that we often have from archaeological sites or process is information on the order of events. These constraints might come from stratigraphy (Harris 1989) or from the definition of parameters describing a process. To introduce a single constraint $t_a < t_b$ between two event parameters we introduce an element in the prior:

$$p_H(t_a, t_b) = \begin{cases} 1, & \text{if } t_a < t_b \\ 0, & \text{otherwise.} \end{cases} \tag{29}$$

The function p_H is the same as the Heaviside function $H(t_b - t_a)$ except in the case where $t_a = t_b$. For multiple constraints in a sequence we define a prior along the same lines:

$$p_H(t_a, t_b, t_c, \dots) = \begin{cases} 1, & \text{if } t_a < t_b < t_c < \dots \\ 0, & \text{otherwise.} \end{cases} \tag{30}$$

Any constraints that we wish to include in a prior model can be expressed as factors of this form.

Two Parameter Processes

Next we will consider processes which can be described by two independent parameters. The most commonly considered of these is the single phase of activity (Buck et al. 1991, 1992, Zeidler et al. 1998, Bronk Ramsey 2009a). Here we have two parameters which describe the phase, a start event t_a and a finish event t_b. Because these describe the limits of the phase they are sometimes referred to as phase boundaries (Bronk Ramsey 2009a). The prior for these boundaries is uniform but we constrain t_a to be before t_b on our time-line. That is the prior is given by:

$$p(t_a, t_b) \propto p_H(t_a, t_b) \tag{31}$$

As there are only two boundary events the prior for the difference between them $s = t_b - t_a$ is uniform over the range $0 - \infty$ which is suitably uninformative. This prior is improper but this will not normally matter in practice. All of the events within the phase then have a normalised prior of:

$$p(t_i \mid t_a, t_b) = \frac{p_H(t_a, t_i, t_b)}{(t_b - t_a)} \tag{32}$$

Thus the overall prior for the model is:

$$p(t) \propto \prod_i \frac{p_H(t_a, t_i, t_b)}{(t_b - t_a)} = \frac{\prod_i p_H(t_a, t_i, t_b)}{(t_b - t_a)^n} \tag{33}$$

where there are n events within the phase.

This is the most widely used type of model for describing archaeological processes. The model has $n + 2$ parameters and an overall scale which we can define as $s = t_b - t_a$. Because the prior contains a factor of s^{-n} this exactly cancels the bias towards longer scales discussed in Section 5.1 above. Thus the prior for the model is scale invariant which is suitably uninformative.

The two parameter approach can be used for other types of process other than the uniform phase: Figure 1 shows some examples implemented within the OxCal package. All of these two parameter process models are effectively a way to mathematically describe a group of events. As an example if we wish to model an exponential rise we can characterise this with two parameters. We have choice over what these parameters are, but to keep the model similar to that for a uniform phase we choose to take a

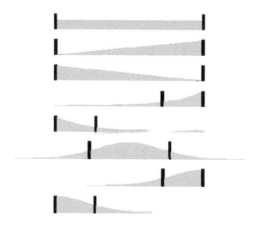

Fig. 1. Range of two-parameter processes or groupings supported in OxCal v4.0 from Bronk Ramsey (2009a).

midpoint, t_a, of the process (that is one time constant before the end) and the end-point, t_b, as being the two event parameters. The overall prior becomes:

$$p(\mathbf{t}) \propto p_H(t_a, t_b) \prod_i p_H(t_i, t_b) \frac{\exp(-(t_b - t_i)/(t_b - t_a))}{(t_b - t_a)} \tag{34}$$

Conceptually the two parameter process is central to all of the more complex models. The reason for this is that in a model which has two independent parameters t_a and t_b each with a uniform unconstrained prior $\sim U(-\infty, \infty)$ the effective prior for both the centroid $(t_a + t_b)/2$ is $\sim U(-\infty, \infty)$ and for the span $s = |t_b - t_a|$ is $\sim U(-0, \infty)$. This property will be referred to here as 'linear scale invariance'.

Constrained Processes

We can put together our two formalisms for constraint and two parameter process to provide a grouped sequence. Most commonly we have a uniform phase of events for which the order is pre-defined (usually by stratigraphic constraint). Adapting the prior from equation 33, where we take $i = 1 \ldots n$, we have:

$$p(\mathbf{t}) \propto \frac{p_H(t_a, t_1, \ldots, t_n, t_b)}{(t_b - t_a)^n} \tag{35}$$

In principle there is no mathematical reason why this cannot be extended to other types of process, though it is hard to see the archaeological situation where this would be relevant.

Multiple Phases

The model for the two-parameter process which describes a single uniform phase can easily be expanded to include the case of multiple phases. Where the phases are entirely independent, the prior is proportional to a product of the priors for the individual phases. Thus if, for example, we have two phases, one of which is from t_a to t_b and another from t_c to t_d, then the prior would be given by:

$$p(\mathbf{t}) \propto \prod_i \frac{p_H(t_a, t_i, t_b)}{(t_b - t_a)} \prod_j \frac{p_H(t_c, t_j, t_d)}{(t_d - t_c)} \tag{36}$$

Here there is no relationship between the phases at all (Fig. 2A). We might also know that the second phase follows the first with an unknown gap (the phases are sequential, as in Fig. 2B). This would give us an additional element to the prior of p_H (t_b, t_c):

$$p(\mathbf{t}) \propto p_H(t_b, t_c) \prod_i \frac{p_H(t_a, t_i, t_b)}{(t_b - t_a)} \prod_j \frac{p_H(t_c, t_j, t_d)}{(t_d - t_c)} \tag{37}$$

Another common interpretation of the archaeological evidence might be that the phases follow directly on from one another (they are abutting as in Fig. 2C). In this

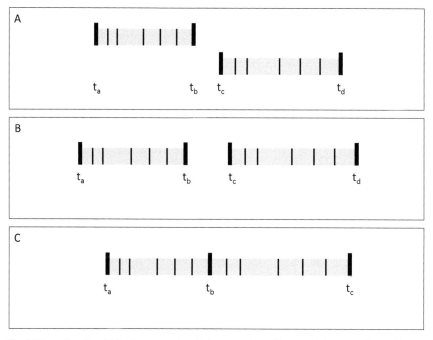

Fig. 2. Examples of multiple phase models with just two phases. Two such phases can be modelled to be independent (A), sequential (B) or abutting (C).

case we have one less boundary parameter because the end of the first phase t_b would be the start of the second phase, which ends at t_c. In this case our prior is:

$$p(\mathbf{t}) \propto \prod_i \frac{p_H(t_a, t_i, t_b)}{(t_b - t_a)} \prod_j \frac{p_H(t_b, t_j, t_c)}{(t_c - t_b)} \tag{38}$$

The first two of these (equations 36 and 37) are effectively four parameter process models, whereas the last (equation 38) is a three-parameter process model.

Now in the case of the dependent phases (examples of which are given in equations 37 and 38), we no longer have linear scale invariance between the earliest and latest boundaries in the model. This is exactly the same problem as we saw for independent events discussed in Section 5.1 above. We can solve this problem by treating all of the inner boundaries in a model as part of a single uniform phase for the chronology as a whole, starting with the first boundary t_a and ending with the last one t_m. Each of these inner boundary events t_k has a normalised prior of:

$$p(t_k \mid t_a, t_m) = \frac{p_H(t_a, t_k, t_m)}{(t_m - t_a)} \tag{39}$$

Thus equation 37 for the sequential phase example (Fig. 3B) becomes:

$$p(\mathbf{t}) \propto p_H(t_b, t_c) \prod_i \frac{p_H(t_a, t_i, t_b)}{(t_b - t_a)} \prod_j \frac{p_H(t_c, t_j, t_d)}{(t_d - t_c)} \frac{1}{(t_d - t_a)^2} \tag{40}$$

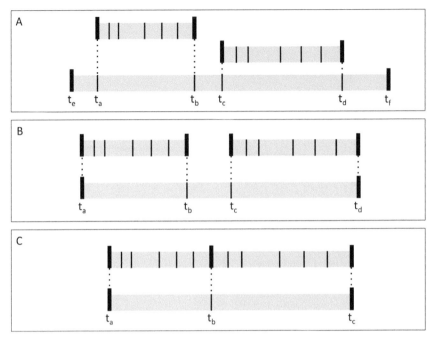

Fig. 3. Examples of multiple phase models with just two phases but modeled using a two parameter process to describe the phase boundaries. The independent (A), sequential (B) or abutting (C) phases should be compared to Fig. 2.

and equation 38 for the abutting phase example (Fig. 3C), becomes:

$$p(\mathbf{t}) \propto \prod_i \frac{p_H(t_a, t_i, t_b)}{(t_b - t_a)} \prod_j \frac{p_H(t_b, t_j, t_c)}{(t_c - t_b)} \frac{1}{(t_c - t_a)} \tag{41}$$

Thus what we have done is to add another two process model at a higher level which shares parameters with the start of the first phase and the end of the last. This process model is used for all of the intervening phase boundaries.

Model Hierarchies

In deriving the multiple parameter process models in the previous section, we have already introduced the notion of a hierarchy in that the process parameters (or boundaries) are taken to be higher level elements of the chronology under consideration. This can in principle be generalised to cover the nesting of processes. As an example we will consider the simple case of two independent phases described by the prior in equation 36. We would introduce a two-parameter (t_e and t_f) process model to cover the four boundary events.

$$p(\mathbf{t}) \propto \frac{p_H(t_e, t_a, t_b, t_f) p_H(t_e, t_c, t_d, t_f)}{(t_f - t_e)^4} \prod_i \frac{p_H(t_a, t_i, t_b)}{(t_b - t_a)} \prod_j \frac{p_H(t_c, t_j, t_d)}{(t_d - t_c)} \tag{42}$$

This model (Fig. 3A) will now be scale invariant again once more which the independent phase model is not on its own. In general parameterising the model for overall linear scale invariance makes most difference when there are a large number of phases in the model, but it is a good principle on which to base model definitions. Within OxCal such model hierarchies can be realised by nesting one set of phase models within another (Bronk Ramsey 2009a).

Now if we look back at the modifications made in equations 40 and 41, they are actually just a special case of the nested model where the start and end boundaries of the outer model are constrained to be the same as the first and last boundaries within the sequence (compare Figs. 2 and 3). Our justification for building models in this way needs ultimately to come from our understanding of human activity as a complex system which goes through periods of equilibrium and change on a range of different temporal scales (Bronk Ramsey 2003).

The Trapezium Process Model

So far we have considered only processes which in their individual elements can be described by two parameters as illustrated in Fig. 1. However, there are some archaeological processes that might better be described with a more complex process model. The case of this kind which has been studied in most detail is the trapezium process model (Karlsberg 2006, Lee and Bronk Ramsey 2012). In this model we assume that the underlying archaeological activity has a period where the activity grows (at a linear rate), a period of stasis, followed by a period of decline (again linear) as shown in Fig. 4. The process can be described by four parameters (Karlsberg 2006), either as the four change points (t_a, t_b, t_c, t_d) under the constraint $t_a < t_b < t_c < t_d$ where the activity follows a prior:

$$p(t) \propto \begin{cases} 0 & \text{for } t < t_a \\ \dfrac{t - t_a}{t_b - t_a} & \text{for } t_a < t < t_b \\ 1 & \text{for } t_b < t < t_c \\ \dfrac{t_d - t}{t_d - t_c} & \text{for } t_c < t < t_d \\ 0 & \text{for } t > t_d \end{cases} \tag{43}$$

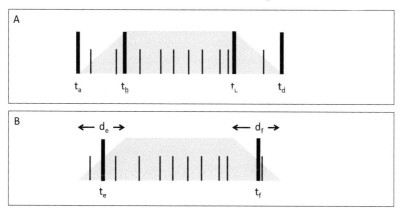

Fig. 4. Formulations of the trapezium model (see text).

Alternatively it can be re-parameterised to two time parameters ($t_c = (t_a + t_b)/2$, $t_f = (t_b + t_c)/2$) which give the centre point of the onset and end of the phase of activity, and two more ($d_e = (t_b - t_a)$, $d_f = (t_d - t_e)$) which give the period for the rise and fall in activity respectively (Lee and Bronk Ramsey 2012). The main point of this parameterisation is for cases where you have a succession of such phases, but it also emphasises the underlying similarity with the two-parameter uniform phase. The issues involved in ensuring linear scale invariance in such a model are more complicated than the uniform phase case and discussed in detail in Lee and Bronk Ramsey (2012).

This type of model lends itself better to changes in cultural activity assumed to be gradual in nature.

Sedimentation

So far the models that we have considered are based either on our understanding of archaeological processes, combines with constraints which normally come from stratigraphy. However, stratigraphic order is only one type of information which can be derived from considering structured deposits. The two most relevant examples for archaeology are tree-rings in wood (or derived harcoal) and sedimentary sequences.

In the case of wood with countable tree-rings we have a very special situation where we know the relationship between different radiocarbon samples measured for radiocarbon. This means that effectively there is only one independent variable in the model (typically taken to be the age of the outer ring). So if t_i and t_{i+1} are the ages of two samples taken from the wood where there are g_i whole rings separating them, then:

$$t_{i+1} = t_i + g_i \qquad \text{for } i < n \tag{44}$$

where there are n samples in total. This simple relationship provides an easy way to 'wiggle-match' radiocarbon sequences measured on wood onto the calibration curve (Bronk Ramsey et al. 2001, Christen 2003). Given that in this particular case, we do know the overall length of the group of dates, and because there is only one independent variable involved, there are no issues of linear scale invariance involved here. There is a somewhat analogous situation in varved lakes and ice-cores, except that in those instances, there is often some uncertainty associated with the gap between samples (Bronk Ramsey 2008). This is rarely directly relevant to archaeology and so will not be discussed further here.

The much more common situation in archaeological contexts is sedimentary deposits including peat formation. In these cases we have stratigraphic ordering, but we also expect the deposition rate to be reasonably regular. This is to be contrasted with the situation in many archaeological sites (urban settlements, cave shelters, etc.) where there may be very sporadic changes in level, sometimes with large increments in level in short periods and at other times erosion or stasis.

For this discussion we will consider the timing of deposition t at a vertical position in the final deposit (after compaction, etc.) of z. In order to simplify the discussion here we will consider the model implications of interpolating between two points in a sequence with timing t_i and t_{i+1} at positions z_i and z_{i+1} respectively. We will consider the timing of the deposition in the range $z_i < z < z_{i+1}$.

At one extreme, if we expect very variable deposition rates, then as above, all we can use is stratigraphic constraint which tells us that:

$$p(t|z) = \begin{cases} \dfrac{1}{t_{i+1} - t_i} & t_i < t < t_{i+1} \\ 0 & \text{otherwise} \end{cases} \tag{45}$$

or this could be included in the prior as a constraint:

$$p(t) \propto p_H(t_i, t, t_{i+1}) \tag{46}$$

However, the implication of this model is that we get no useful chronological information from the depth at all: a sample near the top of this range is just as likely to be similar in age to the bottom as the top. This is really only reasonable if we think that there is only one deposition event between z_i and z_{i+1}.

Another extreme assumption is that the deposition between the two points proceeds perfectly uniformly. In this case only t can be directly calculated from the t_i and t_{i+1} and there are only two independent variable in the model:

$$t = t_i + (t_{i+1} - t_i) \frac{(z - z_i)}{(z_{i+1} - z_i)} \tag{47}$$

In practice, however, no deposition is truly continuous or constant in rate, and so although this might be a convenient mathematical model it is not realistic. What is required is a process model for the deposition itself. One approach is to assume that the deposition is actually governed by an underlying Poisson process with discrete deposition events occurring at random (Bronk Ramsey 2008). For this we assume that there are a certain number of events per unit length given by the parameter k. So if for example our z is measured in meters and k = 1000, this model is like having a simple sequence deposition model with an event every 1 mm. Under this assumption our interpolated age probability for position z is:

$$p(t|z, k) \propto \frac{(t - t_i)^{k(z-z_i)} (t_{i+1} - t)^{k(z_{i+1}-z)}}{(t_{i+1} - t_i)^{k(z_{i+1}-z_i)}} \tag{48}$$

which reaches a maximum when:

$$t = t_i + (t_{i+1} - t_i) \frac{(z - z_i)}{(z_{i+1} - z_i)} \tag{49}$$

as for the uniform deposition case. Indeed the two extreme cases of unstructured deposition and uniform deposition are simply limits of this model in the limit where k tends to zero or infinity respectively (see Fig. 5). The details of this model, including ways to ensure that the overall model is scale in variation, and dealing with the case where the underlying sedimentation is not constant is dealt with in Bronk Ramsey (2008). One difficulty of this model is the need to work our the most appropriate k value. However, this problem can be overcome by allowing k to be a parameter in the model itself, and allow the data to inform us on the uniformity of deposition (Bronk Ramsey and Lee 2013).

The Poisson process is however, only one approach that can be taken. The other most widely used approach is to assume a gamma process (Blaauw and Christen 2005, Haslett and Parnell 2008). In practice, given that all of these models provide a simplified model of the underlying processes, there is a lot to be said for testing several different models to check for the robustness of the findings.

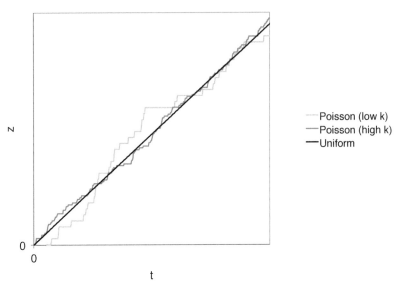

Fig. 5. Comparison of deposition model assumptions from Bronk Ramsey (2008). The Poisson process model allows for variability in deposition rate by assuming that the underlying deposition process is incremental rather than continuous.

Model Implementation

So far we have considered only the mathematical formulation of Bayesian models relevant to archaeological research. The fact that such models have become so widely used only over the last couple of decades, is in large part because they cannot be solved analytically and require Markov Chain Monte Carlo methods to calculate posterior probability densities.

Several statistical packages have been developed which allow the user to use Bayesian models together with chronological data, including BCal (Buck et al. 1999), OxCal (Bronk Ramsey 1995, 2008, 2009a,b), BWigg (Christen 2003), BChron (Haslett and Parnell 2008) and BPeat (Blaauw et al. 2003). Each of these packages have slightly different approaches to the formulation of models and different algorithms for the implementation of the MCMC. However, in many cases the underlying mathematical models are similar, though many packages only implement certain types of model.

Reporting Model Outputs

The main point of any Bayesian modelling is to obtain posterior estimates for the model parameters. Most often this is in the form of marginal posterior density estimates for events within the chronological model. Such outputs are normally given as probability distributions (binned at a suitable resolution), and summarised in terms of highest probability density (HPD) ranges. Unlike many applications, it has been common practice to generate split ranges rather than just a single HPD range. This means that, for example, the 95.4% range is defined as the set of subranges with the shortest combined length which contains 95.4% of the probability in the distribution. The reason that this is common practice is partly as a legacy of the intercept method of calibrating radiocarbon dates, and also because the extra information in these split ranges is often quite useful in interpreting chronologies. However, there are some disadvantages in this approach in that range fragmentation can result from noise in the probability distribution functions derived from MCMC analysis (to an extent which depends on bin-size).

Other information can also be pulled out from Bayesian models of this kind, as part of the MCMC analysis. It is for example possible to generate probability distributions for the first of the group of events, the difference between two events or the covariance between model parameters (Bronk Ramsey 2009a, Bronk Ramsey and Lee 2013). Sometimes this information on the relationship between model parameters can be as important as the age estimates themselves.

Model Output Comparison

Given the output of any analysis is primarily given as probability distributions, it is worth considering, from a mathematical perspective, what the most appropriate methods are for comparing such distributions.

One application of such a comparison might be to compare the model outputs from two different models, or from two runs of the same model to see how similar they are. Supposing we have two different marginal posterior density estimates $p_i(t)$ and $p_j(t)$. For comparison purposes a suitable mathematical approach might be to look for a scalar product (or inner product). As the functions are real, the inner product will require, symmetry, linearity (on either argument) and that the product is positive definite. Since the probability is always positive this is easily achieved as:

$$< p_i, p_j > = \frac{\int p_i(t)p_j(t)dt \int p_i(t)dt \int p_j(t)dt}{\sqrt{\int p_i(t)p_i(t)dt \int p_j(t)p_j(t)dt}} \qquad (50)$$

The properties of this scalar product are:

$$< p_i, p_j > = < p_j, p_i > \qquad (51)$$

$$< ap_i, p_j > = < p_i, ap_j > = a < p_i, p_j > \qquad (52)$$

$$< p_i, p_j > > 0 \qquad (53)$$

$$< p_i, p_i > = \left(\int p_i(t)dt \right)^2 \qquad (54)$$

$$< p_i, p_j > = 0 \quad \text{if } \int p_i(t)p_j(t)dt = 0 \qquad (55)$$

Furthermore in the usual situation in which $p_i(t)$ and $p_j(t)$ are each normalised we have:

$$< p_i, p_j > = \frac{\int p_i(t)p_j(t)dt}{\sqrt{\int p_i(t)p_i(t)dt \int p_j(t)p_j(t)dt}} \qquad (56)$$

$$< p_i, p_i > = 1 \qquad (57)$$

This inner product would be useful in comparing model outputs. One place where it is used is for the convergence measure used in the OxCal program (Bronk Ramsey 1995) which is just $< p_i, p_j >^2$ from equation 56.

The other application of distribution comparison is for evaluating the likelihood distributions in relation to the posteriors. It is important that we do this because if the overlap between the likelihood and posterior distributions is very small it implies that the model does not agree with the observations. In this case we do not expect the distributions to be the same, and the likelihood and marginal posterior are different elements of the model. For this reason the symmetrical inner product is not appropriate. One approach, which is very useful in practice, is the integral of the likelihood over the posterior. For an individual parameter we can compare this integral for marginal density from our full Bayesian model $p_1(t_i|y)$ to that for the marginal density from the null model described above which is $p_0 (t_i|y_i) = p(y_i|t_i)$. The ratio that we get from this:

$$F_i = \frac{\int p(y_i|t_i)p_1 (t_i|\mathbf{y})dt_i}{\int p(y_i|t_i)p_0 (t_i|\mathbf{y})dt_i} = \frac{\int p(y_i|t_i)p_1 (t_i|\mathbf{y})dt_i}{\int p(y_i|t_i)p(y_i|t_i)dt_i} \tag{58}$$

has the property that it is 1 if the marginal posterior is the same as the likelihood. It might be higher than 1 if the marginal posterior is concentrated in the highest portions of the likelihood distribution. This factor gives us the ratio of the likelihood of this particular parameter under the two models and so if the factor is very low it implies that the model is unlikely. This measure can be used to identify outliers and in Bronk Ramsey (1995), the agreement index is $100F_i$. The factor F_i only makes use of the information for one parameter in the model, but the approach can be generalised to provide a factor for the model as a whole:

$$F_{model} = \frac{\int p(\mathbf{y}|\mathbf{t})p_1 (\mathbf{t}|\mathbf{y})d\mathbf{t}}{\int p(\mathbf{y}|\mathbf{t})p_0(\mathbf{t}|\mathbf{y})d\mathbf{t}} \tag{59}$$

Although we might be tempted to use these factors to compare two models (since by taking a ratio between them we eliminate the null model from the equation), it should be noted that they are not formal Bayes factors and in general more constrained (but possibly more realistic models) will often have lower F_{model} values. In particular many models have an $F_{model} < 1$ and yet are certainly more appropriate than the null model which assumes complete independence of all the events.

Model comparison is an area where further research and development is certainly needed.

Conclusion

The application of mathematical methods to this aspect of archaeology has been a very fruitful one. Not only has it been useful in providing quantitative archaeological chronologies, but also in making us think about the very nature of those chronologies and the assumptions that underly them. The process of formalising our approach to chronologies makes us define what we mean by the time-line, which may be very different to the way people perceive age and date in much non-quantitative analysis. It also helps us to think what is an appropriately neutral set of assumptions to apply when we come to study the processes revealed in archaeology. The notion of the two parameter process which has 'linear scale invariance' is something which is central to almost all the approaches discussed here; in the end the reason that we choose this as an approach is that both absolute time and duration of process are central to much that we wish to study through chronology. A hierarchical approach to model building is also something which ultimately comes from treating human activity as an inherently complex process with scales of structure visible at many different levels. The approaches taken in this field also highlight how we think about comparison of results with uncertainty more generally, and specifically in cases where the usual classical statistical approaches (based for example on assumptions of normality) cannot be used.

There are still many areas where further research and ideas are needed. These are particularly in the areas of model comparison where we would like much more powerful tools to allow us to test different ideas through chronological models. A related issue is what to do when we don't have a model of what is going on archaeologically and we just want the data to inform us about the chronology of underlying changes. In such cases the null model is still probably a poor representation of the processes and thus simply looking at summed probability distributions is not a good solution.

There is now a good framework for analysis of chronology using Bayesian models. This provides a very useful basis for archaeologists wishing to develop novel models relevant to their research, and for mathematicians and statisticians interested in extending these methods and improving our understanding of the fundamental nature of the models that we employ.

References Cited

http://c14.arch.ox.ac.uk/oxcalhelp/hlp_analysis_calend.html.

Blaauw, M. and J.A. Christen. 2005. Radiocarbon peat chronologies and environmental change. Journal of the Royal Statistical Society Series C-Applied Statistics 54: 805–816.

Blaauw, M., G.B.M. Heuvelink, D. Mauquoy, J. van der Plicht and B. van Geel. 2003. A numerical approach to C-14 wiggle-match dating of organic deposits: best fits and confidence intervals. Quaternary Science Reviews 22(14): 1485–1500.

Bronk Ramsey, C. 1995. Radiocarbon calibration and analysis of stratigraphy: The OxCal program. Radiocarbon 37(2): 425–430.

Bronk Ramsey, C. 2001. Development of the radiocarbon calibration program OxCal. Radio-carbon 43(2A): 355–363.

Bronk Ramsey, C. 2003. Punctuated dynamic equilibria: a model for chronological analysis. pp. 85–92. *In*: R.A. Bentley and H.D.G. Maschner (eds.). Complex Systems and Archaeology. University of Utah Press.

Bronk Ramsey, C. 2008. Deposition models for chronological records. Quaternary Science Re- views 27(1-2): 42–60.

Bronk Ramsey, C. 2009a. Bayesian analysis of radiocarbon dates. Radiocarbon 51(1): 337–360.

Bronk Ramsey, C. 2009b. Dealing with outliers and offsets in radiocarbon dating. Radiocarbon 51(3): 1023–1045.

Bronk Ramsey, C. and S. Lee. 2013. Recent and Planned Developments of the Program OxCal. Radiocarbon 55(2-3): 720–730.

Bronk Ramsey, C., J. van der Plicht and B. Weninger. 2001. Wiggle matching radiocarbon dates. Radiocarbon 43(2A): 381–389.

Buck, C.E., W.G. Cavanagh and C.D. Litton. 1996. Bayesian approach to interpreting archaeological data. Wiley, Chichester.

Buck, C.E., J.A. Christen and G.N. James. 1999. BCal: an on-line Bayesian radiocarbon calibration tool. Internet Archaeology 7.

Buck, C.E., J.B. Kenworthy, C.D. Litton and A.F.M. Smith. 1991. Combining Archaeological and Radiocarbon Information—a Bayesian-Approach to Calibration. Antiquity 65(249): 808–821

Buck, C.E., C.D. Litton and A.F.M. Smith. 1992. Calibration of Radiocarbon Results Pertaining to Related Archaeological Events. Journal of Archaeological Science 19(5): 497–512.

Buck, C.E. and A. Millard. 2004. Tools for constructing chronologies: crossing disciplinary boundaries. Springer, London.

Christen, J.A. 1994. Summarizing a Set of Radiocarbon Determinations—a Robust Approach. Applied Statistics-Journal of the Royal Statistical Society Series C 43(3): 489–503.

Christen, J.A. 2003. Bwigg: an internet facility for Bayesian radiocarbon wiggle matching. Internet Archaeology 7.

Christen, J.A. and E.S. Pérez. 2009. A new robust statistical model for radiocarbon data. Radiocarbon 51(3): 1047–1059.

Gelfand, A.E. and A.F.M. Smith. 1990. Sampling-Based Approaches to Calculating Marginal Densities. Journal of the American Statistical Association 85(410): 398–409.

Gilks, W.R., S. Richardson and D.J. Spiegelhalter. 1996. Markov chain Monte Carlo in practice. Chapman & Hall, London.

Harris, E.C. 1989. Principles of archaeological Stratigraphy. Academic Press, London.

Haslett, J. and J. Parnell. 2008. A simple monotone process with application to radiocarbon-dated depth chronologies. Journal of the Royal Statistical Society: Series C (Applied Statistics) 57(4): 399–418. URL http://dx.doi.org/10.1111/j.1467-9876.2008.00623.x

Hogg, A.G., Q. Hua, P.G. Blackwell, M. Niu, C.E. Buck, T.P. Guilderson, T.J. Heaton, J.G. Palmer, P.J. Reimer, R.W. Reimer, C.S.M. Turney and S.R.H. Zimmerman. 2013. SHCal13 Southern Hemisphere Calibration, 0–50,000 Years cal BP. Radiocarbon 55(4).

Hua, Q., M. Barbetti and A.J. Rakowski. 2013. Atmospheric Radiocarbon for the Period 1950–2010. Radiocarbon 55(4).

Ivanovich, M. and R.S. Harmon. 1992. Uranium-series disequilibrium: Applications to earth, marine, and environmental sciences. Oxford University Press.

Jones, M. and G. Nicholls. 2002. New radiocarbon calibration software. Radiocarbon 44(3): 663–674.

Karlsberg, A.J. 2006. Statistical modelling for robust and flexible chronology building. Ph.D. thesis, University of Sheffield.

Lee, S. and C. Bronk Ramsey. 2012. Development and Application of the Trapezoidal Model for Archaeological Chronologies. Radiocarbon 54(1): 107–122.

Miles, D.H. 2005. New Developments in the Interpretation of Dendrochronology as Applied to Oak Building Timbers. Ph.D. Thesis, University of Oxford.

Nicholls, G. and M. Jones. 2001. Radiocarbon dating with temporal order constraints. Journal of the Royal Statistical Society Series C-Applied Statistics 50: 503–521.

Reimer, P.J., E. Bard, A. Bayliss, J.W. Beck, P.G. Blackwell, C. Bronk Ramsey, P.M. Grootes, T.P. Guilderson, H. Haflidason, I. Hajdas, C. Hatte, T.J. Heaton, D.L. Hoffmann, A.G. Hogg, K.A. Hughen, K.F. Kaiser, B. Kromer, S.W. Manning, M. Niu, R.W. Reimer, D.A. Richards, E.M. Scott, J.R. Southon, R.A. Staff, C.S.M. Turney and J. van der Plicht. 2013. IntCal13 and Marine13 Radiocarbon Age Calibration Curves 0–50,000 Years cal BP. Radiocarbon 55(4).

Rhodes, E.J., C. Bronk Ramsey, Z. Outram, C. Batt, L. Willis, S. Dockrill and J. Bond. 2003. Bayesian methods applied to the interpretation of multiple OSL dates: high precision sediment ages from Old Scatness Broch excavations, Shetland Isles. Quaternary Science Reviews 22(10-13): 1231–1244.

van der Plicht, J. 1993. The Groningen Radiocarbon Calibration Program. Radiocarbon 35(1): 231–237.

Zeidler, J.A., C.E. Buck and C.D. Litton. 1998. Integration of Archaeological Phase Information and Radiocarbon Results from the Jama River Valley, Ecuador: A Bayesian Approach. Latin American Antiquity 9: 160–179.

15

Modelling the Effects of Post-depositional Transformations of Artifact Assemblages using Markov Chains

Geoff Carver

Introduction

The following is part of a study of archaeological excavation and documentation methodologies, aimed at developing standards and procedures suitable for recording evidence of post-depositional transformations of—and for evaluating the significance these transformations may have had on—the archaeological record. Problems with recording this evidence will be discussed elsewhere. The present study represents an initial exploration of the problems of modelling the effects of post-depositional transformations on archaeological remains, focusing on the potential use of stochastic methods for estimating the probabilities of loss of different artifact materials over time.

First Principles

Archaeology has been defined as "The systematic description or study of human antiquities, esp. as revealed by excavation" (OED 1997). Archaeologists study human antiquities—artifacts—and not the fossil remains of extinct species (as in geology and/or palaeontology). Much archaeological data is revealed by digging up artifacts (or removing them from their matrix; although tangential to the present discussion, the

Broicher Strasse 39b, 51429 Bensberg, Germany.
Email: gjcarver@t-online.de

complex issue of whether this is soil or sediment is worth noting). In contrast to grave-robbers, looters, treasure hunters, and our own antiquarian predecessors, archaeologists aim to be "scientific". Thus, archaeology can also be defined as:

> the application of scientific method to the excavation of ancient objects… it is based on the theory that the historical value of an object depends not so much on the nature of the object itself as on its associations, which only scientific excavation can detect… digging consists very largely in observation, recording and interpretation (Woolley 1961: 18).

Because our discipline has generally fallen short of these ideals, archaeology has also been defined as, "the discipline with the theory and practice for the recovery of unobservable hominid behaviour patterns from indirect traces in bad samples" (Clarke 1973: 17).

These "bad samples" might be expressed in terms of increasingly smaller samples of an original population:

1) The range of hominid activity patterns and social and environmental processes which once existed, over a specified time and area.
2) The sample and traces of these (1) that were deposited at the time.
3) The sample of that sample (2) which survived to be recovered.
4) The sample of that sample (3) which was recovered by excavation or collection (Clarke 1973: 16 [original emphasis]).

However "bad" these samples may be, they are often all we have, and archaeologists have to do as much with them as possible.

For purposes of consistency in the following (and following Clarke's numbering), these samples will be defined as:

1) the *original* population;
2) the *deposit* sample;
3) the *survival* sample;
   and
4) the *recovery* sample.

Clarke linked each of these samples to its own body of theory: a body of theory relating culture to artifacts (*Pre-depositional* theory), another relating to the way artifacts enter the archaeological record (whether they are lost, discarded, etc.: *Depositional* theory), preservation (*Post-depositional* theory) and recovery (*Retrieval* theory). Paraphrasing the first quotation from Clarke, the archaeological process may be defined in terms of using the recovery sample ("the sediments, features, and items of which [the archaeological record] is composed") to infer the otherwise "unobservable hominid behaviour patterns" of the original population (i.e., the "past" many archaeologists claim to study).

Following a systems-analytical approach, each of these samples would be considered separately; but since archaeologists are generally interested only in the cultures which produce Clarke's traces, interpretive "theory" is generally directed only towards the cultures which produce artifacts without considering how cultural processes and/or patterns can be inferred from "bad samples", while Depositional theory is largely restricted to inferences based on what (little) archaeologists know about how artifacts enter the archaeological record. Instead of considering a potentially wide variety of scenarios, archaeologists often follow a mental shortcut, the "Pompeii Premise", an assumption that everything recovered was deposited where it had been used.

The problem is, that—by not taking the depositional process and Clarke's various samples into account—many archaeologists fail to recognize that, "In far too many cases, the evidence used by an archaeologist owes many of its properties, not to the past phenomena of interest, but to various formation processes" (Schiffer 1983: 697).

In fact, "The likelihood that tools, containers, and other goods will remain in their original location of use unmolested is very low" (Cameron 2006: 28).

On the one hand, given limits to budgets and limits to recording technologies, there may be no practical way to fix Clarke's "bad" recovery samples, so archaeologists may have been reluctant to try. It might also be argued that many unreflective or possibly optimistic (naïvely deterministic?) archaeologists fail to understand what archaeology is and/or what the limitations about they themselves can do:

> When some archaeologists are asked to describe what they do, they often say "I study the past." Actually this statement is incorrect.... All that remains of past events is circumstantial evidence. Archaeologists do, however, make observations on events in the present: we record the events connected with excavating the archaeological record; with examining the sediments, features, and items of which it is composed (Binford 2001: 46).

In terms of communication theory, the archaeological "message" is subject to a mix of noise and loss (Clarke's bad "samples"). The problem is that, despite Clarke's claim (1973: 18) that "Archaeology is, after all, one discipline"—archaeology is NOT one discipline in part because of "fragmentation"—"Each subsystem... is developing its own vocabulary which is fast becoming incomprehensible to specialists in other fields" (Hodder 1992b: 27).

I would like to suggest that there are parallels between the archaeological process—deposition, post-deposition and recovery—and the development of archaeology as a discipline; and that archaeology's inability to address the problems of "bad samples" reflects a lack of integration or "fragmentation".

Communication failures within the discipline may help explain Hodder's warning (1992a: 12) against "the danger" of using unspecified "information-processing approaches" in "reducing the meanings of objects to 'bits' of information which are studied simply in terms of their effectiveness in conveying messages," while failing to recognize the important role those "'bits' of information" play in computing, systems analysis, etc.

Although Hodder did not specify which "information-processing approaches" he meant, his description is consistent with those outlined in Claude Shannon's seminal paper, "A mathematical theory of communication" (1948).

Shannon's communication model presents a means for comparing the archaeological "message" which was transmitted to that which was received. Within either an efficient industrial or scientific process, there should be an iterative series of tests to evaluate data quality (Shannon's feedback loop ["correction channel"]). Although there should be tests, for example, to evaluate the degree to which the recovery sample reflects the deposit sample in order to test the efficiency of retrieval methods, and whether or not the margin of error exceeds an acceptable tolerance, archaeologists seem to be sceptical of mathematics. In contrast to the physical sciences which—following Galileo and Newton—use math extensively, archaeology tends to be textual. Given their interest in the past, archaeologists may even be suspected of being opposed to technology and "progress".

In addition to the way small archaeological samples inhibit statistical analysis, the reasons for this scepticism may be historical: in attempting to provide their nascent discipline with firm foundations, 19th century archaeologists (i.e., Pettigrew 1850: 173–174) insisted they had been strongly influenced by the scientific method outlined by Sir Francis Bacon, and Bacon has been criticized for failing to recognize the role math would play on science (cf. Bury 1920: 52, Butterfield 1982: 87, Kuhn 1977b: 48).

This influence may ultimately have proven to be counterproductive; "sometime between 1800 and 1850" (i.e., about the time archaeology was becoming a discipline—what Glyn Daniel [1975: 32] labelled the "Antiquarian revolution"),

> there was an important change in the character of research in many of the physical sciences, particularly in the cluster of research fields known as physics. That change is what makes me call the mathematization of Baconian physical science one facet of a second scientific revolution (Kuhn 1977a: 220).

That such "mathematization" never occurred in archaeology may help explain why statistics are under-utilized in archaeology (typically being used only to test whether or not the sample recovered from any given context is compatible with those recovered from the rest of the site, or whether or not an excavation area is large enough to support valid conclusions), while continued adherence to Baconian principles is suggested by recent, otherwise inexplicable attempts to re-link archaeology and written history via hermeneutics ("archaeology is a text"), rather than to geology.

Transformation Factors

Geology has studied the cycle of erosion, transportation, and deposition of sedimentary particles since the days of Hutton (1726–1797). As a result, geologists and sedimentologists can model depositional processes mathematically, relating the strength of current, divergence of density of the material, particle size, etc.

Although the Huttonian cycle affects all archaeological sites, there is some question as to whether archaeological deposits should be considered as being sediments or soils. Like the question of whether light is a particle or a wave, some compromise may be necessary. Certainly, once deposited, archaeological remains are subject to soil formation processes, and this is reflected in the general term for mixing of soils and sediments, "pedoturbation".

Granted, soils are more complicated than geological sediments, but geological sediments are also subject to post-depositional soil formation processes.

If archaeology was a systematic discipline—and if the import of geological concepts had been more efficient, resulting not in the "corrupted versions of geological hand-me-downs" outlined in "nearly all current textbooks on archaeology" (Harris 1989: xiii)—then it would be normal for archaeologists to follow geological practice by identifying not just the "the main evidences of position" (Petrie 1904: 50)—where Binford's "sediments, features, and items" were found (provenience)—but also the source (provenance [Shannon's "information source"]) and transport media (Shannon's "signal") of archaeological sediments (or "deposits").

Thus, instead of a "Pompeii Premise", archaeologists might be better able to recognize cases which contradict "corrupted versions of geological" theories—Roman material dumped into a medieval cellar, for example—and thus make it easier to recognize the need to identify intrusive postdating or "residual remains" predating the deposit.

The fact that this is not the case reflects the fact that archaeologists generally do not understand geology and, hence, are unable to critically evaluate, for example, the problems of translating geological "laws" of stratigraphy into archaeological contexts.

Erosion is generally a response to elements like air and water, or combinations of the two. Wind can blow away fine sediments, leaving artefacts on the surface. Besides washing soils (and artifacts) downhill, water mixes soil by pushing up air bubbles. Moisture also increases the effects of gravity on the downslope movement and mixing of sediments, causing solifluction and soil creep; this is especially common in saturated soils in periglacial environments or permafrost.

The effects of such erosional and/or transport media may be augmented by thermal causes; frost causes soil to expand, moving artifacts upwards and giving them a vertical orientation, cause soil to crack, etc.

Deposits may also be disturbed by plants or animals. Floralturbation is disturbance by plants: roots can destroy walls and move artifacts, either through growth or when trees fall. Faunalturbation is disturbance by fauna. Two chapters of Darwin's study of worms (1896) documented their effect on archaeological remains; other studies have examined the effects of direct and indirect traces clams, dogs, rabbits, armadillos, etc., leave in the archaeological record. One should also consider the degree to which a given environment is conducive to or inhibits the growth of fungi or bacteria.

In addition to these natural factors, there is cultural transformation due to ploughing, planting and harvesting, irrigation, construction, etc., to consider.

On the other hand, having formed deep under the ocean, geological sediments are largely immune to soil-formation processes: "the very conditions that promote preservation also decree that few organisms, if any, make their natural home in such places" (Gould 1989: 62).

Soil type also plays a role. Seasonal shrinking and swelling of clay due to changes in soil moisture causes cracks to open and close, pushing some artifacts upwards, and allowing smaller artifacts and stones to fall down the cracks. Recognition of this process was particularly important when proving the contemporaneity of artifacts with the bones of extinct fossil species at Brixham cave (cf. Babbage 1859: 68–69, Prestwich 1860: 300).

The materials from which artifacts are made are also important. Table 1 lists environments conducive to the preservation of different materials and Table 2 lists environments where preservation is less likely. One of the aims of the present exercise is to suggest more systematic means for ways of quantifying—giving a probability value to—these labels of "possible" and "unlikely" survival, before testing the statistical significance of the inferred effects.

It should perhaps be noted that since it would be pointless (and generally "unscientific") to speculate on something that was not recovered (Schrödinger's cat is under-utilized in archaeology), no attempt will be made to estimate the amount of material that might have been in a deposit where no evidence of any was found.

Table 1. Burial environments and their action on materials: survival of materials listed is possible (after Watkinson and Neal 1998: 8 [Table 1a]).

Modelling the Effects of Post-depositional Transformations of Artifact Assemblages using Markov Chains

POSSIBLE SURVIVAL OF MATERIAL LISTED					
DAMP BURIAL [Oxygen present]			WATERLOGGED BURIAL [Limited or no oxygen present]		
Acid soil	Neutral or Weak Acid or Alkali	Alkaline soil	Acid soil	Neutral or Weak Acid or Alkali	Alkaline soil
Gold, silver [and their alloys]	Gold, silver, copper, lead, tin, zinc [and their alloys]	Gold, silver, copper, lead [and their alloys]	Gold, silver [and their alloys]	Gold, silver, copper, lead, tin, zinc [and their alloys]	Gold, silver, copper, lead, tin, zinc [and their alloys]
Slag	Iron	Iron	Slag	Iron	Iron
Roman glass	Slag	Bone	Shale	Slag	Slag
Shale	Limestone	Limestone	Wood	Shale	Shale
Amber	Marble	Alabaster	Leather	Limestone	Limestone
Jet	All Ceramics	Marble	Keratin	Alabaster	Alabaster
Porcelain	Glass	All Ceramics	Horn	Marble	Marble
Stoneware	Bone		Amber	Amber	Amber
Earthenware	Shale		Jet	Jet	Jet
			Porcelain	Glass	Bone
			Stoneware	Wood	Wood
			Earthenware	Leather	Leather
			Textiles	Textiles	Textile [vegetable fibres; flax hemp]
			[proteinic fibres: silk & wool]	Horn	Horn
				Bone	Bone
				All Ceramics	All Ceramics

Table 2. Burial environments and their action on materials: survival of materials listed is possible (after Watkinson and Neal 1998: 9 [Table 1b]).

SURVIVAL OF MATERIAL LISTED IS UNLIKELY					
DAMP BURIAL [Oxygen present]			**WATERLOGGED BURIAL** [Limited or no oxygen present]		
Acid soil	Neutral or Weak Acid or Alkali	Alkaline soil	Acid soil	Neutral or Weak Acid or Alkali	Alkaline soil
Copper, lead, tin, zinc [and their alloys]	Tin, zinc [and their alloys]	Tin, zinc [and their alloys]	Lead, tin, zinc [and their alloys]		Lead, tin [and their alloys]
Iron	Textile	Textile	Textile [vegetable fibres]		Textiles [proteinic fibres]
Wood	Wood	Wood	Bone		Horn
Leather	Leather	Leather	Glass		Glass
Horn	Horn	Horn	Terracotta		
Textiles	Alabaster	Alabaster	Limestone		
Bone	Limestone	Limestone	Alabaster		
Limestone					
Alabaster					
Ceramics [low fired]					
Glass [Medieval]					

The following example shows how these factors can combine to alter an archaeological site (in this case, a burial mound):

> It is difficult to reconstruct the appearance of the mound after the burial had been completed. Four factors contributed to its change in later times:

1) The decay of all organic material inside the mound, the wood of the boat (hull) and any equipment placed in it (bilge cover, oars, etc.). The decay of such objects and any other organic grave-goods (wooden chests, for example) would certainly have caused movement of earth within the mound, displacing some of the stones.
2) Rabbits burrowing into the loose soil brought about further destruction.
3) An unknown, but certainly quite considerable, quantity of loose soil must have disappeared from the surface of the mound, or where it was brought to the surface by rabbits, by erosion. In modern conditions, during spells of dry weather, there is a definite tendency to wind erosion, a feature which is aggravated by the scouring properties of heavy rain on any slope.
4) Any stones exposed, either originally or by subsequent erosion, were liable to be used in building field-walls. The remains of a field-wall along the southern edge of the plateau of the hill-fort… lends colour to this point (Bersu and Wilson 1966: 8–9).

If nothing else, it should be obvious that archaeological deposits are subject to a large number of potential post-depositional transformation processes. This complexity has been suggested in the present study by signifying that the past and present theoretical and/or disciplinary contexts in which archaeology is performed parallels the physical contexts (i.e., the materials archaeologists excavate).

Taphonomy

In order to begin approaching such problems systematically, some archaeologists have come to recognize the potential impact of such post-depositional transformation processes as taphonomy. Taphonomy is "the branch of palaeontology that deals with the processes by which animal and plant remains become preserved as fossils" (OED 1997), but its application has extended beyond palaeontology:

> This term was originally coined to refer to the transition of paleontological material from the biosphere to the lithosphere… and, strictly speaking, anything that happens between the death of an animal and the arrival of its bones in the laboratory is the subject of taphonomy…. However, among zooarchaeologists, it has informally come to refer to the study of factors that, by distorting the faunal record, interfere with the use of faunal data to infer prehistoric behavior (Dibble et al. 1997: 629).

Taphonomy provides a useful example because it is particularly well-studied, in large part due to the need to account for the over- and under-representation of bones from certain species and certain types of bones (elements): some bones are preserve better (horse bones, being dense, generally preserve better than bird or fish bones) or easier to identify than others (vertebrae are often classifiable only according to size). Although bone is not the only material affected by post-depositional environments, "For want of a better term, taphonomy is used here to cover an even wider range of natural processes that distort the whole of an archaeological assemblage" (Dibble et al. 1997: 630).

The taphonomic processes illustrated in Fig. 1 are represented as being hierarchical, sequential and deterministic. The flowchart formatting—overly simplistic, like all such schematic representations—implies that a certain sequence of events or processes not only can occur but will occur, that transformation processes act as a sort of filter, obscuring the human behaviour responsible for what was actually deposited in the archaeological record, and that one only needs to factor in some series of variables in order to infer the source population. Thus a sample of bones has been diminished 20% by rodent gnawing (bones provide rodents with a necessary source of calcium), a further 10% by frost, 5% by desiccation, etc.

In reality taphonomic pathways are not necessarily linear: there can be loops as material is continually re-exposed and (re-)deposited, subjected first to rodent gnawing, then to frost and sun, erosion, battering in a streambed, etc. Shannon's communication system—with its feedback loops and probabilistic approach—provides a better model.

Archaeological documentation presents another problem. Since archaeological documentation normally does not record traces of transformation processes, there is rarely any evidence from which to derive the necessary variables.

Archaeologists consequently face two alternatives:

1) Claim that the effects of transformational processes are not quantifiable (i.e., unknowable) and hence speculative;
2) Infer some set of variables to explain or otherwise quantify assemblage loss.

Both of these solutions are unsatisfactory, the first partially because it represents what may largely be considered to be the status quo, and the present chapter obviously would not have been written if its author was satisfied with this state of affairs. The problem is that—without some form of quantification and/or statistical analysis— taphonomy is either assumed to have occurred without taking into consideration the effects this might have had on the assemblage (except, perhaps, for some tentative phrasing in the conclusions), or disregarded altogether (i.e., the "Pompeii premise").

While it may seem that this chapter aims at attaining the second solution in some form, the reality is subtly different. On purely scientific and philosophical grounds it does not seem possible to state with any certainty what any given deposit sample was like based solely on the recovery sample; it seems too speculative to claim that "Here are three bones, whereas originally there were four." What can be argued, and what will be outlined in the following, is something like "Here are three bones; although under certain conditions there might originally have been four—maybe more—it is also possible that the quantity has not changed over time." Instead of variables, it should be possible to assign probability values for the different scenarios—extreme loss, minor loss, no loss at all[1]—resulting from various environmental factors, and then to simulate the cumulative effects these processes may have had over time in order to simulate a deposit sample from which the recovery sample may have been derived.

The present study aims to go beyond mere simulation; the aim is to outline a possible method for identifying contexts which may have been subject to significant

---

[1] Although the archaeological assemblage is transformed through both destruction and movement, the present study deals only with the former.

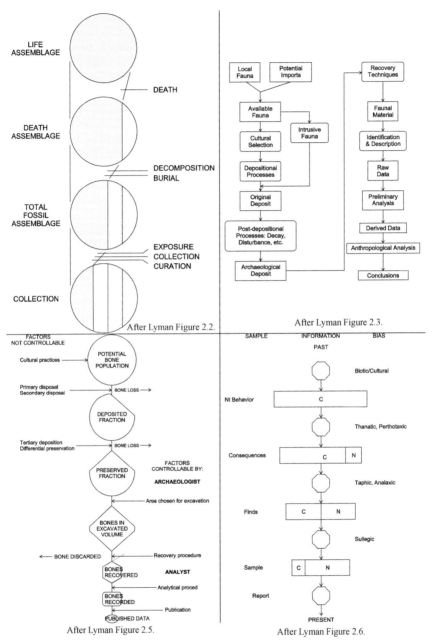

After Lyman Figure 2.2.

After Lyman Figure 2.3.

After Lyman Figure 2.5.

After Lyman Figure 2.6.

Fig. 1. Taphonomic models.

alteration by post-depositional transformations, by testing whether or not the simulated recovery sample differs significantly from the deposit sample under various scenarios.

If there is no statistically significant difference between the sample and the (simulated) population from which it derives (i.e., if the sample is representative), then there is no need to take post-depositional transformations into consideration in the subsequent analysis. If the difference does prove to be significant, then we need to consider its source. Is it due to false premises, extremely unlikely scenarios (the worst of all possible cases), or is it an indication that we need to examine our data more carefully in order to find evidence either supporting or disproving some set of conditions which need to be factored into the equation?

Estimating the Significance of the Taphonomic Process

Archaeological finds (i.e., the recovery sample) are generally summarized in a tabular form:

Table 3. Example of a table showing quantities of recovered artifacts classified by material.

	Ceramic	Bone	Bronze	Iron
1	2	1	3	4
2	1	2	2	3
3	3	4	1	2
4	5	4	2	1

Table 3 can be rewritten as a matrix:

$$\begin{pmatrix} 2 & 1 & 3 & 4 \\ 1 & 2 & 2 & 3 \\ 3 & 4 & 1 & 2 \\ 5 & 4 & 2 & 1 \end{pmatrix}$$

Although less familiar in terms of a medium for representing archaeological data, this matrix better conveys the idea of transformation; that is, the data in the matrix represents the end stage (in this case) of a process. The matrix can then be conceived as bearing a relation to the original population, and the intervening transformations (from original population to deposit and/or survival samples) can then be modelled as a series of mathematical functions:

$$\begin{pmatrix} 3 & 2 & 4 & 5 \\ 1 & 2 & 3 & 4 \\ 4 & 4 & 2 & 3 \\ 5 & 5 & 2 & 3 \end{pmatrix} \Rightarrow \begin{pmatrix} 2 & 1 & 3 & 4 \\ 1 & 2 & 2 & 3 \\ 3 & 4 & 1 & 2 \\ 5 & 4 & 2 & 1 \end{pmatrix}$$

One common means for quantifying transformation processes is to model them as a linear function (that is, the degree of decay is proportional to the length of time spent in the ground, as with radioactive decay [Fig. 2]). Although this analogy may be appealing in part because most archaeologists have a vague understanding of radiocarbon dating, but there is no reason to believe that decay rates remain constant over time. There may have been no significant change; or change may have been cataclysmic: artifacts decaying rapidly shortly after deposition, followed by a period of stasis once they have reached equilibrium with their environment (cf. Watkinson and Neal 1998: 8 [Fig. 3]). Less likely, perhaps: they could have remained in a period of stasis, and decayed shortly before excavation due to, for example, acid rain, global warming, or modern agricultural practices (Fig. 4).

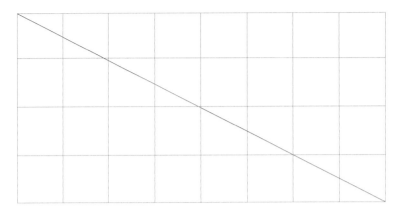

Fig. 2. Gradual, linear decay.

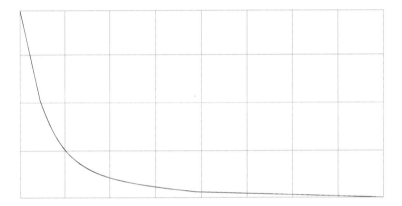

Fig. 3. Catastrophic decay.

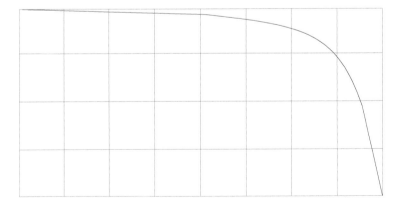

Fig. 4. "Last minute" decay.

The true picture is more likely to have been a sort of "punctuated equilibrium": a fairly cataclysmic decay shortly after deposition (as more volatile materials reacted strongly to their new environment), followed by incremental decay periods in reaction to subsequent events: changes in soil acidity, rising groundwater, particularly deep frost, infestation by a new species of burrowing rodents as a result of changes in climate, land-use, ground cover, etc.

If these factors cannot be identified specifically, their cumulative effects can be modelled randomly; not random in the sense of cause being wholly arbitrary, but in the sense that physicists are able to predict the rate of radiocarbon decay even if they cannot accurately determine which particular atom will decay at any given point in time.

Markov

Thus the effects of post-depositional transformations can be modelled using stochastic methods, which follow some random probability distribution or pattern, thereby allowing behaviour to be analysed statistically but not predicted precisely. The requisites for using stochastic methods are two discreet events (the initial and transformed states), and a probability of transformation from one state to another. In this case, the two events—artifact deposition (the deposit sample) and artifact retrieval (the recovery sample)—are not independent: the recovery sample is related to the deposited, and it is not possible for the retrieved population to be larger than the deposited population.

For the purposes of illustration, a "Markov chain" was chosen as a stochastic method simple both to model and explain; in keeping with attempts to equate archaeology with "text", it is essentially the method Claude Shannon outlined.

Markov chains seem to have been little used in archaeology, and are not found in the indexes to standard introductory works (i.e., Orton 1980, Shennan 1997). One exception (Thomas 1978: 235) listed "Random walks, black boxes, Markov chains, matrix analysis, and so forth" as practices to be avoided. Because this is not referenced, it is unclear which applications (or abuses) the author meant.

Other archaeological applications include the use of Markov chain Monte Carlo (MCMC) methods to interpolate phosphate data from a field survey (Buck et al. 1996: 282). Litton and Buck (1998) used MCMC to reduce the margin of error on ^{14}C dates. A recent example (Bagnoli 2013) used MCMC to simulate the decay of stone inscriptions.

The following section examines Orton's example (2000: 195–196) of deteriorating conditions in a museum collection, which parallels the taphonomic transformation being modelled in the present study. He lists the proportion of finds which are in the given priority categories at the beginning of the survey:

- 40% are initially little,
- 30% are low,
- 20% high,
  and
- 10% urgent.

The probabilities that artifacts in any one of these initial states will change over a given year are:

- 4% (i.e., 40% x 0.10) move from little to low
- 2% (i.e., 40% x 0.05) move from little to high
- 6% (i.e., 30% x 0.20) move from low to high
- 3% (i.e., 30% x 0.10) move from low to urgent
- 5% (i.e., 20% x 0.25) move from high to urgent

These lists can be rewritten as tables—the initial proportions as Table 4, and the probability of transformations as in Table 5—and rewritten as a series of matrices:

Table 4. Priority state expressed as a percentage of the collection.

Little	40
Low	30
High	20
Urgent	10

Table 5. Probabilities of change of state.

		Probability of changing from initial to second state			
		Little	**Low**	**High**	**Urgent**
Initial state	Little	0	0.10	0.05	0
	Low	0	0	0.20	0.10
	High	0	0	0	0.25
	Urgent	0	0	0	1

$$\begin{bmatrix} 40 \\ 30 \\ 20 \\ 10 \end{bmatrix} \Rightarrow \begin{bmatrix} .85 & .10 & .05 & 0 \\ 0 & .70 & .20 & .10 \\ 0 & 0 & .75 & .25 \\ 0 & 0 & 0 & 1.0 \end{bmatrix} \Rightarrow \begin{bmatrix} 34 & 4 & 2 & 0 \\ 0 & 21 & 6 & 3 \\ 0 & 0 & 15 & 5 \\ 0 & 0 & 0 & 10 \end{bmatrix} \Rightarrow \begin{bmatrix} 34 \\ 25 \\ 23 \\ 18 \end{bmatrix}$$

The initial values in the first matrix are transformed by the probabilities in the second matrix to produce the values in the third. Hence, of the 40% of the collection which had a priority rating of little at the beginning, after the first year only 85% remained (100– [.10+.05]), or 34% of the total collection (40% x 0.85) could still be classified as "little". Similarly, over the course of the first year the low value is reduced to 70% of its original value (100– [.30+.20]), but gains the 4% (40% x 0.10) that has changed from "little" to "low".

This process can be repeated for a second year:

- 3.4% (i.e., 34% x 0.10) move from little to low
- 1.7% (i.e., 34% x 0.05) move from little to high
- 5.0% (i.e., 25% x 0.20) move from low to high
- 2.5% (i.e., 25% x 0.10) move from low to urgent
- 5.75% (i.e., 23% x 0.25) move from high to urgent

Table 5 can be rewritten in matrix form:

$$\begin{bmatrix} 34 \\ 25 \\ 23 \\ 18 \end{bmatrix} \Rightarrow \begin{bmatrix} .85 & .10 & .05 & 0 \\ 0 & .70 & .20 & .10 \\ 0 & 0 & .75 & .25 \\ 0 & 0 & 0 & 1.0 \end{bmatrix} \Rightarrow \begin{bmatrix} 28.9 & 3.4 & 1.7 & 0 \\ 0 & 17.5 & 5.0 & 2.5 \\ 0 & 0 & 17.25 & 5.75 \\ 0 & 0 & 0 & 18.0 \end{bmatrix} \Rightarrow \begin{bmatrix} 28.9 \\ 20.9 \\ 23.95 \\ 26.25 \end{bmatrix}$$

And so on until, after about 30 iterations (i.e., 30 years) approximately 99% of the collection will be in urgent need of attention. Since this simple example does not take into account additions to the collections, nor the results of restoration work (both of which would alter the progress of the curves), the condition does not improve, and the applicability to taphonomic modelling—and parallels to "entropy" in Shannon's work (which showed how such changes in direction are easily modelled)—should be obvious.

For present purposes, a major drawback is the fact that—unlike taphonomic analysis, which necessarily moves from a known recovery sample to an unknown initial state of archaeological assemblages (i.e., the deposit sample)—Orton's initial values and probabilities are known. Similarly, Bagnoli's use of MCMC to simulate the decay of stone inscriptions (2013) relied upon a presumed original state and accurate measurements of the annual decay of limestone.

A related problem is the fact that, in contrast to the processes illustrated sequentially in Fig. 1, taphonomic analysis generally moves from a known present *backwards* to an unknown past. Bagnoli's example shows, however, why it seems legitimate to use Markov chains to model probabilities of changes of state not only forwards but also backwards in time. Estimates of the deposit sample are therefore dependent upon accurate estimates of probabilities of decay, which in turn require detailed information regarding the post-depositional environment and any number of contextual factors. Although Bagnoli had this information, difficulties in estimating the original population generally limits archaeological analysis to the level of presence or absence.

The following example addresses the first of these problems.

The use of Markov Chains to Model the Effects of Taphonomic Processes

The recovery environment needs to be considered in terms of whether decay or preservation was likely, or whether it provides any evidence—Clarke's "indirect traces" or Binford's "circumstantial evidence"—of past events. If bones were recovered, they need to be examined in order to determine how well they have been preserved: are there signs of rodent gnawing, exfoliation or root etching? Were they likely to have been effected by groundwater, frost, or an extremely acidic soil? Ceramics need to be compared to those recovered from other contexts in order to help identify whether they were recovered from of primary or secondary deposits, etc.

Once potential environmental factors have been identified, probability values for rate of decay have to estimated for a given interval (year, decade, century). Since the rate of decay is unknown, an element of chance must also be introduced into these estimates: estimated probabilities that a large proportion of bone will decay within any given time period, or that relatively little will decay, etc. The chain will then be run a few thousand times in a program like "R" or WinBugs, the results plotted, and their distribution analysed.

These transformations probably have not been modelled previously in part because of the difficulties associated with quantifying these probabilities. An interim solution is suggested in the following.

There are two main—but interrelated—problems to overcome. The first is the problem of comparing quantities of the recovered sample to those in the original state, and the second is modelling probabilities themselves.

In the Orton example, it was assumed that all of the artifacts having a given priority would have an equal probability of changing state. In reality it seems more likely that the 2% changing from little to high represent a particularly delicate material: organic remains, for example, or ancient glass. Since what is being modelled here are the different survival probabilities for artifacts made of different materials—the fact that organic remains are less likely to be preserved than stone objects—there needs to be a unit of measure which will allow comparison between quantities of different artifact classes.

But what does it mean to say that ceramics constituted 40% of the finds recovered from a single context? In order to use Orton's model, we need to be able to say something like:

- 40% of the finds are ceramic,
- 30% are bone,
- 20% metal,
  and
- 10% glass;

which underlines a number of difficulties. To illustrate this, these percentages are rewritten in tabular form (Table 6), then summarized as numbers of artifacts instead of percentage of an assemblage, and compared at different scales (Table 7), which obviously makes little sense: comparing the number of sherds to pieces of bone is meaningless.

Using the figures provided in Table 3, a reduction from a deposit sample—real or estimated—comprising 40 sherds to a recovery sample of only 34 sherds would be deceptive: would this represent an actual 15% loss due to taphonomic processes or

Table 6. Materials as percentage of an assemblage.

		%
Material	Ceramic	40
	Bone	30
	Metal	20
	Glass	10

Table 7. Comparison of numbers of artifacts.

			Fragments	Units	Types
Material	Ceramic	40	Sherds	Vessels	Fabrics
	Bone	30	Bones	Animals	Species
	Metal	20		Swords	Alloys
	Glass	10	Splinters	Bottles	

could those "missing" sherds have been found if the sampling had been done using a finer sieve (i.e., had some of the sherds only been reduced in size, even as a result of the recovery and post-excavation processes)? And could taphonomic processes be inferred from an increase in the number of sherds (i.e., from an initial count of 40 to a recovered sample of 46) due to breakage (before, during or after excavation)?

And while bone and metals might also break, both their relative rarity and easily recognisable diagnostic features mean that they are often more likely to be reassembled. So two pieces of bone might be recovered, then fitted together to produce one composite; of 40 sherds, 30 might produce 5 larger pieces (large fragments of 2 pots), leaving 10 sherds left over. Thus, although taphonomic theory assumes that the overall quantities of artifacts are reduced over time, the numbers may increase, and may be greatly affected by the availability of post-excavation resources.

It would also be misleading to compare the number of objects of different material classes when modelling taphonomic processes because of the difficulties of comparing numbers of bones to the number of pot sherds or metal artifacts. Under certain environmental conditions whole bones will be destroyed and metal will undergo extensive corrosion, but ceramics are often broken up into increasing numbers of smaller pieces.

The problem is in finding a standard unit of measure which would allow such comparison; an equation which can be expressed formally as a series of ratios:

$$40\alpha:30\beta:20\gamma:10\delta$$

In the Orton example $\alpha = \beta = \gamma = \delta = \%$, but in this case it is unclear how the variables relate to one another:

40 sherds:30 bones:20 pieces of metal:10 splinters
Or:

40 pots:30 animals:20 swords:10 bottles

etc.

Not noted in this analogy is the fact that the Orton example refers to a single collection, and that numerical relations could be expressed as 40 little to 30 low to 20 high to 10 urgent to 1 collection (40:30:20:10:1). Applying this model to the sample recovered from an archaeological excavation for the purpose of modelling the effects of post-depositional transformations would need to convert the single collection to a number of stratigraphic or excavation units (contexts). Thus, although the total assemblage might produce a ratio of 40 ceramics to 30 bones to 20 pieces of metal to 10 pieces of glass to 1 site, it would be more accurate and useful to write this in terms of a site consisting of (for example) 250 units (whatever these might be: standardized excavation units, cubic meters of earth, etc.)—40:30:20:10:250—in order to allow comparison between different time periods (as in the famous Stone, Bronze and Iron Ages, etc.) and different functions (lithic workshop, trash pit, living area, etc.), different environmental conditions (below levels where frost, worms, roots or rodents cause disturbance, above the groundwater, etc.).

While one alternative might be to compare weights of the different parts of the assemblage—of 100 kg of finds recovered, 40 kg (40%) were ceramic, 30% bone, etc.—this does not solve the problem, since differences of density are not taken into account. Some bone (large mammal) is denser than ceramic, and some (fish, bird) is lighter, while metal becomes less dense as it corrodes.

One solution might be to compare volume of artifacts. This solves both the immediate problem of enabling comparison of the relative proportions of material types at time of analysis—comparison which is not possible with any other method (p bones + q sherds = x% of one layer)—it also helps model the deposit sample, since the volume of artifacts recovered from any given context (stratigraphic unit) can then be considered relative to the volume of the context as a whole: comparison to layer volume enables estimating an upper limit on the size of a given assemblage. If—at time of excavation—a layer consists of 100 litres of deposit, of which ceramics compose 40%, then the maximum portion that the ceramics ever could have been is 100% of the original volume. Besides being highly improbable, this will have observable consequences.

Volumes have largely been ignored because they are not conducive to archaeology's traditional two-dimensional recording media (drawings and photos): plans concentrate on areas and spatial distributions, while profiles highlight temporal change. Recording contexts as volumes was also impractical until the advent of electronic surveying equipment and computer-assisted drafting. It is expected that measuring artifacts or contexts in terms of volumes will become standard practice as soon as suitably practical methods—such as 3D laser scanning or Structure Through Motion (STM)—become affordable, and as the wider use of electronic recording and archaeological documentation lead archaeologists away from the two-dimensional planes of drawings and photos to computer documentation.

Although this concept may seem revolutionary, it is actually very practical, and volume analysis would help indicate, for example, the amount of work involved in creating a fill (how many bucket/wagon loads?). For present purposes measuring volumes helps bypass any number of the problems related to estimating the deposit sample by counting the pot sherds or bone fragments in the recovery sample.

Observable Consequences

We will consider a simple model, and start with a ceramic sample of 20 litres retrieved from a context with a volume of 100 litres. It follows that the original ceramic population must have been in the range of 20 to 100 litres. If the same context yields 15 litres of bone and 5 litres of bronze, then the maximum total amount of ceramic which could have been in the original (deposit) population was 100−(15+5); thus the possible range is between 20 and 80 litres. Similar range calculations can be performed for the other materials.

Obviously—since the relative proportions of mineral component (silt, clay and sand) in the "soil" matrix should equal that in the original deposit sample—if the excavated soil matrix is not composed almost entirely of material which might be shown to have once been bones or fired ceramic, then it is highly unlikely that bones or ceramics ever composed 100% of the layer; this will be independently verifiable by soil analysis. If the original population (a deposit sample consisting of a hypothetical maximum of 100 litres of ceramic) has been reduced to 20 litres (20%) by post-depositional transformations, then the remaining 60% of the layer—the "soil" matrix surrounding the retrieved sherds—must consist of transformed ceramic: the remains of sherds which have either been dissolved in acidic soil, or mechanically fragmented to a size not normally recognisable as being an artifact (i.e., fine ceramic dust suitable for grog) by mechanical wasting, frost, etc. Similarly, if bones decayed, then the soil matrix may contain large quantities of bone collagens (and possibly DNA).

The only exceptions are cases where, for example, ground water has leached collagen into another layer; there should be evidence for this, however, and the stratigraphic integrity of the context should be questioned.

Testing Significance

What has been outlined so far is the basic framework for a complex simulation. Although it is not clear whether taking this further would be a worthwhile exercise or not, continuing debate indicates that this problem is taken seriously. The long-held assumption that the recovered and the deposition samples are similar seems to be a necessary part of inferring the "universal laws of human behaviour" archaeologists aim to recover from "indirect traces" or "circumstantial evidence", and is often implicit in the identification of the activity areas analysed using GIS and other spatial tools. Like any assumption, however, this must be tested before it can have any scientific validity. And in this sense, the process of modelling and simulating post-depositional transformations is intended to serve as a control on the interpretive process. Significant differences between the simulated deposit sample and the actual recovery sample should highlight factors contributing to these changes.

This also highlights the need to record evidence of post-depositional transformations (since the process would be meaningless without the possibility of comparing simulated results to something more substantial), and the need to re-examine archaeological theory and practice. Among other things, being constrained by recording media, archaeologists currently conceive of layers/strata in terms of their boundaries—as surfaces—rather than as volumes. While this may be acceptable to some documentation media and 2.5D GIS, it is not good enough for true 4D spatial analysis (or 5D when including probability) or simulating post-depositional movement.

New forms of research, including GIS and data-mining, combined with such recent theoretical developments as contextual analysis, phenomenology and especially "reflexive methods" underline the fact that archaeologists require more from technology than transparency (i.e., metadata) and data access. Archaeologists also need indicators of precision and accuracy, and means for testing data validity. It has long been recognized that "our aim should be for adequate, reliable, and representative data" (Binford 1964: 438). The question is: if—as Clarke claimed—archaeology is a science of bad samples, what can be done to make them better, when method and theory prevent us from achieving this goal.

Conclusions

The solution outlined above is by no means intended to be definitive, but merely offers a "modest proposal" for what could be done, should archaeologists really want it.

Modelling post-depositional transformations and testing their potential significance requires documentation more consistent and detailed than is practical—perhaps possible—with current technology. This does not mean that the necessary technology could not be developed, someday, should there be a demand. Implementation would require a radical shift in thinking, though, not only about how but also why we do archaeology, as shown by the way even the most advanced digital tools are now often used within paper-based paradigm. Drawing styles for profiles and plans that emphasize layer boundaries, for example, ultimately reinforce a superficial understanding of contexts by obscuring the potential effects of post-depositional transformations: the perception of impenetrable boundaries is created simply by drawing a line (cf. Carver 2010), and evidence for anything that might have happened within any given stratum is unknown because it goes unrecorded.

More importantly, questions of data completeness and quality were less important for paper-based archives, in part because bad data cannot travel far. Since it is impossible to predict what will happen to information disseminated in digital form, who will use it, or how it will be used, in the brave new world of tomorrow, the most important question becomes: how "bad" are our samples?

References Cited

Babbage, C. 1859. Observations on the Discovery in Various Localities of the Remains of Human Art Mixed with the Bones of Extinct Races of Animals. Proceedings of the Royal Society of London 10: 59–72.

Bagnoli, P.E. 2013. Dating Historical Rock Art on Marble Surfaces by Means of a Mathematical Model for Natural Erosion Processes. pp. 269–278. *In*: G. Earl, T. Sly, A. Chrysanthi, P. Murrieta-Flores, C. Papadopoulos, I. Romanowska and D. Wheatley (eds.). Archaeology in the Digital Era: Papers from the 40th Annual Conference of Computer Applications and Quantitative Methods in Archaeology (CAA), Southampton, March 26–29, 2012. Amsterdam University Press, Amsterdam.

Bersu, G. and D.M. Wilson. 1966. Three Viking Graves in the Isle of Man. Monograph Series Society for Medieval Archaeology, London.

Binford, L.R. 1964. A Consideration of Archaeological Research Design. American Antiquity 29(4): 425–441.

Binford, L.R. 2001. Constructing Frames of Reference: An Analytical Method for Archaeological Theory Building Using Hunter-Gatherer and Environmental Data Sets. University of California Press, Berkeley.

Buck, C.E., W.G. Cavanagh and C.D. Litton. 1996. Bayesian Approach to Interpreting Archaeological Data. Statistics in practice (Chichester, England). Wiley, Chichester, England.

Bury, J.B. 1920. The Idea of Progress; An Inquiry into its Origin and Growth. Macmillan and Co., London.

Butterfield, H. 1982. The Origins of Modern Science 1300–1800. 2nd ed. Bell & Hyman, London.

Carver, G. 2010. Doku-porn: Visualising Stratigraphy. pp. 109–122. *In*: S. Koerner and I. Russell (eds.). Unquiet Pasts; Risk Society, Lived Cultural Heritage, Re-designing Reflexivity. Ashgate, Farnham, Surrey.

Cameron, C.M. 2006. Ethnoarchaeology and contextual studies. pp. 22–33. *In*: D. Papaconstantinou (ed.). Deconstructing Context: A Critical Approach to Archaeological Practice. Oxbow, Oxford.

Clarke, D.L. 1973. Archaeology: the Loss of Innocence. Antiquity 47(185): 6–18.

Daniel, G.E. 1975. A Hundred and Fifty Years of Archaeology. 2nd ed. Duckworth, London.

Darwin, C. 1896. The Formation of Vegetable Mould, through the Action of Worms, with Observations on their Habits. D. Appleton, New York.

Dibble, H.L., P.G. Chase, S.P. McPherron and A. Tuffreau. 1997. Testing the Reality of a "Living Floor" with Archaeological Data. American Antiquity 62(4): 629–651.

Gould, S.J. 1989. Wonderful Life: The Burgess Shale and the Nature of History. W.W. Norton, New York.

Harris, E.C. 1989. Principles of Archaeological Stratigraphy. 2nd ed. Academic Press, London.

Hodder, I. 1992a. Symbolism, Meaning and Context. pp. 11–23. *In*: I. Hodder (ed.). Theory and Practice in Archaeology. Routledge, London.

Hodder, I. 1992b. Symbols in Action. pp. 24–44. *In*: I. Hodder (ed.). Theory and Practice in Archaeology. Routledge, London.

Kuhn, T.S. 1977a. The Function of Measurement in Modern Physical Science. pp. 178–224. *In*: T.S. Kuhn (ed.). The Essential Tension: Selected Studies in Scientific Tradition and Change. University of Chicago Press, Chicago.

Kuhn, T.S. 1977b. Mathematical versus Experimental Traditions in the Development of Physical Science. pp. 31–65. *In*: T.S. Kuhn (ed.). The Essential Tension: Selected Studies in Scientific Tradition and Change. University of Chicago Press, Chicago.

Litton, C. and C. Buck. 1998. An Archaeological Example: Radiocarbon Dating. pp. 465–480. *In*: W.R. Gilks, S. Richardson and D.J. Spiegelhalter (eds.). Markov Chain Monte Carlo in Practice. Chapman & Hall, Boca Raton, Fla.

Lyman, R.L. 1999. Vertebrate Taphonomy. Cambridge Manuals in Archaeology; Cambridge University Press, Cambridge.

OED. 1997. New Shorter Oxford English Dictionary. Oxford University Press, Oxford.

Orton, C. 1980. Mathematics in archaeology. Collins archaeology; Collins, London.

Orton, C. 2000. Sampling in archaeology. Cambridge Manuals in Archaeology; Cambridge University Press, Cambridge.

Petrie, W.M.F. 1904. Methods and Aims in Archaeology. Macmillan and Co., London.

Pettigrew, T.J. 1850. On the study of Archæology, and the Objects of the British Archæological Association. The Journal of the British Archaeological Association VI: 163–177.

Prestwich, J. 1860. On the Occurrence of Flint-Implements, Associated with the Remains of Animals of Extinct Species in Beds of a Late Geological Period, in France at Amiens and Abbeville, and in England at Hoxne. Philosophical Transactions of the Royal Society of London 150: 277–317.

Schiffer, M.B. 1983. Toward the Identification of Formation Processes. American Antiquity 48(4): 675–706.

Shannon, C.E. 1948. A Mathematical Theory of Communication. Bell System Technical Journal 27(3): 379–423.

Shennan, S. 1997. Quantifying archaeology. 2nd ed. Edinburgh University Press, Edinburgh.

Thomas, D.H. 1978. The Awful Truth about Statistics in Archaeology. American Antiquity 43(2, Contributions to Archaeological Method and Theory): 231–244.

Watkinson, D. and V. Neal. 1998. First Aid for Finds. Third ed. RESCUE/UKIC Archaeology Section, London.

Woolley, L. 1961. Digging up the Past. Penguin, Harmondsworth.

16

Time and Probabilistic Reasoning in Settlement Analysis

Enrico R. Crema

Introduction

Archaeology has a long running research tradition in the study of spatial data (Hodder and Orton 1976, Rossignol and Wandsnider 1992, Bevan and Lake 2013), an endeavour that encouraged the borrowing of statistical methods from other disciplines. Many of these techniques have been adapted to answer existing questions, others fostered new lines of enquiry. Nonetheless the field of spatio-temporal analysis remains surprisingly under-developed in archaeology. Despite theoretical debates on the notion of time (e.g., Bailey 1983, Gamble 1987, Murray 1999, Holdaway and Wandsnider 2008, etc.) and a large number of studies in chronometry (e.g., Buck and Millard 2003, Lyman and O'Brein 2006), spatio-temporal statistics did not experience the same flourishing of ideas observed in spatial statistics and GIS-led analyses.

The distinction between spatial and spatio-temporal analyses is however a delusion as the former can be actually regarded as a special case where time is hold constant. This simplification is dictated by the assumption that, with other things being equal, neglecting the role of this dimension has a minimal influence on the reconstruction of the generative process behind the observed pattern. Most archaeologists would undoubtedly feel uncomfortable with such a statement. However, the great majority of regional analyses in archaeology do not formally consider the role played by time, which is often relegated as a qualitative attribute with little or no analytical function. Yet time plays a fundamental role in several ways. First, most spatial analyses start from the definition of an arbitrary slice of the time continuum, which we can refer to

Universitat Pompeu Fabra, Department of Humanities, 25-27, Ramon Trias Fargas, 08005, Bracelona.

CaSEs - Complexity and Socio-Ecological Dynamics Research Group.

UCL Institute of Archaeology, 31-34 Gordon Square, WC1H 0PY London.
  Email: e.crema@ucl.ac.uk

as *target interval*. *Events* occurring earlier are excluded from analysis, and hence their role in the emergence of the spatial pattern during the *target interval* is not considered. Second, spatial processes occurring within the *target interval* are often assumed to be stationary. Given the extensive duration of these chronological slices, this assumption is in most cases wrong, with the observed pattern being often the cumulative outcome of several distinct spatial processes. This problem is known as *time-averaging* (Stern 1994), a direct consequence of how much of the pattern we observe is a function of the way we subdivided our data (a problem comparable to the modifiable areal unit problem in geography, cf. Openshaw 1984). Third, defining the membership of each constituent unit of the spatial pattern, known as *event*, to a specific time window is not always straightforward. Temporal coordinates of an *event* can be defined by a *time-span* of possible existence delineated by a start and an end point (i.e., *terminus post quem* and *terminus ante quem*). Archaeological *events* have usually fairly large *time-spans* of existence, with low levels of accuracy and precision in their definition, and an undefined probability distribution between boundaries. As a result, the combination of arbitrary time-slicing processes and the uncertainty associated with each archaeological *event* can lead to a rather blurry picture of the past. Despite this, the general practice is to acknowledge their potential impact only outside the analytical stage.

This negligence is surprising, especially considering the large number of quantitative contributions in chronometry (see, e.g., Buck and Millard 2003, chapter 14 this volume). The lack of integration of these works in regional analyses is partly due to a research bias that promoted the usage of advanced statistical techniques to a limited number of case studies. Indeed most of these works rely almost exclusively on radiocarbon dates as their primary data (e.g., Parnell et al. 2008, Bocquet-Appel et al. 2009, Shennan et al. 2013, see also below).

However, the striking majority of archaeological data is not associated with radiocarbon or other forms of scientific dating (see discussions in Bevan et al. 2012 and Crema 2012), and are instead based on categorical definitions (e.g., "Early Bronze Age", "Late Archaic", etc.) inferred from the presence, absence, and frequencies of diagnostic cultural artefacts (Lyman and O'Brien 2006). This is particularly the case of settlement studies, where only few elements (e.g., dwellings, settlements, etc.) are coupled with direct scientific dating, a condition that limits the adoption of available quantitative methods in diachronic studies.

This chapter will first provide an overview of archaeological spatio-temporal analyses restricting its scope to their application within studies of regional settlement pattern. It will then discuss the limits imposed by temporal uncertainty and discuss a solution based on probabilistic reasoning and Monte-Carlo simulation. A case study from prehistoric Japan will then illustrate an example of this approach. The final discussion will review the limits and the potentials of spatio-temporal statistics in archaeology.

Spatio-temporal Analysis and Regional Settlement Study in Archaeology

The last decade saw a marked growth in the number of quantitative and computational studies of regional settlement pattern in archaeology. These cover a wide array of different themes including: political boundaries (Bevan 2011, Stoner 2012) and catchment areas (Ullah 2011); spatial patterning of settlement sites (Palmisano 2013);

distribution and layout of residential features within individual settlements (Crema et al. 2010, Eve and Crema 2014); settlement hierarchy (Crema 2013, Altaweel 2014); inter-settlement interaction (Evans and Rivers 2012); and settlement location (Carleton et al. 2012). These broad themes have been approached using a variety of techniques including point-pattern statistics (e.g., Bevan et al. 2013), model-based approach based on statistical physics (Bevan and Wilson 2013, Davies et al. 2014), and agent-based simulation (Kohler et al. 2007, Crema 2014).

Despite this flourishing of quantitative and computational techniques, the temporal dimension has been rarely approached. The few exceptions proposed in computational archaeology have been primarily focused on the visual representation of time, ranging from the development of stand-alone software packages (e.g., the TIMEMAP project http://www.timemap.net/) to the creation of GIS package plug-ins (Green 2011). Most mathematical and statistical studies have placed their emphasis on specific slices of the time-continuum, focusing on the identification of the generative process behind the observed pattern. This might range from the measurement of a summary statistic to a more formal approach involving the comparison of the observed pattern against precisely defined null models. The latter include simple random patterning of settlement location to complex point process models where both intensity and interaction are defined (Eve and Crema 2014). Others developed simulation-based approach where different assumptions on settlement behaviour have been mathematically defined to provide general theoretical expectations (Crema 2014) or precise and testable predictions (Evans and Rivers 2012, Bevan and Wilson 2013).

Successful examples of spatio-temporal analyses rely primarily on high quality data where temporal uncertainty is either confined to small intervals or quantified by scientific methods capable of assigning probability distributions for the timing of each *event*. One of the best-known examples of high-quality dataset is the spatial distribution of Classic Maya terminal dates, a record signalling the date of erection of the most recent monument at 47 sites dated between the 8th and 9th century AD (Kvamme 1990, Premo 2004). As the cessation of this practice is interpreted as the demise of the elite class and a marker of a profound socio-economic change, the most recent dates incised on these monuments have been used as a proxy of societal 'collapse' and their spatial distribution studied in relation to the extent of local socio-political interaction spheres and the general patterning of large-scale environmental stress. This case study fostered the application of a variety of geostatistical techniques (e.g., trend surface analysis; local and global indices of spatial autocorrelation, etc.) where 'time' was treated as a continuous variable representing the timing of the 'collapse', recorded at sample locations (i.e., archaeological sites).

The increasing availability of datasets with a large number of C14 samples have led to what has been labelled as the "third radiocarbon revolution" (Bayliss 2009:126), an extensive development of bespoke statistical techniques aimed to go beyond chronometry. The most successful example is the reconstruction of past population fluctuations by means of summed probability distributions obtained from calibrated C14 dates (e.g., Collard et al. 2010) sometimes coupled with more sophisticated analyses, such as the identification of possible causal relationship with climatic changes (e.g., using Granger causality analysis, as in Kelly et al. 2013), or an extensive usage of Monte-Carlo methods for detecting statistically significant episodes of population booms and busts, a solution which overcomes many of the problems related to sampling bias and idiosyncrasies in the calibration curve (Shennan et al. 2013, Timpson et al. 2014).

The wide availability of C14 samples has also led to a development of spatio-temporal analyses, such as spatio-temporal kernel density analysis (Grove 2011) and Bayesian hierarchical modelling (as in Onkamo et al. 2012), enabling to cover a wide variety of topics from inferences on the dynamics of reoccupation of north-western Europe at the end of the Ice Age (Blackwell and Buck 2003), to the detection of centres of renewed expansion of European Neolithic (Bocquet-Appel et al. 2009).

The great majority of archaeological data are however based on categorical definition of time that are characterised by higher levels of uncertainty in an undefined form. Dating generally involves the creation of arbitrary sets of artefact types grouped and linked to a spatio-temporal extent (i.e., an archaeological "period", "phase", or "culture") and the definition of memberships to one or more of these. Although this is a relatively straightforward exercise, ambiguities associated with the precise definition of the diagnostic traits, the state of conservation of the artefacts, or any form of subjective judgment of the membership introduces a significant degree of uncertainty (Bevan et al. 2012).

Settlement archaeology has an additional layer of uncertainty. *Events* are not directly dated, and their chronology is usually indirectly inferred from the dating of artefacts that are found in association with the target object. For example, the date of a residential feature is often derived from dates of objects recovered from its habitation floor. Thus, the dating process inherits the uncertainty of individual artefact dating, and also introduces the problem of how these objects are associated with the target *event* (Dean 1978). This process of association requires the definition of a depositional model, which specifies how objects are discarded and physically located in their archaeologically recovered contexts. This introduces further levels of uncertainty that cannot be overcome by simply improving the accuracy of artefact dating. A number of quantitative solutions have been proposed in the literature (e.g., Buck et al. 1992, Buck and Sahu 2000, Bellanger et al. 2008, Roberts et al. 2012), but the wide variety of depositional processes do not allow the development of unique universal solutions.

Settlement archaeology thus poses a highly challenging field for pursuing spatial analysis. The high quality temporal data provided by scientific dating does not provide a statistically sufficient sample size for most settlement studies, and does not solve the intricacies of indirect dating. On the other hand the relative dating used in most archaeological datasets imposes an arbitrary slicing of the time-continuum into unequally sized "phases" and "periods", with the temporal uncertainty of each *event* being rarely reported in formal fashion. These constraints in the archaeological data impede the adoption of existing spatio-temporal analysis for investigating regional settlement pattern, and highlight the importance in developing appropriate solutions.

Temporal Uncertainty, Probabilistic Reasoning, and Monte Carlo Simulation

Measuring Uncertainty

We can formalise the problems described in the previous section by defining a space-time comprising a set of *events* e, each with spatial coordinates s_1 and s_2 and a temporal coordinate t. If we ignore the spatial extent and the temporal duration of the *events*, we can characterise e as a spatio-temporal point pattern within a three-dimensional space.

However, the temporal uncertainty of most archaeological *events* does not allow us to assign a single value for *t*, which needs to be substituted by a probability distribution τ. This can be directly obtained from chemical or physical analysis (e.g., radiometric dating, thermo luminescence dating, etc.), or more frequently expressed in terms of membership to one or more archaeological periods based on a variety of methods (e.g., typological analysis, indirect dating, etc). In the latter case, τ is expressed as a vector of chronological periods P_1, P_2, ..., P_L with length *L*, each associated with a probability of membership. Thus low and high chronological uncertainty can be measured in terms of entropy; the distribution $τ_A = \{0,0,0,1,0\}$ will have the lowest, while $τ_B = \{0.2,0.2,0.2,0.2,0.2\}$ will have the highest level of uncertainty. This *event*-wise uncertainty can be quantified using Shannon's entropy value (Shannon 1948), as follows:

$$U_e = -\sum_{i=1}^{L} P_i \, ln \, P_i \qquad (1)$$

where U_e is bounded between 0 and $-L(L^{-1}lnL^{-1})$.

Chronological uncertainty is ideally an intrinsic and independent property of each *event*, and consequently all periods should be associated with similar levels of uncertainty. However, in many cases some chronological periods exhibit higher uncertainty than others due to intrinsic properties of the diagnostics used in the dating process. Bevan et al. (2012) explored this structural property of uncertainty in typological dating by devising two numerical indices. The first measures the overall uncertainty of each period:

$$U_j = \frac{\sum_{i=1}^{n} min \, (P_{ij}, max_{k \neq j}(P_{ik}))}{\sum_{i=1}^{n} P_{ij}} \qquad (2)$$

where the uncertainty U_j for the period *j* is the sum of the minimum between the probability P_{ij} associated with each *event* *i* and the highest probability assigned to the *event* to any other period (P_{ik}), divided by the sum all probabilities of all *events* for the specific period. Low values (with a minimum of 0) of U_j suggest that *events* associated with period *j* have diagnostic independence (i.e., *events* have generally higher probabilities), while high values (with a maximum of 1) indicate the opposite. The second index measures instead the pair wise uncertainty between two periods:

$$U_{jk} = \frac{\sum_{i=1}^{n} min \, (P_{ij}, P_{ik}) \cdot 2}{P_{ij} + P_{ik}} \qquad (3)$$

where U_{jk} is the ratio between the sum of overlaps between periods *j* and *k* for each event *i*, and the maximum possible overlap between the two periods. U_{jk} ranges between 0 indicating complete independence (no shared probabilities) and 1 indicating that all *events* share the same probabilities for the two periods.

Bevan and colleagues (2012) applied their indices to pottery sherds collected from an intensive survey in the Greek island of Antikythera. They showed how their data was characterised by an extremely wide range of overall uncertainty (between 0.091 to 0.961) and a pair wise uncertainty exhibiting higher values between abutting phases.

Whilst in many cases the vector of probabilities associated with each event is independent, and hence not influenced by the knowledge of any other event, there are some situations where this condition is not met. This is the case when the *events* in question are features that exhibit a stratigraphic relationship, from which we can infer the temporal topology between *events*. This additional layer of knowledge can considerably improve the accuracy of the dating process; a line of research that has been explored by several archaeologists applying a Bayesian inferential framework (e.g., Buck et al. 1992, Ziedler et al. 1998).

Probabilistic Reasoning and Monte Carlo Simulation

Given the nature of our dataset, the application of standard spatial statistics is hindered by two issues. Firstly, if we seek to examine the spatial pattern for a specific chronological period *P* we need to formally integrate the uncertainty associated with each event. Secondly, comparisons between different archaeological periods need to take into account differences in their absolute duration.

One solution to the first problem is the adoption of weighted analysis (Crema et al. 2010, Grove 2011), whereby, for a given period, *events* with lower uncertainty have a lower contribution to the computation of the summary statistic. Whilst this approach allows the straightforward application of many techniques, the resulting statistics hide the underlying uncertainty of the raw data. An example can illustrate this. Consider two sets of *events*—A and B—with identical spatial coordinates but different temporal definitions. For a given period *j* all *events* of A have a probability of 0.95, while all *events* of B have a probability of 0.01. If we measure the mean centre of distribution of the two sets we will obtain the same spatial coordinates, though clearly the degree of confidence we would assign to the two summary statistics should be different, given the differences in their intrinsic uncertainty.

An alternative solution, that returns the uncertainty in the final summary statistic, can be achieved with the following three steps:

1) compute the probabilities π_1, π_2, ..., π_m for each of the *m* possible spatial configuration expected from the data in a given period *P*;
2) calculate the relevant summary statistic S_1, S_2, ..., S_m for each of these permutations;
3) obtain the probability of a given interval of the summary statistic $[S_\alpha, S_\beta]$ by summing the probabilities of all permutations where the $S_\alpha \leq S \leq S_\beta$ is met.

Figure 1a illustrates an example of this workflow with an artificial dataset representing the location of 12 archaeological sites, each associated with their probability of existence for a given temporal interval. The number of possible permutations *m* is given by L^n, where *L* is the number of possible states (archaeological periods) and *n* is the number of *events*. Since we are dealing with a single chronological period, we are interested only in two possible states (the target period *P* and all other periods), and hence $m = 2^{12}$. The example illustrated in Fig. 1 shows the average distance to nearest neighbour and the Clark and Evans Nearest Neighbour Index (NNI, Clark and Evans 1954) for three possible spatial configurations (Fig. 1b~d), and the probability estimate of defined intervals for both summary statistics (Fig. 1e and 1f).

The algorithm illustrated above provides the exact probability of any summary statistic, which quantitatively accounts for the combined uncertainty of all *events* within a dataset. However, *m* (the number of permutations) grows exponentially with the number of *events*, making the computation of the summary statistics intractable for

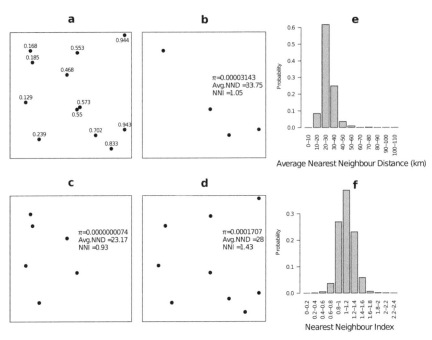

Fig. 1. An illustration of a spatial analysis with probabilistic reasoning: a) point data with probability of existence assigned to each event; b~d) three possible permutations of dataset with corresponding probability of occurrence (π), the average distance to the nearest neighbour (Avg. NND), and Clark and Evan's nearest neighbour index (NNI); e) exact probabilities for different bins of Avg. NND; and d) exact probabilities for different bins of NNI.

most datasets. One way to solve this problem is to sample from all possible permutations using the Monte-Carlo method (Crema et al. 2010, Crema 2012, Bevan et al. 2013, Yubero et al. In Press):

1) simulate k possible spatial configurations of the observed set of *events*;
2) calculate relevant summary statistics $S_1, S_2, ..., S_k$ for each of the k simulations;
3) estimate the probability of a given interval $[S_\alpha, S_\beta]$ by calculating the proportion of k satisfying the condition $S_\alpha \le S \le S_\beta$.

Figure 2 presents a scatter plot of the estimated probability of the condition NNI >1 using the same artificial data of Fig. 1 against different values of k (number of Monte Carlo simulations). The graph shows how, by increasing the number of simulations, the estimated probability approaches the true probability asymptotically (here shown as a dashed line). Acceptable values for k depend strongly on the strength of the pattern, the nature of the summary statistics, and the uncertainty associated with each event. Consequently, the only way to determine the optimal number of k is to plot the variation of summary statistic and determine whether the estimated probability becomes stable. Given the large values of m one might suspect that k should be similarly high, but empirical studies using this method show how this is not necessarily the case (see Crema et al. 2010).

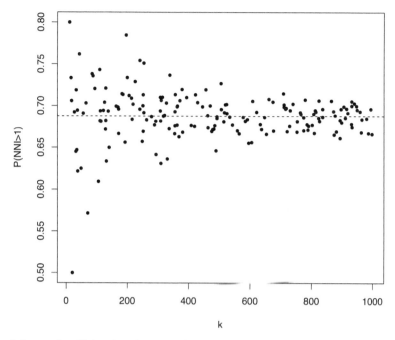

Fig. 2. Scatter-plot of *k* (number of Monte-Carlo simulations) against the probability of obtaining an NNI greater than 1. The horizontal dashed line shows the true probability.

The second issue, i.e., biases derived by comparing archaeological periods with different durations, can be solved by slicing the time-continuum into equally sized *time-blocks* (cf. with the notion of *chronon* in temporal databases, e.g., Snodgrass 1992) and extracting the probability of existence for each of them. Since these analytical units do not correspond with boundaries of archaeological periods, we need to define a continuous probability distribution of each event within their archaeological phases. This is not possible in most cases, and hence the most conservative approach is to adopt a uniform probability distribution (i.e., an equal prior) following the principle of insufficient reason (see also the application of aoristic analysis in archaeology; Johnson 2004, Crema 2012). Given that cultural traits often exhibit an initial expansion, a peak period, and a decline in popularity, adopting a "flat" probability distribution might however generate a systematic bias in the probability distribution. A unimodal shape might in fact have a closer resemblance to the true underlying probability distribution (Roberts et al. 2012, Manning et al. 2014), an assumption that plays a pivotal role in seriation techniques (Kendall 1971), and has also been integrated as a prior (using a trapezoidal distribution) in the Bayesian analysis of archaeological periods (Lee and Bronk-Ramsey 2012). However, most relative chronologies provide only the *terminus ante* and *post quem*, limiting the application of these alternative models to cases where required parameters can be estimated from the empirical record.

Case Study: Continuity and Discontinuity in the Settlement Pattern of Eastern Tokyo Bay (Japan) during the Mid-5th Millennium BP

The method described in the previous section offers a simple workflow for analysing regional settlement data using conventional spatial and spatio-temporal analyses that are widely available in the literature. Here, a case study from the Jomon culture of Japan is illustrated to showcase limits and potentials of the proposed approach. Detailed discussion on the case study, as well as a set of analyses covering a larger temporal span can be found elsewhere (Crema 2013). The choice of the case study was dictated by three compelling reasons. First, Japanese archaeology offers one of the most impressive settlement record known for prehistoric hunter-gatherers (Habu 2004), with a continuously growing dataset sustained by large scale rescue projects. These excavations provide an excellent sample for settlement studies, with multiple scales of observations ranging from the regional distribution of sites to the spatial layout of residential features within individual settlements. Second, Jomon hunter-gatherers produced large quantities of ceramic materials, which are often recovered on the floor of residential features, providing an economic (compared to radiometric methods) and effective form of indirect dating. Third, existing studies suggest that Jomon culture experienced several episodes of dramatic changes in demography and subsistence orientation (Imamura 1999), providing an ideal context where we anticipate possible changes in regional settlement pattern.

The study area is a 15 × 15 km square located on the Eastern shores of Tokyo Bay in Central Japan. This area is renowned for the high number of archaeological sites during the Middle (5500–4420 cal BP) and the Late (4420–3320 cal BP) Jomon periods, with the transition between the two characterised by a sharp decline in the number of residential features coupled with major changes in settlement hierarchy (Crema 2013). The present study re-examines these macro-scale trends by focusing on the continuities and discontinuities in the Jomon residential patterning during the second half of the Middle Jomon period.

The dataset comprise 934 pit houses recorded from 120 Jomon sites identified within the study area. Each residential unit has been associated with one or more archaeological phases based on published typological analysis of diagnostic pottery recovered from the habitation floor. These relative chronological assignments have then been converted into temporal intervals defined by an absolute start and end date, using the sequence proposed by Kobayashi (2004).

To reduce the potential bias generated by differently sized chronological phases, the interval between 5000 and 4400 cal BP has been sliced into 6 phases (**I**: 5000–4900; **II**: 4900–4800; **III**: 4800–4700; **IV**: 4700–4600; **V**: 4600–4500; and **VI**: 4500–4400 cal BP). Thus, the temporal definition of each residential unit has been described by a vector of eight probability estimates: 6 phases defined for this study (**I~VI**), and two dummy phases aggregating the probability of existence before (**Pre-I**) and after (**Post-VI**) our temporal scope. Probability estimates have been retrieved assuming a uniform distribution within each archaeological phase (see Fig. 3), while stratigraphic relationships between overlapping residential units have been recorded in order to ensure that the temporal topology between pit houses were conserved during the simulation process.

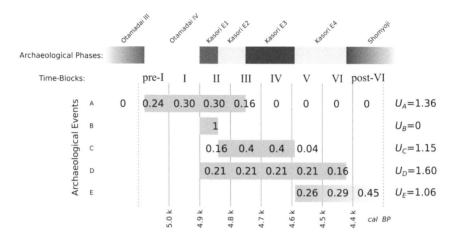

Fig. 3. Workflow for defining the probability of existence of each event within each phase. Each event is defined by equal membership to given archaeological periods, which has been translated into an absolute time-span of possible existence (light grey rectangles). The proportion of time-span within each time-block is then translated into probabilities. Longer time-spans have larger individual uncertainty quantified by U_e.

The distribution of event uncertainty (Fig. 4a) shows a strong peak in intermediate values, with less than 10% of data having $U_e = 0$, and no cases where the entropy value is the highest (shown here as a dashed vertical line). The overall uncertainties of the six phases show in most cases intermediate values (Fig. 4b), with the exception of phase **II** (4900–4800 cal BP), which exhibits a peak value of 0.9. This suggests that probability estimates attributed to *events* in this period are in most cases lower than other dominant phases (i.e., pit houses rarely have their highest probability of existence during this phase). The analysis of the pair wise uncertainty (Table 1) confirms the high uncertainty associated with phase **II**, with a high value of U_{jk} (=0.75) with the previous phase **I**. On the other hand, Table 1 also indicates that most of the uncertainty is distributed between immediately adjacent phases, indicating that this is rarely shared across more than two consecutive phases.

Given the large number of possible permutations ($m = 8^{934}$), computing the probability of each possible spatial configuration is not a viable option, and hence the Monte Carlo simulation approach has been adopted. Convergence test indicated that 1,000 iterations are sufficient to obtain a stable estimate of the probabilities of relevant summary statistics.

Previous analyses (see Crema 2013) on a larger dataset have indicated two key findings. First, settlement hierarchy exhibits a strong fluctuation, with an alternation between phases with few large sites and many smaller settlements, and episodes of a more uniform distribution of sizes. Figure 5 depicts a standardised rank-size plot for all 1,000 iterations with the hypothetical Zipfian (i.e., settlements of rank r have size Sr^{-1}, with S being the size of the largest settlement; Zipf 1949) drawn as a dashed line. During phase **I**, the observed rank-size distribution has a steeper line compared to the one expected by the Zipfian distribution. In the subsequent phases the size distribution becomes increasingly less hierarchical until phase **IV**, when the opposite trend starts. This fluctuation can be measured using the A-coefficient (Drennan and Peterson 2004,

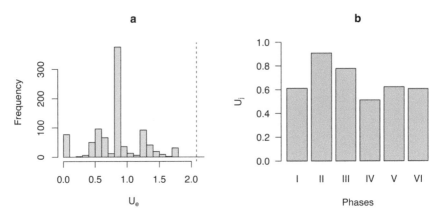

Fig. 4. Distribution of event uncertainty (a) and the overall uncertainty of each phase (b).

Table 1. Matrix of pair wise uncertainty.

	Pre-I	I	II	III	IV	VI	Post-VI
I	0.29						
II	0.15	0.37					
III	0.07	0.22	0.75				
IV	0.04	0.1	0.18	0.22			
V	0.03	0.08	0.08	0.08	0.39		
VI	0.01	0.03	0.03	0.03	0.09	0.4	
Post-VI	0.33	0.03	0.03	0.03	0.09	0.19	0.29

Crema 2013), a summary statistic that measures the amount of deviation from the Zipfian distribution. The index returns negative values for strong hierarchy, values close to zero for patterns conforming to Zipf's law, and positive values for more uniform distributions. Despite the effects of temporal uncertainty, the distribution of A-coefficients (Fig. 5) shows clear Gaussian shape and small variances for all periods.

Second, the total number of residential units shows an initial growth, followed by a sharp decline between phases **V** (4600–4500 cal BP) and **VI** (4500–4400 cal BP), with an average rate of decline at 0.68 pit houses per year. Figure 6 shows the box-plot of the five transitions of interests, with whiskers indicating the minimum and the maximum rates of change among the 1,000 simulation runs. The first four transitions all show positive median values, albeit high levels of uncertainty can be observed in the transition from phase **IV** (4700–4600 cal BP) to **V** (4600–4500 cal BP). The last transition suggests instead a sharp decline in the number of residential units, with a complete absence of simulation runs with positive rates of change.

These results indicate continuous changes in the settlement pattern, whereby the steady growth in residential density from phase **I** to **V** is coupled with shifts in hierarchy. However, these global statistics beg the question as to whether the variation in the settlement hierarchy was the result of a differential growth rate between different settlements or the outcome of a sharp discontinuity characterised by the formation of new settlements and the demise of others. More generally, these global statistics do not show whether temporal variations in residential density exhibit any form of

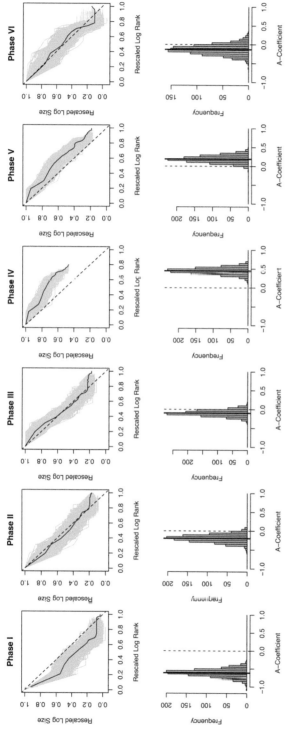

Fig. 5. Upper row: standardised rank size plot of the settlement sizes. Light-grey lines indicate individual runs of the simulation and the solid black line represents the mean trend. Lower row: distribution of the *A*-coefficient.

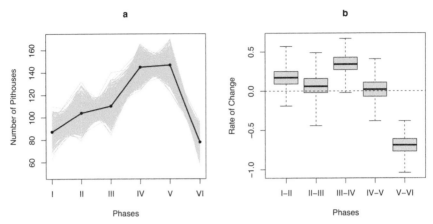

Fig. 6. Variation in the overall number of residential units (a) and corresponding rates of change (b).

spatial structure. If the fluctuations in the rank-size curves are the result of a mixture of residential growth and decline, are these opposite trends spatially clustered? If so, are they the result of local changes in the environment inducing the abandonment or colonisation of specific areas?

The temporal continuity of settlement occupation can be assessed using a variant of the Spatio-Temporal Join-Count Statistic (*st-JCS*), originally developed by Little and Dale (1999). The rational of the method can be summarised as follows. For a given transition from i to j, we first compute J_{ij}, the number of locations that are occupied by at least one residential unit in both periods. We then permute the occupied locations of both periods for n times, and generate a distribution of expected number of joins $\hat{J}_{ij}$ given a null hypothesis of random settlement locations. Statistical significance for $J_{ij} > \hat{J}_{ij}$ will provide a simple assessment of the continuity in the choice of residential sites. Figure 7 shows the distribution of the p-values obtained from 1,000 permutations test conducted for each of 1,000 simulated spatio-temporal patterns. The results strongly support the presence of continuity in occupation for all transitions except for the one between phase **III** and **IV**, where less than 15% of the runs had a significance level below 0.05. Interestingly, the lack of continuity[1] in the this key transition matches the highest change in the settlement hierarchy as seen in Fig. 5, when the mean A-coefficient shifts from slightly negative values (indicating weak hierarchy) to positive values (suggesting a more uniform distribution of settlement sizes; Fig. 5), and the period of highest growth in the overall residential density (Fig. 6).

As mentioned earlier, the global analysis depicted in Fig. 6 obscures the presence of any spatial variation in the rates of change of residential density. Figure 8 illustrates this problem by plotting the frequency distribution of rates of change between consecutive phases for each location, excluding all cases where sites were unoccupied in both phases. The histograms show how positive and negative rates of change co-exist in all transitions, indicating how the box plot in Fig. 6 hides a mixture of opposite trajectories in the change of residential density.

---

[1] Notice that the transition however does not exhibit any evidence of significant discontinuity (i.e., $J_{ij} < \hat{J}_{ij}$).

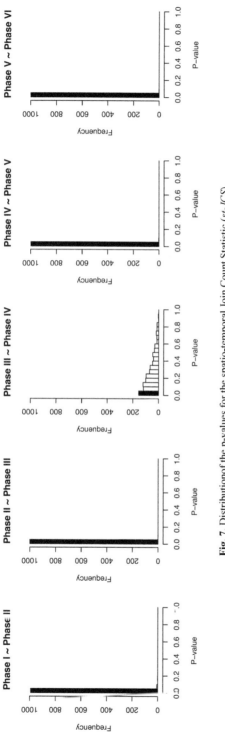

Fig. 7. Distribution of the p-values for the spatio-temporal Join Count Statistic (*st-JCS*).

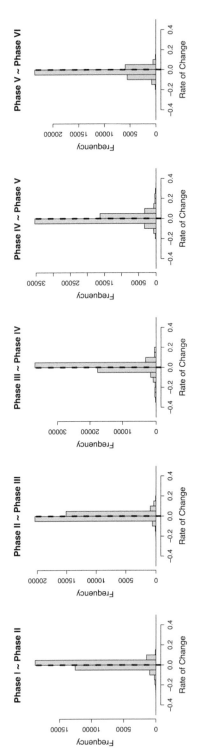

Fig. 8. Distribution of rates of site level changes in residential density.

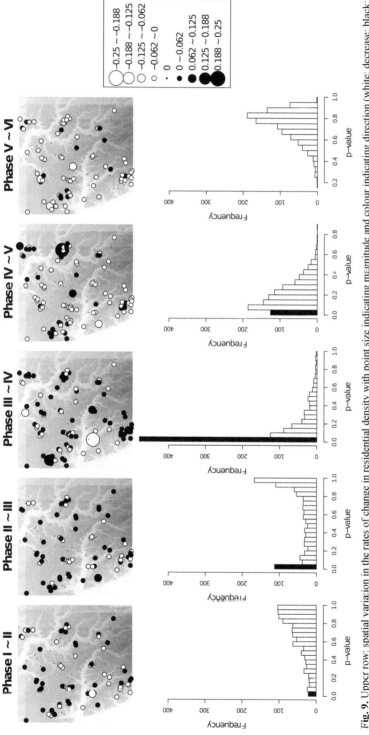

Fig. 9. Upper row: spatial variation in the rates of change in residential density with point size indicating magnitude and colour indicating direction (white: decrease; black: increase). Lower row: distribution of p-values for Moran's I spatial autocorrelation test. Frequencies of runs with p<0.05 are highlighted in black.

The spatial heterogeneity in the temporal variation of residential density can also be visualised as bubble plots (Fig. 9), with point size indicating the mean magnitude of change, and the colour indicating whether the trend is positive (black) or negative (white). All transitions displayed on Fig. 9 show a mixture of locations with opposite trajectories of changes in residential density, suggesting how the overall variation in the number of pit houses was characterised by a complex spatial structure. We can formally examine this by evaluating whether the rates of change in residential density exhibit spatial autocorrelation. Statistically significant results would suggest that local differences in the underlying environment might be a potential driver in the evolution of settlement pattern. The histogram on the lower row in Fig. 9 shows the distribution of p-values of Moran's I spatial autocorrelation test (Moran 1950), with a distance radius set at 2.5 km. The results suggest how the highest number of significant positive autocorrelation is observed during the transition from phase **III** and **IV**, when ca. 50% of the simulations had a p-value below 0.05. This corresponds to the highest overall growth rate in the number of residential units (Fig. 6), the largest change in settlement hierarchy (Fig. 5), and the only transition where *st-JCS* did not exhibit significant continuity in site occupation (Fig. 7). These results suggest that this period was characterised by a major change in the settlement patterning, with the concurrent abandonment of key areas and the colonisation of others, perhaps driven by episodes of local resource depletion (Toizumi and Nishino 1999). The transition from stage **V** and **VI** is instead characterised by the smallest number of runs with significant p-values. The location of sites with positive rates of change seems to be mixed with those showing the opposite pattern rather than being confined in specific areas.

Discussion and Conclusion

The analyses presented in the case study highlight both the benefits and the limitations of the Monte-Carlo approach proposed here. Despite the presence of relatively high levels of temporal uncertainty in the archaeological data and number of possible permutations, a sample of 1,000 simulation runs was sufficient to generate robust outputs in the summary statistics that can help understand the settlement evolution in the case study area. The method enabled diachronic comparisons using an array of statistics that have been already used in archaeology (e.g., Spatial autocorrelation, Rank-Size Analysis, *A*-coefficient) as well as the introduction of methods that has not been applied before (e.g., spatio-temporal Join Count Statistic).

Yet these techniques also highlighted intrinsic problems that are present in this and other datasets. The temporal slicing and the definition of arbitrary temporal blocks of 100 years have undoubtedly facilitated the comparison between different settlement configurations, but at the same time showed the intrinsic limits of the archaeological record for answering specific research questions. For example, during the transition from phase **V** to **VI**, a lack of spatial autocorrelation and the co-presence of sites with opposite rates of change in residential density have been noticed. This does not necessarily indicate that these opposite trends were concurrent, as one process might have occurred earlier than the other. In other words, the available data does not provide enough information to determine whether the spatial process within the two time-blocks was stationary or non-stationary, hindering part of our understanding of the settlement history of the region. This is not an intrinsic problem of the proposed method, which instead highlights the limits of the available data.

In fact, the dataset used in the present case study highlights other, deeper problems in the archaeological record that limit the straightforward adoption of probabilistic technique proposed here. For example, the precise definition of the chronological boundaries of the archaeological phases adopted in the case study (based on Kobayashi 2004) hides the underlying uncertainty associated with these chronologies. Archaeologists are aware of the fuzzy nature of these boundaries and that uniform distribution does not correctly represent the assumption of a unimodal shape in the probability of existence of *events* within an archaeological phase. However, these crucial data are rarely available, and when they are present, they are retrieved only within narrow contexts where advanced statistical techniques can be coupled with robust chronological proxies. In the present case study, this enforced the use of a uniform probability distribution, which, at the current state of archaeological knowledge, represents our best approximation.

The lack of interest in defining and reporting uncertainties in chronometric data is perhaps the key reason why the development of spatio-temporal statistic in archaeology is still in its infancy. There is a wide misconception amongst archaeologists that uncertainty is a negative aspect of their research, a weakness of our record that should be kept hidden when possible. This attitude can lead to a substantial loss of information, and a reduced effort in attempting to describe, quantify, and ultimately overcome the problem of uncertainty. Crisp boundaries with no explicit definition of the underlying probability distribution are the preferred chronological description of archaeological phases, and quantitative approaches to the problem are seen as an unnecessary complication with no practical use (but see Manning et al. 2014). This trend is perhaps due to the lack of methods that can benefit from the quantification of archaeological uncertainties. This chapter is an attempt to illustrate that archaeological research can highly benefit from a stronger awareness of uncertainty, and that a development of this research agenda is highly desirable.

Acknowledgments

I would like to thank Juan Barceló and Igor Bogdanovic for their editorial work and for inviting me to contribute on this volume.

References Cited

Altaweel, Mark. (In press) Settlement Dynamics and Hierarchy from Agent Decision-Making: A Method Derived from Entropy Maximization. J Arch. Metod. doi:10.1007/10816-014-9219-6

TIMEMAP project http://www.timemap.net/.

Bailey, G.N. 1983. Concepts of Time in Quaternary Prehistory. Annu. Rev. Anthropol. 12: 165–192.

Bayliss, A. 2009. Rolling out revolution: using radiocarbon dating in archaeology. Radiocarbon 51: 123–147.

Bellanger, L., R. Tomassone and P. Husi. 2008. A Statistical Approach for Dating Archaeological Contexts, J. Data. Sci. 6: 135–154.

Bevan, A. 2011. Computational models for understanding movement and territory. pp. 383–394. *In*: V. Mayoral Herrera and S. Celestino Pérez (eds.). Tecnologías de Información Geográfica y Análisis Arqueológico del Territorio, Anejos de Archivo Español de Arqueología, Mérida.

Bevan, A., J. Conolly, C. Hennig, A. Johnston, A. Quercia, L. Spencer and J. Vroom. 2012. Measuring Chronological Uncertainty in Intensive Survey Finds, Archaeometry 55: 318–328.

Bevan, A., E. Crema, X. Li and A. Palmisano. 2013. Intensities, Interactions and Uncertainties: Some New Approaches to Archaeological Distributions. pp. 27–52. *In*: A. Bevan and M. Lake (eds.). Computational Approaches to Archaeological Space, Left Coast Press, Walnut Creek.

Bevan, A. and M. Lake (eds.). 2013. Computational Approaches to Archaeological Space, Left Coast Press, Walnut Creek.

Bevan, A. and A. Wilson. 2013. Models of Settlement Hierarchy Based on Partial Evidence. J. Archaeol. Sci. 40: 2415–2427.

Blackwell, P.G. and C.E. Buck. 2003. The Late Glacial human reoccupation of north-western Europe: new approaches to space-time modelling, Antiquity 77: 232–240.

Bocquet-Appel, J.-P., S. Naji, M.V. Linden and J.K. Kozlowski. 2009. Detection of diffusion and contact zones of early farming in Europe from the space-time distribution of 14C dates. J. Archaeol. Sci. 36: 807–820.

Buck, C.E., C.D. Litton and A.F.M. Smith. 1992. Calibration of Radiocarbon Results Pertaining to Related Archaeological *Events*, J. Archaeol. Sci. 19: 497–512.

Buck, C. and A. Millard (eds.). 2003. Tools for Constructing Chronologies: Crossing Disciplinary Boundaries, Springer Verlag, London.

Buck, C. and S.K. Sahu. 2000. Bayesian models of relative archaeological chronology building. Appl. Stat. 49: 423–440.

Carleton, W.C., J. Conolly and G. Iannone. 2012. A locally-adaptive model of archaeological potential (LAMAP), J. Archaeol. Sci. 39: 3371–3385.

Collard, M., K. Edinborough, S. Shennan and M.G. Thomas. 2010. Radiocarbon evidence indicates that migrants introduced farming to Britain, J. Archaeol. Sci. 37: 866–870.

Clark, P. and F. Evans. 1954. Distance to nearest neighbor as a measure of spatial relationships in populations. Ecology 35: 445–453.

Crema, E.R., A. Bevan and M. Lake. 2010. A probabilistic framework for assessing spatio-temporal point patterns in the archaeological record, J. Archaeol. Sci. 37: 1118–1130.

Crema, E.R. 2012. Modelling Temporal Uncertainty in Archaeological Analysis, J. Archaeol. Method. Th. 19: 440–461.

Crema, E.R. 2013. Cycles of change in Jomon settlement: a case study from Eastern Tokyo Bay, Antiquity 87: 1169–1181.

Crema, E.R. 2014. A simulation model of fission-fusion dynamics and long-term settlement change. J. Archaeol. Method. Th. 21: 385–404.

Davies, T., A. Wilson, A. Palmisano, M. Altaweel and K. Radner. 2014. Application of an entropy maximizing and dynamics model for understanding settlement structure: the Khabur Triangle in the Middle Bronze and Iron Ages, J. Archaeol. Sci. 43: 141–154.

Dean, J.S. 1978. Independent Dating in Archaeological Analysis. Adv. Archaeol. Method. Th. 1: 223–255.

Drennan, R.D. and C.E. Peterson. 2004. Comparing archaeological settlement systems with rank-size graphs: a measure of shape and statistical confidence, J. Archaeol. Sci. 31: 533–549.

Evans, T.S. and R.J. Rivers. 2012. Interactions in Space For Archaeological Models, Adv. Complex Syst. 15: 1150009.

Eve, S. and E.R. Crema. 2014. A house with a view? Multi-model inference, visibility fields, and point process analysis of a Bronze Age settlement on Leskernick Hill (Cornwall, UK), J. Archaeol. Sci. 43: 267–277.

Gamble, C. 1987. Archaeology, geography and time, Prog. Hum. Geogr. 11: 227–246.

Green, C.T. 2011. Winding Dali's clock: the construction of a fuzzy temporal-GIS for archaeology Archaeopress, Oxford.

Grove, M. 2011. A Spatio-Temporal Kernel Method for Mapping Change in Prehistoric Land-Use Patterns, Archaeometry 53: 1012–1030.

Habu, J. 2004. Ancient Jomon of Japan, University of Cambridge Press, Cambridge.

Hodder, I. and C. Orton. 1976. Spatial Analysis in Archaeology, Cambridge University Press, Cambridge.

Holdaway, S. and L. Wandsnider (eds.). 2008. Time in archaeology: time perspectivism revisited, University of Utah Press, Salt Lake City.

Imamura, K. 1999. Jomon no jitsuzo wo motomete, Yoshikawakoubunkan, Tokyo (In Japanese).

Johnson, I. 2004. Aoristic Analysis: seeds of a new approach to mapping archaeological distributions through time. pp. 448–452. *In*: F. Ausserer, W. Börner, M. Goriany and L. Karlhuber-Vöckl (eds.). [Enter the Past] the E-way into the Four Dimensions of Cultural Heritage: CAA2003. BAR International Series 1227. Archaeopress, Oxford.

Kelly, R.L., T.A. Surovell, B.N. Shuman and G.M. Smith. 2013. A continuous climatic impact on Holocene human population in the Rocky Mountains, Proc. Natl. Acad. Sci. USA 110: 443–447.

Kendall. D.G. 1971. Seriation from abundance matrices. pp. 215–252. *In*: F.R. Hodson, D.G. Kendall and P. Tautu (eds.). Mathematics in the Archaeological and Historical Sciences, Edinburgh, Edinburgh University Press.

Kobayashi, K. 2004. Jomonkenkyu no shinsiten: tanso14nenndaiokutei no riyo, Rokuichishobo, Tokyo (In Japanese).

Kohler, T.A., C.D. Johnson, M. Varien, S. Ottman, R. Reynolds, Z. Kobti, J. Cowan, K. Kolm, S. Smith and L. Yap. 2007. Settlement Ecodynamics in the Prehispanic Central Mesa Verde Region. pp. 61–104. *In*: T.A. Kohler and S.E.v.d. Leeuw (eds.). The Model-based Archaeology of Socionatural Systems, SAR Press, Sante Fe.

Kvamme, K.L. 1990. Spatial autocorrelation and the Classic Maya collapse revisited, J. Archaeol. Sci. 17: 197–207.

Lee, S. and C. Bronk-Ramsey. 2012. Development and applications of the trapezoidal model for archaeological chronologies. Radiocarbon 54: 107–122.

Little, L.R. and M.R.T. Dale. 1999. A method for analysing spatio-temporal pattern in plant establishment, tested on a Populus balsamifera clone, J. Ecol. 87: 620–627.

Lyman, R.L. and M.J. O'Brien. 2006. Measuring Time with Artefacts: A History of Methods in American Archaeology, University of Nebraska Press, Lincoln.

Manning, K., A. Timpson, S. College, E.R. Crema, K, Edinborough, T. Kerig and S. Shennan. In Press. The chronology of culture. A comparative assessment of European Neolithic dating methods, Antiquity.

Moran, P.A.P. 1950. Notes on Continuous Stochastic Phenomena. Biometrika 37: 17–23.

Murray, T. 1999. Time and Archaeology, Routledge, London.

Onkamo, P., J. Kammonen, P. Pesonen, T. Sundell, E. Moltchanova, M. Oinonen, M. Haimila and E. Arjas. 2012. Bayesian spatiotemporal analysis of radiocarbon dates from eastern fennoscandia, Radiocarbon 54: 649–659.

Openshaw, S. 1984. The Modifiable Areal Unit Problem, Geobooks, Norwich.

Parnell, A.C., J. Haslett, J.R.M. Allen, C.E. Buck and B. Huntley. 2008. A flexible approach to assessing synchroneity of past *events* using Bayesian reconstructions of sedimentation history, Quat. Sci. Rev. 27: 1872–1885.

Palmisano, A. 2013. Zooming Patterns Among the Scales: a Statistics Technique to Detect Spatial Patterns Among Settlements. pp. 348–356. *In*: G. Earl, T. Sly, A. Chrysanthi, P. Murrieta-Flores, C. Papadopoulos, I. Romanowska and D. Wheatley (eds.). CAA 2012: Proceedings of the 40th Annual Conference of Computer Applications and Quantitative Methods in Archaeology (CAA), Palla Publications, Southampton.

Premo, L. 2004. Local spatial autocorrelation statistics quantify multi-scale patterns in distributional data: An example from the Maya Lowlands, J. Archaeol. Sci. 31: 855–866.

Roberts, Jr., J.M., B.J. Mills, J.J. Clark, W.R. Haas Jr., D.L. Huntley and M.A. Trowbridge. 2012. A method for chronological apportioning of ceramic assemblages. J. Archaeol. Sci. 39: 1513–1520.

Rossignol, J. and L. Wandsnider (eds.). 1992. Space, time, and archaeological landscapes, Plenum Press, New York.

Shannon, C.E. 1948. A mathematical theory of communication. The Bell System Technical Journal 379–423, 623–656.

Shennan, S., S.S. Downey, A. Timpson, K. Edinborough, S. Colledge, T. Kerig K Manning and M.G. Thomas 2013 Regional population collapse followed initial agriculture booms in mid-Holocene Europe, Nat Commun. DOI:10.1038/ncomms3486.

Snodgrass, R.T. 1992. Temporal Databases. pp. 22–64. *In*: A.H. Frank, I. Campari and U. Formentini (eds.). Theories and Methods of Spatio-Temporal Reasoning in Geographic Space, Springer-Verlag, Berlin.

Stern, N. 1994. The implications of time-averaging for reconstructing the land-use patterns of early tool-using hominids, J. Hum. Evol. 27: 89–105.

Stoner, W. 2012. Modeling and Testing Polity Boundaries in the Classic Tuxtla Mountains, Southern Veracruz, Mexico. J. Anthropol. Archaeol. 31: 381–402.

Timpson, A., S. Colledge, E. Crema, K. Edinborough, T. Kerig, K. Manning, M.G. Thomas and S. Shennan. 2014. Reconstructing regional population fluctuations in the European Neolithic using radiocarbon dates: a new case-study using an improved method. J. Archaeol. Sci. 52: 549–557.

Toizumi, T. and M. Nishino. 1999. Jomonkouki no miyakogawa muratagawaryouiki-kaizukagun. Chibakenbunkazaisentaa KenkyuKiyo 19: 151–171 (In Japanese).

Ullah, I.I. 2011. A GIS method for assessing the zone of human-environmental impact around archaeological sites: a test case from the Late Neolithic of Wadi Ziqlâb, Jordan J. Archaeol. Sci. 38: 623–632.

Yubero-Gómez, M., X. Rubio-Campillo and J. Lopez-Cachero (In press). The study of spatiotemporal patterns in integrating temporal uncertainty in late prehistoric settlements in northeastern Spain, Anthropological and Archaeological Science.

Ziedler, J.A., C.E. Buck and C.D. Litton. 1998. Integration of Archaeological Phase Information and Radiocarbon Results from the Jama River Valley, Ecuador: A Bayesian Approach, Lat. Am. Antiq. 9: 160–179.

Zipf, G.K. 1949. Human Behavior and the Principle of Least Effort. Cambridge: Harvard University Press.

17

Predictive Modeling and Artificial Neural Networks: From Model to Survey

Luca Deravignone,[1,*] *Hans Peter Blankholm*[2] and *Giovanna Pizziolo*[3]

Introduction

Luca Deravignone

A main characteristic of archaeology is that we have to deal not only with high complexity phenomena like culture, society, and economic aspects of the past, but also with imprecise and heterogeneous datasets. The archaeological record is generally incomplete, depending on the quality of conservation, even when accurately documented. When performing quantitative analyses it is like always working on samples. The archaeological record is extremely heterogeneous and therefore trying to understand history from material culture appears as an extremely complex matter.

If we look beyond the archaeological excavation, the analysis of the territory and surrounding landscape always yields a lot of precious information and data. In particular the study of settlement patterns by the use of spatial analysis, with its high level of objectivity and synthesis, represents a useful means for the study of this kind of phenomena. However, the theories on these subjects have reached such a high level of complexity that new methodologies have been considered (Barceló 2009). Acknowledging the non linearity of problems pertaining to the study of human

[1] Department of European, American and Intercultural studies, Sapienza University of Rome, P.le Aldo Moro 5, 00185 Roma, Italy.
  Email: luca.deravignone@uniroma1.it
[2] Institute of Archaeology and Social Anthropology, University of Tromsø, 9037 Tromsø, Norway.
  Email: hans.peter.blankholm@uit.no
[3] Department of History and Cultural Heritage, University of Siena, Via Roma 56, 53100 Italy.
  Email: pizziolo@unisi.it
* Corresponding author

processes, recent years have seen the use of Artificial Neural Networks (ANN) and other adaptive systems techniques have spread within the archaeological community.

Aside from referring the explanation of how ANNs work to dedicated articles (e.g., Abdi et al. 1999, see also Barceló, this volume, Chapter 1), it is sufficient to say that they are based on abstract analogies with the human brain, the foremost being the capacity of learning by looking at examples. Among the several types of ANNs that exist, the approach presented here is based on Feed-forward Neural Networks, a supervised method that involves some inputs (variables) and a desired output (information). Here the training process is based on the recurring feed of the training pattern to the ANN until the weights of the connections between the neurons reach a combination suitable for relating the input and the output. Once the ANN is trained (i.e., it has "learned the problem"), the user can present new records as inputs and the net will generate an output according to the previous training.

The basic assumption of this approach is that archaeological sites may be conceived as entities located in a territory, characterized by the variables of the place where they are located. This is especially true for settlement sites since the reasons involved in the choice of one place instead of another are usually far from being random. The variables (see section below) can be environmentally, socially or economically oriented, or related to many other aspects. So what if we use these variables as input features, and presence/absence of an archaeological site as the final output? A good starting point is that this methodology allows us to use not only continuous variables that can be easily quantified (i.e., distances or degrees) as inputs, but also arbitrary and different scales or Boolean values. Concurrently we can conceive the output of the net as a model for that settlement, which is important not only for using it to produce predictive maps, but also for the analysis of the settlement pattern itself. The system also allows several other operations; for example, it is possible to train a net on a certain territory, time frame, or settlement type and then apply the model to another one in order to highlight and analyze similarities or differences.

Regarding the practical aspects of the methodology, the inputs derive from raster layers related to the different variables used, and the output itself will be a raster map. On this map the value of a single cell indicates how much that cell satisfies the input-output relationship previously learned by examples. If our aim is to predict site locations, we can also look at the result as a predictive map.

When we began to develop the methodology (Deravignone and Macchi 2006), we decided to release the developed tools under Open Source license and for different GIS platforms and operating systems.

For this method, the tools needed for performing spatial analyses with the use of ANNs are: a GIS platform, a neural networks simulator, and specific utilities that allows connection of these two environments. A detailed description of the whole process and used tools can be found in the "Grosseto Predictive Modeling Method" (GPMM) manual (Deravignone et al. 2013).

As indicated, one of the most important characteristics of ANNs is their ability to analyze non-linear processes and their flexibility regarding the examination of different case studies. An ANN, in fact, can combine a basic level of abstraction while keeping all the characteristics of a "standard" quantitative approach. Another of the strong points of this method, as also explained before, is that an ANN can learn how to solve a certain problem merely by observing some examples. This means that the analysis can be performed without external preconceptions or subjective influences like the *a priori* application of a certain theory or model.

This aspect is at the base of the training phase and at the same time allows using data derived from samples, making ANNs very useful for archaeological purposes. Moreover, this method allows visualizing the evolution of settlement systems by observing and comparing results diachronically in time or even synchronically in different territories.

In terms of research design, there is no need for weighting variables since, by its own nature, ANN takes into consideration how much every single variable is influencing the result. This means that if a variable has a very similar value for all records, it will automatically be under-weighted and *vice versa*.

However, a major problem may be represented by the "black box" aspect of the procedure. Even if different ways of analyzing the process are available, the ANN procedure does not allow one to see what really happens during the processing phase and this is of some importance, especially during the interpretation of results.

The first attempt using this procedure was on medieval hilltop settlements, castles, of southern Tuscany, Italy. This choice was made especially because of the huge database made by the Archaeology Department at the University of Siena (Francovich and Milanese 1989), counting 1500+ records from the region. This made it possible to use a source that had already been extensively studied quantitatively and for its spatial patterning, and at the same time to have a useful and strong base to create training patterns and test patterns (see case study below).

After several phases of analyses and testing, both in the laboratory and in the field (Deravignone 2011), a further step was necessary: to consider another case study in order to see how the methodology would work with a very different research question, in a different environment, with different site types and a different time frame.

The right occasion was offered by the second author. His project concerned the study of Neolithic and Early Metal age pit-dwellings in Senja, North Norway. A survey had already taken place so a basic sample was ready for analyses. In turn, the authors conducted more extensive studies and analyses (see Senja study below).

Archaeological ANN and Variables

Giovanna Pizziolo

Before we proceed to our cases, we need to elaborate on the modeling and use of variables in ANN analyses in general and in our examples in particular.

When building a model, or a system of analysis, variables play a fundamental role and affect the results obtained and/or the direction of interpretation. ANNs offer an independent means to analyze and classify the landscape, but the rules that the black box provides are dependent on the variables you put into the system. In short, the results may change according to the number and type of variables you use to "train" the system.

In general, if "To be able to read the landscape is to be able to abstract the complexity of the landscape into those variables or quantities that are needed" (Zubrow 2003), it is clear that variables are a crucial part of landscape archaeology research. A vivid debate on which kind of variables are valuable for the investigation of settlement strategies of the past has been going on for decades (Verhagen and Whitley 2012). A specific focus on this question arose with the introduction of GIS into landscape archaeology. GIS allows one to consider in a handy way variables such as geomorphology, geology, hydrology, and vegetation cover, and in an efficient way all the morphological information derived by Digital Elevation Models (DEM). The latter may be used to calculate altitude, slope, aspect, and other terrain variables such as watersheds, or erosion/accumulation areas. These analytical opportunities have been widely exploited in landscape archaeology studies and in some cases, due to their uncritical application, they also have been considered as a risk for ending up with interpretational environmental determinism of patterns or to produce just "obvious relationships" (Gaffney et al. 1995). Furthermore, GIS has been used to calculate cost-distance or view shed areas (Van Leusen 1999) with the aim to introduce more "human" variables into the interpretation of archaeological landscapes. In particular, the cost-distance provides a better understanding of movement in a three-dimensional space compared to a simple calculation of linear distance. Moreover, new important parameters have been introduced for the analysis of movement and its relationship with landscape (Llobera 2000) which demonstrates the complexity of dealing with these matters. Other attempts to introduce socio-cultural variables to the study of archaeological landscapes come from the predictive modeling of Kamermans et al. (2009). The studies on accessibility, visibility and settlement continuity (Verhagen et al. 2013) in particular, seem useful for predictive research which uses both inductive and deductive methods.

The need to deal with "the qualitative" through GIS emerged from a diverse range of applications in social science. The possibility to use different types of qualitative and quantitative variables is a further advantage as demonstrated by the "mixed-methods approach" developed to study the distribution of socio-cultural phenomena (e.g., Jung and Elwood 2010).

We will not discuss these issues in depth; we only wish to highlight that the discussion on environmental and cultural and/or social variables involved in the study of archaeological landscape is still open and naturally evolves with the introduction of new methods and tools. See, as an example, the role of statistical analysis as it has been highlighted in Woodman and Woodward (2002) or the role of non-parametric statistical tests to establish the significance of each variable in predicting the likelihood of site presence within a regression approach. In this perspective the use of ANNs offer interesting possibilities as it is heuristic and helps avoiding a judgmental approach in modeling. The methodology allows one to deal with variables by evaluating their reciprocal relationships and by considering them more as proxies which shows indirect indications. This ability is motivating when dealing with landscapes which by themselves are basically constituted by reciprocal relationships between man and the environment in a synchronic and diachronic perspective. In fact ANNs "may be used to extract patterns that are too complex to be noticed by either humans or other computer techniques" (Zubrow 2003) and may thus reveal hidden relationship among land characteristics and settlement patterns. The inspiring result of ANN applications is that analyses related to landscape archaeology may be appreciated throughout a map format.

Following the above, we decided to use sets of variables which mainly pertain to the geographical characteristics of the territories for our case studies. A first attempt to introduce variables more related to the cultural sphere has been carried on by the introduction of cost-distance calculation to natural resources.

It needs to be considered, however, that we are dealing with information extracted from present day landscapes and, aside from geological data, we cannot refer in detail to palaeoenviromental reconstructions of landscape transformations as, for example, changes related to coastline displacements or palaeohydrological conditions. The latter is particularly important in Arctic areas where dramatic changes took place from prehistory to the present time. Almost the same group of variables are used for our two case studies, but as they differ in terms of geographical and chronological context they will be commented upon separately.

Medieval Tuscan Castles

The analytical details on this case study are presented below; here we focus on particular aspects of variable selection belonging to the region and the use of ANN. The variables are related to different types of issues and obtained from different sources. The geographical and topographical representativity is satisfactory across the region.

Elevation is extracted from the DEM (cell size 50 m and 10 m, respectively). It should be emphasized that the study area covers a wide range of geomorphological characteristics which vary among interior mountain relief, smooth hills and coastal plains. The absolute elevation above sea level makes sense when compared to other variables.

Slope in degrees provides information to contextualise the terrain variability. The combination of elevation and slope may offer hints on the three-dimensional morphologies defining classes like hilltop, hillside, or plain positions without having to perform a supervised pre-classification. The variable may help to study the spatial changes of the location of castles through time. Moreover, the slope gradient around a site may indicate its level of accessibility/defence.

Aspect is another variable used to understand settlement strategies, for example, site location according to sun radiation.

Distance from rivers. Fresh water availability is a basic need for human life. This variable may be related to this fundamental necessity or to possible communication routes, in particular as a proxy variable to relate sites to valley corridors.

Distance from sea. The distance from sea may indicate whether a site belongs to the inner part of Tuscany or to the coastal belt.

Distance from dioceses introduces a specific socio-cultural factor to the environmental—geographical settings. The Diocesi played a fundamental role in the religious-political-administrative organization of medieval communities in Tuscany. The spatial relationship with centres of political power is a variable that can be highlighted by ANN analysis.

Cost-distance from rivers offers a way to calculate terrain variability, avoiding the possibly misleading information of Euclidean distance. This is particularly relevant when adding human factors to analysis; introducing the perspective of moving across the terrain. This variable increase the possibility to explore the above mentioned "valley corridors".

Cost-distance from sea may help exploring morphology variation of the coastal belt.

Cost-distance from dioceses. In this case it seems appropriate to refer to "landscapes of power" that relate to human movement across a given administrative territory.

Landsat TM5 satellite imagery is a source that introduces a wide range of opportunities. It is composed of five different bands which may reflect several characteristics of the terrain. This can provide information on soil-types, hydrology, or land use.

Comments and Further Steps

Taking into consideration the large area under study we cannot plan to further increase the accuracy of data. However some improvements may be achieved if we can better define, in spatial terms, the socio-cultural factors that played a role in the subdivision of territories of "power" during medieval time. It could be interesting eventually to test the role assumed by diocese boundaries when compared to the distance from churches.

Moreover, other variables extracted from Landsat imagery, such as different types of vegetation cover, may detect bias elements and consequently help to gain better control of the data's representativity.

Senja Pit-Dwellings (North Norway)

Specific information on the Island of Senja is presented below. Here we introduce the variables and their connotations within the geographical and chronological framework of the coastal prehistoric settlements of Senja island.

The variables are partly related to the geological characteristics, and partly derived from the DEM (cell size 100 m).

Elevation allows one to identify chronological trends in predictive configurations. Over the past 12 millennia dramatic changes occurred in relative sea-levels in Norway. In general, the higher a site occurs in the terrain the older it is. However, the sites being coastal and shore-bound also mean that the coastline and surrounding terrain was different from what it is today. In most cases, the terrain from the coastline contemporary with the site and down to the present coastline would have been submerged.

Slope is considered a strong variable that is related to the ability of people to settle and use the land around their sites. It is generally assumed that humans did not live or build houses on ground with more than 10–15 degrees of slope. Slope can also be used to identify edges or flat areas which, again, may reveal flat-topped or slightly round-topped fossil beaches with a high potential for the presence of prehistoric sites.

Fine-grained marine deposits and coarse-grained marine deposits have revealed that coastal pit-dwellings, almost without exception, are situated on one of the above stated types of deposit which are easy to dig into and are characterized by good drainage.

Aspect can help to investigate if there was preponderance for facing the sun (see above) or for finding shelter. This variable may also provide new insights on what sites faced, beyond the fact they were coastal and related to the sea.

Distance from sea may indicate if there were a particular configuration of sites with a particular set of characteristic in existence and, if so, for what behavioral reason. Moreover, the distance from the present day coastline may also be of interest as it can provide proxies, or indirect information, on the depth of water when the sea level was above that of today's.

Distance from rivers is another basic element in land/seascape archaeology as fresh water plays a basic role in the organization of daily human life. North Norway has almost omnipresent freshwater resources. However, we have also considered streams greater than 1 m in width that may have been important for fishing.

Cost distance from sea. Similar to the Distance from sea, this variable is a proxy for shallowness of the sea facing the site. These variables are the best options we have to add some human "parameters" into our set of data (see Medieval Tuscan Castles case study). In the Senja study the cost-distance could be addressed to explore something one is not readily aware of, i.e., were short distances and gentle slopes a must? Did people avoid steep sided and bottomed bays?

Cost distance from river. As the cost distance from sea, this can provide information on the walking distance to water resources.

Beyond the geological information, other variables selected for the ANN application can be considered as "standard" variables to be used in landscape archaeology research and have been chosen in several other predictive research studies that have used other methods of analysis (Graves 2011). This may not seem particularly original, but means that the results of the ANN application can be compared with other methodologies.

Critical Points and Further steps

Reservations to the spatial accuracy of our input data, which do not allow individualizing some morphological features which characterise, when observed at a large scale of analysis, the local landscape, can be addressed. For example, fossil beaches, nowadays characterised by narrow and relatively low ridges located at different elevations along the coastal belt, may represent important landscape features in the development of prehistoric settlement strategies. Unfortunately, however, such units are not necessarily highlighted by a 100 m DEM grid. Other important variables related to the local topography of the coastal fjords are not readily available and must be acquired, but may be worth considering. It is also important to consider that coastal settlements are related to the exploitation of the sea and that much of the traffic was across water and not necessarily land. So, presence of sheltered bays and natural harbours must have been of interest to people in the past. Isthmuses, ness, or the "degree of shelter" that a bay can offer are other types of topographical features that we could add in future analyses.

The GPMM and Medieval Tuscan Castles

Luca Deravignone

Castles and their settlement patterns represent one of the main themes of medieval archaeology in Mediterranean Europe and beyond because of their importance as precious indicators of the study of medieval society, politics, and economy. The region of interest for conducting these analyses was initially constituted by the south-eastern part of Tuscany, Italy, formed by the Provinces of Grosseto, Siena and Arezzo with a particular focus on castles settled before 1150 AD (Fig. 1).

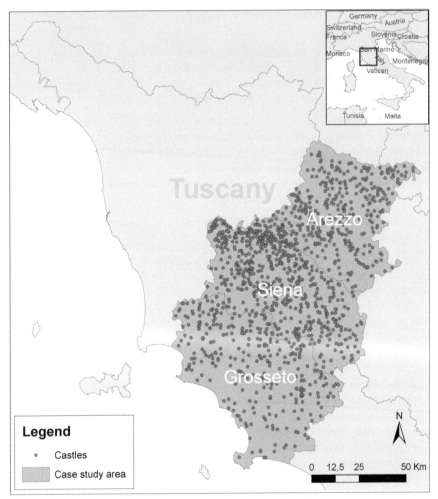

Fig. 1. Italy and Tuscany with castle case study.

From the beginning a dual approach was contemplated: prediction and the study of settlement pattern. In the first case somewhat heterogeneous patterns, constituted by records of different areas and periods, were used, while in the second only castles from a certain territory or time period. This was done especially to highlight possible differences and similarities, continuities and discontinuities, both in a diachronic and synchronic approach.

The choice of variables is presented in the Variables section (above). The purpose was to reach a wider description and understanding using variables that represent, in a more comprehensive way, the qualities and features of each archaeological site.

First Applications

The first results showed almost all hilltops as high probability areas, making the analyses useless since they were representing almost 50 percent of the total study area. Evidently, an improvement was necessary. A first step forward was based on the assumption that not only the site location but also its surrounding areas were important. For medieval settlements, as well as for other kinds of sites, the characteristics of surroundings represent a primary piece of information. This is based on the fact that rural settlements are not only residences for human population, but also productive units integrated with the territory.

This insight was operationalized by performing a simple raster shift. New analyses added 8 shifts (North, North-East, etc. around the compass) relative to selected variables, especially those concerning the geomorphology. These shifts were performed at different distances and showed empirically that the best results were obtained between 100 and 300 m.

The difference between the results of the two analyses, one performed using just straight morphological variables, and another which also included the surrounding values, was clearly visible, particularly when looking at the huge decrease of the high probability areas (Deravignone and Macchi 2006).

An additional improvement was the use of Landsat TM5 satellite imagery as new variables. With a resolution of 30 m this is not suitable for shape recognition, but is advantageous for other important purposes. Certain bands, like infrared, give information that is not even visible to the human eye, thus adding information on, for example, the humidity or even the "quality" of soil.

In this case the differences with the previous analyses were clearly marked and it allowed exclusion of at least 80 percent of our case study area, thus making the task of isolating high probability areas much easier and focused. After a reclassification of the results, it was clear that all areas above the 98 percent value should be considered as high probability areas.

On comparing the results it is clear how the topographical morphology of the site and the proximity to main resources (water, raw materials, etc.) are certainly important considerations for this kind of analyses. The significance of the test was initially measured in the laboratory using GIS. The resulting raster values were re-imported in test patterns constituted by randomly excluded castles (10 percent), left outside the analyses only for testing purposes. The resulting chart became a valid instrument and proved useful for determining the success of the analysis and several runs showed that almost 2/3 of the castles were above 80 percent probability (Fig. 2).

Diachronic and Synchronic Approaches for Settlement Studies

As mentioned above, the ANN approach, as a numerical model, may be applied in another territory in order to measure the similarities or differences between two settlement patterns (training a network on Grosseto province, for example), and then used to analyze the Siena area. In this case the link or association between the two archaeological areas was based on a fully integrated model of the settlement system with its specific environment. In these analyses, however, even if they provided for a

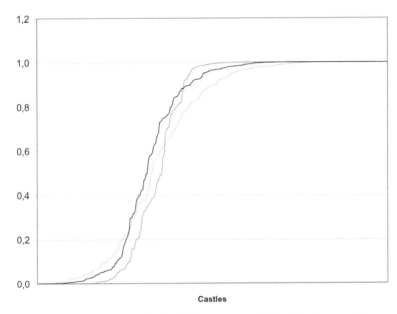

Fig. 2. Laboratory tests results for 3 different runs on Medieval Castles case study.

number of ideas, the interpretation of results was particularly difficult, mainly because of the black box problem.

The same can be stated about analyses conducted with the same approach, but using different time frames instead of using different areas. The castles were divided into time-frames of 25 years, each of them constituting a training pattern. In this case the only easily visible difference, when looking at the resulting raster maps, was that, going further up in time and using the same number of records, there was an increase of high probability areas.

In the Field Tests and Surveys

A necessary further step was to test the analyses in the field. The aim was to choose an area with little information about archaeological sites due to, for example, a high level of forestation and the related impossibility of performing a standard survey. The area of the old Volterra diocese, in particular the Berignone wood area, characterized by very dense vegetation, was selected (Deravignone 2009).

After performing the analyses relative to the old diocese area, the high probability portions were exported into a GPS device in order to be able to find them during survey. This led to the discovery of several new sites, many of them with elevated features or walls (Fig. 3). Although it is still premature to relate these sites as hilltop settlements, it is a good fact because the pottery found made it possible to date all the sites from XI to around the end of XIII century, the same period used in the training process.

In this case it was not necessary to re-do the analyses with the inclusion of the new results. This was so because the number of records was so high (see above) that we were able to create many independent test patterns from them.

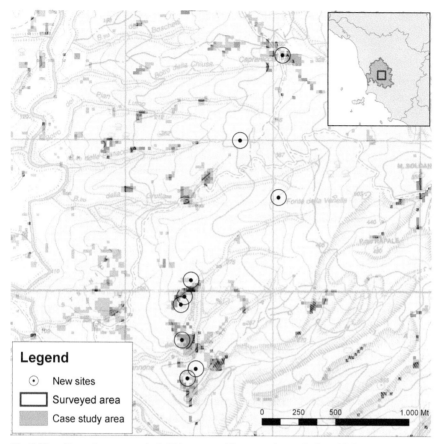

Fig. 3. New sites and survey results in Volterra area. The squared zones indicate high probability areas.

Some Final Considerations

The development of the method led us towards more ambitious goals: by the use of ANN and GIS it is possible to detect variables that may characterize underlying invisible relationships between territory and settlement patterns.

It is also interesting to see the numerous analogies between the high probability areas and aerial photography anomalies. Overlapping the anomalies database compiled by the University of Siena with the resulting probability maps, how many points that are matching very high probability areas and that have not yet been surveyed, is clearly visible.

Another conclusion is that resolution is fundamental for a good and significant outcome of the analysis. This is especially important for morphologic variables like slope and altitude, where, depending also on where the pixel is located, the result can change a lot. The first examples were made with 50 m pixels, but using a 10 m scale, for selected smaller areas, a great increase in precision was noticed. Thus, whenever possible we suggest using the highest resolution available.

The GPMM and Neolithic and Early Metal Age Pit-Dwellings in Senja Island, North Norway

Hans Peter Blankholm

One of the most characteristic traits of Neolithic (ca. 4.500–1.800 BC) and Early Metal Age (ca. 1.800–0 BC) coastal settlements in North Norway is sites with pit-dwellings. Topographically and geologically they are generally situated on fossil, raised beaches in bays, on ness, or on isthmuses.

Finnmark county, bordering Russia, has for decades been the hub of pit-dwelling research (e.g., Schanche 1994) whereas research in Tromsø county to the south has only recently begun (e.g., Blankholm 2009). Still, the number of sites in this region appears too low to allow for more nuanced insights into geographical, topographical and economic choices, and into cultural material variability and social organization. It is thus important to increase both the number of sites and geographical and topographical representation.

To achieve these ends a pilot-project was initiated in 2009 using 1.586 km^2 Senja island and 128 km^2 of the southwestern-most part of the neighbouring Kvaløy island (Fig. 4) as a case and the neural-network-based GPMM as the primary tool.

Senja has a wild, mountainous outer (western) side facing the Atlantic, and a mild and lush inner (eastern) side. In course of the pilot project, surveys were carried out in cooperation with my co-authors.

The tactical goal for the Senja prediction was to increase the probability of finding pit-dwelling sites and simultaneously to reduce the area that needed to be covered by surveys. However, given the large size of the island and the complex logistics involved in the Arctic, priority was given to the latter, but not at the expense of neglecting probabilities; rather, experience has shown that sites are often found in cells contiguous, or very close to, high-probability cells. At the same time priority was given to increase geographical representation, i.e., to search for sites in areas where previously none, or only few, had been found.

Application of the GPMM

This application is based on 44 "original" registered sites which, with a few uncertain exceptions, lie below the 20 m contour. One procedure that could have been used was to either systematically survey along the coast below the 20 m contour; an area of roughly 105 km^2, approximately 6 percent of the total investigated area, or only to survey those areas below the 20 m contour covered by Holocene (fossil) beach deposits—a favorite location for pit-dwellings. The latter would have reduced the area to be covered to around 50 km^2 or about 55 km^2 less than the first option.

However, digitized maps from the Norwegian Mapping Authority (www.statkart. no) of the topography and geology provides for more sophisticated analysis including variables that would not normally be considered, but which may throw more light on the choice of settlement location in the past and given a powerful GIS can also be handled with analytical consistency. Nine variables which we considered relevant for the study were chosen for the prediction (see Archaeological ANN and Variables section). Moreover, to allow for sites to occur in "unexpected" locations (i.e., places where sites are not usually looked for), it was decided to include all areas less than 1 kilometer from the coast instead

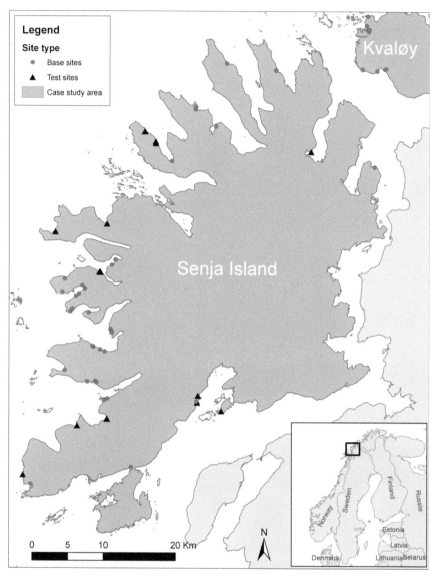

Fig. 4. Norway and Senja island with original and test sites. The lack of sites on eastern, central Senja is most likely due to a combination of the effects of agriculture and limited surveys.

of only the area below the 20 m contour. The grid-size was set at 100 m, which gives a satisfactory level of resolution considering the lay of the land and the quality of the maps.

Field surveys were conducted in 2009 and 2010, respectively, covering a total of 7.86 km^2 (corresponding to 1.3 percent of the total area within the 1 km buffer from the coast or 7.5 percent of the area below the 20 m contour), and targeting high-potential areas in general and areas with few or no previous finds in particular in order to increase general representativity around the island (Fig. 5). A total of 18 new pit-dwelling sites were found during the surveys. This field survey had to be conducted anyway; however

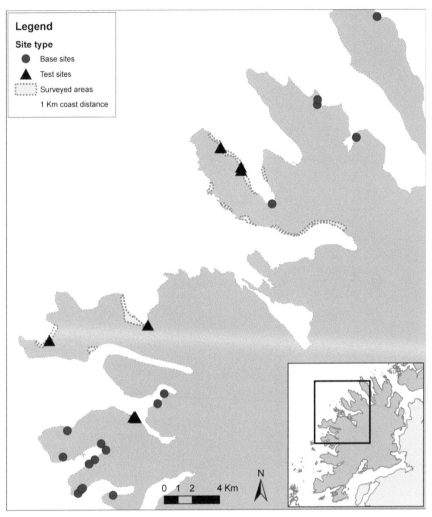

Fig. 5. Senja. Surveyed areas.

all along the way we improved on our research design and found that the surveyed area would be adequate to investigate the quality of our predictions.

First Application

The average probability score for the 44 original sites is 62 percent. While this may be satisfactory by itself in a predictive analysis context, it may, however, be more informative to look at the number of original and test-sites, respectively, assigned to cells with probabilities higher than 75 percent (the critical inflection point in the analysis graph of probabilities which allows for a heuristically chosen cut-off point). Twenty-eight original sites (64 percent) and seven test-sites (39 percent) fall into this category (Fig. 6). However, the relatively low score for the test-sites is more than compensated

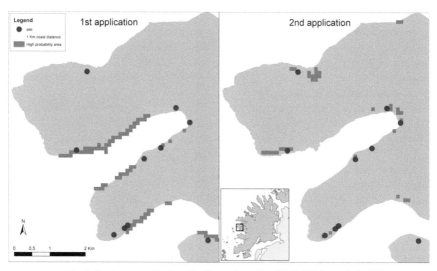

Fig. 6. Senja. 1st and 2nd application—close-up of probabilities and sites.

for by area reduction. The area with probabilities higher than 75 percent and less than 1 km from the coast is 37 km^2 (or 6 percent of the total area of 597 km^2 within the 1 km buffer), which is 68 km^2 less (or only 35 percent) of what would have been required to be covered or sampled by a traditional "below the 20 m contour" survey or 13 km^2 less (or 26 percent) in relation to one that would only have surveyed marine deposits. The intersection of surveyed area with areas of probabilities higher than 75 percent is 9.79 percent, in retrospective, and in all practicalities a 10 percent sample.

Thus, seen in relation to the aims for the application, the GPMM proved very successful.

Second Application

It is considered "good practice" in predictive modeling to feed sites found during surveys back into a new analysis in order to see if improvements, or deeper insights into the variability, can be achieved. When merging the new-found sites with the original ones, the new total is 62. The average probability is 34 percent, which is considerably lower than the 62 percent for the first application (Fig. 6). The number of sites assigned to cells with a probability higher than 58 percent is 28 (45 percent). The area of cells with a probability higher than 58 percent and less than 1 km from the coast is 60 km^2 (10 percent). This is 23 km^2 more to be sampled and surveyed than for the first application, but only 57 percent of what would have been required in a standard "below the 20 m contour" analysis (and 10 km^2 or 20 percent for one on geology and below 20 m). However, as the focus is on the 1 km buffer (and not only the area below the 20 m contour) the method is still considered very useful for reaching the aims of the study. The reason why the average probability dropped and the area increased is likely to be a very common one; as the number of sites increases, so does the range of variability, but at the risk of reaching lower probability scores (see Archaeological ANN and Variables section). In fact, as noted above, only seven out of 18 test-sites had high scores in the first run and it is likely that it is the variability added by the 11 remaining sites that contributes most to the lower general probability. It is also to be considered that the priority given to improve geographical representation (rather than to go solely for potentially high-probability areas) may have contributed to the lower general probability in the second application. The intersection of surveyed area with areas of probabilities higher than 58 percent is 8.9 percent—also in retrospective and in all practicalities a 10 percent sample.

In order to explore more fully the structure of the data and to gain insight into which of the variables should be retained for future analysis, a PCA was run on the probability scores. With 76 percent of the variance explained on the first three components, the result was as expected. The variables slope, aspect, cost-distance from sea, and cost-distance from river are closely interrelated and partly auto-correlated and form a group of their own. Given the general nature of the topography and the strong dependency of the two latter variables on the slope, this is not surprising. Future predictions probably should exclude the two "cost-variables". The two types of marine deposits are in juxtaposition, but do not otherwise seem to lead to differential results. It is suggested that they be lumped together into a single variable for further analysis. The two "distance" variables do seem to be of some importance and should be retained whereas altitude does not seem to play any role and should be left out. New variables should be added however. Given the hindsight of ground-thrusting during the surveys, variables, such as isthmus (essentially giving a sheltered harbour) and degree of shelter would seem very promising for future predictions (see Archaeological ANN and Variables section).

In the present case the emphasis was on 1) to investigate location relative to a wider set of locational variables (affordance variables) in an analytical consistent way, 2) to enable the discovery of coast-bound sites not necessarily tied solely to marine deposits below the 20 m contour, and not least, 3) to reduce the area necessary to be sampled for survey for new sites. For all three purposes the GPMM proved very useful.

Conclusions

Luca Deravignone
Hans Peter Blankholm
Giovanna Pizziolo

In some respects predictive modeling is as much an art as a science. A substantial amount of qualitative assessments goes into the decisions of, for example, choice of grid-placement, cell-size, variables, test-areas and techniques, and the weighing of options and final decisions naturally influence the outcome. It is also so that all methods and techniques must be judged on their individual ability to deliver information of relevance for the achievement of stated goals and ultimately for behavioral interpretation (Blankholm 1991).

In the present chapter the emphasis was on 1) to investigate locations relative to a wider set of locational variables (affordance variables) and socio-economic/cultural variables in an analytical consistent way, 2) to reduce the area necessary to be sampled for survey for new sites, and 3) to try new ways of analyzing ancient settlements patterns in diachronic and synchronic ways. For all purposes the GPMM proved successful and demonstrates its versatility, robustness, and analytical appeal vis-à-vis other methods and techniques for prediction. With the results from our case-studies we feel optimistic about future developments, and are also looking forward to expand our case studies with other sites types and test the results with new databases.

References Cited

www.statkart.no The Norwegian Mapping Authority

Abdi, H.D. Valentin and B. Edelman. 1999. Neural Networks. Newbury Park CA, USA, Sage, Sage University Series.

Barceló, J.A. 2009. The birth and historical development of computational intelligence applications in Archaeology. Archeologia e Calcolatori 20: 95–109.

Blankholm, H.P. 1991. Intrasite Spatial Analysis in Theory and Practice. Aarhus University Press. Århus.

Blankholm, H.P. 2009. Rapport over prøveutgravning av Grindvollen, Tuft 4, Ts.nr. 11776, ID 18030, Juni 2009. Institute of Archaeology and Social Anthropology, University of Tromsø. Norway.

Deravignone, L. 2009. Comprendere le dinamiche insediative con l'aiuto dell'intelligenza artificiale: un caso di studio. pp. 187–192. *In*: G. Macchi Jánica Geografie del Popolamento. Università degli Studi di Siena, Edizioni dell'Università, Siena.

Deravignone, L. 2011. Esperienze di un approccio multi-metodologico per lo studio delle antiche reti insediative. pp. 207–211. *In*: A. Di Blasi Il futuro della geografia: ambiente, culture, economia. Atti del XXX Congresso Geografico Italiano, Pàtron Editore, Bologna.

Deravignone, L. and G. Macchi. 2006. Artificial Neural Networks in Archaeology. Archeologia e Calcolatori 17: 121–136.

Deravignone, L., H.P. Blankholm and J.I. Kleppe. 2013. Grosseto Predictive Modeling Manual for ArcGIS, http://www.researchgate.net/publication/258283865_Grosseto_Predictive_Modeling_Manual_for_ArcGIS.

Francovich, R. and M. Milanese (eds.). 1989. Lo scavo archeologico di Montarrenti e I problemi dell'incastellamento medievale. Esperienze a confronto. Archeologia Medievale 16.

Gaffney, V., Z. Stančič and H. Watson. 1995. The impact of GIS in Archaeology: a personal perspective. pp. 211–229. *In*: G. Lock and Z. Stančič (eds.). Archaeology and Geographical Information Systems, Taylor & Francis, London.

Graves, D. 2011. The use of predictive modeling to target Neolithic settlement and occupation activity in mainland Scotland, Journal of Archaeological Science 38(3): 633–656.

Jung, J. and S. Elwood. 2010. Extending the Qualitative Capabilities of GIS: Computer-Aided Qualitative GIS, Transactions in GIS 14(1): 63–87.

Kamermans, H., M. van Leusen and P. Verhagen (eds.). 2009. Archaeological Prediction and Risk Management. Leiden: Leiden University Press.

Llobera, M. 2000. Understanding movement: a pilot model towards the sociology of movement. pp. 65–84. *In*: G. Lock (ed.). Beyond the Map. Archaeology and Spatial Technologies, Amsterdam: IOS Press/Ohmsha.

Schanche, K. 1994. Gressbakkentuftene i Varanger. Boliger og sosial struktur rundt 2000 f.Kr. Unpublished Ph.D. thesis. University of Tromsø, Norway.

Van Leusen, M. 1999. Viewshed and Cost Surface Analysis Using GIS (Cartographic Modeling in a Cell-Based GIS II). pp. 215–223. *In*: J.A. Barceló, I. Briz and A. Vila (eds.). New Techniques for Old Times. CAA 98. Computer Applications and Quantitative Methods in Archaeology. Proceedings of the 26th Conference. BAR International Series 757.

Verhagen, P. and T.G. Whitley. 2012. Integrating Archaeological Theory and Predictive Modeling: a Live Report from the Scene. Journal of Archaeological Method and Theory 19: 49–100.

Verhagen, P., L. Nuninger, F.-P. Tourneux, F. Bertoncello and K. Jeneson. 2013. Introducing the Human Factor in Predictive Modeling: a Work in Progress. pp. 379–388. *In*: G. Earl, T. Sly, A. Chrysanthi, P. Murrieta-Flores, C. Papadopoulos, I. Romanowska and D. Wheatley (eds.). Proceedings of the 40th Conference on Computer Applications and Quantitative Methods in Archaeology Southampton, 26–30 March 2012, Amsterdam University Press, Amsterdam.

Woodman, P.E. and M. Woodward. 2002. The use and abuse of statistical methods in archaeological site location modeling. pp. 39–43. *In*: D. Wheatley, G. Earl and S. Poppy (eds.). Contemporary Themes in Archaeological Computing, Oxbow Books, Oxford.

Zubrow, E.B.W. 2003. The Archaeologist, the Neural Network, and the Random Pattern: Problems in Spatial and Cultural Cognition. *In*: M. Forte and P.R. Williams (eds.). The Reconstruction of Archaeological Landscapes through Digital Technologies. Italy-United States Workshop. Oxford (BAR S1151).

18

Spatial Cluster Detection in Archaeology: Current Theory and Practice

Benjamin Ducke

Point Distributions and Clusters

Archaeologists spend a great amount of their time pondering maps of *points*, so-called *distribution maps* of sites or artefacts, and trying to make sense of them (Fig. 1; see also Fig. 2 and Fig. 8, upper left). The mapped points are, of course, *manifestations* of something much more complex and interesting. And these hidden *processes* behind the points are the real phenomena that are of academic interest. When attempting inference from indirect evidence, such as point distributions, the use of formal methods is of particular importance. A rigid analytical approach helps the human interpreter avoid jumping to premature conclusions or reading too much into the data. Indeed, the eagerness of the human brain to identify *patterns*, i.e., recurring regularities, in visual information seems hard to control and even harder to replace with a more objective approach. The following sections will discuss mathematical methods for the detection of *clusters*, a type of pattern that is characterised by locally increased densities. Such clusters are assumed to be indicative of higher frequencies of human activities, i.e., they represent "hotspots" of social processes (note that this deviates from the use of the term "cluster" within the context of multivariate clustering analysis as discussed by Mucha et al. in Chapter 9).

More often than not, the observed or assumed presence of clusters provides the sole point of departure for reasoning about the processes behind the points. Accordingly, the challenge of cluster detection has been a frequent topic of spatial analysis in archaeology that made its first widely known appearance in Hodder and Orton (1976)

Deutsches Archäologisches Institut, Hohenzollerndamm 150-151, Haus 2, 14199 Berlin.
  Email: benducke@fastmail.fm

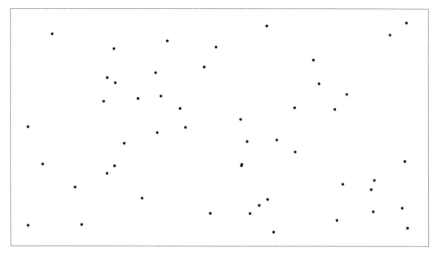

Fig. 1. A plain distribution map of 50 points. The question at hand is whether and where this data contains any clusters, i.e., local concentrations of points. The reader is invited to use his or her own visual judgement and then read the conclusions of this text.

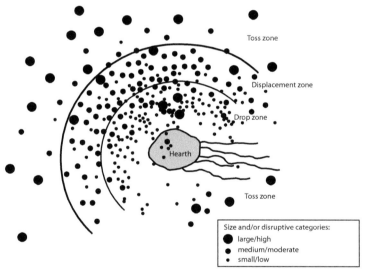

Toss zone

Displacement zone

Drop zone

Hearth

Toss zone

Size and/or disruptive categories:
● large/high
● medium/moderate
· small/low

Fig. 2. Model of litter dispersion and size distribution around a central hearth as formulated by Stevenson (1985): Smaller litter immediately drops to the floor ("drop zone"), whereas larger pieces are deliberately tossed aside ("toss zone"). Redrawn from Stevenson 1985, Fig. 10.

and culminated in the seminal work by Blankholm (1991). Since then, the number of published cluster detection methods have greatly increased across the whole spectrum of "spatial disciplines" (geography, ecology, forensics and epidemiology, to name just a few), but so have the volumes of data to be analysed, as well as the expectations of the capabilities of modern software for cluster detection. Therefore, it seems only appropriate to provide a concise exposure of current spatial cluster detection for the archaeological readership, illustrating the most important concepts and capabilities of modern, computationally intense methods.

Before delving into mathematical and technical details, a few underlying principles must be explained. A particularly powerful approach to scientific data analysis is the *model-based* approach. Compared to the "free style interpretation" approach, a good model significantly increases explanatory power. A common example is the evidence provided by maps of scattered artefacts on an excavated site. By observing the spatial locations of a number of artefacts (or other remains), archaeologists wish to infer the nature of the anthropogenic processes behind them and a model can greatly help them achieve this. Thus, Stevenson (1985) provides a model of what type of artefact scatter to expect at various distances from a fireplace (Fig. 2), while Stapert (2003) takes this approach one step further by adding the assumption that a fireplace within an enclosed space should result in slightly different depositional patterns. It is worth taking a moment to reflect on what such relatively simple models can actually achieve: Based on nothing more than the spatial distributions of artefacts, it becomes possible to test whether architectural features once existed on a site that *does not have preserved direct evidence* of such features.

Often, however, such elegant, preconceived models will not be available. In fact, archaeologists must consider themselves lucky if there is any *a priori* knowledge at all about what to expect from the data, such as an approximate number and size of "hotspots" or even the vague locations of focal points around which clusters can be expected to manifest. In addition, the creation of any explanatory model, no matter how refined, must also start with much simpler observations, frequently nothing more than the locations of interesting *events* (in the following, the word "event" will be used synonymously with "point", in accordance with the common terminology of spatial statistics), such as an offering being deposited, a coin lost or a settlement built.

Terminology

At this point, a short terminological review is necessary. The following word usages are reductive and may not fully overlap with their everyday meanings:

- An *event* is a point on a map that represents the location of any discrete object or phenomenon of interest.
- A *process* is any naturally occurring or anthropogenic agent capable of physically manifesting itself as a series of events.
- A *pattern* is the characteristic manifestation of a series of events that can, at least in theory, be repeated. Analytically speaking, the most important aspect of a pattern is that it can be associated with a specific process.
- A *cluster* is a local concentration of events. The two fundamental properties of a cluster are (a) that it can be delineated and (b) that there is a significantly higher event density within the borders of a cluster than on its outside.
- *Clustering* is the general (global) tendency of a process to manifest itself in clusters.
- A *cluster detector* is any quantitative method capable of locating and delineating clusters in a dataset. To achieve this, a valid cluster detector must either assign each point to a cluster or declare it as "noise".

- Finally, a *model* (in this narrow context) is an understanding or idea of the processes behind observed point patterns. In an even narrower context, the term *cluster model* denotes the "idea of what constitutes a cluster", as used by a specific cluster detector.

These basic terms and associated concepts are commonly found in the literature on spatial statistics. They also extend quite naturally from the domain of spatial analysis into that of spatio-temporal analysis. However, the analysis of space-time data is fraught with its own challenges that cannot be discussed in depth here (see Crema et al. 2010 for a state-of-the-art study on the subject).

Pitfalls

Ironically enough, the very simplicity of point distributions is accompanied by some of the toughest analytical problems, many of which are subtle and frustratingly common, yet do not have a simple solution. There is a structural commonality behind most of the pitfalls in spatial analysis that lies in the fact that all observations and results of computations are *scale-dependent*. Just how this problem manifests itself with seemingly endless variety has been the subject of dedicated treatises (see Lock and Molyneaux 2006). The following manifestations of scale-dependency are prevalent enough to have been identified and given common names across disciplines (for further illustrations and explanations from a geographer's point of view, see O'Sullivan and Unwin 2003: 26–40):

1. The *MAUP* (modifiable areal unit problem) expresses the fact that the results of spatial statistics depend on the extent of the study region. The archaeological excavation is a perfect example: Excavated trenches act as analytical windows into the complete data source. Through these windows, one could for example perceive a set of postholes as regularly spaced entities, whereas the fact that they form part of a larger cluster would only become visible if more of the site was excavated. There really is no "solution" to the MAUP. One should wisely choose the scale of analysis and rigorously adhere to this choice when collecting data, formulating research aims and identifying processes of interest.
2. *Edge effects* are strongly related to the MAUP and are owed to the fact that many spatial statistics become unreliable towards the edges of the study region. For example, all cluster detectors discussed in the following use some notion of distance between points to decide whether they are part of a specific cluster (e.g., see the discussion of the DBSCAN algorithm and its parameter ε). However, distance measurements are problematic for points near the edges of the study region, because there could be other, closer points just beyond that border, for which the analysis would be blind. Edge effects are typically more pronounced for "arbitrary" regions such as excavated trenches that do not represent "natural" boundaries for the underlying processes. In such cases, an option of last resort is to simply set the study region to be somewhat smaller than the actual extent of data. The data in the remaining, outer buffer can then be used to correct for edge effects in computations, but must never be used for inference and interpretation.
3. In archaeology, "smearing" (otherwise known as *non-stationarity*) describes the effects of post-depositional perturbation of artefact locations by processes such as soil erosion or ploughing, on the observed distributions. This is an extremely

tough-to-control problem on many sites, as the displacement processes must be identified, and at least partly accounted for, prior to meaningful spatial analysis.

4. Point processes are influenced by background structures and effects. The existence of hard, physical boundaries that block the movement of material can lead to accumulation of remains against these boundaries, which will create "meaningless" (in relation to the actual anthropogenic processes) clusters. Networks (such as roads between towns) also have a strong influence on the structure of spatial activities, as does the distribution of natural resources. One may think of these examples as specific cases of the more general concept of *covariates*, i.e., background variables that shape the spatial manifestations of point processes in a more or less complicated fashion, often referred to as *trends*. Once identified, such background variables can be catered for in the spatial analysis by systematically removing any trends from the data. Methods for detecting clusters in the presence of networks have been published by Okabe et al. (2009), for instance.

For the purposes of this chapter, the intention behind exposing the above problems is merely to stress that one of the most fundamental recent achievements in the field of spatial statistics has been progress in understanding the true magnitude of the challenge, rather than in introducing novel computational methods (although there is no lack of the latter, either). While modern software implementations of statistical methods may offer some options for dealing with problems like edge effects, such automatisms can never be blindly relied on, and neither should the output of any cluster detector naïvely be accepted as a "final" or "optimal" result.

The Barmose Site as a Benchmark for Cluster Detectors

For simplicity's sake, the following discussion will remain at the archaeological site level, looking at the distribution of artefacts across relatively small, excavated areas. Nevertheless, all principles and methods also apply to any other scale of data analysis (for a study at landscape scale, see, e.g., Bevan 2012).

More than two decades ago, Blankholm (1991) published his book on "intrasite spatial analysis", in which he presented a systematic comparison of a large number of mathematical methods, testing each one for its ability to detect clusters in artefact distribution maps. One of two benchmark sites for his study was Barmose I, an Early Mesolithic camp site from South Zealand, Denmark (Fig. 3; see Blankholm 1991, pp. 183 for more details), that had preserved traces of activity areas and a full spectrum of artefacts. Based on these remains, Blankholm formulated an archaeological expectation of the site's cluster structure and then proceeded to test how well this structure was revealed by the different methods, some of which (in particular *k*-means) are still popular, while others have (perhaps for good reasons) disappeared into obscurity. The fact that the site data, including all artefact positions, was published as an appendix of the book makes it possible to revisit Barmose I and see how well current cluster detection methods fare in the "Barmose challenge". The latter could be phrased as: "Is there a cluster detector capable of revealing the structure of Barmose's activity areas, using only the indirect evidence of artefact locations?"

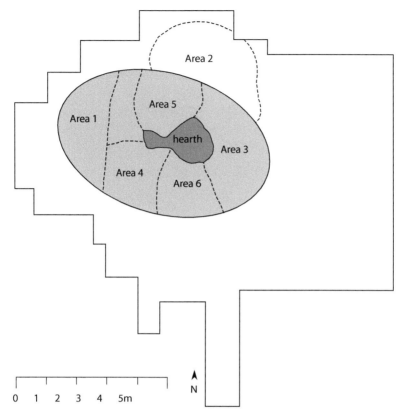

Fig. 3. Excavated surface area of the Early Mesolithic Barmose I site (Zealand, Denmark). The distribution of artefacts and other remains suggests a central hearth area with several activity areas around it. Redrawn after Orton 2004, Fig. 3.

Methods for Spatial Cluster Detection

With regard to Barmose I, the most significant limitation to the systematic comparison of cluster detectors is the one so often encountered in archaeology: There is no certainty about the original appearance of the site's activity areas. Therefore, there is also no strictly defined number or shape of clusters that would constitute a "correct" output for any given algorithm (although the number of plausible cluster solutions is certainly limited by many constraints). This problem emphasizes that, in archaeological practice, the usefulness of a cluster detector is defined as much by its properties as by the solutions that it produces. More specifically, the following should be true for a useful cluster detector:

- Its mathematical core should be simple, so that archaeologists can understand why a certain input leads to a certain output,
- it should assume as little as possible about the clustering structure (i.e., the number, sizes and locations of clusters),

- it should be robust against noise, i.e., it should just ignore the odd scattering of sherds around the actual clusters, and it should not assume that every point is indeed part of a cluster,
- it should either assign each point to a cluster or classify it as noise,
- and finally, it should always produce the same result when given the same input data.

Unfortunately, no single method has yet emerged that would meet all of the above criteria. The best existing option is a combination of tools to tackle the problem from different angles.

Kernel Density Estimates

An alternative to treating points and clusters as *discrete* entities is to look at the *continuous* measure of point *density*, i.e., the number of points within a given area. This is most prominently represented by *kernel density estimation* (KDE), a method that is both robust and sensitive. It is widely available in GIS and makes an excellent exploratory tool that can be used to reveal structure in the data before subjecting it to a "proper" cluster detector. KDE works by moving a circular search window, the so-called "kernel", across the entire study area (see O'Sullivan and Unwin 2003: 85–88 for illustrated details). In the simplest case, the kernel has a fixed radius, the so-called "bandwidth" that needs to be specified by the analyst (KDE with variable bandwidth will not be discussed here, as it leads to results that are exceedingly hard to interpret). At each position of the study area, the density of points within the kernel will be computed as a probability estimate, usually based on a Gaussian distribution, that lends more weight to points closer to the centre of the (Gaussian) kernel. The result is a continuous and smooth estimation of density across the entire study area (see Fig. 4).

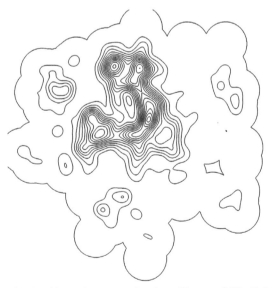

Fig. 4. A KDE of artefact densities on the excavated surface of Barmose I (Fig. 3). In this example, the Gaussian kernel has a bandwidth of 25 cm (compare with Fig. 8, upper left).

Being a probability-based method, KDE cannot be used to make definitive conclusions about the locations or shapes of individual clusters. However, it is also a very sensitive method, capable of revealing even the most subtle local changes in density. The key to the successful use of KDE as an exploratory technique is in choosing the "correct" bandwidth. This can be done by experimenting with a number of different values or, in the absence of any more specific knowledge, by first subjecting the data to a method capable of suggesting interesting bandwidths, such as the one discussed next.

Ripley's K and L

Whereas cluster detection as discussed here aims at *local* effects, more often than not, some *global* understanding of the data's properties is also required. One of the most immediate problems in any spatial analysis is the "proper" choice of scale. In the context of cluster detection, this translates into plausible expectations for the bandwidth of a KDE or the average size of expected clusters; as required to validate the results of cluster detection and to provide initial parameter values ("starting points") for many algorithms. In some cases, such expectations may be derived from knowledge about the processes involved. For example, one would expect the size of clusters of flint flakes formed around a working area to correspond with the typical radius of debris "fallout" from a person chipping away at a block of raw material.

Alternatively, Ripley's *K* (also known as "k hat" or *k* function: Ripley 1976, see also Conolly and Lake 2006: 166–168) can help reveal interesting clustering effects across a large spectrum of scales. Ripley's *K* is a very sensitive, graphical method for discovering global clustering structure at different scales. Its working principle is that of a typical "scan statistic", i.e., a computationally intense procedure that examines the data across all interesting scales (search radii) and points out the most significant results (for more details and an illustration, see O'Sullivan and Unwin 2003: 92–95). Formally, the function "k hat" is defined as

$$\hat{K}(s) = \lambda^{-1} n^{-1} \sum_{i \neq j} I(d_{ij} < s)$$

with λ (lambda) being the average point density within a maximum search radius of *s*. The total number of analysed points is *n*, and d_{ij} is the distance between each individual pair of points. For any search radius, clustering is indicated by the observed average distance between point pairs falling significantly below the expected distance. Significance is established by comparison with a large number of randomly generated points. The search radius is increased in small steps, and *s* is re-computed for each new radius. The resulting function graph shows the search distance on the horizontal (X) axis and the result for *k* on the vertical (Y) axis (Fig. 5). There is one line that represents the actual *k* values and two more that represent the confidence envelope for the assumption of *complete spatial randomness* (CSR). If the line of the *k* values falls in between the two enveloping lines, then there is no significant difference from randomness. For those distances, where the *k* values are outside the envelope, the following applies:

1. If the *k* values fall *above* the envelope, then significant *clustering* is indicated,
2. and if they fall *below* the envelope, then significant *dispersion* is indicated.

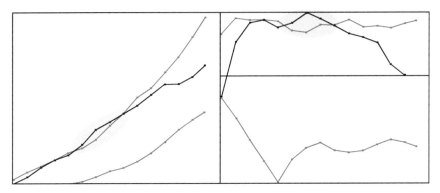

Fig. 5. Plots of K (left) and L (right) functions for the burins at Barmose I, computed for search radii with a step-wise increase of 10 cm. The black curve represents the data; the grey curves are the envelope of simulated random point processes. Significant clustering is indicated for bandwidths between 60 and 90 cm (grey areas). Redrawn from Orton 2004, Fig. 1.

Ripley's K is a versatile, multi-scale analysis tool, and there are many modifications to it that make it even more useful in practice. One of them is the L function, a normalised reprojection of K that is easier to read and can be directly compared between different datasets. As an example, Fig. 5 shows the plots of k and l values for one specific type of artefact (burins) from Barmose I, as published by Orton (2004).

K-means as a Spatial Cluster Detector

The k-means algorithm is a stock method of multivariate data analysis. It is robust and easy to understand; two characteristics that largely explain its popularity. In archaeological analysis, k-means has been applied to the problem of spatial cluster detection by simply feeding it the X and Y coordinates of recorded artefacts (e.g., Koetje 1987). The k-means algorithm defines a cluster A as the set of all points that lie closer to the fictitious centroid ("central point") of A than to that of any other cluster's centroid. The locations of the k centroids must be determined from the point data itself. This is achieved by an iterative approximation scheme (compare with Fig. 6):

1. Start with k randomly placed centroids.
2. Assign each point to its nearest centroid to form the initial clusters.
3. Re-set each centroid to the mean coordinates of the points in its cluster.
4. Repeat 2 and 3, until re-setting the centroids will no longer change the clusters.

The above scheme immediately reveals one of k-means' greatest drawbacks: It is necessary to know the number of clusters (k) in advance. This is a requirement that will be hard to meet in most archaeological scenarios. After all, where would such knowledge come from, given the typical point distribution map or artefact scatter? However, even if it were possible to know k (or even a limited range of plausible k values for trial-and-error) in advance, there are other, equally severe, limitations that stem from k-means' centroid-based cluster model. The spatial manifestation of the k-means solution is always a *complete*, non-overlapping partition of space that is the equivalent of a Voronoi diagram (see Fig. 8, upper right). This output bears little geometric resemblance to the

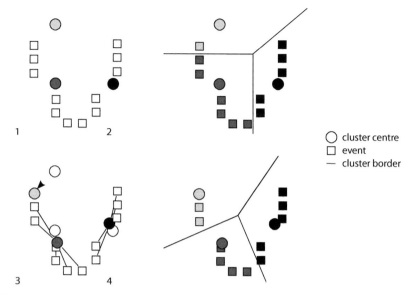

Fig. 6. Graphical illustration of the k-means algorithm with $k = 3$. See text for details. Adapted from Wikipedia entry "k-means clustering".

actual shapes of the activity areas that the clusters ought to represent and it does not allow for the exclusion of "noisy points" from the cluster areas. In summary, k-means is not well suited for use as a spatial cluster detector. Fortunately, there are alternative methods that address its weaknesses.

EM Clustering

The e*xpectation-maximization* (EM) algorithm is a probabilistic alternative to the k-means algorithm that provides a less rigid clustering solution, allowing for noise and overlapping clusters (Dempster et al. 1977, see Scott and Hillson 1988 for an archaeological application). It assumes that the distances of points from the centres of their respective clusters follow a normal, i.e., Gaussian distribution (an assumption that is usually met, given a sufficient number of data points for each cluster). The global model for the entire dataset is therefore a mixture of Gaussian distributions, each one with unknown parameters mean and variance that have to be estimated by the EM algorithm. Just as in the k-means case, the number of clusters, k, must be given. Starting with random distribution parameters for all k clusters, the EM algorithm will then iteratively optimise the assignment of the points to the distributions of the individual clusters. This is followed by an optimisation of the distribution parameters to fit the latest cluster assignments. Every pass through the data involves an expectation-maximization step for all clusters:

1. Expectation step: estimate parameters mean and variance for each cluster from the points assigned to it.
2. Maximization step: re-assign points to clusters after re-estimating parameters from updated clusters.

3. Repeat from step 1, until there are no longer any significant changes to the clusters and their distributions.

The expectation step uses a so-called *Q*-function, and the maximization step uses a maximum likelihood function, for which the details are provided by Wu (1983) and Hastie et al. (2001), respectively. A glance at the output of the EM algorithm (Fig. 8, lower left) shows some substantial improvements over the output of *k*-means. The probable locations, sizes and orientations of the clusters are represented by enclosing ellipses, with centroids and radii determined by the means and variances of the underlying point coordinate distributions. In addition, the cluster boundaries are "fuzzy", with points towards the edges classified as noise, and they are allowed to partially overlap, adding a more realistic notion of uncertainty to the result. However, the EM cluster representation is still rather abstract and does not provide more than a rough idea of the geometrical properties of the activity areas that are presumably represented by the clusters. In addition, there is still the problem of having to specify *k* in advance, and the initial randomization of the cluster distributions leads to significant fluctuations in results when the EM algorithm is run repetitively on one and the same input data.

DBSCAN

DBSCAN ("density-based spatial clustering of applications with noise") is an example of a current, computationally intense method that was designed with spatial data analysis in mind. DBSCAN uses a *density-based* cluster model that has some very advantageous properties in the context of spatial cluster detection. It is robust against noise, there is no need to predetermine the number of clusters, and the detected clusters can have complex, including concave, shapes.

The complete DBSCAN algorithm is a little too complex to be exposed here (for details see the original publication by Ester et al. 1996), so that an explanation of its core principles will have to suffice. Essentially, DBSCAN assigns points to clusters, based on the principle of *density-reachability* and *density-connectivity* (Fig. 7). Two points are determined to be *density-connected* if they are linked via a series of *core points*, each one being no further than a distance threshold ε (epsilon) from the next. A point is classified as a core point if it is surrounded by a minimum number (*minPts*) of other points that are all within the distance threshold of the core point. Finally, a cluster is defined as the set of all points that are density-connected via the same core points. All other points are classified as noise.

As opposed to *k*-means and EM clustering, DBSCAN does not require the number of clusters to be known in advance. Instead, it takes two parameters that should be easier to supply in most situations: the search distance ε and the minimum number of points (*minPts*) that constitute an "acceptable" cluster. While ε has direct influence on the shape and size of the detected clusters, the second parameter decides which points are classified as noise. From an archaeological point of view, choosing values for ε and *minPts* amounts to making informed statements about the expected spatial range and intensity of the processes behind the clusters.

One slight complication to keep in mind is that ε represents a lower distance threshold rather than a simple, fixed bandwidth (as in the case of KDE). A more significant limitation of DBSCAN lies in its poor performance with *inhomogeneous* data, where one would expect to find clusters of strongly varying densities. To address

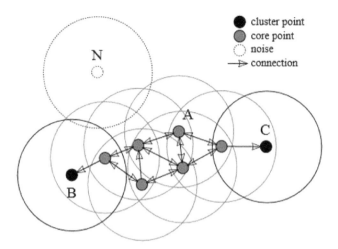

Fig. 7. Graphical illustration of the DBSCAN algorithm's working principles (see text for details): Points in A are core points that are both close to neighbouring points and surrounded by a sufficient number of other core points. Points B and C are density-reachable and therefore part of the same cluster as A. N is neither a core point nor density-reachable and therefore classified as noise. The circles correspond to the distance threshold ε (epsilon). Adapted from Wikipedia entry "DBSCAN".

these concerns, the OPTICS ("ordering points to identify the clustering structure") algorithm allows to find good values for ε, while at the same time being able to handle inhomogeneous data (for details see Ankerst et al. 1999).

Comparison of Outputs

Having discussed the properties of three different cluster detectors from a theoretic point, it is now time to let the proverbial "rubber hit the road" and compare their performances in practice. To this end, both *k*-means and EM clustering where parametrised with *k*=6, in accordance with the anticipated number of activity zones on the Barmose I site (see Fig. 3). The parameter settings ε=30 cm and *minPts*=10 were found by first holding *minPts* constant at a (subjectively chosen) value of "10" and varying ε within close range of the clustering scale suggested by the results for Ripley's *K* that were previously published by Orton (2004; see also Fig. 5; note that said *k* values apply to burins only, whereas this example includes all types of artefacts).

Looking at the side-by-side comparison in Fig. 8, it immediately becomes apparent that the three detectors produce greatly different results. As expected, the output of *k*-means is only minimally related to the expected spatial structure of the site. While the assignment of the points to the clusters does appear valid (prior expectations about the clusters aside), the actual cluster shapes, as represented by the lines of the Voronoi diagram, do not. The most apparent problem is the complete partitioning that also includes scattered points which should rather be classified as noise and does not represent a good approximation of the actual cluster shapes.

The output of EM clustering is more realistic, insofar as that the clusters allow for some overlap and noise. In fact, if one considers only the elliptical envelopes

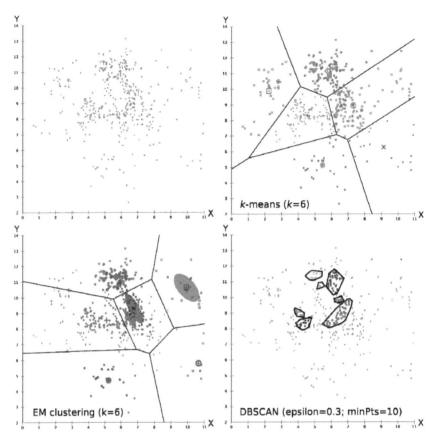

Fig. 8. Distribution of artefacts on Barmose I (upper left); outputs of *k*-means, EM clustering and DBSCAN. Compare with proposed site structure shown in Fig. 3. Produced with ELKI 0.6b2 (http://elki.dbs.ifi.lmu.de/).

(representing the one and two sigma ranges of the underlying probability distributions) of the clusters, and once more disregards any *a priori* expectations of the site's structure, then the EM output shown in Fig. 8 does represent *one* objectively valid solution. Due to the stochastic nature of the EM algorithm, however, this particular solution will not be exactly reproducible, and there is little justification why it should be preferred over any other solution that might arise when re-running the EM algorithm on the same data.

By contrast, the output of DBSCAN fits the expected site's structure much better, with the number of detected clusters almost identical to the expected count (seven instead of six) and all clusters compactly outlined and neatly arranged around the central hearth, as anticipated by the nature of the activity areas. As should be expected from the "smearing" effects on an archaeological site, a large part of the data cannot be assigned to any cluster and was classified as "noise" by DBSCAN. This solution is deterministic and perfectly reproducible.

Aspects of Practical Application

The multivariate cluster detection problem is not identical to the spatial one. Multivariate data analysis aims to condense complex data into a much lower number of synthetic variables that represent dominant trends. Space, on the other hand, is a fixed, physical frame of reference and not just another data dimension. Consequently, the shape of spatial clusters has *meaning*, providing important clues about the nature of the processes behind the observed point distributions. A cluster detector with a spatial cluster model, such as DBSCAN, will therefore inevitably produce more useful results than one with a non-spatial model, such as the classic k-means. Neither of these methods makes any assumptions about the statistical nature of the processes behind the point distributions. They are strictly exploratory methods, useful for revealing interesting structure in the data, but incapable of providing statistical evidence to support the validity of any given result. Some detectors, such as EM clustering, are highly unlikely to produce identical results when run consecutively on the same input data.

Cluster Detection in Artefact Distributions

In the absence of more specific *a priori* knowledge about the expected clustering structure, a good approach is to first explore the data's characteristics using tools such as Ripley's K and KDE, and to use the insights gained from this to parametrise a cluster detector. There are, of course, many other methods for determining parameters, such as KDE bandwidth settings, automatically, but one should never expect any of them to perform reliably for all input data.

In addition, keeping in mind the pitfalls of spatial data analysis in the context of artefact distribution, one should be aware of the following necessities:

- Consider the effect of the excavation's trench layout on the observable patterns,
- try to control the time dimension by looking at "slices" of artefact distributions, that represent individual occupation layers,
- determine the influence of any hard boundaries or other features of the site that could structure the spatial distributions of the material remains,
- and try to further disentangle the artefact distributions by looking at artefact types separately, as well as in relation to each other (see Orton 2004 on how to apply Ripley's K for this purpose).

Software for Cluster Detection

Concerning available software implementations of the methods discussed here, one of the best options is the free and open source research software ELKI (http://elki. dbs.ifi.lmu.de/), which implements an enormous array of cluster detection algorithms, including the ones discussed here. Counter-intuitively, spatial statistics (not to be confused with geostatics, i.e., the theory and practice of kriging interpolation!) are *not* a strength of current GIS. Sometimes, basic exploratory tools such as k-means, Ripley's K and KDE are available, but the more useful methods for discrete cluster detection, such as DBSCAN, are absent. The prime source of software implementations of both established and cutting edge research in spatial statistics remains the open source R project (http://www.r-project.org); see also Bivand et al. (2008).

Notes on 3D and 4D Cluster Detection

With deeply stratified sites and the availability of digital hardware to measure artefact locations in 3D, one might wonder about the possibility to extend the concepts and methods of 2D cluster detection into 3D space. However, such an extension is subject to the "curse of dimensionality": Cluster detection in 3D is usually much harder than in 2D, even though it seems like a matter of simply adding one additional Z coordinate to the input data. In the archaeological context, the "curse" makes itself felt in the way artefacts are deposited vertically. As a general rule, they do not spread out in the vertical dimension as much as in the horizontal (just like stratigraphic layers), so that the Z dimension appears "squeezed" in comparison to the X and Y dimensions, and it can be difficult to compensate for this in a method like KDE with a single bandwidth setting. To make matters worse, the degree of "packing" of artefacts may vary across the site.

This problem is greatly exaggerated in 4D analysis, where the naïve approach would be to simply add another t coordinate for the time dimension. However, since there is no simple common measure for distances in both space and time, it is not possible to run algorithms, such as the ones discussed here, on 4D data and obtain valid results. Indeed, *integrated* space-time data analysis is a challenging problem that is only now being understood in sufficient depth (see Cressie and Wikle 2011). Established methods for space-time cluster detection derive mostly from the field of epidemiology, where they are used to detect significantly increased occurrences of diseases in space and time (e.g., Kulldorff and Nagarwalla 1995, Jacquez 1996). However, none of these methods are capable of revealing structure in the data in the way that, e.g., DBSCAN is able to provide clues about the actual shapes of clusters.

Conclusions

Clusters characterise anthropogenic processes in space and time, and across all scales, from individual action to the organisation of entire settlement systems. Cluster detection is a seemingly simple task with no readily available, simple solution. This conundrum is perhaps best explained by the very simplicity of the input data itself. The search for clear structure in something as simple as a set of points is hindered by too many degrees of freedom. No mathematical approach, no matter how refined, will ever change this. On the other hand, the capabilities of current algorithms, such as DBSCAN, are very encouraging.

The more *a priori* knowledge exists (such as about the number, focal points or estimated diameters of clusters) the easier the task of parametrising algorithms will become. However, it is still necessary to mitigate the risk of being deceived by the suggestive visual nature of point distributions and reading subjective and arbitrary "patterns" into the input data. As an example, Fig. 1 contains a depiction of the manifestation of a completely random point process. The obvious problem is that even random processes do suggest some degree of clustering, albeit not a statistically significant one. Thus, even if one does not accept arguments such as transparency and reproducibility in favour of a mathematical approach: The deceptive nature of point distributions really mandates such a formal approach.

Without the ability to answer basic structural questions, such as the one about clustering, no further insights may be gained, no keen interpretations may be made and indeed no interesting research questions formulated. Given that clusters and their detection play such a pivotal role for archaeological reasoning in space and time, it is somewhat surprising that the discipline has not produced another systematic survey of related methods since Blankholm's now-classic book (Blankholm 1991). In this short treatise, only three current cluster detectors were discussed, in order to highlight the possibilities and limitations of current algorithms. New cluster detectors are being published on a regular basis, and it would be futile to attempt to cover even a representative sample here.

The adoption into archaeology of many more concepts and methods from key research fields, such as point pattern analysis, is currently hindered by the heavy handed style of fundamental works, such as Ripley (1981). Nevertheless, interest in "hard core" spatial analysis is currently seeing a revival, as expressed by a series of recent publications by Bevan and colleagues (Bevan 2012, Crema et al. 2010, Baddeley et al. 2013). The latter provide higher level entry points for interested archaeologists, including the essential topics of statistical reasoning with temporal data, probability-based inference and significance testing with point distributions. It seems a safe bet that such advanced research will provide archaeology's next hotbed for fundamental research and innovation in data analysis, including, perhaps, the next generation of cluster detection methods.

References Cited

ELKI (http://elki.dbs.ifi.lmu.de/).

R project (http://www.r-project.org).

Ankerst, M., M. Brunig, H.-P. Kriegel and J. Sander. 1999. OPTICS: Ordering Points Tt Identify the Clustering Structure, in ACM SIGMOD international conference on Management of data (New York: ACM Press), S. 49–60.

Baddeley, A.J., R. Turner, J. Mateu and A. Bevan. 2013. Hybrids of Gibbs Point Process Models and Their Implementation, Journal of Statistical Software, 55/11 (http://www.jstatsoft.org/v55/i11).

Bevan, A. 2012. Spatial Methods for Analysing Large-scale Artefact Inventories, Antiquity 86: 492–506.

Bivand, R., E.J. Pebesma and V. Gomez-Rubio. 2008. Applied Spatial Data Analysis with R (New York: Springer).

Blankholm, H.P. 1991. Intrasite Spatial Analysis in Theory and Practice (Aarhus University Press).

Conolly, J. and M. Lake. 2006. Geographical Information Systems in Archaeology (Cambridge University Press).

Crema, E., A. Bevan and M.W. Lake. 2010. A probabilistic framework for assessing spatio-temporal point patterns in the archaeological record, Journal of Archaeological Science 37/5: 1118–1130.

Cressie, N. and C.K. Wikle. 2011. Statistics for Spatio-Temporal Data (Hoboken, NJ: Wiley).

Dempster 1 Dempster, A.P., N.M. Laird and D.B. Rubin. 1977. Maximum-Likelihood from incomplete data via the EM algorithm, Journal of the Royal Statistical Society B (Methodological) 39/1: 1–38.

Ester, M., H.P. Kriegel, J. Sander and X. Xu. 1996. A density-based algorithm for discovering clusters in large spatial databases with noise. pp. 226–231. *In*: E. Simoudis, J. Han and U.M. Fayyad, Hg. Proceedings of the Second International Conference on Knowledge Discovery and Data Mining (KDD-96) (Palo Alto, California: AAAI Press).

Hastie, T., R. Tibshirani and J. Friedman. 2001. The EM algorithm. pp. 236–243. *In*: T. Hastie, R. Tibshirani and J. Friedman (eds.). The Elements of Statistical Learning (New York: Springer).

Hodder, I. and C. Orton. 1976. Spatial analysis in archaeology (Cambridge University Press).

Jacquez, G.M. 1996. A k nearest neighbour test for space-time interaction, Statistics in Medicine 15: 1935–1949.

Koetje, T.A. 1987. Spatial Patterns in Magdalenian Open Air Sites from the Isle Valley, Southwestern France, BAR International Series 346 (Oxford: Archaeopress).

Kulldorff, M. and N. Nagarwalla. 1995. Spatial disease clusters: detection and inference, Statistics in Medicine 14: 799–810.

Lock and Molyneaux. 2006. Lock, G. and Molyneaux, B.L., Confronting Scale in Archaeology: Issues of Theory and Practice (New York: Springer).

O'Sullivan, D. and D. Unwin. 2003. Geographic Information Analysis, Hoboken, NJ, John Wiley & Sons, Inc.

Okabe, A., T. Satoh and K. Sugihara. 2009. A Kernel Density Estimation Method for Networks, its Computational Method, and a GIS-based Tool, International Journal of Geographical Information Science 23/1: 7–32.

Orton. 2004. Point pattern analysis revisited, Archeologia e Calcolatori XV; 299–315.

Ripley, B.D. 1976. The second-order analysis of stationary point processes. Journal of Applied Probability Theory 13: 255–266.

Ripley, B.D. 1981. Spatial Statistics (Hoboken, NJ: Wiley).

Scott, W.A. and S.W. Hillson. 1988. An application of the EM algorithm to archaeological data analysis. pp. 43–52. *In*: S.P.Q. Rahtz, Hg., CAA 88. Computer Applications and Quantitative Methods in Archaeology 1988 (Oxford: Archaeopress, 1988).

Stapert. 2003. Towards Dynamic Models of Stone Age Settlements. pp. 5–15. *In*: S.A. Vasil'ev, O. Soffer, J. Kozlowski (Hrsg.): Perceived Landscapes and Built Environments. The cultural geography of Late Paleolithic Eurasia (=BAR Int. Series 1122).

Stevenson. 1985. The Formation of Artifact Assemblages at Workshop/Habitation Sites: Models from Peace Point in Northern Alberta. American Antiquity 50/1: 63–81.

Wu, C.F.J. 1983. On the Convergence Properties of the EM Algorithm, Annals of Statistics 11, 1: 95–103.

19

Non-Euclidean Distances in Point Pattern Analysis: Anisotropic Measures for the Study of Settlement Networks in Heterogeneous Regions

Joan Negre Pérez

Introduction

In the statistical analysis of spatial point patterns, stationarity is often assumed to mean that the spatial point process has constant intensity and uniform correlation depending only on the lag vector between pairs of points (Møller and Toftaker 2012). In other words, it is assumed that the correlation between the elements of a spatial distribution is a function of the Euclidean distance between them. This framework has been vastly used in Spatial Analysis to describe settlement processes, taking into account a homogenous and undifferentiated surface that is easy to generalise. These assumptions fail when we consider the historical and economical dynamics that took place in space. This failure arises from the fact that the possibility of movement constraints have hardly been taken into account. Social action is supposed to have been performed in a simple and homogeneous Euclidean surface where only straight line distances are considered.

Based on this simplification, geographers have been able to build a series of overly abstract (and hardly empirically reliable) models that pretend to analyse economic behaviour. For example, five of the most important geographical models have been

Laboratori d'Arqueologia Quantitativa, Departament de Prehistòria, Facultat de Filosofia i Lletres, Universitat Autònoma de Barcelona, 08193 Bellaterra (Barcelona) Spain.

Centro Austral de Investigaciones Científicas, Consejo Nacional de Investigaciones Científicas y Técnicas, B. Houssay 200, 9410 Ushuaia (Tierra del Fuego) Argentina.
Email: negreperez@gmail.com

modelled assuming this premise. These are, in chronological order, Von Thunen's model of agricultural land use; Weber's model of industrial location; Christaller's central place formulation; the gravity model of spatial interaction and Hägerstrand's model of the geographical spread of innovation (Tobler 1993: 1). We can observe some examples of this trend in the works that analyse the influence of cities and territorial settlements in the emergence of spatial dependence (Fujita 1989, Krugman 1991). This urban phenomenon is one of the most revisited topics by historical sciences, adding new types of socio-spatial relationships, such as concentration, accessibility and spatial interaction (Barceló et al. 2002: 45–46). Further, new analytic variables relating mainly to communication networks, their costs and benefits are also being introduced in modern research scenarios (Labbé et al. 1995). In some instances, the development of these models has been closely linked to the needs of economists, who have contributed to bring up their limitations and integrate new interaction factors, as in the case of shipping costs brought up by Alfred Weber at the beginning of 20th century. Despite the efforts, some of these new models of the location of economical and historical processes still assume a homogeneous space with no interaction with the social sphere.

Our approach is based on the idea of Social Space, in the sense that relationships and transformations performed by social environment on the spatial container of social action should be considered in terms of an articulated social construct. It is this ability to influence social action and to be influenced by the social environment what generates the heterogenic and anisotropic characterisation of physical and social space. In other words, social interactions generate directional dependence in the spatial variability of places where social action is (was) performed. Thus, this issue is addressed through developing new methodological approaches in order to analyse historical settlement patterns assuming some kind of spatial heterogeneity, this is a topographical dependence, as a working hypothesis. Usually, the more local the fixed focus, the less spatial variability. This is fundamentally an approach that would be useful at macro-scale. We consider, therefore, that distances between two synchronous archaeological sites in a geographical environment are socially and spatially dependent and cannot be analysed using Euclidean metrics. That is, movement on this type of surfaces presents an anisotropic or directionally dependent behaviour. For example, it is assumed that is not as difficult to travel between these two points when they are in a flat and firm area as when the path must pass through a very hilly mountainous area. Under geometric anisotropy, the variance of any spatial process is the same in all directions, but the strength of the spatial autocorrelation is not (Schabenberger and Gotway 2005: 46). Consequently, the main goal of this chapter is to propose certain possible solutions to implement non-Euclidean measures in the study of a spatial structure such as a settlement pattern.

Non-Euclidean Distances in Spatial Statistics: An Archaeological Perspective

Spatial Statistics provides a set of tools for putting the analysis of coordinated data into practice. Those tools enable us to extract as many information as possible from a set of locations, providing lots of specific knowledge about the diachronic transformations of a society. We approach here the analysis of temporal changes in the structure of a spatial distribution of archaeological observables, what gives us an idea of social dynamics in the past. Going from archaeological data to historical explanation is possible on the basis that any historical process has a spatial dimension. In this way, space is a key point in the internal relations of a social formation, so that whatever happens, it happens in space and time (Wegener 2000: 3). One of the most obvious premises which can be deduced from this axiom is the commonly accepted First Law of Geography, according to which near events are more related than distant ones (Tobler 1970: 234). But, which are these events? Are they an accurate description of the spatial structure of a particular society?

Our starting point is the idea of *habitat,* considered to be a result of the interaction between the social and the spatial spheres of human behaviour. In this way, the simplified spatial structure of this occupation network can be quantified in terms of the point pattern distribution of the system nodes, that is, the settlements. Traditionally, Euclidean metrics have been used as the most common measure of distance between *habitats* for spatial statistics purposes. In this particular kind of geometry, the distance between two points p and q is calculated as the length of the line segment connecting them ($\overline{pq}$) when $p = (p_1, p_2, ..., p_n)$ and $q = (q_1, q_2, ..., q_n)$. For a Euclidean n-space R^n the applied formula is:

$$d(p, q) = \sqrt{(q_1 - p_1)^2 + (q_2 - p_2)^2 + \cdots + (q_n - p_n)^2} = \sqrt{\sum_{i=1}^{n}(q_i - p_i)^2}$$

Such a measure assumes the implicit validity to spatial homogeneity. On the opposite, we should focus our investigation on archaeological cases of point pattern analysis in which anisotropy is the rule. We propose to address this issue presenting a non-Euclidean adaptation of the well-known Ripley's K-function in order to implement a more reliable tool for detecting deviations from spatial homogeneity. An improvement is expected by using this alternative measure, because it is better suited to situations in which spatial processes do not follow the standard properties of a Euclidean space, as it happens usually in Archaeology.

Alternative measures to Euclidean distances have been largely tested in Archaeology to understand the optimal paths between different settlements. One of the most usual alternatives to straight metrics in our discipline are cost-weighted distance models, implemented in most GIS software packages. From a methodological perspective, the main goal of this kind of approaches is to define the least cost path to reach a known point from each cell location in the original raster dataset. The cost-weighted distance algorithms, therefore, calculate the length of the irregular vectors formed by a spatial distribution using the shortest weighted distance; this is the path with least accumulated travel cost. An inter-point cost-weighted distance matrix allows understanding the spatial correlation between the settlements in a more trustworthy way, represented as a point pattern distribution.

There is a large variety of algorithms to calculate cost-weighted surfaces. The basic purpose of these is to assign an impedance value to each cell of a raster layer, that is, the ease with which it can be crossed. The trouble is that such value is extremely dependent on the assumptions made by the researcher. From the simplest models, based on the percentage of slope, to the most complex that integrate a large number of interrelated variables, the final result is a grid in which each cell has a value that allow us to measure relative distances based on the costs of movement (Conolly and Lake 2006: 221–224).

Several modelling algorithms have been proposed to shape a well-fit cost surface over the last decades, and most of the discussion about them is focused in the range of the variables used as inputs (Marble 1996, Llobera 2000, Bell and Lock 2000, van Leusen 2002). However, Wheatley and Gillings noticed that in all of them, social relationships had not been taken into account (Wheatley and Gillings 2002: 155–156). Only physical factors were considered, while the presence of roads, the influence of settlement size, or other social and cultural indicators had been left out. On the other hand, I argue for more complex models in which five basic considerations should be highlighted:

1. The directional dependence or anisotropy of human movement across any surface.
2. The heterogeneity of geomorphologic features limiting the mobility. For instance, the relevance of the hydrological network.
3. The possibility of using different alternative means of transport and road networks.
4. In the case of a sedentary society, the uneven distribution of population across the landscape, with levels of demand and specific consumption propensities, that implies diverse kinds of gravitational models of settlement with differentiated levels of attraction and/or repulsion.
5. Economic activities are diversified and they may affect the capacity and possibilities for movement around.

Formally, the resulting cost-weighted model can be defined as an f function which describes for each cell of our model a series of real, positive values with respect to the difficulty to go through them, that is, its cost-weighted density. Therefore, the cost-weighted movement dx to the point x is $f(x)dx$. From this function the cost of any path in A can be calculated as the integration of every cell in the model which is gone through (Muñoz 2012: 55). Thus, the cost of an α path between points $s_1, s_2 \in A$ will be

$$\int_0^1 f(\alpha(t)) \; \alpha'(t)dt$$

Once the use of relative distances proves to be an appropriated choice to solve the least cost path between enumerated settlements, it must be proved that this approach can also be applied to geostatistical or spatiostatistical functions. In any case, anisotropically cost-weighted distances maintain the same general properties than their metric counterparts (Waller and Gotway 2004: 321):

- non-negativity $(d(x, y) \geq 0)$
- symmetry $(d(x, y) = d(y, x))$
- triangle inequality $(d(x, z) \leq d(x, y) + d(y, z))$

Sometimes, geographical publications assume the satisfaction of all axioms except of symmetry owing to the effects of directional dependence, considering the pattern of distances as a quasimetric. Nevertheless, most studies on the subject highlight the low variability of friction for a given cell whatever the directivity, often confused as direction, taken (Llobera 2000: Fig. 2; van Leusen 2002: VI, 7).

Despite the fact that the application of cost-weighted distances in this kind of functions involves a work of adaptation, the results might be, in some cases, statistically non relevant. The more homogeneous is the surface under study, the less significant are the changes with respect to the use of Euclidean measures. To obtain mathematical validity, not just argumentative, results obtained must be checked with the positive definition of the covariance matrix of the observations. This condition requires that for any n number, set of locations $s_1, \ldots, s_n$ and complex set of coefficients $\alpha_1, \ldots, \alpha_n$, the R function verify the next relationship:

$$\sum_{i,1=1}^{n} \overline{\alpha}_i\, \alpha_j R(d(s_i, s_j))$$

where d represents the cost-weighted distance between their arguments (Muñoz 2012: 201). Ultimately, and through verification of the above validation factors, a functional model can be described. Although it cannot always be used as a form of statistical proof, it can provide the best linear unbiased prediction.

The physical reality in which human activities took place in the past should not be considered in terms of an absolute geometric container, and for this reason, it is too simple to generalise spatial phenomena formalisation with Euclidean metrics (Kitchin 2009). The importance of particular space contexts or surroundings in the understanding of each case study must be taken into consideration, assessing the need of using relative distances instead of Euclidean ones. Intuitively, it is noted that the more heterogeneous the space, the more statistical variability of underlying spatial variables. The importance of a large range of factors on analysing spatial human patterns, such as physical, social or economic ones, oblige us to reconsider the methods that have been used traditionally. Even though the use of cost weighted distances is not a new argument (Smith 1989), the problems related with its implementation have prevented its diffusion among social disciplines. Thus, in this chapter I propose a methodological approach in order to implement a spatial homogeneity descriptor, based on an anisotropic modification of the standard Ripley's K-function.

Implementing Non-Euclidean Distances in Ripley's K-Function

Ripley's K-function, $K(h)$, is a widely used statistic to detect clustering or inhibition in point process data (Ripley 1976). It is commonly used as a test, where the null hypothesis is that the point process under consideration is a homogeneous Poisson process and the alternative is that the point process exhibits clustering or inhibitory behaviour (Veen and Schoenberg 2006: 293). Its introduction in social disciplines can be linked to the limitations of previous point pattern analysis methods regarding the assessment of local and multi-spectral phenomena, that is, second order processes. This function combines distance and frequency methods in order to identify not only dispersion rates but the measure of them in an increasingly range of scales. Its capability to explain spatial dynamics and overtake classical scale and confine problems is the major reason why it has become one of the most applied tools for studying settlement patterns (Macchi 2009: 237, Bevan and Conolly 2006). Formally, the function is described by Ripley as:

$$\hat{K}(h) = \frac{A(\sum k(x, y))}{N^2}$$

In this notation, A represents the area of the study region, $k(x, y)$ the number of elements included within a circle centred at x with a radius y, and N the total amount of elements in the settlement network under analysis (Ripley 1981: 158–159). The application process of the Ripley's K-function is based in a relterative measure of frequency in n buffers for a permanent radius increase. The value of the estimator is finally compared with the null hypothesis of Complete Spatial Randomness, also known as homogeneous Poisson process, determined by the formula $\hat{K}(h) = \pi \times r^2$. For visual analysis purposes, Ripley's K-function is usually transformed as a linearised representation of $\hat{K}(h)$:

$$\hat{L}(h) = \left(\frac{\hat{K}(h)}{\pi}\right)^{\frac{1}{2}}$$

The method has its own limitations. The main one is the assumption of a Euclidean simplified space. Another key limitation corresponds to goodness of fit when the null hypothesis is not a stationary Poisson model. In this case, research has tried to introduce several weighted versions of the K-function, such as those applied to evaluate competing models for the spatial background rate for California earthquakes (Veen and Schoenberg 2006: 294).

In the case of implementing cost-weighted distances in Ripley's K-function, a direct use of them in the general equation has proved to be methodologically unsuitable. Cost-weighted buffers and inter-distances matrix must be manually calculated for each entity in our spatial distribution, and it is necessary to repeat the operation for the diverse spatial distributions used as reference. This handmade work took, in previous experimental tests with a cost-weighted adaptation of Nearest Neighbour algorithm, several months to be implemented. In the specific case presented by Ripley's K-function calculations, the whole task seemed to be unachievable in the current research scenario.

Considering this, the approach we present here conceives the implementation problem inversely, modelling a new spatial distribution based in its pairwise cost-weighted distance matrix, and represented in a pseudo-Euclidean framework. In its basis, it is a transformation problem to be solved with ordination techniques, such as non-metric multidimensional scaling (MDS). Specifically, these methods iteratively seek a good-fit solution to the general task of embedding Euclidean coordinates to a set of similarity relations between the objects. Unlike other ordination methods, MDS makes few assumptions about the nature of the data, becoming well suited for a wide variety of data. Neither linear nor modal relationships are assumed for the data matrix; any distance measure between data points is allowed, unlike other methods which specify particular measures, such as covariance or correlation in principal components analysis, or the implied chi-square measure in detrended correspondence analysis (Holland 2008). Moreover, because MDS operates on dissimilarities, no statistical distribution assumptions are necessary.

There is, however, a condition to be fulfilled by the data in a MDS analysis: non-negativity, symmetry and triangle inequality assumptions of measured distances. Cost-weighted distances between archaeological settlements satisfy all these conditions, as has already been mentioned. Therefore, we have created a $n \times n$ symmetrical matrix of all pairwise cost-weighted distances. The multidimensionality of this matrix is reduced to a two-dimensional solution applying a Multidimensional Scaling standard algorithm (MDS), whose final results constitute a 2D point distribution, whose clustering or inhibition can be measured using the standard form of the Ripley's K-function.

In this way, the original spatial coordinates of archaeological settlements are transferred to a different 2D arrangement in such a way that original spatial differences are preserved in the new configuration, which is based on cost-weight assumptions. In this work, Paleontological Statistics Software (PAST) has been used as working environment. In it, the MDS algorithm may converge on a different—but equal— solution in each run which is actually a sequence of 11 trials, from which the one with smallest stress is chosen (Hammer 2013: 82). Calculations are based on the work of Taguchi and Oono (2005) for very large data sets as a maximally unsupervised data mining device in the context of bioinformatics.

For testing the final stress of the new cost-weighted 2D configuration, Euclidean distances between points are regressed against the original input matrix and the predicted ordination distances for each pair of samples is calculated. In a perfect ordination, all distances would fall exactly on the regression, that is, MDS 2D final configuration would match the rank-order of distances in the original cost-weighted distance matrix. The goodness of fit of the regression is measured based on the sum of squared differences between ordination-based distances and the distances predicted by the regression. This goodness of fit is used as stress test and can be calculated in several ways, with one of the most common being Kruskal's Stress:

$$Kruskal\ Stress = \sqrt{\frac{\Sigma_{h,i}\,(d_{hi} - \hat{d}_{hi})^2}{\Sigma_{h,i}\,d_{hi}^2}}$$

where d_{hi} is the ordinated distance between samples h and i, and $\hat{d}$ is the distance predicted from the regression (Holland 2008). Once the empirical spatial distribution is re-scaled, it is possible to run the Ripley's K-function on the sample and obtain the first results, from which we must consider the size of the flat surface of the analysis to be taken into account in further theoretical analysis in the same study area.

Complete Spatial Randomness Hypothesis in Heterogeneous Regions

Complete Spatial Randomness (CSR) describes a point process whereby point events occur within a given study area in a completely random fashion. In classical Ripley's K-function method, this kind of distribution is usually represented as a Homogenous Spatial Poisson Process which fits the null hypothesis ($H_0 : \hat{K}\,(h) \sim CSR$). As Ripley's K algorithm is a function of distance and density, the $\hat{K}\,(h)$ value for an isotropic (non-directionally dependent) analysis of the random point pattern might be solved as a function of the frequency in each framed radius through this formula:

$$\hat{K}(h) = \pi r^2$$

This stochastic distribution allows us to use its values as a benchmark, in which $\hat{K}(h) > \pi r^2$ implies that the average value is greater than 1 then the average neighbour density is greater than the average point density on the studied domain. Therefore, the points are aggregated. Inversely, $\hat{K}(h) < \pi r^2$ indicates the opposite situation, a spatial distribution where the points are dispersed.

Nevertheless, when spatial dependence, this is the influence of topography in the analysis of the point pattern analysis, is taken into consideration, the isotropic Spatial Poisson Process cannot be accepted as a reliable null hypothesis. This situation implies the necessity of designing a new CSR index based on the above-mentioned premises. We have calculated manually the index in terms of the $\hat{K}\,(h)$ values for several random distributions composed by the same number of entities as the in the data input. At least five random samples must be created and analysed following the process described in the previous pages in order to finally obtain a set of linearised distributions in a defined topographical area. The mean and variance of those random configurations is used for testing the null hypothesis in the modified Ripley's K-function, and for obtaining the confidence interval where "random" variation is expected. We understand by "random variation" the range of variability of the null hypothesis, which is defined in terms of spatial independence of the point pattern regarding the surface under study. That is, the cost-weighted distances do not affect the choice of the settlement place and it is not observed any kind of spatial correlation between entities and the irregularity of the surface (topography). As a regional-scale method, this probability interval will vary from case to case, so that it must be independently recalculated for each region under analysis. Simulations such as Montecarlo methods cannot be used in these problems, forcing to laborious calculations in order to obtain a few examples for comparing purposes.

A Case Study: Rural Settlements in a Riverside-Mountainous Region

The case study we presented here is located in the current Catalan regions of Montsià and BaixEbre (Tarragona), at the northeast of Iberian Peninsula. These territories coincide with the medieval boundaries fixed by the feudal administration in 1149 AD, just after the fall of the Muslim city of Tortosa and its territory. A central axis of this area is defined by the lower course of the Ebro River, delimiting a geomorphologic unity identified as a tectonic depression and oriented northwest-east. The width of the valley varies from 2 to 4 km, and the sedimentology indicates a silt composition of the fluvial terraces, formed throughout the Holocene. It is around these spaces where most of the agrarian production was located in medieval times owing to the important fertility of these soils that were capable of producing a large number of good quality crops.

Regarding the deltaic plain in the estuary of the river, its formation began about 6,000 years ago, once the eustatic rise became lower than the rate of sedimentation. Since this period, distinct deltaic lobes have succeeded each other, but the current deltaic system configuration of the Ebro is barely a thousand years old. Despite the fact that this

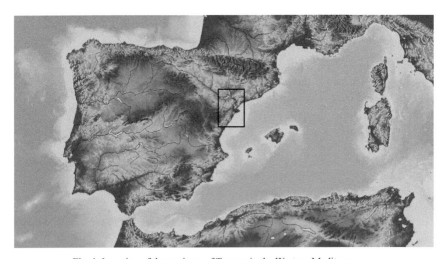

Fig. 1. Location of the territory of Tortosa in the Western Mediterranean.

plain was already formed in the 11th century, the truth is that no permanent cultivation or settlement process could be developed in this muddy area until the Late Middle Ages.

We focus our analysis on the study of the settlement pattern of this territory during the last century of the Islamic dominion of Tortosa (c. 1050–1150). The fall of this region at the close of 1148 supposed the end of the feudal expansion initiative in this area, which began as a part of the Christian Crusades project about fifty years before. This event signified the end of more than four centuries of Islamic presence in that region and the beginning of a complex process of change in the settlement structures. Chronologically, this period has been chosen because of the vast volume of written data generated by Christian feudal agents after the conquest regarding the previous organisation of this territory. In addition, long archaeological survey campaigns were carried out in this area, clarifying and fixing the different places mentioned by written sources. Thanks to all this data, it is possible to reconstruct with a high degree of confidence where the human habitat was and how this region was spatially, economically and socially organised.

Fig. 2. Location of rural settlements along the lower course of the Ebro River.

Methodologically, the first step to be taken is the calculation of the $\hat{K}(h)$ values referring to the empirical spatial distribution under analysis. In this case, the 30 settlements seem to be located along the river axis and they do not show apparently any particular pattern. We have built a cost-weighted inter-distance matrix, and rescaled this multidimensional configuration applying a non-metric Multidimensional Scaling to obtain a new 2D configuration of points on which Ripley's K-function can be performed. For interpretative purposes, the results will be linearised through $\hat{L}(h)$ transformation.

To define a null hypothesis of spatial independence and randomness, we have created 5 different configurations of 30 points randomly dispersed across this territory.

Their respective local cost-weighted distance matrices have been rescaled (MDS) and $\hat{L}(h)$ values have been calculated. The confidence interval of the null hypothesis spatial independence assumption has been defined using the maximum and minimum values for each iteration, using the graphic result as a probability rank of the null hypothesis.

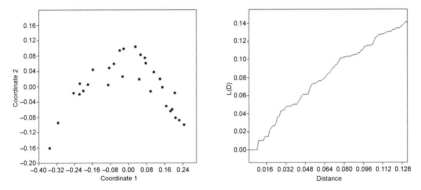

Fig. 3. Rescaled Spatial Distribution and Ripley's K-function results.

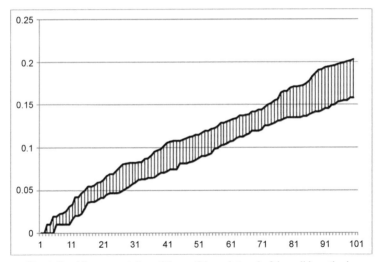

Fig. 4. Graphic representation of the confidence interval of the null hypothesis.

Finally, both empirical and theoretical results have been compared to infer the pattern shown by the settlement network under study. If the values of the empirical $\hat{L}(h)$ are located under the confidence interval, the null hypothesis is rejected, and a global dispersion mechanism is assumed as alternative hypothesis. On the other hand, if the values of the empirical $\hat{L}(h)$ are located over the interval of confidence, the null hypothesis is also rejected, and an aggregation social mechanism is inferred as alternative hypothesis. In this specific case, it can be observed a clear inhibition process tending to a dispersion pattern at all scales. This trend is clearer at higher scales. In any case, our results cannot be read as direct measures of spatial distance but scale ranks.

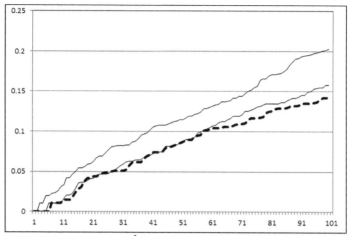

Fig. 5. Comparison between empirical $\hat{L}(h)$ and the null hypothesis interval of confidence.

To understand these results in a historical perspective, we should translate the notion of social aggregation and disaggregation regarding in Ripley's K-function terms. In spatial statistics, both aggregation and disaggregation patterns are different sides of the same coin, a statistical process of overdispersion. It implies the presence of greater variability in a dataset of points than would be expected based on a given simple statistical model, in this case, the above-stated Complete Spatial Randomness one. In other words, the study of spatial overdispersion in historical settlement networks is one of the most valuable indexes to analyse their spatial structure.

This issue has been largely addressed by most of the geographers and historians who tackled the importance of point pattern analysis from a social viewpoint. Although researchers have tried to describe settlement patterns as dispersed, random or clustered, it is hard to decide objectively which of these types of patterns prevailed in a given area (Kariel 1970: 124). There are still some researchers who brag about themselves and their capability to analyse a spatial pattern with a mere glance at a map. They are the same people who relegate any quantitative analysis to a more or less esoteric field. Ripley's K-function has become in the past decade one of the best tools that deal with these kinds of problems, even though some limitations has been remarked in this text. Thus, its function is to allow the researches to measure and describe precisely the degree of spatial overdispersion shown by a sample of settlements in a more reliable way.

Social structures tend to act over the settlement, transportation and productive networks in an extremely sound way. Consequently, the transformations shown by these can be formally analysed as a result of a series of historical and political decisions and they might also become a precise indicators of social change. A remarkable theoretical approach to the study of rural settlement was made by John C. Hudson at the end of the sixties. In this work, the author proposed a series of spatial processes, similar to those defined for plant ecology, which would explain changes in settlement distribution over time. Three phases were synthesised: colonisation, by which the occupied territory of a population expands; spread, through which settlement density increases and distance between sites decrease; and competition, defined in terms of the process having generated regularity in the settlement pattern (Hudson 1969: 365).

This general model has repeatedly proved to be useful in the analysis of rural spatial patterns, but its implementation depends on the confidence degree of the overdispersion analysis of the point pattern under study. Thus, the first phase of the model might be correlated with a randomness pattern of the spatial distribution, owing to the few entities already present in the exploitation processes of the territory. On the other hand, clustering or aggregation dynamics should be taken into account as spatial indicators of the spread phase. In this period, settlement multiplies and tends to concentrate themselves in a more restricted area, given rise in some cases to cooperation strategies. Finally, the latter phase proposed by Hudson requires a more inhibited spatial pattern in order to facilitate work dynamics and the exploitation of productive areas without competition between the settlements.

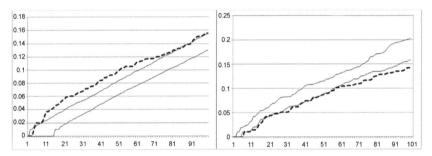

Fig. 6. Comparison between Euclidean-based Ripley's K-function and Rescaled Ripley's K-function.

In the proposed case study, cost-weighted transformed Ripley's K-function suggests a clear tendency towards disaggregation at all scales whereas the non-transformed function show high spatial aggregation levels. We consider that this latter explanation is erroneous because Euclidean distances between medieval settlements do not represent the time needed to arrive from one site to another. Consequently, the results shown by the traditional Ripley's algorithm mislead the reading of the historical process. The real distances between settlements were at that time too high to maintain the communal strategies suggested by the non-transformed function. On the contrary, our results present a clear stage of competition between settlements on both the left and right riversides. That strategy might be linked to the intervention of the City, as a headquarter of a regionalised State, in organising and managing the most important productive spaces. The construction of important hydraulic infrastructures along the riverbanks reinforces this hypothesis as well (Negre 2013: 380–392). From a geographical viewpoint, we should remark the importance of the river as a source of directivity and heterogeneity in spatial variation. Euclidean-based approaches gave equal value to distances between sites on the same riverbank than to sites placed in opposite sides of the river, and this is wrong from the real point of view. The river cannot be crossed in absence of an appropriate bridge or necessary equipment (boats). On the opposite, cost-weighted distances tend to present as nearest neighbours sites on the same side of the river.

This fact can be detected in the differences between the Euclidean configuration of 2D spatial points and the Cost-weighted MDS configuration. It occurs due to the influence of the river as a demarcation between both riverbanks, insulating some settlements from the apparently near ones.

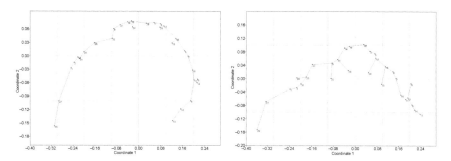

Fig. 7. Spanning-Tree Comparison between Euclidean-based and Cost-weighted.

Conclusions

Throughout this text we have tried to offer an alternative to spatiostatistical methods based on the traditional assumption of homogeneous or Euclidean spaces. A Ripley's K-function of the MDS transformation of a cost-weighted distance matrix is an appropriated solution, if we recalculate the null hypothesis based on the idea of cost-weighted distances among random series of points in the same geographical area as studied, with the same topography and geomorphological barriers. This technique provides a more exact insight of the spatial structure, taking into account topographical reality and the possibility of attractor points. Although it is true that the implementation of this approach is far from being automated, the results are very clear about the differences with the standard method based on the Euclidean assumption. The higher the irregularity of the studied area, that is the higher the topographical variability, the more needed are these new ways of calculating the degree of over dispersion within the configuration of points.

Thus, the geographical surface cannot still be measured as a plane and inert container for social processes. We stand for presenting simpler investigations in terms of content, but with much more accurate results. In that way, we consider that this new approach should be progressively integrated in GIS software packages, preferably open source.

Acknowledgments

The present work benefited from the input of Juan Antonio Barceló, Head of the Quantitative Archaeology Laboratory and Professor at the Autonomous University of Barcelona, and Giancarlo Macchi, Assistant Professor at the University of Siena, who provided valuable comments to the undertaking of the research presented here.

References Cited

Barceló, J.A., G. Pelfer and A. Mandolesi. 2002. The origins of the city. From social theory to archaeological description. Archeologia e calcolatori 13: 41–63.
Bell, T. and G. Lock. 2000. Topographic and cultural influences on walking the Ridgeway in later prehistoric times. pp. 85–100. *In*: G. Lock (ed.). Beyond the Map: Archaeology and Spatial Technologies. IOS Press, Amsterdam.

Bevan, A. and J. Conolly. 2006. Multi-scalar approaches to settlement pattern analysis. pp. 217–234. *In*: G. Lock and B. Molyneaux (eds.). Confronting Scale in Archaeology: Issues of Theory and Practice. Springer, New York.

Conolly, J. and M. Lake. 2006. Geographical Information Systems in Archaeology. Cambridge University Press, Cambridge.

Fujita, M. 1989. Urban economic theory. Land use and city size. Cambridge University Press, Cambridge.

Hammer, Ø. 2013. Paleontological Statistics Version 3.0. Reference Manual. University of Oslo, Oslo.

Holland, S. 2008. Non-metric Multidimensional Scaling (MDS). Technical Report. University of Georgia, Athens.

Hudson, J. 1969. A location theory for rural settlement. Annals of the Association of American Geographers 59: 365–381.

Kariel, H.G. 1970. Analysis of the Alberta settlement pattern for 1961 and 1966 by nearest neighbor analysis. Geografiska Annaler 52(2): 124–130.

Kitchin, R. 2009. Space II. pp. 268–275. *In*: R. Kitchin and N. Thrift (eds.). International Encyclopedia of Human Geography. Elsevier, Oxford.

Krugman, P. 1991. Increasing Returns and Economic Geography. J. Polit. Econ. 99/3: 483–499.

Labbé, M., D. Peeters and J.F. Thisse. 1995. Location on Networks. pp. 551–624. *In*: M. Ball, T. Magnanti, C. Monma and G. Nemhauser (eds.). Handbooks of Operations Research and Management Science: Networks. North-Holland, Amsterdam.

Llobera, M. 2000. Understanding movement: a pilot model towards the sociology of movement. pp. 65–84. *In*: G. Lock (ed.). Beyond the Map: Archaeology and Spatial Technologies. IOS Press, Amsterdam.

Macchi, G. 2009. Spazio e misura. Introduzioneaimetodigeografici-quantitativiapplicatiallo studio deifenomenisociali. Edizionidell'Università, Siena.

Marble, D.F. 1996. The human effort involved in movement over natural terrain: a working bibliography. Technical Report: Ohio State University, Department of Geography.

Møller, J. and H. Toftaker. 2012. Geometric anisotropic spatial point pattern analysis and Cox processes. Research Report Series, Department of Mathematical Sciences, Aalborg University.

Muñoz, F.M. 2012. Geoestadística en regiones heterogéneas con distancia basada en el coste. PhD Thesis, University of Valencia, Valencia.

Negre, J. 2013. De Dertosa a Turtusa. L'extrem oriental d'al-Tagr al-A'là en el context del procésd'islamitzaciód'al-Andalus. PhD Thesis, Universitat Autònoma de Barcelona, Bellaterra.

Ripley, B. 1976. The second-order analysis of stationary point processes. J. Appl. Probab. 13: 255–266.

Ripley, B. 1981. Spatial Statistics. Wiley, New York.

Schabenberger, O. and C. Gotway. 2005. Statistical methods for spatial data analysis. Chapman & Hall/CRC Press, Boca Raton.

Smith, T.E. 1989. Shortest-path distances, an axiomatic approach. Geographical Analysis 21(1): 1–31.

Taguchi, Y. and Y. Oono. 2005. Relational patterns of gene expression via non-metric multidimensional scaling analysis. Bioinformatics 21: 730–740.

Tobler, W. 1970. A computer movie simulating urban growth in the Detroit region. Economic Geography 46: 234–240.

Tobler, W. 1993. Three presentations on geographical analysis and modelling. Technical Report. pp. 93-1, University of California Santa Barbara, Santa Barbara.

vanLeusen, P.M. 2002. Pattern to process: methodological investigations into the formation and interpretation of spatial patterns in archaeological landscapes. PhD Thesis, University of Groningen, Groningen.

Veen, A. and P. Schoenberg. 2006. Assessing Spatial Point Process Models Using Weighted *K*-functions: Analysis of California Earthquakes. Lecture Notes in Statistics 185: 293–306.

Waller, L. and C. Gotway. 2004. Applied Spatial Statistics for Public Health Data. Wiley, Hoboken.

Wegener, M. 2000. Spatial models and GIS. pp. 3–20. *In*: S. Fotheringham and M. Wegener (eds.). Spatial Models and GIS: New and Potential Models. Taylor & Francis, London.

Wheatley, D. and M. Gillings. 2002. Spatial technology and archaeology. The archaeological applications of GIS. Taylor & Francis, London.

20

Lattice Theory to Discover Spatially Cohesive Sets of Artifacts

Michael Merrill

Introduction

Galois Lattice

The Galois lattice (Birkhoff 1967) is recommended for the analysis of two-mode social networks (Freeman and White 1994, Duquenne 1991). Mohr et al. provide a good summary of the two-mode property as it relates to the lattice: "A Galois lattice, however, has the special property of representing two orders of information in the same structure such that every point contains information on both logical orders simultaneously" (Mohr et al. 2004: 10). A Galois lattice can be viewed as the unfolding of the structure of multidimensional, two-mode binary data. In the lattice-based method of this paper, one mode is a set of n artifact types $A = \{a_1, a_2, \cdots a_n\}$ and the other mode is the set $C = \{c_1, c_2, \cdots c_m\}$ of m cliques, each of which comprises a subset of three or more of the artifact types in set A. A membership relation $I \leq AxC$ links these two sets (see Merrill and Read 2010). When an artifact type a_i belongs to a specific clique c_j, it follows that $(a_i, c_j) \in I$ and 0 otherwise. The membership relations between artifact types and cliques can therefore be represented by a nxm binary matrix M, where for any element m_{ij} in M, $m_{ij} = 1$ if $(a_i, c_j) \in I$ and 0 otherwise. The mathematics needed to unfold and graphically show the complete two mode structure of the clique-artifact type binary matrix as a Galois lattice are beyond the scope of this chapter. The mathematical details of the theory, representation, as well as examples of applications of Galois lattices in the analysis of two mode binary data are available elsewhere (Davey and Priestley 2002, Duquenne 1991, 1999:419–428, Wille 1982, 1984).

School of Human Evolution and Social Change, Arizona State University, Tempe, Arizona, USA.
  Email: mlmerril@asu.edu

Outline of a Lattice-Based Spatial Analytic Method

Figure 1 outlines the steps in a lattice-based method for intrasite spatial analysis. It is important to recognize that in this method a spatial similarity measure based on average distances is computed but the spatial similarity measures used in the analysis are probabilities. For each pair of types, say Type A (represented by m 2D points) and Type B (represented by n 2D points), a measure (see Mielke and Berry 2001) of their spatial similarity is computed. The spatial measure is converted to a probability for Type A by comparing the observed spatial measure of Type A to that for all possible permutations ($\frac{(m+n)!}{m!n!}$) of the $m + n$ points, where m of the points are Type A and n are Type B in each permutation. Step 1 involves computing a matrix of p-values that provides a spatial similarity measure for each pair of artifact types. Details for the method of computing the matrix of p-values (called delta P-values in Merrill and Read 2010) for the pairwise spatial similarity for the distribution of artifact types are presented elsewhere (see Step 1 in Merrill and Read 2010: 423–425). The probabilities relate to the distribution of the measure of spatial similarity to the distribution of these values, hence when two pairs of artifact types have the same measure of spatial similarity, each of the spatial similarity values may come from different distributions and so the probabilities for these spatial measures will be different.

The matrix of delta P-values is then used to construct a graph and identify spatially cohesive sets of artifact types or cliques in the graph. A cut-off value of 0.05 is used in keeping with the fact that the delta P-values are probabilities and typical statistically based cutoff values for significance levels are 0.05 or 0.01. It should also be noted that, like cutoff values for statistical significance, a change in the cutoff value would have an effect on the adjacency matrix (which is used to construct the graph in Step 2, Fig. 1). The goal of this method is to look for a clear pattern, though, not one dependent on whether the cutoff value is 0.05 or 0.01.

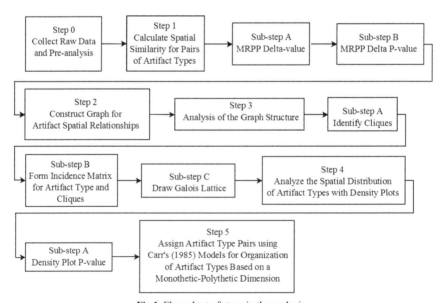

Fig 1. Flow chart of steps in the analysis.

In this method, a clique (Fig. 1; Step 3A) is a set S of artifact types such that each type T in S has a spatial distribution similar to the spatial distribution of any other type T* in S. Intuitively, a clique, is a set of *spatially cohesive artifact types*. Cliques may overlap; that is, one artifact type can be in more than one clique. The lattice structure is determined by the cliques (Fig. 1; Step 3B). Density plots (Fig. 1; Step 4) are computed as bivariate kernel density estimates (or KDEs). In the application of the lattice method later in the chapter, I use the sum of asymptotic mean square error (SAMSE) pilot plug-in bandwidth selection method to compute bivariate density estimates (Duong 2007, Duong and Hazelton 2003). The KDEs are graphically displayed as contour plots with the upper 25, 50 and 75 percent contours. These contours are the boundaries of the highest density regions of a sample (Bowman and Foster1993, Hyndman 1996).

In this chapter I add a test statistic (Duong et al. 2012, Schauer et al. 2010) to Sub-step A of Step 4 (Merrill and Read 2010; Fig. 1), which (given estimates of two steady state probability density functions, f_A and f_B, determined by the 2D spatial coordinates of the artifacts in Lattice Chains A and B) is used to test the null hypothesis $H_0 : f_A = f_B$. If the null hypothesis is rejected then the spatial distribution of the intrasite areas of concentration (as determined by the density plots in Step 4) of Lattice Chains A and B are significantly different. More details of Step 5 and the remaining steps of the lattice-based analysis are provided in the original description of the method (Merrill and Read 2010).

Application of the Lattice-Based Spatial Analytic Method to a Large Data Set

CA-LAn-803 is a typical Early period Phase X (King 1990: 28–31) residential site on the coastline of Malibu, California (Figs. 2 and 3), that was occupied some time during 6000–4500 B.C. It is located on a slightly sloped mesa adjacent to the Pacific

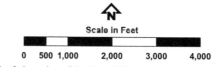

Fig. 2. Location of CA-LAn-803.

Fig. 3. 1949 aerial photograph of the Puerco Mesa area in Malibu, California. Large ellipse indicates approximate location of LAn-19 and smaller ellipse the location of LAn-803. LAn-19 was probably first occupied after 4000 B.C. and LAn-803 was probably used somewhat earlier (ca. 5500–4000 B.C.). Photograph provided by Dr. Chester King.

Color image of this figure appears in the color plate section at the end of the book.

Coast Highway (Fig. 3), which overlooks the Pacific Ocean. Previous observations of the intrasite patterning of artifacts in Early period (and later) sites similar to LAn-803 (Gamble 1983, Holmes 1902) provide a baseline of expectations for the intrasite spatial patterning of some kinds of artifacts in these sites. For example, Gamble's 1983 analysis showed that groundstone, heavy hammers, and choppers were most often found in houses, in the case of the Pitas Point Site (CA-VEN-27) near Ventura, California, which was occupied long after the Early period, but has many of the same kinds of artifacts found in Early period sites. CA-LAn-1107, an Early period site in Malibu, California provides a spatial arrangement of surface artifacts for comparison with LAn-803. Many flakes in this site, possibly associated with male outdoor activities such as butchery (Gamble 1983), were present in small and discrete areas, downslope from the area of the site where shell and groundstone were found (King 1990). A very similar spatial patterning of surface artifacts exists in other Early period Malibu sites, such as LAn-451 and LAn-803 (Fig. 4). These observations provide the basis for what I call the gendered dichotomy of house and outdoor activity areas in Early period Chumash residential sites.

The Gendered Dichotomy of Household and Outdoor Activity Areas in Chumash Early period Residential Sites

Certain kinds of artifacts, such as manos, metates, and heavy hammerstones typically concentrate in household areas in Early period sites within the Chumash cultural region of southern California (King and Merrill 2002). Food processing, cooking, and

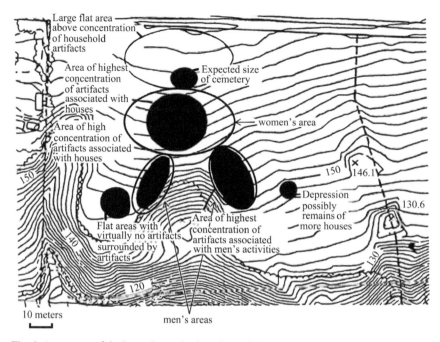

Fig. 4. Assessment of the internal organization of LAn-803 (King and Merrill 2002). LAn-803 is a medium sized Early Period Phase x (ca. 5500–4000 B.C.) settlement in Malibu, California. There is little available information concerning the organization of settlements from this time period in southern California.The different site areas shown on this map are based on field observations of over 700 artifacts on the surface of the site.

groundstone manufacture are activities strongly associated with household areas in Chumash sites from this time period. Ethnohistorically, the household and domestic activities are associated with Chumash women (Gamble 1983: 124).

Based on field observations made by Chester King and other archaeologists (see previous section), specific types of artifacts such as chert flakes, burins, gravers, and domed scrapers consistently concentrate in what may be outdoor areas that are spatially separate from house locations in Early period Chumash residential sites. Butchery, wood, bone, and hide working are some of the activities associated with these kinds of artifacts (Keeley 1978). Ethnohistorically, the areas outside the household (and activities conducted there) are associated with Chumash men (Gamble 1983).

In 1997, Dr. Chester King and I digitized the positions of over 700 surface artifacts in LAn-803 from a map we had made, just prior to the destruction of the site by development. We also constructed a typology for these artifacts using digitized images I made from a video we had taken of the artifacts during the mapping process. This provided the raw data for an earlier spatial analysis (Merrill 2002) that contributed to the development of the lattice-based methodology used in this study. In this chapter, I use a simplified version of our original LAn-803 artifact typology. Also, only 460 out of slightly over 700 finds could be assigned to a type class and the remaining artifacts that were not typed because they were chunks of lithic material without obvious signs of use wear, retouch, flake removals, or other forms of human modification. Based on the consistency and lack of diagnostic characteristics of the unassigned types, the potential loss of information from excluding these types is probably not significant. Future analysis of the digitized artifact images (using methods provided by Read 2007) may significantly refine the typology, and (in particular) the typology of the chert flakes (which are the most abundant type in the sample). The refined and potentially more complete sample should then be re-analyzed and the results compared with those from the analysis in this chapter. Figure 5 is a scatterplot of the two-dimensional spatial distribution of the artifact types I use in this analysis.

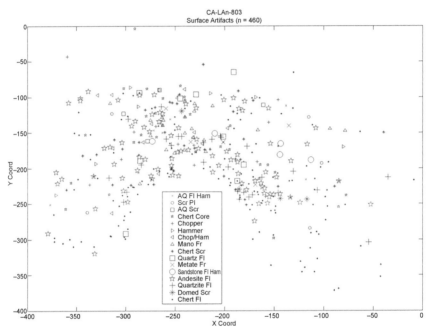

Fig. 5. Plot of the typed LAn-803 surface artifacts used in this analysis. Distance in both axes is in feet (0 to 400 feet for both axes).

Lattice-Based Analysis of the LAn-803 Surface Artifacts

Step 1 of Fig. 1 produces Table 1; Step 2 produces a graph, and Step 3 results in Table 2 (which is needed, since mathematicians know about adjacency matrices, but not archaeologists), Tables 3 and 4, and Fig. 6. Details on the construction of a Galois lattice from Table 4 are provided elsewhere (Merrill and Read 2010). For example, consider the only chain in the LAn-803 lattice, which contains domed scrapers and chert flakes. In this chain (as in all other chains in the lattice), artifact types which belong to the largest number of cliques (Table 5, Fig. 6) correspond to the node at the bottom of the chain and artifacts associated with successive nodes moving up the chain belong to increasingly fewer cliques.

Table 5 consists of the distinct and complete chains of artifact types in Fig. 6. It is clear that there are two less artifact type lattice chains in Table 5 than cliques in Table 3 and that the observed distributions of the number of artifact types (which are

Table 1. MRPP P-values for the sixteen LAn-803 artifact types. See Step 1 (Fig. 1); also see Merrill and Read 2010 for more information on MRPP P-values (probabilities, not averages of the spatial measures).

Artifact Type	1	2	3	4	5	6	7	8	9	10	11	12	13	14	15	16
1	0	0.821	0.81	0.966	0.073	0.923	0.901	0.008	0.458	0.86	0.345	0.275	0.151	0.027	6.30E-04	0.883
2	0.821	0	0.681	0.648	0.005	0.48	0.411	0.003	0.062	0.263	0.124	0.023	0.022	9.80E-04	4.50E-05	0.433
3	0.81	0.681	0	0.573	0.043	0.774	0.616	0.002	0.104	0.463	0.093	0.074	0.039	3.30E-03	4.70E-04	0.623
4	0.966	0.648	0.573	0	0.359	0.947	0.872	0.059	0.873	0.884	0.76	0.875	0.739	0.232	0.076	0.741
5	0.073	4.80E-03	4.30E-02	3.60E-01	0	0.47	0.184	0.217	0.1	0.096	0.017	0.484	0.089	0.21	0.07	0.134
6	0.923	0.48	0.774	0.947	0.47	0	0.927	0.135	0.907	0.956	0.1074	0.502	0.828	0.712	0.758	0.57
7	0.901	0.411	0.616	0.872	0.184	0.927	0	0.03	0.558	0.646	0.825	0.548	0.234	0.068	6.80E-03	0.908
8	8.00E-03	3.00E-03	2.00E-03	0.059	0.217	0.135	0.03	0	0.036	0.338	1.80E-03	0.211	0.056	0.284	0.216	0.059
9	0.458	0.062	0.104	0.873	0.1	0.907	0.558	0.036	0	0.669	0.044	0.214	0.769	0.109	1.70E-04	0.792
10	0.86	0.263	0.463	0.884	0.096	0.956	0.646	0.038	0.669	0	0.086	0.262	0.302	0.033	2.10E-04	0.815
11	0.345	0.124	0.093	0.76	0.017	0.107	0.825	0.002	0.044	0.086	0	0.237	3.60E-03	1.00E-03	2.40E-07	0.45
12	0.275	0.023	0.074	0.875	0.484	0.502	0.548	0.211	0.214	0.252	0.237	0	0.241	0.405	0.056	0.489
13	0.151	0.022	0.039	0.739	0.089	0.828	0.234	0.056	0.769	0.302	3.60E-03	0.241	0	0.124	2.00E-05	0.389
14	0.027	9.80E-04	3.30E-03	0.232	0.21	0.712	0.068	0.284	0.109	0.033	0.001	0.405	0.124	0	0.085	0.119
15	6.30E-04	4.50E-05	4.70E-04	0.076	0.07	0.758	6.80E-03	0.216	1.70E-04	2.10E-04	2.40E-07	0.056	2.00E-05	0.085	0	0.027
16	0.883	0.433	0.623	0.741	0.134	0.57	0.908	0.059	0.792	0.815	0.45	0.489	0.389	0.119	0.027	0

Table 2. Adjacency matrix constructed from the matrix in Table 1, using a cut-off of 0.05. This is used to construct a graph of the pairwise spatial relationships of artifact types and to detect the cliques in the graph. See Steps 2 and 3 (Fig. 1). Also see Merrill and Read 2010 for more detail.

Artifact Types	No.	1	2	3	4	5	6	7	8	9	10	11	12	13	14	15	16
Chopper	1	0	1	1	1	1	1	1	0	1	1	1	1	1	0	0	1
Chopper/Hammer	2	1	0	1	1	0	1	1	0	1	1	1	0	0	0	0	1
Hammer	3	1	1	0	1	0	1	1	0	1	1	1	1	0	0	0	1
AQ Flaking Hammer	4	1	1	1	0	1	1	1	1	1	1	1	1	1	1	1	1
Sandstone Flaking Hammer	5	1	0	0	1	0	1	1	1	1	1	0	1	1	1	1	1
Scraper Plane	6	1	1	1	1	1	0	1	1	1	1	1	1	1	1	1	1
AQ Scraper	7	1	1	1	1	1	1	0	0	1	1	1	1	1	1	0	1
Domed Scraper	8	0	0	0	1	1	1	0	0	0	0	0	1	1	1	1	1
Chert Scraper	9	1	1	1	1	1	1	1	0	0	1	0	1	1	1	0	1
Chert Core	10	1	1	1	1	1	1	1	0	1	0	1	1	1	0	0	1
Mano Fragment	11	1	1	1	1	0	1	1	0	0	1	0	1	0	0	0	1
Metate Fragment	12	1	0	1	1	1	1	1	1	1	1	1	0	1	1	1	1
Andesite Flake	13	1	0	0	1	1	1	1	1	1	1	0	1	0	1	0	1
Quartzite Flake	14	0	0	0	1	1	1	1	1	1	0	0	1	1	0	1	1
Chert Flake	15	0	0	0	1	1	1	0	1	0	0	0	1	0	1	0	0
Quartz Flake	16	1	1	1	1	1	1	1	1	1	1	1	1	1	1	0	0

represented as vectors x and y in R script) in the clique [x<-c(7,8,9,9,9,9,9,10)] and lattice chain [y<-c(4,5,5,7,7,8)] sets may have very different distributions (Tables 3 and 5). A two-sample Kolgomorov-Smirnov (or K-S) test [using the R function ks.test(x,y) and a function written in R code to compute the permutation p-value of this statistic (Rizzo 2008)] was used to test the null hypothesis that x and y are from the same distribution. Both the approximate (because of ties) p-value ($p = 0.04226$) of the standard two-sample K-S test and a simulation result for the permutation p-value (mean p-value = 0.019538, min p-value = 0.008, and max p-value = 0.035 for 1,000 computations of the permutation p-value) reject the null hypothesis ($\propto = 0.05$). This supports the idea that the order of the lattice structure acts as an additional noise filter (Merrill and Read 2010).

Kernel Density Estimate Plots and P-values

The bivariate kernel density estimate (KDE) plots used here provide an objective way to visualize the distribution of sets of spatially cohesive artifact types for a range of different densities, as well as an unbiased means for estimating the boundaries of their distribution within a site. In the bivariate KDE plots, the area(s) of highest (peak) artifact density are readily identified by 75 percent contours (Duong 2007).

Table 3. Cliques detected in the graph using a mathematical algorithm. See Merrill and Read 2010 for more detail. Cliques are spatially cohesive sets of artifacts.

Clique 1	Clique 2	Clique 3	Clique 4	Clique 5	Clique 6	Clique 7	Clique 8
Chopper	AQ Flaking Hammer	Chopper	Chopper	AQ Flaking Hammer	AQ Flaking Hammer	Chopper	Chopper
AQ Flaking Hammer	Sandstone Flaking Hammer	Hammer	Hammer	Sandstone Flaking Hammer	Sandstone Flaking Hammer	Chopper/Hammer	Chopper/Hammer
Sandstone Flaking Hammer	Scraper Plane	AQ Flaking Hammer	AQ Flaking Hammer	Scraper Plane	Scraper Plane	Hammer	Hammer
Scraper Plane	AQ Scraper	Scraper Plane	Scraper Plane	Domed Scraper	Domed Scraper	AQ Flaking Hammer	AQ Flaking Hammer
AQ Scraper	Chert Scraper	AQ Scraper	AQ Scraper	Metate Fragment	Metate Fragment	Scraper Plane	Scraper Plane
Chert Scraper	Metate Fragment	Chert Scraper	Chert Core	Andesite Flake	Quartzite Flake	AQ Scraper	AQ Scraper
Chert Core	Andesite Flake	Chert Core	Mano Fragment	Quartzite Flake	Chert Flake	Chert Scraper	Chert Core
Metate Fragment	Quartzite Flake	Metate Fragment	Metate Fragment	Quartz Flake		Chert Core	Mano Fragment
Andesite Flake	Quartz Flake	Quartz Flake	Quartz Flake			Quartz Flake	Quartz Flake
Quartz Flake							

Table 4. 16 x 8 Presence/Absence Matrix for the Galois lattice shown in Fig. 6. The rows of the table are LAn-803 artifact types and the rows are the cliques or (in this case) spatially cohesive sets of artifact types in Table 3.

Artifact Type		Clique 1	Clique 2	Clique 3	Clique 4	Clique 5	Clique 6	Clique 7	Clique 8
Chopper	1	1	0	1	1	0	0	1	1
Chopper/Hammer	2	0	0	0	0	0	0	1	1
Hammer	3	0	0	1	1	0	0	1	1
AQ Flaking Hammer	4	1	1	1	1	1	1	1	1
Sandstone Flaking Hammer	5	1	1	0	0	1	1	0	0
Scraper Plane	6	1	1	1	1	1	1	1	1
AQ Scraper	7	1	1	1	1	0	0	1	1
Domed Scraper	8	0	0	0	0	1	1	0	0
Chert Scraper	9	1	1	1	0	0	0	1	0
Chert Core	10	1	0	1	1	0	0	1	1
Mano Fragment	11	0	0	0	1	0	0	0	1
Metate Fragment	12	1	1	1	1	1	1	0	0
Andesite Flake	13	1	1	0	0	1	0	0	0
Quartzite Flake	14	0	1	0	0	1	1	0	0
Chert Flake	15	0	0	0	0	0	1	0	0
Quartz Flake	16	1	1	1	1	1	0	1	1

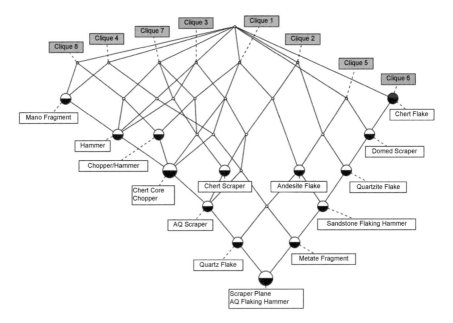

Fig. 6. Minimal representation of the lattice generated from the presence/absence matrix in Table 4. This simplified representation helps to visually identify the six distinct and complete artifact type chains in the lattice (Table 5).

Color image of this figure appears in the color plate section at the end of the book.

Table 5. All of the distinct and complete lattice chains of artifact types in Fig. 6.

Lattice Chain #1	Lattice Chain #2	Lattice Chain #3	Lattice Chain #4	Lattice Chain #5	Lattice Chain #6
AQ Fl. Ham.	AQ Fl. Ham.	AQ Fl. Ham.	AQ Fl. Ham.	AQ Fl. Ham.	AQ Fl. Hammer
Scraper Plane	Scraper Plane	Scraper Plane	Scraper Plane	Scraper Plane	Scraper Plane
Quartz Fl.	Quartz Fl.	Quartz Fl.	Quartz Fl.	Metate Fr.	Metate Fr.
AQ Scraper	AQ Scraper	AQ Scraper	Andesite Fl.	Sandstone Fl. Hammer	Sandstone Fl. Hammer
Chert Core	Chert Core	Chert Scraper		Andesite Fl.	Quartzite Flake
Chopper	Chopper				Domed Scraper
Chopper/ Ham.	Hammer				Chert Flake
	Mano Fr.				

The lattice-based method is also capable of identifying overlapping structure in site organization, which previous analytical methods applied to intrasite spatial analysis cannot (Merrill and Read 2010). It is clear that all of the KDE plots (and therefore the artifact sets corresponding to each of the lattice chains) have areas of highest concentration (75 percent contour) in the "house" area of the site (Figs. 7 and 8). In contrast, only the artifact sets corresponding to Lattice Chains # 4, 5 and 6 have areas of highest concentration (75 percent contour) in the right "outdoor" activity area in Fig. 4. Or, equivalently, the artifact sets corresponding to Lattice chains # 1, 2, and 3 only concentrate (75 percent contour) in the "house" area, and the artifact sets corresponding to Lattice Chains # 4, 5 and 6 have significant concentrations in both the "house" and "outdoor" activity areas, with a much more spatially extensive concentration of the artifact set of Lattice Chain # 6 in the right outdoor activity area (Figs. 4, 7 and 8). Also, the artifact sets of Lattice Chains # 4 and 5 have a similar representation (in terms of the area enclosed by the 75 percent contour) in both the "house" and right "outdoor" activity areas (Fig. 8, top). The major difference between Lattice Chains # 4, 5 and 6 (which are on the right side of the lattice, see Fig. 6) is that Lattice Chains # 4 and 5 are the only sets with andesite flakes (which concentrate in the house area), and Lattice Chain # 6 is the only set with quartzite flakes, domed scrapers, and chert flakes, which are most dense in the outdoor activity sub-areas of the site.

This suggests that andesite flakes possibly were preferred for household related activities (such as food preparation and the cutting of fibers for basketry manufacture) and/or were curated there. Domed scrapers, and quartzite and chert flakes were possibly selected for outdoor activities that may have included butchery and woodworking. Empirical studies comparing the use efficiency and wear patterns of andesite, quartzite, and chert flake tools in specific tasks, along with ethnoarchaeological research may help to explain the observed differences in the spatial distribution of flakes made from different materials in LAn-803.

The KDE test p-values (Table 6) show that the null hypothesis ($H_0 : f_A = f_B$, where f_A and f_B are respective estimates of the steady state probability density function of the spatial distribution of Lattice Chains A and B in LAn-803) is rejected for Lattice Chain # 6 compared pairwise to Lattice Chains # 1 and 2 ($\alpha = 0.05$). I interpret these results as strong evidence that Lattice Chains # 1 and 2 are "tool kits" used mostly in

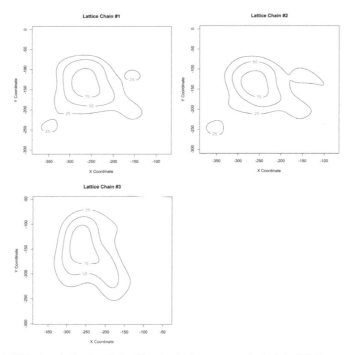

Fig. 7. KDE plots for Lattice Chains #1 to 3, which concentrate in the LAn-803 "house area".

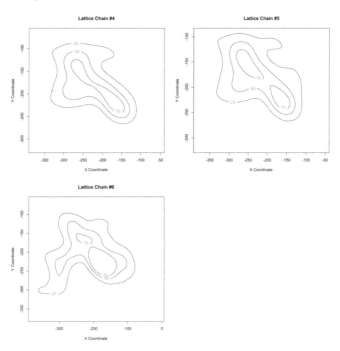

Fig. 8. KDE plots for Lattice Chains # 4, 5, and 6, which concentrate in the LAn-803 "house area" and largest "outdoor activity area".

Table 6. KDE test p-values. LC = Lattice Chain.

	LC 1	LC 2	LC 3	LC 4	LC5
LC 2	0.8297304				
LC 3	0.6714231	0.5017339			
LC 4	0.1182128	0.08514293	0.9418796		
LC 5	0.09585501	0.08503078	0.7998196	1	
LC6	1.560951e-05	3.020392e-06	0.07736818	0.1498393	0.1481896

household activities and Lattice Chain # 6 is a "tool kit" most often used in outdoor types of activities. Interestingly, the null hypothesis is not rejected for the pairwise comparison of Lattice Chains # 3, 4, and 5 with any of the other Lattice Chains (Table 6). This may suggest that Lattice Chains # 3, 4, and 5 are general purpose "tool kits" used equitably in both house and outdoor activities, which require scraping tools and a lightweight hammerstone.

Carr Models of Artifact Type Organization

Carr defines a depositional set in a manner that captures the effect all formation and disturbance processes have on the spatial distribution of artifacts: "A depositional set may be thought of as a mathematical set, the organization of which is the end product of structural transformations (archaeological formation and disturbance processes) operating upon a previously structured set or (in terms of the spatial distribution of some sets of artifacts) activity sets organized by human behavior" (Carr 1989: 117).

Carr develops models of artifact organization in relation to site formation processes. Different models relate to different formation processes and different use patterns of spatially associated artifact pairs. For example, Carr Model 3 is associated with simpler formation processes than Model 6, which suggests that there is less variability in the ways pairs of spatially cohesive artifact types (whose spatial structure is consistent with Model 3) are used in relation to one another within a site. Artifact pairs whose spatial arrangement corresponds to Model 6 were probably used in a greater variety of ways. I provide two examples (one associated with the house area and the other with outdoor activities) from LAn-803 to illustrate the identification and use of the Carr models for relating spatial patterning of artifact types to past human behavior.

I use the following procedure to assign pairs of artifact types in lattice chains to Carr models. First, KDE plots (75 percent contours) are used to estimate the range of x-y coordinates of artifact types in sub-areas of the site most strongly associated with specific lattice chains. Groups in the sub-area(s) are identified with a cluster analysis of the x-y coordinates and plotted. Then the spatial pattern of each artifact type pair within the discrete sub-area(s) of the site associated with the lattice chain can be assigned to one of Carr's models of organization of artifact types along a monothetic-polythetic continuum, by identifying the following characteristics of discrete groups consisting of three or more artifact types within the sub-area (Carr 1989:336, Merrill and Read 2010:430). This is accomplished by answering the following series of questions. (1) Do the sets of groups in the sub-area always have both types of artifacts (globally monothetic) or does at least one group have only one artifact type (globally polythetic)? (2) If one artifact type occurs in each of the groups, does the other type also occur, and vice versa (symmetrical co-arrangement)? (3) If not, what is the relative density of one artifact type compared to the other in each of the groups (magnitude of asymmetry)? (4) Does this vary between groups? (5) Does the same artifact type always have the highest frequency in each of the groups (the same direction of asymmetry in each of the groups), or does this vary? (6) Does one artifact type in each of the groups always have the other artifact type for a nearest neighbor and vice versa (locally monothetic set of artifact pairs)? If not, the arrangement of artifact pairs in the sub-area is locally polythetic.

In the lattice-based intrasite spatial analysis of LAn-803 surface artifacts, chert cores and choppers only occur in Lattice Chains #1 and 2 (Table 5), and are strongly linked spatially to the "house area" (see Fig. 4). A hierarchical cluster analysis and two-dimensional spatial plot of the chert cores and choppers from the house area (Fig. 9) suggests three clusters (groups). Since one of the groups (Group 3) contains only choppers, this corresponds to the condition unique to Carr Models 5 and 6, specifically: "Asymmetry within groups taken to the extreme, where one type does not occur in some groups" (Carr 1989:341). Since Group 3 consists of only choppers, and Groups 1 and 2 are dominated by cores (Fig. 9, right), this shows there are differences between the groups in the directions of their asymmetries (Carr 1989: 341), which corresponds to Carr Model 6 (which is characterized by variable organization among spatially associated artifact types, Carr 1989: 352). Also, Group 3 (Fig. 9, right) contains only choppers. This may suggest that in Early period Chumash residential sites, choppers were sometimes used in household activities that were spatially segregated from where chert cores were kept or used. Interestingly, in the largest area of the site associated with male outdoor activities (Fig. 4), and where Lattice Chain # 6 has a high (75 percent contour) concentration (Fig. 8), there is only one AQ flaking hammer. This may suggest that flake production did not often occur in this area, and possibly more often in house areas. Since chert cores and AQ flaking hammers are both spatially associated in the house area (see Lattice Chains #1 and 2), chert flakes (in particular) may have been produced primarily in the LAn-803 house area, and (in contrast) most often used outdoors away from houses (as suggested by the strong spatial correspondence of Lattice Chain # 6 and the largest "men's activity area", see Figs. 4 and 8).

Chert and quartzite flakes, only occur in Lattice Chain #6 (Table 5), which (as discussed earlier) concentrates in the largest area of the site associatedwith outdoor (and male) activities (Fig. 4). In this case, three groups are identified by a cluster analysis of the spatial locations of chert and quartzite flakes in this area (Fig. 10). All three clusters have both chert and quartzite flakes (Fig. 10, right). The magnitude of asymmetry varies between the groups: 4:1 (Groups 1 and 3) and 19:6 (Group 2) see Fig. 10, right. The direction of asymmetry does not vary between groups, since chert flakes predominate in all three groups (Fig. 10, right). Based on these and other criteria (see Carr 1989:341, Table 13.2), the spatial patterning of chert and quartzite flakes in this case corresponds to Carr Model 3, which is also both locally and globally monothetic.

The organizational assumptions of Model 3 suggest that chert and quartzite flakes may have often been used together in outdoor activities (which may include butchery, woodworking, and other tasks requiring slicing, cutting, and scraping). Additional research is needed to understand more precisely what kinds of activities chert and quartzite flakes were applied in Early period Chumash residential sites.

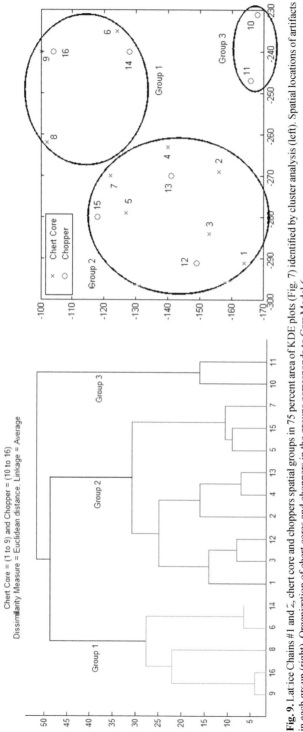

Fig. 9. Lattice Chains #1 and 2, chert core and choppers spatial groups in 75 percent area of KDE plots (Fig. 7) identified by cluster analysis (left). Spatial locations of artifacts in each group (right). Organization of chert cores and choppers in the groups corresponds to Carr Model 6.

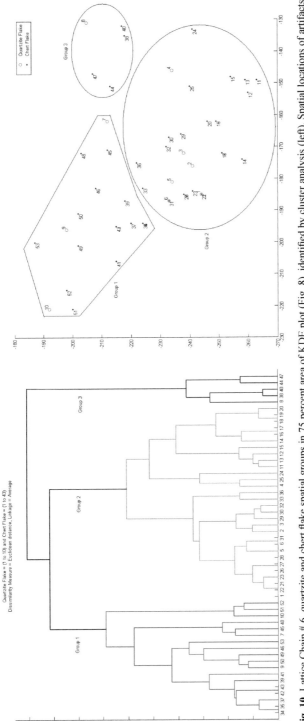

Fig. 10. Lattice Chain # 6, quartzite and chert flake spatial groups in 75 percent area of KDE plot (Fig. 8), identified by cluster analysis (left). Spatial locations of artifacts in each group (right). Organization of quartzite and chert flakes in the groups corresponds to Carr Model 3.

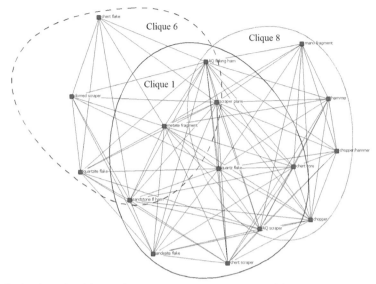

Fig. 11. Graph produced from adjacency matrix (Table 2) of the spatial relationships of the sixteen LAn-803 artifact types. There are eight cliques in the graph (Table 3), of which three are outlined with beziergons.

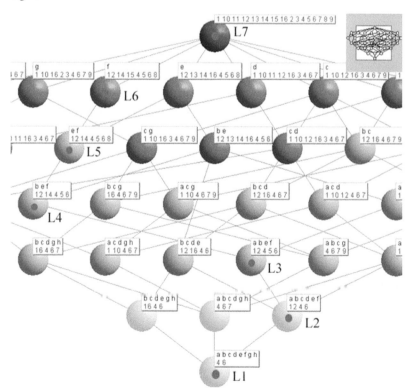

Fig. 12. Visual representation of fully labeled Lattice Chain #6 (Table 6) with a portion of the Galois lattice unfolded from the presence/absence matrix in Table 4. Each node in the chain has a dot.

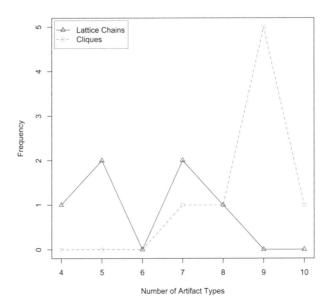

Fig. 13. Frequency distribution of the number of artifact types in cliques and lattice chains (Tables 3 and 5, respectively).

Conclusion

In this chapter, I provide a specific example of how lattice theory is useful for studying the complex spatial organizational structure of artifacts and ecofacts in archaeological sites. Hopefully this will generate interest in developing new applications of lattice theory to uncover spatial patterning and other structural relationships in archaeological sites that may otherwise not be observed. For example, an extension of the method presented in this chapter to three-dimensional space is one possibility.

Acknowledgements

Dr. Chester King (Topanga Anthropological Consultants) provided a historical aerial photograph of LAn-803.

References Cited

Birkhoff, G. 1967. Lattice Theory, 3rd ed. American Mathematical Society, Providence, Rhode Island.

Bowman, A.W. and P. Foster. 1993. Density based exploration of bivariate data. Statistics and Computing 3: 171–177.

Carr, C. 1989. Alternative Models, Alternative techniques: variable approaches to intrasite spatial analysis. pp. 302–473. *In*: C. Carr (ed.). For Concordance in Archaeological Analysis: Bridging Data Structure, Quantitative Technique and Theory. Waveland Press, Inc., Prospect Heights, Illinois.

Davey, B.A. and H.A. Priestley. 2002. Introduction to Lattices and Order. Cambridge University Press, Cambridge.

Duong, T. 2007. ks: Kernel density estimation and kernel discriminant analysis for multivariate data in R. Journal of Statistical Software 21: 1–16.

Duong, T., B. Goud and K. Schauer. 2012. Closed-form density-based framework for automatic detection of cellular morphology changes. PNAS 109: 8382–8387.

Duong, T. and M.L. Hazelton. 2003. Plug-in bandwidth matrices for bivariate kernel density estimation. Journal of Nonparametric Statistics 15: 17–30.

Duquenne, V. 1991. On the core of finite lattices. Discrete Mathematics 88: 133–147.

Duquenne, V. 1999. Latticial structures in data analysis. Theoretical Computer Science 217: 407–436.

Freeman, L.C. and D.R. White. 1994. Using Galois lattices to represent network data. Sociological methodology 23: 127–146.

Gamble, L. 1983. The organization of artifacts, features, and activities at Pitas Point: a coastal Chumash village. Journal of California and Great Basin Anthropology 5: 103–129.

Holmes, W.H. 1902. Anthropological Studies in California. Government Printing Office, Washington D.C.

Hyndman, R.J. 1996. Computing and graphing highest density regions. The American Statistician 50: 120–126.

Keeley, L.H. 1978. Preliminary microwear analysis of the Meer Assemblage. pp. 73–86. *In*: F.V. Noten (ed.). Las Chasseurs de Meer (Dissertations Archaeologicae, 18), De Tempel, Brugge, Belgium.

King, C.D. and M. Merrill. 2002. Significance of Ahmanson Ranch Archaeological Sites. Topanga Anthropological Consultants, Topanga, California.

King, C.D. 1990. Evolution of Chumash Society: A Comparative Study of Artifacts Used for Social System Maintenance in the Santa Barbara Channel Region before A.D. 1804. Garland Publishing, Inc., New York.

Merrill, M. 2002. A spatial analysis of mapped surface artifacts in several early period sites in Los Angeles and Ventura Counties. pp. 87–114. *In*: C.D. King (ed.). Significance of Ahmanson Ranch Archaeological Sites. Topanga Anthropological Consultants, Topanga.

Merrill, M. and D. Read. 2010. A new method using graph and lattice theory to discover spatially cohesivesets of artifacts and areas of organized activities in archaeological sites. American Antiquity 75: 419–451.

Mielke, P.W. and K.J. Berry. 2001. Permutation methods: a distance function approach. Springer, New York.

Mohr, J.W., M. Bourgeois and V. Duquenne. 2004. The Logic of Opportunity: A Formal Analysis of The University of California's Outreach and Diversity Discourse. University of California, Berkeley.

Read, D. 2007. Artifact classification: A conceptual and methodological approach. Left Coast Press, Walnut Creek, California.

Rizzo, M.L. 2008. Statistical Computing with R. Chapman & Hall/CRC, Taylor & Francis Group, Boca Raton, Florida.

Schauer, K., T. Duong, K. Bleakley, S. Bardine, M. Bornens and B. Goud. 2010. Probabilistic density maps to study global endomembrane organization. Nature Methods 7: 560–566.

Wille, R. 1982. Reconstructing lattice theory: an approach based on hierarchies of concepts. *In*: I. Rival (ed.). Ordered Sets. Reidel Publishing Company, Dordrecht-Holland, Boston, Massachusetts.

Wille, R. 1984. Lattice diagrams of hierarchical concept systems. International Classification 11: 77–86.

Beyond Mathematics: Modeling Social Action in the Past

21

Gradient Adaptive Dynamics Describes Innovation and Resilience at the Society Scale

Carsten Lemmen

Introduction: Simulating Prehistoric Societies

Simulating a complex system like a human society faces an inherent challenge: On the one hand, a simulation requires an underlying model, i.e., a target-oriented and simplified mapping of the system (Stachowiak 1973). On the other hand, the observations we make about societies, e.g., about the foraging-farming transition (Boserup 1965), are a result of complex interaction patterns between many agents possessing a multitude of individual strategies, preferences, and options to act (Shanks and Tilley 1987). No model captures the full complexity of a society, by definition. No simulation can reproduce the full spectrum of observations in a society, by definition.

Observing Stachowiak's (1973) definition, the target-orientation in modeling means to reduce the complexity of a system such that certain aspects of the modeled system are described that help to answer a specific set of (research) questions. The challenge in modeling is to isolate these few important aspects and to describe the processes that govern their evolution.

Many processes in a society are more easily described and measured not at the group level, but at the individual group member level. Variously termed individual-based, agent-based, or multi-agent modeling, behavioral rules are specified for individual group members; the behavior of the group is then an emergent property of the many actions of and interactions between individuals (Grimm 1999, Acevedo et al. 2004). In societal simulations, individuals are allowed to learn, or to adapt their individual strategies. They form a complex adaptive system, which is often characterized by a

C.L. Science Consult, Lüneburg, Germany, and Institut für Küstenforschung Helmholtz-Zentrum Geesthacht, Germany.
Email: science@carsten-lemmen.de

larger resilience to perturbation at the group level compared to the individual level (Roberts et al. 2002, Lansing 2003, Redman and Kinzig 2003, Peeples et al. 2006).

A different, top-down approach explicitly describes group properties. The contribution of individuals and their properties to this group is described by probability distributions; properties of the individuals are aggregated in the group (Picard and Franc 2001, Witten and Richardson 2003). In archaeology, this approach has often been used to describe the dispersal of gene pools, cultural traits, and of languages in space, and is usually formulated as a reaction-diffusion equation that describes local (reaction) and spatial (diffusion) properties (Davison et al. 2009, Ackland et al. 2007, Isern and Fort 2012).

In genetics, this approach was introduced as adaptive dynamics by Metz and colleagues (Metz et al. 1992, Abrams 1997, Dieckmann and Law 1996), going back to earlier work by Fisher (1930). When genetically encoded traits influence the fitness of individuals, that gene's prevalence in a population changes. Adaptive dynamics describes the change of the probability of the trait in the population by considering its mutation rate and its fitness gradient, i.e., the marginal benefit of changes in the trait for the (reproductive) fitness of the individual. To ecological systems, this metaphor was first applied by Wirtz and Eckhardt (1996) and Vincent et al. (1996) independently and later extended to cultural traits of human communities by Wirtz and Lemmen (2003). Because many traits are usually involved in (socio-)ecological applications, the term gradient adaptive dynamics (GAD, the vector of traits follows its marginal benefit gradient) was introduced.

In this chapter, we revisit the gradient adaptive dynamics approach, build a simple analytical model of a GAD for describing the foraging-farming transition, and shortly highlight results of case studies using GAD in archaeology.

Gradient Adaptive Dynamics

Gradient adaptive dynamics describes the coevolutionary development of population density p and one or more growth-rate influencing traits x from an aggregated community-average perspective over time t. The population is composed of n members $i \in \{1...n\}$ with relative frequency p_i/p and growth rate r_i:

$$r_i = \frac{1}{p} \cdot \frac{dp_i}{dt} = r_i(x_i, \ldots, E(t)). \tag{1}$$

where x_i is the ith growth-influencing trait; the specific growth rate can also depend on other variables and time-dependent environmental conditions $E(t)$.

For timescales much longer than demographic timescales, equilibrium populations ($\frac{d}{dt}p_i = 0$) can be assumed such that the system is in an optimal, fully adapted, state. The gradient adaptive dynamics approach is useful when the timescale of adaptation is of the same order or faster than the timescale of the population dynamics.

Since $\sum_i^n p_i/p = 1$ is a probability density function, the mean of a quantity x is calculated as

$$\langle x \rangle = \sum_{i=1}^{n} \frac{x_i p_i}{p}. \tag{2}$$

The change of $\langle x \rangle$ with time was introduced by Wirtz and Eckhardt (1996) as the effective variable ansatz (EVA) to describe adaptation; it assumes that, within an individual, x_i is constant and that $\langle x \rangle$ is changed only by a change in the traits' distributions across the population, i.e., $\frac{d}{dt}x_i = 0$ for all x_i:

$$\frac{d \langle x \rangle}{dt} = \langle x \cdot r \rangle - \langle r \rangle \cdot \langle x \rangle \tag{3}$$

Using a Taylor expansion of growth rate about $x = \langle x \rangle$, the above equation can be reformulated in terms of the $(k+1)^{st}$ central moment, of which the first summand is zero. Neglecting moments higher than $k=2$, the temporal change of $\langle x \rangle$ is

$$\frac{d \langle x \rangle}{dt} = \sigma_x^2 \cdot \frac{\partial}{\partial x}r(\langle x \rangle) + v_x\sigma_x^3 \frac{\partial^2}{\partial x^2}r(\langle x \rangle), \tag{4}$$

where $\sigma_x^2 = \langle (x - \langle x \rangle) \rangle$ denotes the second central moment (variance) of x, and v_x describes the skewness (third central moment normalised with cube of σ) of x. The full derivation is given by equations 7–9. Essentially, the population averaged value of a trait changes at a rate which is proportional to the gradient of the fitness function evaluated at the mean trait value.

If the probability distribution of x is known, the variance and sknewness can be deduced; for example, in a common case where $\langle x \rangle$ expresses a variable with uniform distribution in the range $[x_1; x_2]$, the variance is $\sigma_x^2 = (\langle x \rangle - x_1)(x_2 - \langle x \rangle)$. In other cases, variance and skewness have to be specified explicitly. The third moment may not be necessary in some cases and the approximation can be truncated after the second order term; if not, specific closure terms have to be derived (Wirtz and Eckhardt 1996, Merico et al. 2009).

The system of ordinary differential equations formed by the growth equation 1 and the adaptive dynamics equation 4 for one or several mutually independent characteristic traits $\langle x_1 \rangle, \langle x_2 \rangle, \dots$ forms the GAD. Hereinafter, and when not required for understanding of the equations, I shall leave out the angular brackets to denote the mean of a quantity for better readability.

GAD for the Foraging-Farming Transition

The GAD approach is implemented in the Global Land Use and technological Evolution Simulator (GLUES, Lemmen et al. 2011) for the foraging–farming transition. This simulator mathematically resolves the dynamics of population density and three population-averaged characteristic sociocultural traits $x = \{T_A, T_B, C\}$: technology T_A, share of agropastoral activities C, and economic diversity T_B. These are defined for preindustrial societies as follows:

1. Technology T_A is a trait which describes the efficiency of food procurement—related to both foraging and farming—and improvements in health care. In particular, technology as a model describes the availability of tools, weapons, and transport or storage facilities. It aggregates over various relevant characteristics of early societies and also represents social aspects related to work organization and knowledge management. It quantifies improved efficiency of subsistence, which is often connected to social and technological modifications that run in parallel.

An example is the technical and societal skill of writing as a means for cultural storage and administration, with the latter acting as a organizational lubricant for food procurement and its optimal allocation in space and among social groups.

2. A second model variable *C* represents the share of farming and herding activities, encompassing both animal husbandry and plant cultivation. It describes the allocation of energy, time, or man-power to agropastoralism with respect to the total food sector.

3. Economic diversity T_B resolves the number of different agropastoral economies available to a regional population. This trait is closely tied to regional vegetation resources and climate constraints. A larger economic diversity offering different niches for agricultural or pastoral practices enhances the reliability of subsistence and the efficacy in exploiting heterogeneous landscapes.

The adaptive coevolution of the food production system $\{T_A, T_B, C\}$ and population *p*, implement the conceptual model that was, e.g., proposed by Boserup (1981, p. 15): "The close relationship which exists today between population density and food production system is the result of two long-existing processes of adaptation. On the one hand, population density has adapted to the natural conditions for food production [...]; on the other hand, food supply systems have adapted to changes in population density."

Productivity and Growth

The Neolithic transition is characterized by changes in subsistence intensity (SI). Subsistence intensity describes a community's effectiveness in generating consumable food and secondary products; this can be achieved based on an agricultural (with fractional activity *C*) and a hunting-gathering life style (with fractional activity $1 - C$). SI is dimensionless and scaled such that a value of unity expresses the mean subsistence intensity of a hunter-gatherer society equipped with tools typical for the mature Mesolithic and living in an affluent natural environment.

$$SI = (1 - C) \cdot \sqrt{T_A} + C \cdot T_B \cdot T_A \cdot TLI \qquad (5)$$

The agricultural part of SI increases linearly with T_B and with T_A: The more economies (T_B) there are, the better are sub-regional scaled niches utilised and more reliable returns are generated when annual weather conditions are variable; the higher the technology level (T_A), the better the efficiency of using natural resources (by definition of T_A). While a variety of techniques can steeply increase harvests of domesticated species, analogous benefits for foraging productivity are less pronounced and justify a less than linear dependence of the hunting-gathering calorie procurement on T_A. We use a square root formulation, which satisfies $\sqrt{T_A} < T_A$ since T_A is generally larger than unity.

We introduce an additional temperature constraint (TLI) on agricultural productivity which considers that cold temperature could only moderately be overcome by Neolithic technologies. This limitation is unity at low latitudes and approaches zero at permafrost conditions.

The domestication process is represented by T_B, which is the number of realised agropastoral economies. We link T_B to natural resources by expressing it as the fraction *f* of potentially available economies (PAE) by specifying $T_B = f \times PAE$, where the latter corresponds to the richness in domesticable animal or plant species within a specific region.

Food Productivity and Human Growth Rate

The growth rate r of a regional population relative to its size is mainly controlled by its subsistence intensity SI:

$$r = \mu \cdot (\text{FEP} - \gamma\sqrt{T_A p}) \cdot (1 - \omega T_A)$$

$$\cdot \; \text{SI} - \rho \cdot p \cdot e^{-T_A/T_{A\text{lit}}}, \tag{6}$$

with growth rate parameters μ and ρ. The subsistence intensity's contribution to growth is modulated by environmental utility (FEP), societal impacts on the environment $(-\gamma\sqrt{T_A p})$, and organisational losses within a society $(1 - \omega T_A)$. As technology advances, more and more people neither farm nor hunt: Construction, maintenance, administration draw a small fraction ω of the workforce away from food-production. The impact on the environment is modeled as a function of population density and technology (IPAT, Ehrlich and Holdren 1971) on the utility of the environment (FEP, see Lemmen et al. 2011 for details). The loss term is mediated by technologies (T_A, with $T_{A\text{lit}} = 12$), which mitigate, for example, losses due to disease.

Case Studies

The GAD approach has been used in archaeology for a variety of research questions revolving around the foraging-farming transition that occurred during the last 10,000 years: there, the relationship between humans and their environment underwent a radical change from mobile and small groups of foraging people to sedentary extensive cultivators and on to high-density intensive agriculture modern society (Bogaard 2004); these transitions fundamentally turned the formerly predominantly passive human user of the environment into an active component of the Earth system.

The most striking global impact is only visible and measurable during the last 150 years (Crutzen and Stoermer 2000, Crutzen 2002); much earlier, however, the use of forest resources for metal smelting from early Roman times and the medieval extensive agricultural system had already changed the landscape (Barker 2011, Kaplan et al. 2009); global climate effects of these early extensive cultivation and harvesting practices are yet under debate (Lemmen 2010, Kaplan et al. 2011, Stocker et al. 2011, Ruddiman 2013), while the chicken-and-egg problem of whether innovation or population growth came first is not yet settled (Boserup 1965, Lemmen 2014).

Population Pressure and Innovation

Is innovation in the foraging-farming transition the result of population pressure, or is rapid population growth the result of innovations? This question separates Malthuslan (technology driven, an "invention-pull view of population history" Lee 1986) from Boserupian (population driven) economic traditions. These traditions were developed out of different approaches, where the first was based on theoretical assumptions about the growth of population versus resources, and the latter was based on field observations (Malthus 1798, Boserup 1965). What would happen if we analyzed a model such as the one presented in this chapter for the relationship between population pressure (i.e., the difference between capacity and population density) and innovation (i.e., the change in technology $\frac{d}{dt}T_A$)?

The ecological formulation of the model based on the concept of capacity indicates that it is in essence Neomalthusian (Richerson and Boyd 1998, Boserup 1981), and thus should follow an invention-pull dynamics on population growth. Quite the opposite is observed in simulations: innovation is greatest when population pressure is high, just as in Boserup's ethnographic observations. Only the analytical investigation of the relation between population pressure and innovation shows that indeed innovation leads the dynamics, in contrast to the macroscopic view (Lemmen 2014). The GAD model teaches us that it is plausible to assume Malthusian principles behind Boserupian observations.

Transitions to Agriculture

How long does the transition take? This is difficult to measure in the radiocarbon record because of the poor or missing temporal resolution in many sites. A possible measure could be to take the interval between the earliest and the latest radiocarbon date associated with the presence of agropastoral activities in a larger region. For Europe, the analysis of 631 dates in the Neolithic radiocarbon dataset by Pinhasi et al. (2005), distributed in 66 regions across the continent, yields transition durations between 300–1200 years (Lemmen and Wirtz 2014). In simulations employing the GAD approach, the authors found a similar transition time (of 250–900, defined as the 5–95% duration interval).

Resilience to Climate Change

Was the spread of agropastoralism from the Fertile Crescent throughout Europe influenced by rapid climatic shifts? Environmental pressures may have induced migrations (Childe 1942, Weninger et al. 2009), or may have created open space for inmigration (Berger and Guilaine 2008). The establishment of permanent cultivation areas and infrastructures was likely favored by the otherwise stable Holocene climate (van der Leeuw 2008).

On the other hand, reduced mobility after investments in settlement infrastructure may have increased the sensitivity of novel farmers to environmental alterations (Janssen and Scheffer 2004). Individual climate events have repeatedly been blamed for societal demise (Weiss et al. 1993, DeMenocal 2001, Weninger et al. 2005), but the causal relationship is hotly debated (Erickson 1999, Coombes and Barber 2005).

In 2003, Wirtz and Lemmen argued that a continentally different propensity of the climate system to environmental variation should have delayed the establishment of farming systems in some world regions compared to others. They later showed (Wirtz et al. 2010) how climate during the Holocene became, e.g., increasingly variable in western North and South America. With a GAD ansatz, we tested the effect of climate events on the foraging-farming transition in Europe (Lemmen and Wirtz 2014): we find that in this world region not migration triggers, but production failures impacted the dynamics of few regions; the majority of European regions, however, was not affected at all by climate change events. The authors argue that climate events may not have been as important for early sociocultural dynamics as other (endogenous) factors.

Carbon Emissions

When did humans change to alter global climate? Ruddiman (2003, 2013) argues for a start of human interference by carbon dioxide some 8000 years ago, much earlier than the industrial revolution. At that time, the first widespread agricultural areas appeared in China, the Mediterranean, and central Europe.

Foremost, the transition from hunting-gathering subsistence to agropastoral life style had a great impact on our Earth system, or as M. Zeder (2008) states, "domesticates and the agricultural economies based on them are associated with radical restructuring of human societies, worldwide alterations in biodiversity, and significant changes in the Earth's land forms and its atmosphere".

Neolithic agropastoral subsistence for the first time required long-term removal of forest to create space for settlements, crops, or animals; additionally, timber was harvested for construction and as fuel. Simulations with the GAD model introduced in this chapter yield for any place on Earth the population density p and the life style C, which can be converted to a crop land demand per person with estimates by Gregg (1988). In forested areas, demand for crop area is equivalent to deforested area, former trees are mostly replaced by annual herbs. Lemmen (2010) demonstrates how to obtain a carbon emission estimate from the crop demand and calculates the prehistoric carbon emissions of world regions. Their value of 29 Gt of carbon emissions between 9500–2000 BC is small compared to current emissions of about 9 Gt per year (Friedlingstein et al. 2010), based on the same population dataset, however, Kaplan et al. (2011)—using a more detailed carbon cycle model—derived much higher prehistoric emissions of up to 357 Gt by AD 1850.

Conclusion

Gradient adaptive dynamics (GAD) is an established tool in many ecological fields outside archaeology, when the dynamic statistical properties of a group consisting of adaptive members are explicitly described. In societal modeling, this tool is relatively recent, and has not been used to its full potential yet. While delivering a consistent, parsimonious and computationally efficient way to describe complex adaptive systems, its greatest challenge lies with the identification of important traits and with the formulation of the trait-dependent growth dynamics.

I demonstrated an example formulation of a societal GAD approach, and a variety of applications, ranging from economic theory to the quantification of carbon dioxide emissions. The adaptive formulation resulted in resilient simulated societies, where external disturbances do not have a major destructive impact on the simulated system. On the other hand, the usage of simulated population and life style for determining land use effects indicates larger impacts of prehistoric people on their environment than is generally assumed.

A. Appendix

A.1. Derivation of trait dynamics

The change of $\langle x \rangle$ with time t is

$$
\begin{aligned}
\frac{d\langle x \rangle}{dt} &= \frac{d}{dt}\left(\sum_{i=1}^{n} \frac{x_i p_i}{p}\right) = \sum_{i=1}^{n} \frac{d}{dt}\left(\frac{x_i p_i}{p}\right) \\
&= \sum_{i=1}^{n}\left[\frac{p_i}{p}\cdot\frac{dx_i}{dt}\right] + \left[x_i\frac{d}{dt}\left(\frac{p_i}{p}\right)\right] \\
&= \sum_{i=1}^{n} x_i \cdot \left(\frac{1}{p}\frac{dp_i}{dt} + p_i\frac{d(p^{-1})}{dt}\right) \\
&= \left(\sum_{i=1}^{n} \frac{x_i r_i p_i}{p}\right) - \frac{dp}{dt}\cdot\left(\sum_{i=1}^{n} \frac{x_i p_i}{p^2}\right) \\
&= \langle x \cdot r \rangle - \langle r \rangle \cdot \langle x \rangle,
\end{aligned}
\tag{7}
$$

assuming that within an individual x_i is constant and that $\langle x \rangle$ is changed only by a change in the traits distribution across the population, i.e., for all i, $\frac{d}{dt}x_i = 0$.

Taylor expansion of relative growth rate about $x = \langle x \rangle$ yields the expression

$$
r(x) = \sum_{k=0}^{\infty} \frac{1}{k!}\frac{\partial^k r}{\partial x^k}(\langle x \rangle) \cdot (x - \langle x \rangle)^k.
\tag{8}
$$

This expansion can be used to reformulate equation 3, equation 7 in terms of the $(k+1)^{\text{st}}$ central moment.

$$
\begin{aligned}
&\langle r \cdot x \rangle \quad - \quad \langle r \rangle \cdot \langle x \rangle \\
&= \sum_{k=0}^{\infty} \frac{1}{k!}\frac{\partial^k r}{\partial x^k}(\langle x \rangle) \quad \cdot \quad \left\langle x\cdot(x-\langle x \rangle)^k\right\rangle - \langle x \rangle \\
&\quad \cdot \sum_{k=0}^{\infty} \frac{1}{k!}\frac{\partial^k r}{\partial x^k}(\langle x \rangle) \quad \cdot \quad \left\langle (x-\langle x \rangle)^k\right\rangle \\
&= \sum_{k=0}^{\infty} \frac{1}{k!}\frac{\partial^k r}{\partial x^k}(\langle x \rangle) \quad \cdot \quad \left[\left\langle x\cdot(x-\langle x \rangle)^k\right\rangle\langle x \rangle\right. \\
&\quad \left.\cdot \left\langle (x-\langle x \rangle)^k\right\rangle\right] \\
&= \sum_{k=0}^{\infty} \frac{1}{k!}\frac{\partial^k r}{\partial x^k}(\langle x \rangle) \quad \cdot \quad \left\langle (x-\langle x \rangle)(x-\langle x \rangle)^k\right\rangle \\
&= \sum_{k=0}^{\infty} \frac{1}{k!}\frac{\partial^k r}{\partial x^k}(\langle x \rangle) \quad \cdot \quad \left\langle (x-\langle x \rangle)^{k+1}\right\rangle
\end{aligned}
\tag{9}
$$

In this derivation, I made use of the identity $\langle r(x) \rangle = r(\langle x \rangle)$ and of the linear transformation $\langle \sum x \rangle = \sum \langle x \rangle$.

The first summand ($k = 0$) of this series is zero, since $\langle x - \langle x \rangle \rangle = 0$. If all higher moments (> 3) are neglected, we arrive at the approximation

$$
\frac{d \langle x \rangle}{dt} = \langle x \cdot r \rangle \quad - \quad \langle r \rangle \cdot \langle x \rangle
$$
$$
= \sigma_x^2 \cdot \frac{\partial r}{\partial x}(\langle x \rangle) \quad + \quad v_x \sigma_x^3 \frac{\partial^2 r}{\partial x^2}(\langle x \rangle)
$$

(10)

Table 1. Symobls and variables used in the text and equations (in order of appearance in text). A useful parameter set is $\mu = \rho = 0.004$, $\omega = 0.04$, $\gamma = 0.12$, $\delta T_A = 0.025$, $\delta T_B = 0.9$; a variable $\delta_C = C(1 - C)$; and initial values for $p_0 = 0.01$, $T_{A,0} = 1.0$, $T_{B,0} = 0.8$, and $C_0 = 0.04$.

Symbol	description	unit	typical range
p	population density	km^{-2}	> 0
x	growth-influencing trait		> 0
t	time	a	9500–1000 BC
r	specific growth rate	a^{-1}	
E	environmental constraints		
$\langle \cdot \rangle$	mean / first moment of ·		
σ^2, δ	variance		
v	skewness		
T_A	technology trait		> 0
T_B	economic trait		> 0
C	labor allocation trait		$]0; 1[$
SI	subsistence intensity		
TLI	temperature limitation		$[0; 1]$
FEP	food extraction potential		$[0; 1]$
ω	administration parameter		
γ	exploitation parameter		
μ	fertility rate	a^{-1}	
ρ	mortality rate	a^{-1}	

Acknowledgments

This study was partly funded by the Swiss National Science Foundation (SNF, grant ACACIA). GLUES is free and open source software and can be obtained from http://www.sf.net/p/glues. I thank K.W. Wirtz for initiating and supporting the work on GLUES.

References Cited

Abrams, P.A. 1997. Evolutionary responses of foraging-related traits in unstable predator prey systems, Evol. Ecol. pp. 673–686.

Acevedo, M., B. Callicott and M. Ji. 2004. Coupled Human and Natural Systems. A Multi-Agent Based Approach, North.

Ackland, G.J., M. Signitzer, K. Stratford and M.H. Cohen. 2007. Cultural hitchhiking on the wave of advance of beneficial technologies, Proc. Natl. Acad. Sci. USA 104(21): 8714–8719, doi: 10.1073/pnas.0702469104.

Barker, G. 2011. Archaeology: The cost of cultivation, Nature 473(7346): 163–4.

Berger, J. and J. Guilaine. 2008. The 8200 cal bp abrupt environmental change and the Neolithic transition: A Mediterranean perspective, Quaternary International.

Bogaard, A. 2004. The nature of early farming in central and south-east Europe. Doc. Praehist. 31(4): 49–58.

Boserup, E. 1965. The conditions of agricultural growth, Aldine, Chicago, 124 pp.

Boserup, E. 1981. Population and technological change: A study of long-term trends, University of Chicago Press Chicago.

Childe, V.G. 1942. What happened in history, Pelican/Penguin, Harmondsworth, 256 pp.

Coombes, P. and K. Barber. 2005. Environmental determinism in Holocene research: causality or coincidence? Area 37(3): 303–311.

Crutzen, P.J. 2002. Geology of mankind, Nature 415, 23.

Crutzen, P.J. and E.F. Stoermer. 2000. The Anthropocene, Glob. Chang. Newsl. 41.

Davison, K., P.M. Dolukhanov and G. Sarson. 2009. Multiple sources of the European Neolithic: Mathematical modeling constrained by radiocarbon dates, Quat. Int. 44: 1–17.

DeMenocal, P.B. 2001. Cultural responses to climate change during the late Holocene, Science. 292(5517): 667–673.

Dieckmann, U. and R. Law. 1996. The dynamical theory of coevolution: a derivation from stochastic ecological processes, J. Math. Biol. 34(5–6): 579–612.

Ehrlich, P.R. and J.P. Holdren. 1971. Impact of population growth, Science. 171(977): 1212–1217.

Erickson, C.L. 1999. Neo-environmental determinism and agrarian "collapse" in Andean prehistory, Antiquity 73(281): 634–642.

Fisher, R.A. 1930. The Genetical Theory of Natural Selection, Clarendon, Oxford.

Friedlingstein, P., R.A. Houghton, G. Marland, J. Hackler, T.A. Boden, T.J. Conway, J.G. Canadell, M.R. Raupach, P. Ciais and C. Le Quéré. "Update on CO_2 Emissions." Nature Geoscience 3, no. 12(2010): 811–12.

Gregg, S. 1988. Foragers and farmers:population interaction and agricultural expansion in prehistoric Europe, University of Chicago Press.

Grimm, V. 1999. Ten years of individual-based modeling in ecology: what have we learned and what could we learn in the future? Ecol. Modell. 115: 129–148.

Isern, N. and J. Fort. 2012. Modeling the effect of Mesolithic populations on the slowdown of the Neolithic transition. J. Archaeol. Sci. 39(12): 3671–3676, doi:10.1016/j.jas.2012.06.027.

Janssen, M.A. and M. Scheffer. 2004. Overexploitation of Renewable Resources by Ancient Societies and the Role of Sunk-Cost Effects. Ecol. Soc. 9(1): 6.

Kaplan, J.O., K.M. Krumhardt and N. Zimmermann. 2009. The prehistoric and preindustrial deforestation of Europe. Quat. Sci. Rev. 28(27-28): 3016–3034, doi:10.1016/j.quascirev.2009.09.028.

Kaplan, J.O., K.M. Krumhardt, E.C. Ellis, W.F. Ruddiman, C. Lemmen and K. Klein Goldewijk. 2011. Holocene carbon emissions as a result of anthropogenic land cover change. The Holocene 21(5): 775–791, doi:10.1177/0959683610386983.

Lansing, J.S. 2003. Complex Adaptive Systems. Annu. Rev. Anthropol. 32(1): 183–204, doi: 10.1146/annurev.anthro.32.061002.093440.

Lee, R. 1986. Malthus and baserup: A dynamic synthesis. pp. 96–130. *In*: D. Coleman and R. Schofield (eds.). The State of Population Theory: Forward from Malthus. B. Blackwell.

Lemmen, C. 2010. World distribution of land cover changes during Pre- and Protohistoric Times and estimation of induced carbon releases, Géomorphologie: Relief, Proc., Environ. 4(2009): 303–312.

Lemmen, C. 2014. Malthusian Assumptions, Boserupian Response in Transition to Agriculture Models. *In*: Ester Boserup's Legacy on Sustainability, M. Fischer-Kowalski et al. (eds.). Human-Environment Interactions 4, Springer.

Lemmen, C. and K.W. Wirtz. 2014. On the sensitivity of the simulated European Neolithic transition to climate extremes, J. Archaeol. Sci. 51: 65–72, doi:10.1016/j.jas.2012.10.023.

Lemmen, C., D. Gronenborn and K.W. Wirtz. 2011. A simulation of the Neolithic transition in Western Eurasia, J. Archaeol. Sci. 38(12): 3459–3470, doi:10.1016/j.jas.2011.08.008.

Malthus, T.R. 1798. An Essay on the Principle of Population, Cambridge University Press.

Merico, A., J. Bruggeman and K.W. Wirtz. 2009. A trait-based approach for downscaling complexity in plankton ecosystem models, Ecol. Modell. 220(21): 3001–3010, doi: 10.1016/j.ecolmodel.2009.05.005.

Metz, J.A.J., R.M. Nisbet and S.A.H. Geritz. 1992. How should we define fitness for general ecological scenarios. Trends Ecol. Evol. 7: 198–202.

Peeples, M.A., C.M. Barton and S. Schmich. 2006. Resilience Lost: Intersecting Land Use and Landscape Dynamics in the Prehistoric Southwestern United States. Ecol. Soc. 11(2): 22.

Picard, N. and A. Franc. 2001. Aggregation of an individual-based space-dependent model of forest dynamics into distribution-based and space-independent models. Ecol. Modell. 145: 69–84.

Pinhasi, R., J. Fort and A.J. Ammerman. 2005. Tracing the origin and spread of agriculture in Europe, Public Libr. Sci. Biol. 3(12): e410, doi:10.1371/journal.pbio.0030410.

Redman, C.L. and A. Kinzig. 2003. Resilience of past landscapes: resilience theory, society, and the longue durée, Conserv. Ecol.

Richerson, P. and R. Boyd. 1998. Homage to Malthus, Ricardo, and Boserup: Toward a general theory of population, economic growth, environmental deterioration, wealth, and poverty, Human Ecology Review 4: 85–90.

Roberts, C.A., D. Stallman and J.A. Bieri. 2002. Modeling complex human environment interactions: the Grand Canyon river trip simulator, Simulation 153: 181–196.

Ruddiman, W.F. 2003. The anthropogenic greenhouse era began thousands of years ago, Clim. Change, pp. 261–293.

Ruddiman, W.F. 2013. The Anthropocene. Annu. Rev. Earth Planet. Sci., doi:10.1146/annurev-earth-050212-123944.

Shanks, M. and C. Tilley. 1987. Re-constructing Archaeology: Theory and Practice, New Studies in Archaeology, 320 pp. Cambridge University Press, Cambridge, United Kingdom.

Stachowiak, H. 1973. Allgemeine Modell theorie, Springer, Wien.

Stocker, B.D., K. Strassmann and F. Joos. 2011. Sensitivity of Holocene atmospheric CO¡sub¿2¡/sub¿ and the modern carbon budget to early human land use: analyses with a process-based model, Biogeosciences 8(1): 69–88, doi:10.5194/bg-8-69-2011.

van der Leeuw, S.E. 2008. Climate and Society: Lessons from the Past 10000 years, Ambio 37(14): 476–482.

Vincent, T.L.S., D. Scheel, J.S. Brown, T.L. Vincent, T.A, Naturalist and N. Deo. 1996. Trade-Offs and Coexistence in Consumer-Resource Models: It all Depends on what and where you Eat 148(6): 1038–1058.

Weiss, H., M.-A. Courty, W. Wetterstrom, F. Guichard, L. Senior, R.H. Meadow and A. Curnow. 1993. The Genesis and Collapse of Third Millennium North Mesopotamian Civilization, Science(80-.) 261(5124): 995–1004, doi:10.1126/science.261.5124.995.

Weninger, B., E. Alram-Stern, L. Clare, U. Danzeglocke, O. Jöris, C. Kubatzki, G. Rollefson and H. To-dorova. 2005. Die Neolithisierung von Südosteuropa als Folge des abrupten Klimawandels um 8200 Cal B.P., in Clim. Var. Cult. Chang. Neolit. Soc. Cent. Eur. 6700-2200calBC, RGZM-Tagungen, vol. 1, edited by D. Gronenborn. pp. 75–118. Römisch-Germanisches Zentralmuseum, Mainz, Germany.

Weninger, Bernhard, Lee Clare, Eelco Rohling, Ofer Bar-Yosef, Utz Böhner, Mihael Budja, Manfred Bundschuh, et al. "The Impact of Rapid Climate Change on Prehistoric Societies during the Holocene in the Eastern Mediterranean." Documenta Praehistorica 36 (2009): 7–59. doi:10.4312/dp.36.2.

Wirtz, K.W. and B. Eckhardt. 1996. Effective variables in ecosystem models with an application to phytoplankton succession, Ecological Modeling 92: 33–53.

Wirtz, K.W. and C. Lemmen. 2003. A global dynamic model for the neolithic transition. Clim. Change 59(3): 333–367.

Wirtz, K.W., G. Lohmann, K. Bernhardt and C. Lemmen. 2010. Mid-Holocene regional reorganization of climate variability: Analyses of proxy data in the frequency domain, Palaeogeogr. Palaeoclimatol. Palaeoecol. 298(3-4): 189–200, doi:10.1016/j.palaeo.2010.09.019.

Witten, G. and F. Richardson. 2003. Competition of three aggregated microbial species for four substrates in the rumen. Ecol. Modell. 164(2-3): 121–135, doi:10.1016/S0304-3800(02)00383-6.

Zeder, M.A. 2008. Domestication and early agriculture in the Mediterranean Basin: Origins, diffusion, and impact. Proc. Natl. Acad. Sci. USA 105(33), 11597-11604. doi:10.1073/pnas.0801317105.

22

Two-dimensional Models of Human Dispersals: Tracking Reaction-Diffusion Fronts on Heterogeneous Surfaces[#]

Fabio Silva[1,][*] and *James Steele*[2]

Introduction to Mathematical Models of Human Dispersal: Fisher-Skellam-KPP

Reaction-diffusion models are often used to describe the spatial and temporal dynamics of populations under the influence of natural or cultural selection (where a species, gene, or cultural innovation expands its geographical range, typically because of some advantage over competitors in the means of capturing energy and converting it into a larger population size through self-reproduction, e.g., Murray 2002, 2003, Okubo and Levin 2001, Kandler and Steele 2009). In ecology they provide a way of translating assumptions or data about movement, mortality, and reproduction of individuals, at a local scale, into global conclusions about the persistence or extinction of populations and the coexistence of interacting populations (Cantrell and Cosner 2003).

Historically, R.A. Fisher developed the now classic Fisher model for the spread of an advantageous genetic mutation (Fisher 1937). The same reaction-diffusion system

[1] Institute of Archaeology, University College London, 31–34 Gordon Square, London WC1H 0PY, UK and, School of Archaeology, History and Anthropology, University of Wales Trinity Saint David, Lampeter Campus, Ceredigion SA48 7ED, UK.

[2] Institute of Archaeology, University College London, 31-34 Gordon Square, London WC1H 0PY, UK.

Email: j.steele@ucl.ac.uk.

[*] Corresponding author: fabio.silva@ucl.ac.uk.

[#] Parts of this chapter have previously appeared in our review paper on dispersal models for archaeologists (Steele 2009), and in our methodological paper introducing the Fast Marching method to an archaeological audience (Silva and Steele 2012).

was also investigated simultaneously to Fisher's paper by Kolmogorov et al. (1937), with somewhat greater mathematical detail. Thus in ecology the usual reference is to the 'Fisher equation', or to the 'Fisher-Skellam model', while in mathematics the usual reference is to 'Fisher-KPP' or 'KPP-Fisher' or simply to the 'KPP' equation. Using such a model that incorporated the combined effects of selection and dispersal, Fisher showed that after a certain allele is established in a population, it will spread in a wave-of-advance-like pattern with a velocity that is proportional to the selective advantage of the allele and the length scale of the population's diffusive interaction. J.G. Skellam used a similar framework in theoretical studies of population dispersal (Skellam 1951) wherein he established the relationship between the random walk as the description of individual movements of members of a biological species, and the diffusion equation as the description of the dispersal of the population as a whole (Cantrell and Cosner 2003). Skellam was particularly interested in reaction-diffusion models, which he used to study the spread of muskrats in Central Europe. He showed that predictions of the reaction-diffusion framework fit the actual data very well (Skellam 1951). Parenthetically, we may note that this model has also been studied in a coupled system for two interacting chemicals, an activator and an inhibitor, with the reaction term given by their reaction kinetics, in a highly influential paper on biological pattern formation by Turing 1952; cf. Murray 2012.

Understanding large-scale human dispersals requires us to model two components, for which this model system is an effective approximation. The population growth process constitutes the reaction term, and some measure of the sum of individual spatial displacements constitutes the diffusion term (in cases where more than one population is involved, we must also consider the nature of the interaction). The basic system can be described as:

$$\frac{\partial n}{\partial t} = f(n,\ \alpha,\ K) + D\nabla^2 n$$

where $n(\mathbf{r},t)$ denotes the population density at time t and at position $\mathbf{r} = (x,y)$. This system has two components: a nonlinear population growth (or 'reaction') term and a linear population dispersal (or 'diffusion') term. $f(n,\alpha,K)$ is the population growth function which in the Fisher-KPP equation is taken to follow the logistic growth law proposed by Verhulst (1838) and widely used in theoretical population biology (Murray 1993). This function describes a self-limiting density-dependent population increase and is given by $f(n,\alpha,K) = \alpha n(1 - \frac{n}{K})$ where α is the intrinsic maximum population growth rate and K is the carrying capacity, related to local environmental factors. D is a diffusion constant which specifies the mean spatial dispersal rate of individuals between birth and reproduction, and ∇^2 is the Laplacian operator which redistributes the population from regions of higher density into those of lower density (for simplicity we shall use the Laplacian operator in the equations in this chapter; many recent authors however use an integral formulation that allows variably-shaped dispersal kernels). In general individuals will move from their birthplace a distance λ during their generation time τ. The square of this distance will in general be proportional to the time available (this is a standard random walk result); the constant of proportionality is the diffusion constant $D = \frac{\lambda^2}{2d\tau}$, where d is the number of dimensions in the system being modeled, typically, 1 or 2 (Einstein 1905). The width of the wave front region, over which the population changes from a high to a low density, can be shown to be dependent on D and α and to have an intrinsic spatial scale $\xi \sim \sqrt{\frac{D}{\alpha}}$. It can be shown that the speed v at which this wave front travels is also related to D and α, tending asymptotically to approach $v = 2 \sqrt{D\alpha}$ (Fisher 1937, Kolmogorov et al. 1937).

It is no trivial matter to resolve the detailed demographic processes involved in prehistoric dispersals. Much effort has been expended estimating speeds of spatial population expansion, because the relation between the speed v and the product $D\alpha$ implies that front propagation rate is determined by population growth and diffusion rates (and that by dating the arrival of a population at successive locations in space, one can therefore reconstruct the underlying demography). Archaeologists have found that radiocarbon dating can be used to obtain sufficiently precise estimates of the age of cultural events at different spatial locations for this purpose, although there are many subsidiary methodological and sampling issues which need to be addressed before the results can bear close examination (e.g., Steele 2010). With the advancement of radiocarbon measurement and calibration techniques, we are seeing the application of accurate and precise dating to increasingly remote episodes of human dispersal during and subsequent to the last ice age.

For hunter-gatherer dispersals, early applications of the Fisher-KPP model to Palaeolithic dispersal problems include Young and Bettinger (1995) and Steele et al. (1996, 1998), neither of which focus specifically on front speeds (due partly to limits on the accuracy and precision of available archaeological dates). Studies focused on front propagation speed as estimated from radiocarbon dates (first arrival times) include Mellars' (2006) attempts to estimate the speed of spread of anatomically-modern humans into Europe prior to the last glacial maximum; Housley et al. (1997) and Fort et al.'s (2004) studies of the rate of subsequent late glacial recolonization of northern Europe as the ice receded; and Hamilton and Buchanan's (2007) study of the diffusion of Clovis spear point technology in late glacial North America. For the spread of farming, the classic study was by Ammerman and Cavalli-Sforza (e.g., 1984); their work introduced the F-KPP system to archaeologists, who typically now refer to this variant as the 'wave of advance' model. Ammerman and Cavalli-Sforza fitted a linear regression curve to dates and distances from Jericho, finding a mean rate of spread of about 1 km yr[-1] and finding this to be consistent with the front propagation speed predicted for this reaction-diffusion system with ethnographically-derived values for the demographic terms. More recently, the radiocarbon record has been revisited by Gkiasta et al. (2003) and by Pinhasi et al. (2005). Pinhasi et al. (2005) fitted a linear regression curve to dates from a set of 735 archaeological sites in Europe and the Near East using various origins and two possible distance measures, and found an average rate of spread for the Neolithic transition in the range of 0.6–1.3 km yr[-1]. Pinhasi et al. (2005) found that this spread rate was consistent with predictions from a slightly modified F-KPP model of demic diffusion, taking anthropological estimates of 2.9–3.5% annually for α and a mean dispersal rate $D = 1400$–3900 km[2]/generation, with 29–35 years for the generation time τ. There are increasing numbers of other examples of archaeological applications of this model (see, e.g., Davison et al. 2006, Patterson et al. 2010, Baggaley et al. 2012).

Tracking the Dispersal Front Using Fast Marching Methods, as an Alternative to Solving the Differential Equations

Increasingly, archaeologists seek to analyse models of dispersal in a two-dimensional, geographically-explicit formulation. To solve the Fisher-KPP equation in a discrete time and space lattice implementation requires discretization techniques. In our own earlier work (Steele et al. 1996, 1998), we approximated time differentials at particular sites by finite differences (Press et al. 1986):

$$\frac{dn(\mathbf{r}, t)}{dt} \approx \frac{n(\mathbf{r}, t + \Delta_t) - n(\mathbf{r}, t)}{\Delta_t}.$$

Typically we used $\Delta_t = 1$ year. Space differentials were similarly approximated by finite differences:

$$D\nabla^2(\mathbf{r}_0) = \mathrm{h}^{-2} \sum_\alpha w_\alpha D'_\alpha [n(\mathbf{r}_\alpha) - n(\mathbf{r})],$$

where for a given position $\mathbf{r}_0$ the sum is taken over nearest neighbour sites $\mathbf{r}_\alpha$ on the lattice, and where the lattice size is h. There are two types of neighbour sites: those along the lattice axes and those along the diagonals. The sum is weighted appropriately with parameters w_α; this parameter is typically $\frac{2}{3}$ for sites α along the lattice axes and $\frac{1}{6}$ along the diagonals. The effective diffusion parameter D'_α, appropriate to motion between the sites $\mathbf{r}_0$ and $\mathbf{r}_\alpha$, is given by $D'_\alpha = \sqrt{D(\mathbf{r}_\alpha)D(\mathbf{r}_0)}$. In practice in any given simulation in our own early study, only two values of D were used: $D = D_0$ and $D = 0$, the latter representing the fact that the particular cell is inaccessible. We encountered issues of numerical stability as a function of the grid cell resolution: our discretized model gave accurate results so long as the natural length scale in this equation, $\xi = \sqrt{\frac{D}{a}} > h$, where h is the cell size (length). Otherwise the simulated velocity was faster than that predicted analytically. A similar stability issue arises with excessively coarse time discretization. This problem of global domain discretization error due to excessive coarsening of the space and time steps may be exacerbated when solving the equations with cell-specific parameter values, and large changes in values between adjacent cells. Solving this by increasing the mesh refinement can impose large costs in terms of the required computational resources, although adaptive mesh techniques can help to alleviate this.

Mathematical modellers often choose to avoid these implementation issues by analysing some version of the reaction-diffusion system analytically as a continuous system of PDEs, with front propagation dynamics explored numerically in one spatial dimension; this enables them to derive estimated front speeds for a given parameter set which can then be compared with a mean front speed estimated empirically (from measurements on the archaeological record). However, archaeologists may often wish to know what the effects on dispersal might be of the specific geographical structure of some relevant area of the real world, and to answer this a different approach is often required. There is some parallel here with debates around the merits of classical mathematical approaches versus agent-based models in other anthropological applications. The classical approach yields clear analytical or numerical solutions, but at the expense of realism, while the geographically-explicit approach yields realistic solutions but at the expense of some opacity about the sensitivity of the overall pattern to modelled local features: if there is no explicit experimental criterion for discerning how much geography is the 'right amount' to include, we are at risk of over-fitting. How much geography do we really need to consider, to reconstruct the broad dynamic of an archaeologically-recorded dispersal? Here, at any rate, we shall assume that if modelling is to progress our understanding of the large-scale archaeological record, geography *does* matter.

Helpfully, an alternative approach exists which does not require the solving of the discretized RD equations, but simply the tracking of the displacement of the front; and this approach lends itself extremely well to modelling the effects of geographical heterogeneity. Front propagation in two dimensions can be modelled as a shortest path problem. The most familiar application of front propagation techniques in Geographical Information Systems is to combustion fronts in wildfire modelling, where realistic local variation in fuel load and other factors can be incorporated. However, many other geographically-posed front propagation problems can be studied using shortest path techniques. In GIS applications, a front at a particular time point t_x can be conceptualised as the isoline joining all points in the domain which are reached in a given time $t_x - t_0$ by their shortest path from a given origin point (where the front begins to propagate at t_0 and where velocity of movement is defined by units of cost, i.e., time as a cost function of distance). Complete calculation of the shortest paths (valued in cost lengths) in a domain yields a cumulative cost surface. Cumulative cost surfaces are typically calculated in GIS packages using Dijkstra's algorithm (Dijkstra 1959), which solves the single-source shortest-path problem when all edges have non-negative weights. This algorithm starts from the source and works outwards, starting at each iteration with the cell with the lowest cumulative cost value among the cells that the front has already reached and adding this to the values of the cells in its neighbourhood. The neighbourhood is typically defined on a regular grid or raster as including the cells that are reachable in a single step in a Rook's pattern (4-cell neighbourhood), a Queen's pattern (8-cell neighbourhood) or a Knight's pattern (16-cell neighbourhood).

In our own work, we use a closely-related approach known as the 'Fast Marching' method (Sethian 1996), implemented in MATLAB. Fast Marching algorithms (Sethian 1998, 1999) were developed initially for the stable and economical computation of evolving interfaces in front propagation research (e.g., crystal growth and flame propagation; Sethian 1987). Such algorithms involve the computation of finite differences to solve a boundary-value partial differential equation known as the Eikonal equation, and, because of that, are highly stable with a computational complexity independent of the initial conditions (weights, number of source points, etc.) and dependent only on the pixel size of the domain space.

Dijkstra's algorithm with a graph-based (i.e., grid cell neighbourhood) update is prone to introduce artefactual 'staircasing' into least cost paths. The Fast Marching method instead overcomes these constraints by replacing the graph update with a local resolution of the Eikonal equation (in our case, by a second-order finite-difference approximation; Silva and Steele 2012, Eq. 4). This produces a more accurate treatment of the underlying continuous spatial surface. Instead of the Dijkstra update algorithm, with D_j the next cell to be evaluated according to the above iteration rule:

$$D(j) = min(dx + W(j), dy + W(j));$$

we use an Eikonal update:

$$\Delta = 2*W(j) - (dx-dy)^2;$$
$$\quad if \, \Delta > = 0$$
$$\qquad D(j) = (dx + dy + \sqrt{\Delta})/2;$$
$$else$$
$$\qquad D(j) = min(dx + W(j), dy + W(j))$$

where W is the neighbourhood metrical weight. The most time consuming aspect of Dijkstra-like approaches is the management of the list of cells for which cumulative costs have been computed. The Fast Marching method streamlines this process by singly focusing on the narrow band that encloses the propagating front, and using only known upwind values to estimate the cumulative costs.

FMM is a technique that computes the arrival time of an expanding front at each point of a discrete lattice or grid (Sethian 1998). In our context, the expanding fronts are the boundaries separating a 'colonized' region from the outside. The method tracks the evolution of these fronts by assigning to each grid point (pixel) the time at which the front hits the grid point. The FMM hence requires time to be uniquely defined, and thus only applies to cases where the front is uniformly expanding, in a direction perpendicular to the front. Front motion is described by the Eikonal equation, where T is the arrival time of the expanding front and v is its speed, when it crosses the grid point (x,y):

$$\|\nabla T(x,y)\| v(x,y) = 1$$

The FMM works outwards from an initial condition, the source of the expanding front, by selecting a set of grid points in a narrow band around the existing front. This band then "marches forward", freezing the values just calculated and bringing in new ones into its structure. Key to the method's success is in the selection of which narrow band grid point to update. The algorithm developed by Sethian takes advantage of the fact that all upwind neighbouring grid points will yield a value for the arrival time which cannot be smaller than any of the Frozen points, i.e., the points that are already within the region delimited by the front interface. The narrow band grid point with lowest T value can then be selected, frozen and its neighbours brought into the narrow band. Only neighbouring grid points that are already frozen are used in the finite difference calculation of the arrival time.

Solving the Eikonal equation, to calculate the arrival time at each grid point, is done by finite-difference approximations, the simplest of which is the first-order one:

$$1/v(x)^2 = \begin{cases} \max(T(x,y) - T(x+1,y), T(x,y) - T(x-1,y), 0)^2 \\ + \max(T(x,y) - T(x,y+1), T(x,y) - T(x,y-1), 0)^2 \end{cases}$$

where $T(x,y)$ is the arrival time calculated at the selected grid point (x,y), and grid points $(x+1,y)$ and $(x-1,y)$, $(x,y+1)$ and $(x,y-1)$ are the neighbouring points along the horizontal and vertical axis respectively. The maximum is taken precisely to select only the upwind neighbour values to be used in the calculation. A second-order approximation of the Eikonal equation, yielding results which are considerably more accurate, was also proposed (Baerentzen 2000):

$$1/v(x)^2 = \begin{cases} \max\left(\frac{3T(x,y) - 4T(x-1,y) + T(x-2,y)}{2}, -\frac{3T(x,y) - 4T(x+1,y) + T(x+2,y)}{2}, 0\right)^2 \\ +\max\left(\frac{3T(x,y) - 4T(x,y-1) + T(x,y-2)}{2}, -\frac{3T(x,y) - 4T(x,y+1) + T(x,y+2)}{2}, 0\right)^2 \end{cases}$$

It is also common in Fast Marching implementations to express the metrical weight as a function of the local speed of the propagating front, thus outputting an arrival time surface. This allows us to model dispersals over heterogeneous domains. For each individual cell, W_j can be multiplied by a friction factor, boosting or inhibiting the speed with which the front will propagate locally. Implementation of this generalization requires the construction of friction raster layers covering the computational domain, which can then be used as a lookup array by the FM algorithm.

Case Study 1: Dispersal of Farming in Neolithic Europe

To illustrate these methods we have applied them to a dataset of radiocarbon dates related to the dispersal of farming in Neolithic Europe (Silva and Steele 2014). Radiocarbon dates for the earliest Neolithic occupation from the earliest-dated levels of 765 sites in the Near East, Europe, and Arabia had previously been collated from four pre-existing online databases by Pinhasi et al. (2005, Table S1). Using a set of archaeological sites as possible dispersal origin points (note that this is a heuristic device to anchor the spatial modelling, and should not be taken literally), Pinhasi et al. (2005) compared the values for Pearson's correlation coefficient for calibrated age and distance in the above sample of 735 dated sites using Great Circle distances, and shortest path distances along land- and near-offshore based dispersal routes. *Çayönü* in southern Turkey was the archaeological site whose location as an origin gave the best fit for the terrain-dependent shortest path lengths. We checked and confirmed the Great Circle distances from *Çayönü* using the set of site coordinates given by Pinhasi et al. (2005); the associated correlation coefficient value gave us a measure of fit for the baseline (no geography) model ($r = -0.793$; $\rho = -0.751$).

The statistical methodology used to estimate trends in earliest observed arrival dates as a function of distance from the assumed origin, involved fitting regression models (reduced major axis, cf. Steele 2010) to sets of paired values of site dates (mean calibrated radiocarbon ages, calibrated in OxCal using INTCAL09) and distances to sites from the assumed origin. Using regression slopes to estimate average front speeds is established practice in the literature. This enabled us to estimate the mean speed of dispersal (using the regression slope coefficient), and the proportion of the variation in arrival times that was accounted for by that trend (using the correlation coefficient). We estimated (using the correlation coefficient) best-fitting speeds of dispersal in different directions as a function of habitat, with coasts, rivers, and major ecoregions all being given individual values for their possible effects on rates of spread. Cost-distances from the origin to each dated site retained for our analysis were calculated using the FM algorithm with variable costs for traversing the various geographical features, with friction-cost parameter sets estimated both from existing scenarios in the literature and by systematically sampling the wider parameter space using a Genetic Algorithm.

For our base maps we classified polygons representing the Mediterranean and other coastal corridors; the Mediterranean, temperate forest and other present-potential biomes; the Danube-Rhine river corridor; and polygons defining inaccessible areas with a 1,100 m altitudinal cut-off. These were obtained using public-domain GIS map layers (rivers from ESRI World Rivers shapefile, elevations from the ETOP05 Digital Elevation Model, and biomes from the Terrestrial Ecoregions of the World shapefile compiled by Olson et al. 2001) and projected into a Lambert Azimuthal Equal Area projection centred at 45 N, 45 E using GRASS GIS (GRASS Development Team 2012).

To determine the optimal set of reclassification rules (friction factors) for these geographical features, we analysed the radiocarbon dataset in relation to several existing hypotheses in the literature, as well as using an unconstrained GA search. In all models only land was treated as colonisable, subject to a near-offshore coastal buffer of 45 kms which represents the bridging potential of maritime transport. Except for the baseline model (where distances were calculated by spherical geometry using standard software), we used the Fast Marching method to derive shortest-path distances on a cost-surface given by the model's constraints. Although different centres might yield higher correlation coefficients for the different models, we were interested in exploring the effects of adding biogeography and therefore, for direct comparison, we used *Çayönü* as the nominal source of dispersal in all models.

Our results demonstrate the added explanatory value of a geography-rich model. Consistent with earlier results (Pinhasi et al. 2005), Models 1 and 2 (Great Circle and land-based distances, respectively) are able to account for up to 60% of the variation in dates in this dataset, based on the adjusted R^2 values. Among our more geography-rich models, a literature-derived one based on a scenario from Davison et al. (2006), and which stipulated both an altitudinal cut-off and a linear decrease in front propagation rate with latitude, yielded an improved linear association between date and cost distance, with a front speed in southern parts of the domain of 1.05 km/yr decreasing to 0.525 km/yr in more northerly latitudes. This model accounted for 63.4% of the variation in dates, which was an improvement on Models 1 and 2 after correcting for numbers of free parameters. Meanwhile the best-fit model obtained by the GAs yielded a yet-higher value for adjusted R^2, and explained 66.5% of the variation in the dates, which is proportionally a 10% increase in explanatory power over the geography-free or near-geography-free Models 1 and 2. The relative friction weights recovered by the GA search suggest an important accelerating role for a northern Mediterranean coastal dispersal corridor, a significant but less marked accelerating role for the Danube/Rhine corridor, and a decelerating effect of forested biomes away from these corridors, most markedly in the higher-altitude temperate coniferous forest biome. Future work should address demographic interpretations of these geographical effects. Further details and illustrations can be found in the original publication (Silva and Steele 2014).

Case Study 2: Dispersal of Farming in the Bantu-Speaking Regions of Sub-Saharan Africa

To illustrate these methods further, we have also applied them to a dataset of radiocarbon dates related to the dispersal of farming in the Bantu-speaking regions of sub-Saharan Africa from an origin in the grasslands of southern Cameroon (Russell et al. 2014). In an earlier study, we had modelled the interface between two converging fronts in this region to illustrate the crystal growth algorithm (Silva and Steele 2012, see next Section), since archaeologists have long favoured a 'deep split' model, with an Eastern population stream spreading along the northern margin of the rainforest to reach the inter-lacustrine region of East Africa, and a Western population stream spreading from corridors along the western coast of and along the major rivers of the equatorial rainforest. Such archaeological models are widely cited, and seen by many archaeologists as supporting a parallel 'deep split' in the radiation of the Western and Eastern Bantu languages; but they need to be evaluated in relation to physical anthropological, genetic and linguistic data, as well as continuing archaeological discoveries. Some recent phylogenetic work in linguistics and genetics does not find support for such a 'deep split'. Instead these studies find support for some version of a 'pathway through the rainforest' scenario, with the Eastern Bantu language clade radiating much later in time.

In our initial paper (Silva and Steele 2012) we did not build geographical features into our model beyond the land/sea distinction. In our more recent study (Russell et al. 2014) we addressed this limitation. We used an archaeological database containing geographically referenced radiometric determinations that by their association with archaeological material (most commonly pottery) are interpreted by the excavator/ archaeologist as marking the first arrival of Bantu language speaking farmers to an area. Data were collected from those countries in sub-Saharan central, eastern and southern Africa where Bantu languages are spoken today. More details on the database can be found in Russell et al. (2014).

The spatial domain was constructed using appropriate GIS raster and vector files. To define land/sea boundaries we used a present-day world coastlines map, projected using the Lambert Azimuthal Equal Area projection (centred at 10°S, 25°E). This is an appropriate projection for the domain of interest, which is predominantly tropical, with a north-south orientation. To define land cover classes we used the biomes in the 2004 version of the Terrestrial Ecoregions of the World shapefile, compiled by Olson et al. (2001), with limited further aggregation of biome types. In addition, the Congo and Zambezi rivers, and their major tributaries, were taken from the ESRI World Rivers shapefile. These two drainage basins were considered separate features to enable the Congo to be a corridor through the rainforest if needed. Other major African rivers were not considered relevant for this initial study.

For our modelling, which requires an approximate origin point, we chose a point in northwest Cameroon at 5°51'N, 10°4'E, close to the site of Shum-Laka (the oldest site in the database). As in the previous case study, the statistical methodology used to estimate trends in earliest observed arrival dates as a function of distance from the assumed origin involved fitting regression models to sets of paired values of site dates (mean calibrated radiocarbon ages, calibrated in OxCal using INTCAL09) and distances to sites from the assumed origin. This again enabled us to estimate the mean speed of dispersal (using the regression slope coefficient), and the proportion of the variation in arrival times that was accounted for by that trend (using the correlation coefficient). We again estimated (using the correlation coefficient) best-fitting speeds of dispersal in different directions as a function of habitat, with coasts, rivers, and major ecoregions all being given individual values for their possible effects on rates of spread. Cost-distances from the origin to each dated site retained for our analysis were again calculated using the FM algorithm with variable costs for traversing the various geographical features, with friction-cost parameter sets estimated both from existing scenarios in the literature and by systematically sampling the wider parameter space using a Genetic Algorithm.

As in the previous case study, our results made it clear that geography affected dispersal rates: we found effects of corridors, barriers, and of different habitat types. Our GA-optimized results did not support the 'Deep Split' model, and further emphasize the importance of accurate geographical reconstruction, with a key role found for a now-vanished late Holocene savanna corridor through the equatorial rainforest. Further details and illustrations can be found in the original publication (Russell et al. 2014).

The Modified Fast Marching Method for Multiple Converging Fronts

We have further introduced a generalization of this approach to model multiple competing fronts with different origin locations, onset times and propagation rates, where each cell in a grid is populated by the descendants of one or other source population accord to a first arrival rule (Silva and Steele 2012). To model numerically the location and shape of the interface between two converging fronts requires slightly different methods. We use the methodology of spatial tessellation known as Voronoi diagrams (Okabe et al. 2000). These diagrams provide a sectioning of the spatial domain in which the sections are determined by the distances to a previously specified set of points known as generators or sources. An ordinary Voronoi diagram tessellates space into sections that contain all points whose distance to that section's generator point is smaller than to all other sources. This assumes that all sources are identical, which is to say that they all have the same weight when calculating the distances, but this is not always the case and more complicated, weighted Voronoi diagrams account for such variability.

To apply such geometrical methods in modelling multiple converging fronts we first, again, reduce the dynamics of the reaction-diffusion process to a single parameter, front propagation speed (v), which as noted above is asymptotically constant for the Fisher-KPP system with normal (non-anomalous) diffusion. Front speed then becomes a multiplicative weight in computation of the Voronoi diagram. We also allow for different times-of-initiation of front propagation from different foci, with the offset between the earlier and later-initiated processes being treated as an additive weight. The combination of multiplicative and additive weights yields a compoundly weighted Voronoi diagram, but we also impose the constraint that fronts cannot propagate across areas already 'colonized' from a competing source. This must then be modelled as a shortest-path compoundly weighted Voronoi diagram (Okabe et al. 2000, the term 'shortest path' indicates that the front must be propagated around barriers and other obstacles) using the kinds of algorithms which has been developed for the computation of spatial crystal growth processes (where growth rates and/or onset times vary between seeds; Kobayashi and Sugihara 2001), or for the computation of collision-free paths for pursuit evasion (Cheung and Evans 2007).

To locate the final position of the interface, an analytical solution exists for the multiplicatively weighted crystal-growth problem with two converging fronts (Kobayashi and Sugihara 2001). In such cases the interface will have the form of a closed logarithmic spiral. However, for compoundly-weighted cases and for cases with more than two converging fronts, analytical solutions are not available and may be axiomatically unobtainable. Numerical methods are therefore required. The most widely used techniques again involve Fast Marching algorithms (Sethian 1998, 1999).

The method used here extends work on crystal growth where the final locations of interfaces between crystals depend on the times of onset of growth from each individual seed, and the crystals' subsequent individual rates of growth. Modelling is complicated by the fact that the spatial extent of pre-existing and growing crystals constrains the area into which other crystals can expand. This yields the multiplicatively weighted shortest-path, or crystal-growth, Voronoi diagram (Kobayashi and Sugihara 2001). For purely additively weighted diagrams subject to this constraint, where crystals begin growing at different times but all grow at a constant rate, a solution can be found in the Kolmogorov-Johnson-Mehl-Avrami model (e.g., Fanfoni and Tomellini 1988).

Kobayashi and Sugihara (2001) modified the original FMM algorithm, just described, in order to accommodate for the fact that each region internal to a diffusive front (a crystal in their context) acts as an obstacle to all other fronts. The result of this is known as the shortest-path Voronoi Diagram. They were also interested in different speeds for different crystals. Their modifications can be summarized as follows:

- To each grid point is assigned a region ID (crystal name) representing the ordinal number of the diffusive process to which the grid point belongs to;
- Whenever the wavefront with ID *n* reaches a gridpoint, this gridpoint's ID is changed to *n*;
- At the initialization stage:
  - all source points, the generators of the diffusive processes, are attributed an ID number, as above;
  - the same for the grid points that are first neighbours of the sources, i.e., the point that are one grid point apart from the source points;
  - all other points on the grid are initialized to have null ID;
  - Frozen is initialized to be the set of all source points;
- For the calculation of the arrival time at each grid point:
  - The gridpoint's ID is read and the speed v_{ID} is used in [Eq. 4];
  - only upwind neighbours that are also included in the same crystal, i.e., that have the same ID, are used in solving [Eq. 4];

With these modification the FMM outputs a tessellation of the domain in which each grid point belong to a crystal with a certain ID, which has been interpreted as the crystal-growth Voronoi diagram.

To allow for the inclusion of additive weights, which correspond to a time offset in the sourcing of the different processes, another modification was implemented on top of the above mentioned ones. The developed algorithm was implemented in MATLAB® (MathWorks Inc., Natick, MA). Our solution was to separate processes with different time offsets into different grids, henceforth called layers, where the wavefronts were allowed to evolve independently. Each layer can contain more than one process as long as they are all sourced at the same time. To ensure interaction between all processes, all layers share the same ID and Frozen matrices, thus knowing which grid points are already occupied by another process on a different layer. It is on the ID layer, the 'master layer', that the spatial domain is tessellated into the final Voronoi diagram. The expanding wavefronts are allowed to evolve on each layer, as described above, one time-step at a time, to ensure that all layers evolve at equal intervals. Because of the different speeds for each wavefront, the arrival times for each process have to be normalized and a "master time" created that works across all layers (for, e.g., time on the second layer will need to be added a constant factor equal to the first time offset in order to match the time on the first layer). This "master time" is the one used to keep track of the insertion and evolution of different layers. This implementation is fully described below.

Layered Crystal-Growth Fast Marching Algorithm

Step 1—Initialization

1. Cover the spatial domain with grid points. Set initial conditions for all diffusive processes: source coordinates, front propagation speeds, and initiation time offsets.
2. Select the processes to be initiated on the first layer (all those with a null time offset parameter).

Step 2—Inserting Sources in First Layer and First Loop

1. Initialize Frozen to be the set of source points for the processes selected in 1.2. Set arrival time T to zero for these source points and set their ID to be an identifying number for each process.
2. All points neighbour to the source points that are not already Frozen are added to the neighbour list (the narrow band) and their distances calculated using [Eq. 4].
3. Repeat until first time offset is reached or until number of neighbour points is zero.
4. Retrieve the pixel from the neighbour list that has the smallest arrival time value.
5. Add it to Frozen and set its ID to be the region number of its source process.
6. All neighbour points to this pixel that are not already Frozen or belong to another region (i.e., have another ID) are added to the neighbour list and their arrival times calculated.

Step 3—Inserting Sources in Other Layers and Main Loop

This step should be taken whenever new processes are to be generated. When the arrival time reaches the first time offset value, loop (2.3) breaks and a new layer needs to be initialized. Repeat while there are layers to be added:

1. If the gridpoints of the sources to be added do not already belong to a region, repeat 2.1 and 2.2 for the new sources.
2. Until the number of neighbour points is zero repeat 2.3 for each layer, one time-step at a time, i.e., repeat 2.3 for layer 2 until t = t0 + 1, then repeat 2.3 for layer 1 until t = t0 + 1, then repeat 2.3 for layer 2 until t = t0 + 2, etc., where t is the "master time" and t0 the first time offset.

The value of this approach is that it enables modelling of multiple competing fronts, as for example where two comparable innovations appear at different locations and diffuse through neighbouring populations (with their final distribution reflecting some first-arrival rule, see, e.g., Silva et al. 2014). A comparable problem in genetics has recently been considered by Ralph and Coop (2010).

Discussion and Conclusion

We have now outlined a method of modelling front propagation that is numerically very stable and computationally very fast, and which facilitates consideration of the effects of geographical heterogeneity. The compromise made is that the solving method does not incorporate the underlying Fisher-Skellam demographic model (although the parameters for this can be estimated from the front speed). In fact, however, this might be seen as a strength of the new approach. The original Fisher-Skellam-KPP equation can be modified to introduce further complexities, such as density-dependence, advection, competition, or non-Fickian diffusion (Steele 2009). Different combinations of such terms and parameters can result in similar outcomes to the level of resolution available in the archaeological record, making it impossible to distinguish the underlying process by comparison to the radiocarbon record alone. Indeed, the Fisher-KPP approach assumes that demography is the underlying factor behind a dispersal process—the *prime mover*. In archaeology, this is not always true as, for example, the dispersal of pottery might simply indicate innovation diffusion without a dispersing population behind it (cf. Silva et al. 2014). Several different internal dynamics can lead to the same front propagation velocity, and thus result in the same observable outcomes. This is a long-standing debate in archaeology and evidence other than radiocarbon-based front speed needs to be looked at to discern between alternative possibilities.

In other disciplines the Fast Marching Methods, or their big brother Level Set Methods, have been used to solve geometric problems (Sethian 1989), track the dynamics of fluids (Yu et al. 2003), segment medical imaging (Malladi et al. 1995), model tumor dynamics (Hogea et al. 2005), estimate shortest-paths in robotics (Kimmel and Sethian 1996), model seismic velocity in geophysics (Cameron et al. 2007) and model molecular pathways in chemical reactions (Haranczyk and Sethian 2009). The present applications do not exhaust their potential within archaeological and anthropological problems either, and they may find application in other problems requiring efficient computations of shortest paths and of landscape tessellation.

Several other enhancements to the FMM algorithm of interest to archaeologists are possible. Firstly, we might want to model the effects on population dispersals of time-dependent (as well as space-dependent) friction surfaces, since in the timescales involved in large scale dispersals the landscapes can change. Climate, environmental and sea level changes are some of the most important diachronic processes that can affect dispersals in the *longue durée*. These can change the available landmass (e.g., rising sea levels would reduce the domain area), create or destroy barriers and corridors (e.g., glaciers, exposed land bridges), or affect the structure of biogeographical patches (e.g., biomes), thus dynamically constraining dispersal routes. Large and abrupt (catastrophic) change events might also be implemented in the FMM, since these (e.g., volcanic eruptions) can also affect dispersals. A second potential FMM modification would be the implementation of complex interactions between competing populations. This would allow modelling dynamic interactions, such as assimilation and delayed substitution, at the interface between populations (e.g., between advancing Neolithic farmers and the indigenous hunter-gatherers), but again in a geographically explicit two-dimensional modelled domain.

Acknowledgements

We thank Tim Sluckin and Thembi Russell for their collaborations in various phases of this work. Tim steered the early development of a finite difference implementation of the Fisher-KPP equation to model Paleoindian dispersals, while Thembi led in the development of our project modelling the spread of farming into southern Africa.

References Cited

Ammerman, A.J. and L.L. Cavalli-Sforza. 1984. The Neolithic Transition and the Genetics of Populations in Europe. Princeton University Press, Princeton.

Baerentzen, J.A. 2000. On the Implementation of Fast Marching Methods for 3D Lattices, Technical Report IMM-REP-2001-13, DTU.IMM.http://www2.imm.dtu.dk/pubdb/views/publication_details.php?id=841.

Baggaley, A.W., G.R. Sarson, A. Shukurov, R.J. Boys and A. Golightly. 2012. Bayesian inference for a wave-front model of the neolithization of Europe. Physical Review E 86(1): 016105.

Cameron, M.K., S.B. Fomel and J.A. Sethian. 2007. Seismic velocity estimation using time migration velocities, Inverse Problems 23: 1329.

Cantrell, R.S. and C. Cosner. 2003. Spatial Ecology via Reaction-Diffusion Equations (John Wiley & Sons, Ltd.).

Cheung, W. and W. Evans. 2007. Pursuit-Evasion Voronoi Diagrams in 11. In Voronoi Diagrams in Science and Engineering, 2007. ISVD'07. 4th International Symposium pp. 58–65.

Davison, K., P. Dolukhanov, G.R. Sarson and A. Shukurov. 2006. The role of waterways in the spread of the Neolithic. Journal of Archaeological Science 33(5): 641–652.

Dijkstra, E.W. 1959. A note on two problems in connexion with graphs. Numerischemathematik 1(1): 269–271.

Einstein, A. 1905. Über die von der molekularkinetischen Theorie der Warmegeforderte Bewegung von in ruhenden Flussigkeitensuspendierten Teilchen. Ann. Phys. (Leipzig) 17: 549–560.

Fanfoni, M. and M. Tomellini. 1988. The Johnson-Mehl-Avrami-Kolmogorov model—a brief review, Nuovo Cimentodella Societaltaliana di Fisica. D 20(7-8): 1171–1182.

Fisher, R.A. 1937. The wave of advance of advantageous genes. Ann. Eugenics 7: 355–369.

Fort, J., T. Pujol and L.L. Cavalli-Sforza. 2004. Palaeolithic population waves of advance. Cambridge Archaeological J. 14: 53–61.

Gkiasta, M., T. Russell, S. Shennan and J. Steele. 2003. Neolithic transition in Europe: the radiocarbon record revisited. Antiquity 77: 45–61.

GRASS Development Team. 2012. Geographic Resources Analysis Support System (GRASS) Software, Version 6.4.2. Open Source Geospatial Foundation. http://grass.osgeo.org.

Hamilton, M.J. and B. Buchanan. 2007. Spatial gradients in Clovis-age radiocarbon dates across North America suggest rapid colonization from the north. Proceeding of the National Academy of Sciences of the United States of America (PNAS) 104(40).

Haranczyk, M. and J.A. Sethian. 2009. Navigating Molecular Worms Inside Chemical Labyrinths. Proceedings of the National Academy of Sciences 106: 21472–21477.

Hogea, C.S., B.T. Murray and J.A. Sethian. 2005. Simulating complex tumor dynamics from avascular to vascular growth using a general Level Set method. J. Mathematical Biology 53(1).

Housley, R.A., C.S. Gamble, M. Street and P. Pettitt. 1997. Radiocarbon evidence for the late glacial human recolonisation of northern Europe. Proceedings of the Prehistoric Society 63: 25–54.

Kandler, A. and J. Steele. 2009. Innovation diffusion in time and space: effects of social information and of income inequality. Diffusion Fundamentals 11(3): 1–17.

Kimmel, R. and J.A. Sethian. 1996. Fast Marching Methods for Robotic Navigation with Constraints, CPAM Report 669. Univ. of California, Berkeley.

Kobayashi, K. and K. Sugihara. 2001. Crystal Voronoi Diagram and Its Applications (Algorithm Engineering as a New Paradigm).数理解析研究所講究録1185: 109–119. http://hdl.handle.net/2433/64634.

Kolmogorov, A.N., I.G. Petrovskii and N.S. Piskunov. 1937. A study of the diffusion equation with increase in the quantity of matter, and its application to a biological problem. Bulletin of Moscow University, Mathematics Series A 1: 1–25.

Malladi, R., J.A. Sethian and B. Vemuri. 1995. Shape Modeling with Front Propagation: A Level Set Approach, IEEE Trans. on Pattern Analysis and Machine Intelligence 17(2).

Mellars, P. 2006. A new radiocarbon revolution and the dispersal of modern humans in Europe. Nature 439: 931–935.

Murray, J.D. 2002. Mathematical Biology I: an Introduction. Springer-Verlag, 3rd edition.

Murray, J.D. 2003. Mathematical Biology II: Spatial Models and Biomedical Applications. Springer-Verlag, 3rd edition.

Murray, J. 2012. Why are there no 3 headed monsters? Mathematical Modeling in Biology. Notices of the AMS 59(6): 785–795.

Murray, J.D. 1993. Mathematical Biology, 2nd edition. Springer-Verlag, Berlin.

Okabe, A., B. Boots, K. Sugihara and S.N. Chiu. 2000. Spatial Tessellations: Concepts and Applications of Voronoi Diagrams. 2nd ed. Chichester: John Wiley & Sons.

Okubo, A. and S.A. Levin. 2001. Diffusion and Ecological Problems: Modern Perspectives. Springer-Verlag, 2nd edition.

Olson, D.M., E. Dinerstein, E.D. Wikramanayake, N.D. Burgess, G.V. Powell, E.C. Underwood and K.R. Kassem. 2001. Terrestrial ecoregions of the world: a new map of life on earth. BioScience 51: 933–938.

Patterson, M.A., G.R. Sarson, H.C. Sarson and A. Shukurov. 2010. Modelling the Neolithic transition in a heterogeneous environment. Journal of Archaeological Science 37(11): 2929–2937.

Pinhasi, R., J. Fort and A.J. Ammerman. 2005. Tracing the origin and spread of agriculture in Europe. PLoSBiol 3: e410.

Press, W.H., B.P. Flannery, S.A. Teukolsky and W.T. Veterling. 1986. Numerical Recipes: The Art of Scientific Computing. Cambridge University Press, Cambridge.

Ralph, P. and G. Coop. 2010. Parallel adaptation: One or many waves of advance of an advantageous allele? Genetics 186: 647–668.

Russell, T., F. Silva and J. Steele. 2014. Modelling the spread of farming in the Bantu-speaking regions of Africa: an archaeology-based phylogeography. PLoS ONE 9(1): e87854. DOI: 10.1371/journal. pone.0087854.

Sethian, J.A. 1987. Numerical methods for propagating fronts. *In*: P. Concus and R. Finn (eds.). Variational Methods for Free Surface Interfaces. Springer-Verlag, New York.

Sethian, J.A. 1989. A review of recent numerical algorithms for hypersurfaces moving with curvature-dependent speed, Journal Differential Geometry 31: 131–161.

Sethian, J.A. 1996. A fast marching level set method for monotonically advancing fronts. Proceedings of the National Academy of Sciences 93(4): 1591–1595.

Sethian, J.A. 1998. Adaptive Fast Marching and Level Set methods for propagating interfaces. Acta Math. Univ. Comenianae Vol. LXVII 1: 3–15.

Sethian, J.A. 1999. Level Set Methods and Fast Marching Methods: Evolving Interfaces in Computational Geometry, Fluid Mechanics, Computer Vision, and Materials Science. 2nd ed. Cambridge Monographs on Applied and Computational Mathematics. Cambridge University Press.

Silva, F. and J. Steele. 2012. Modeling boundaries between converging fronts in prehistory. Advances in Complex Systems 15(01n02).

Silva, F. and J. Steele. 2014. New methods for reconstructing geographical effects on dispersal rates and routes from large-scale radiocarbon databases. Journal of Archaeological Science 52: 609–620. DOI: 10.1016/j.jas.2014.04.021.

Silva, F., J. Steele, K. Gibbs and P. Jordan. 2014. Modeling spatial innovation diffusion from radiocarbon dates and regression residuals: the case of early Old World pottery. Radiocarbon 56(2): 723–732.

Skellam, J.G. 1951. Random dispersal in theoretical populations. Biometrika 38: 196–218.

Steele, J. 2009. Human dispersals: mathematical models and the archaeological record. Human Biology 81: 121–140.

Steele, J. 2010. Radiocarbon dates as data: quantitative strategies for estimating colonization front speeds and event densities. Journal of Archaeological Science 37(8): 2017–2030.

Steele, J., J. Adams and T. Sluckin. 1998. Modelling Paleoindian dispersals. World Archaeology 30: 286–305.

Steele, J., T. Sluckin, D. Denholm and C. Gamble. 1996. Simulating hunter-gatherer colonization of the Americas. *In*: H. Kamermans and K. Fennema (eds.). Interfacing The Past. Analecta Praehistorica Leidensia 28: 223–227.

Turing, A.M. 1952. The chemical basis of morphogenesis. Phil. Trans. Roy. Soc. B 237: 37–72.

Verhulst, P.F. 1838. Notice sur la loique la population pursuit dans son accroissement. Corresp. Math. Phys. 10: 113–121.

Young, D. and R. Bettinger. 1995. Simulating the global human expansion in the Late Pleistocene. Journal of Archaeological Science 22: 89–92.

Yu, J.-D., S. Sakai and J.A. Sethian. 2003. A coupled level set projection method applied to ink jet simulation, Interfaces and Free Boundaries 193(1): 275–305.

23

The Sustainability of Wealth among Nomads: An Agent-Based Approach

J. Daniel Rogers,[1,*] *Claudio Cioffi-Revilla*[2] *and Samantha Jo Linford*[3]

Introduction

One of the core objectives of social science in general, and of archaeology in particular, is to explain the emergence and development of complex social systems—i.e., social systems with status inequality and government by non-kin-based authority. In this chapter, we implement an empirically calibrated, spatial agent-based model (ABM) as a tool for studying why and how wealth differentials and associated social inequalities are generated and sustained over multiple generations. We also describe the theoretical foundation, formal strategies, and examples of the mathematical and computational approaches needed to develop complex ABMs.

Recent research on inequality and the sustainability of wealth among mobile pastoralists (nomads) has challenged older interpretations. Earlier ethnographic and historical research emphasized the egalitarian nature of pastoralism and the inability to sustain wealth due to environmental vulnerability and marginalization within developing nation-states. In the 1990s, basic interpretations of egalitarianism were reevaluated to

---

[1] Department of Anthropology, NHB 112, Smithsonian Institution, Washington, DC 20013.
  Email: rogersd@si.edu

[2] Center for Social Complexity and Department of Computational Social Science, George Mason University, Fairfax, VA 22030.
  Email: ccioffi@gmu.edu

[3] Department of Anthropology, University of California, Santa Cruz, 1156 High Street, Santa Cruz, CA 95064.
  Email: linford.samantha@gmail.com

* Corresponding author

incorporate a broader understanding of differences based on gender, age, skills, and prestige, whether or not within a stratified or ranked social context (Flanagan 1989). For pastoral nomads, observations of environmental and cultural marginality were tied to the perception that social organization was primarily tribal, relatively egalitarian, and organized around less formal social hierarchies. Borgerhoff Mulder et al. (2009, 2010: 37) combined new theoretical hypotheses and ethnographic data to document mechanisms for intergenerational wealth transmission, effectively contradicting earlier egalitarian theories. Related studies have also reanalyzed cross-cultural interpretations of the sustainability of wealth differences among hunter-gatherers and other small-scale societies (Bowles et al. 2010, Charles and Hurst 2003). These newer studies hold two principal implications for the long-term study of social change: first, the mechanisms of wealth maintenance provide general insights into how social inequalities were sustained and transformed over time in the emergence of complex social systems; secondly, pastoral-based pathways to social complexity both expand and complicate theories on the origins of early states and empires.

> **Definition 1 (Pastoralist Wealth W; Borgerhoff et al. 2010: 37).** Aggregation of material wealth (X), relational wealth (Y), and embodied wealth (Z), where *material wealth* includes herds and accumulated goods such as jewelry, tents, domestic tools, and other property; *relational wealth* is represented by the symbolic and social capital of obligations and prestige that families accumulate by spending other forms of wealth; and *embodied wealth* is the accumulated knowledge of everything from grazing conditions on different landscapes to regional politics. Formally, the norm of pastoralist wealth can be defined as
>
> $$\|W\| = (X^{*2} + Y^{*2} + Z^{*2})^{1/2},$$
>
> where X^*, Y^*, and Z^* are standardized positive values (e.g., strictly positive z-scores) of each wealth component.

This chapter explores two basic questions affecting wealth transmission:

1. *What is the role of different degrees of social controls in the maintenance of all forms of wealth?*
2. *How vulnerable is wealth to external weather events?*

Our empirical frame of reference for this study is herders living in Mongolia in a social environment dominated by groups structured on kinship and other social ties. Within the model there are no large-scale political systems controlling local actions, beyond that of the lineage/clan observed in the simplest chiefdoms. Conflict is not part of the model. The social context of the model is similar to the period of initial social complexity around the beginning of the Bronze Age in eastern Inner Asia (ca. 3000 B.C.E.), prior to the formation of early states and empires (Rogers 2012).

The history of pastoralism in Inner Asia and elsewhere clearly documents emergent and sustained wealth and social inequalities, deeply enmeshed in the formation of complex societies (Honeychurch 2015). Prior to 200 B.C.E., in the antecedent Bronze and Early Iron Ages (ca. 3000 to 200 B.C.E.), archaeological evidence exists for hereditary leadership and substantial wealth differentials in societies generally described as complex or super chiefdoms (Frachetti 2008, Kradin 2006). The burial mounds of Central Asia (*kurgans*) and Mongolia (*khirigsuur*) provide the principal evidence for substantial wealth accumulation by central elites, and presumably the institutionalized intergenerational transmission of wealth among ruling lineages.

Theoretical Framework:
Inequality and Origins of Social Complexity

Formally, we use the following definitions of inequality, sustainability, social complexity, and related concepts.

Definition 2 (Wealth Inequality *G*). Uneven distribution of wealth (material, relational, or embodied) in a given society. Operationally, *G* is defined by the Gini coefficient:

$$G = \frac{\sum_{i=1}^{n} \sum_{j=1}^{n} |w_i - w_j|}{2 \, n^2 \mu}, \tag{1}$$

where $x_i = \|W\|$ (as in Definition 1) denotes the wealth of the *ith* household, *n* is the number of households, and μ is the mean household wealth (Dorfman 1979).

Interestingly, wealth inequality and the Gini coefficient are related to the Pareto power-law of wealth distribution and its scaling exponent by the equation

$$G = \frac{1}{2b - 3}, \tag{2}$$

where *b* is the scaling or Pareto exponent (Cioffi-Revilla 2014: 165, Kleiber and Kotz 2003: 35). Wealth distribution for households is log-normal, not Paretian. Note that equation 2 is also a scaling law. Conversely,

$$b = \frac{1}{2G} + \frac{3}{2}, \tag{3}$$

which goes to complex infinity as $G \to 0$ (perfect equality). Empirical values of *G* therefore yield power-law equivalent values of *b* and vice versa, based on equations 2 and 3.

Definition 3 (Wealth Sustainability). Ability of an individual or household to acquire and retain wealth across generations; trans-generational acquisition and duration of wealth. Operationally, wealth sustainability is measured in terms of time duration *T* measured between the initial acquisition of wealth at some time τ_0 and loss of wealth at some later time τ'.

Note that wealth sustainability is a probabilistic compound event, based on acquisition and retention of assets (tangible or intangible) as sequential and jointly necessary conditions (Cioffi-Revilla 2014: 177), not a simple deterministic outcome. As such, it is subject to a set of formal principles of social complexity. Theoretically, we assume that *T* is a continuous random variable (c.r.v.) with a set of observed realizations $\{t_1, t_2, t_3, ..., t_m\} \in T$ defined by probability functions for density $p(t)$, cumulative density $\Phi(t)$, intensity (hazard rate) $H(t)$, and complementary cumulative density $S(t)$, among the most common probability functions associated with *T*. Wealth duration *T* is discrete in the simulation.

Definition 4 (Social Complexity). Extent to which a given society is governed through non-kin-based relations of authority. Ordinal levels of social complexity include kin-based societies (ground state or level 0), simple chiefdom (local leaders exercise authority), complex chiefdom (regional leaders and local confederates), state (specialized institutions or public administration), and empire (multinational society). A recent overview of data, measurement, and formal theories of the emergence of initial social complexity is provided in Cioffi-Revilla (2014: chs. 5 and 7). The formation of institutionalized hierarchies of authority is typically the social context in which leadership strategies take shape as a result of canonical processes involving, for example, collective action (Cioffi-Revilla 2005, Rogers and Cioffi-Revilla 2009).

Computational Methodology

Agent-based models are increasingly used in all fields of science to explore complex interactions (Kohler and van der Leeuw 2007, Takadama et al. 2010). Computational agent-based modeling methodology consists of the following main six phases of development (from motivation to analysis). The simulation model is called Households World, abbreviated as HHW, as summarized below. Detailed information on design, coding strategies, and algorithms are available in earlier publications (Cioffi-Revilla et al. 2007, Cioffi-Revilla et al. 2010, Rogers et al. 2012).

1. *Motivation:* The core purpose of the study aimed at answering the research questions stated in the Introduction section of this chapter, on the effects of social control, extreme weather events, and social strategies on initial emergence and sustainability of wealth inequality in pre-state nomadic societies of Inner Asia (namely Mongolia and surrounding regions).
2. *Design:* HHW is designed to replicate a broad range of behaviors pertaining to pastoralist households living in a kinship structured social and economic landscape referenced to that of Inner Asia during the Early Bronze Age. HHW includes pastoralist households and their herds, situated in a biophysical landscape affected by weather. The landscape is endowed with biomass for herds to subsist. The overall dynamics are as follows: Weather affects biomass vegetation, which affects herding throughout the annual seasons, which, in turn, affects the movement of households across the steppe. In time, households' herds change in size and location as households congregate in camps as they undergo nomadic migrations in interaction with their herds.
3. *Implementation:* HHW was implemented in 2010, using the MASON toolkit, version 10. The model is written in Java, with graphic facilities to portray multivariable time series, histograms, and maps of the Inner Asia region. Agent decision-making was implemented using ethnographic data from written sources and field observations in Mongolia.
4. *Verification:* HHW was verified using standard procedures for spatial computational ABMs (Cioffi-Revilla 2010: 242, 297), including code walkthrough, unit testing, profiling, debugging, multiple long runs, and parameter sweeps, among others. The current version operates without any known bugs.
5. *Validation:* Several procedures were utilized to test HHW's validity. They included pattern-matching on histograms and related quantitative and qualitative distributions, time-series data, and movement-time relations in model output

data, among others. For example, HHW generates log-normal household wealth distributions similar to those known from ethnographic and historical data (Erdenebaatar 2009, Flores Ettinger and Linford 2013).

6. *Analysis:* Three experiments were conducted, focusing on the effects of (1) social control, (2) marriage strategies, and (3) extreme weather events on initial emergence and sustainability of wealth inequality.

In HHW, households belong to lineages and clans, households have friends, they remember their ancestors, and they have children who marry and begin new households. Households usually obey kinship norms and clan rules, and each day they evaluate the landscape and take their herds out to graze. They tend to camp with friends and relatives in a group similar to the *khot aul* as seen historically and in modern Mongolian pastoralism (Bold 1996). HHW has two socially direct forms of wealth transmission: intergenerational (marriage payment) and direct assistance to kin households in financial trouble (bailout). Here, we ran the simulation for a period of 500 yr. The first 135 yr of the simulation is a calibration phase during which population continues to grow. Effectively, the relatively stable time span of the simulation represents the yr from 135 to 500—the equivalent of over 18 generations.

This study utilizes the concept of a Standard Stocking Unit (SSU) for the purpose of comparability in herd numbers (Humphrey and Sneath 1999: 77). Herd dynamics, including birth, growth, consumption, and death rates are derived from a variety of rangeland studies, including results from the Kherlen Bayaan-Ulaan Grassland Research Center, and ethnographic sources (Begzsuren et al. 2004, Cribb 1991, Redding 1981).

Households and herds 'live' on specific landscapes. This study utilized the Egiin Gol landscape (Honeychurch and Amartuvshin 2007), consisting of a 100 x 100 km area in north central Mongolia (NW corner is at 49° 56' N and 102° 46' E). Households cannot leave the landscape and there is no trade, social, or political interaction beyond the boundaries. Topography and ground cover vegetation are modeled at a scale of 1 sq km. Vegetation variability is based on the Normalized Difference Vegetation Index (NDVI) calculated from a five yr (1995–2000) mean of contemporary biomass (Hansen et al. 1998, 2000). Monthly NDVI rasters are based on atmosphere corrected bands in 500 m resolution. Aggregate biomass is rendered in 14 land cover types with exponential regressions calculated to produce approximations of edible biomass. Biomass coefficients are based on Kawamura et al. (2005) and verified through a variety of rangeland research in Mongolia (Bedunah et al. 2006, Batbuyan 1997).

For better or worse, weather happens. A widely recognized challenge for pastoralists is the vulnerability of herds to extreme weather (but also disease, predators, and theft). In HHW, weather is reflected through the impact of seasonal changes, winter storms called *dzud*, and droughts. These weather patterns and events affect vegetation abundance and growth rates, and therefore the availability of pasture. An approximated frequency and duration of *dzuds* and droughts was derived from 20th century weather and livestock statistics along with descriptive accounts from specific years and storms (National Statistical Office 2001, 2005: 173, Batima et al. 2008: 77, Begzsuren et al. 2004: 792). With the functional time span of the simulation set at 365 yrs (yr 135 to 500) it was estimated that 67 noteworthy weather events were likely to have occurred. Event timing was randomized within the appropriate seasons over the 365 year span.

Experiments and Results

Three simulation experiments were conducted to answer the central question—what are the factors that most affect wealth maintenance over the course of generations? Considering the length of time involved, the relative longevity of kinship lineages was the best measure of wealth maintenance. Three specific measures were used: longevity of individual lineages T, mean longevity of all lineages, and mean herd wealth for individual w_h and aggregated lineage groups w_a.

Experiment 1: Effects of Social Control. Comparisons were conducted between runs of the simulation, first with a higher degree of centrally controlled residence and kinship rules, followed by runs in which individual households and local group camps independently chose where and when to move and with whom to associate. The central question in this experiment asks: how does central control of kinship and mobility affect wealth sustainability? In a variety of real world and modeling studies, the maintenance of mobility is recognized as necessary for success of individual pastoralist families (Barnard and Wendrich 2008, Fernández-Giménez et al. 2012, Galvin et al. 2008, Kerven et al. 2006, Rogers et al. 2012). Greater mobility should improve the success of individual families, but this depends on the local environment.

Results of Experiment 1 are illustrated in Figs. 1a and b, showing human population dynamics with (2a) and without (2b) centrally controlled social rules. Figures 2a and b (run 81) describes the corresponding herd population dynamics. At the start of the simulation, human populations are small and continue to grow for several generations. Both sets of simulation runs show that at approximately 50,000 d (yr 135), the rate of population growth declines as the landscape nears capacity. The first severe weather events were introduced in both simulation sets at 93,126 d (256 yr). Comparing Figs. 1 and 2 shows that allowing lineage leadership to increase control resulted in significantly denser population than when households and camps are more independent (Fig. 2). Also, mean number of animals per lineage was lower when leadership control increased. However, mean household wealth over the entire 500 yr is similar under either lineage-level social control (mean = 273 animals) or local autonomy configurations (mean = 269 animals). The basic differences between the two simulations are summarized in Table 1.

The overall success of lineages was very different under the two levels of centralized control. Figure 3a and b shows the differences. In Fig. 3a, relatively restrictive social controls significantly limit the longevity of lineages. From the point at which the landscape was fully populated (approximately 140 yr [50k d]) to the end of the simulation (500 yr [180,060 d]), no lineage survived. By contrast, the absence of centralized control shown in Fig. 3b shows the much greater longevity of lineages. Of 151 lineages at time 140 yr there were 14 (9%) surviving until 500 yr. This is a small proportion, but still remarkable given the length of time involved. The mean herd wealth of the households in the 14 long-lived lineages is compared to those that did not survive in Table 2. This table also shows Gini coefficients calculated for the two groups. Long-lived lineages were wealthier at the beginning and end of the sequence, although the Gini coefficient shows that individual households were much less wealthy by the end of the sequence.

Experiment 2: Effects of Bailout and Marriage Strategies. Within the model there are only two properties that are unquestionably directed towards the maintenance and transmission of material wealth—bailout and marriage payment. Bailout is the sharing of a small number of animals for subsistence purposes if kin households fall below the

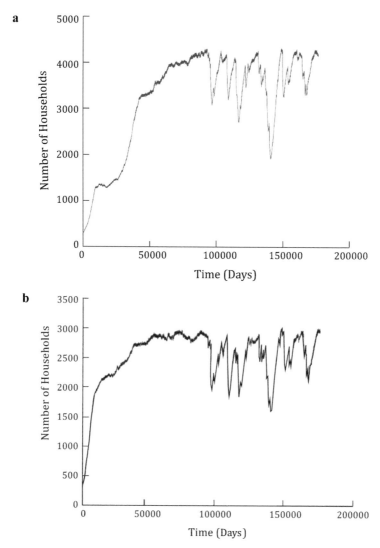

Fig. 1. (a) Human population levels under centrally controlled social rules; (b) human population levels with independent households (Run 80 series).

60 animal sustainability level at any point during the year. Anthropologically there are multiple kinds of marriage payments, including bridewealth, dowry, bride-service, gift exchange, token payments, and sister exchange (Goody 1973). In the simulation, no distinction is made between different forms of exchange; marriage payment is defined in a general sense as the transfer of herd animals from either of the parent households to the newly established household.

Whether parent households give more or less to an offspring household should affect the sustainability of lineages. In Experiment 1 the simulation was run with a 30% transfer of herd animals from one parent household to the daughter household.

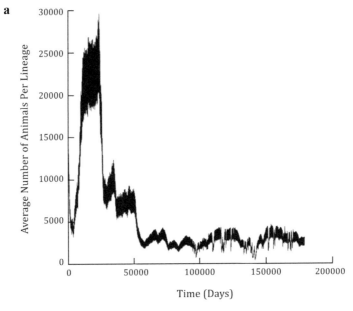

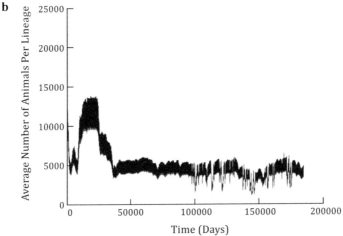

Fig. 2. (a) Herd population levels under centrally controlled social rules; (b) herd population levels under independent households (Run 81 series).

Table 1. Summary of results for two versions of simulation, one with centralized social controls implemented and one with local autonomy.

	Centralized Control	**Autonomous Households**
Human Population	High	Low
Herd Population	High	Low
Household Wealth	Moderate	Moderate
No. of Lineages	High	Low
Mean Herd Size per Lineage	Low	High

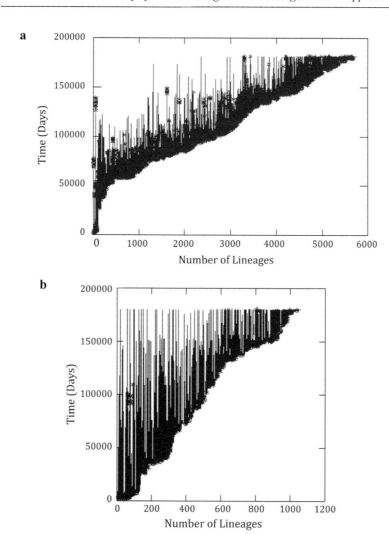

Fig. 3. (a) Lineage longevity under centrally controlled social rules (Run 80 series); (b) lineage longevity under independent households (Run 81 series).

Table 2. Statistics comparing the Longevity Lineages against the shorter lived lineages (Run 81 series).

Time (days)	Calculations:	Longevity Group (14 Lineages)	Other Group (137 Lineages)
Beginning: 49900	Mean herd wealth:	7818	4677
	t-value comparison:	0.0007	
	Gini Coefficient:	0.3755	0.2454
End: 180060	Mean herd wealth:	5695	4306
	t-value comparison	0.0247	
	Gini coefficient:	0.1623	0.2883

Additional runs of the simulation were conducted with the transfer rate reduced to 15%, but not less than the minimum survival level of 60 animals. Additionally, the effects of bailout versus marriage payment on lineage longevity were also compared. Runs of the simulation were conducted with the bailout function turned off and the marriage payment transfer rate set to 30% and with the converse, bailout turned on and marriage payment set to 0.00%.

Results of Experiment 2 show that both bailout and marriage payments had a very significant impact on the longevity of lineages. Eliminating either the bailout or marriage payment options resulted in no lineages surviving to the end of the simulation and an actual reduction in the number of lineages (Run 88 series). If the bailout was maintained, and the marriage payment reduced to 15%, rarely did any lineage survive to the end of the simulation. Within the parameters of the simulation, both marriage payment and bailout are necessary for long-term lineage survivability.

Experiment 3: Effects of Extreme Weather. Herd wealth is especially vulnerable to extreme weather. The simulation does not incorporate conflict, political shifts, or the economics of regional markets, but weather events are certainly a part of the environment module. Ethnographic studies have shown that households with less wealth in Inner Asia fare poorer in extreme weather events than their wealthier neighbors (Cribb 1991: 32, Fernández-Giménez et al. 2012: 7). An analysis was conducted on a series of four events in a specific run of the simulation (Run 82). In yr 262 (96987 d) a late spring snowstorm killed a large percentage of herds, resulting in an initial decline of 267 households (Fig. 4). Additional storms further reduced the number of households, with an eventual total loss of 710. Population levels did not totally recover for 39 yr (>14000 d).

To analyze the results from Experiment 3, the top wealth quartile was compared with the bottom quartile, before and after the series of weather events. In the wealthiest quartile 24 of 43 lineages survive the events and maintained their wealth. In the bottom quartile 24 of 42 lineages survived, 20 of which actually become wealthier. While 56% of the wealthiest lineages were able to maintain their position, it was unexpected that

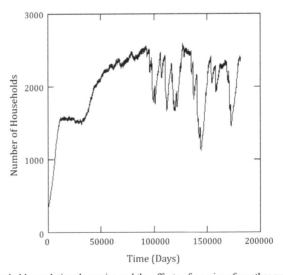

Fig. 4. Household population dynamics and the effects of a series of weather events (Run 82).

almost the same percentage (57%) of poor households also survived, and grew their wealth. Extreme weather events did not spell doom for the poor lineages and may even have opened up opportunities for herd growth in the less populated landscape.

Discussion

Simulation Results

Over time, kinship affiliations within a clan/lineage may become so attenuated that relationships are no longer recognized and the social group splits and new lineages emerge. This is far more likely to happen when one or more lineage households die out without producing daughter households. The timing of new lineage emergence is documented in the simulation (Fig. 3). When strict biological kinship affiliation is used single lineages rarely exist over the course of several hundred years. Logically, it is the exception to this pattern that illustrates the unexpected success of lineages as seen in Experiment 1. In the HHW model, contemporary Mongolian pastoralist kinship organization was used as the starting point, as illustrated in Fig. 5. The Mongolian information is based on interviews conducted with herder families in the Egiin Gol region of northern Mongolia (Erdenebaatar 2009). In Mongolia, kinship affiliation is normally not recognized beyond three generations vertically or beyond second cousins horizontally.

Although ethnographic sources for detailed wealth data in Inner Asia are rare, it is useful to compare one example with the simulation results. Vreeland (1957) conducted interviews with Khalkha Mongols living in the United States, but originally from the Narobanchin territory of Inner Mongolia (northern China). Although this is certainly secondary information, Vreeland accumulated enough wealth data to allow construction of the histogram in Fig. 6a. The Narobanchin pastoralist data shows a very common

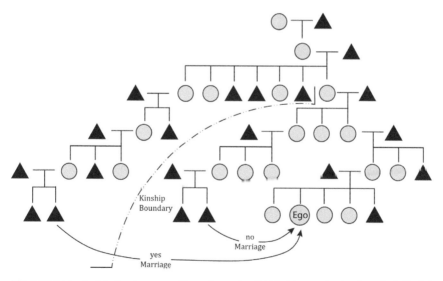

Fig. 5. Exogamy model of contemporary Mongolian pastoralist kinship organization from the Egiin Gol region (based on draft illustration by William Honeychurch and Diimaajayn Erdenebaatar).

a

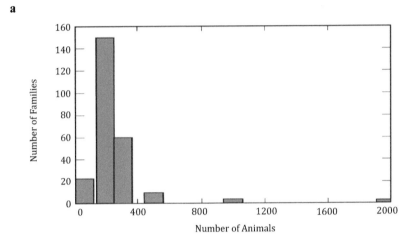

b

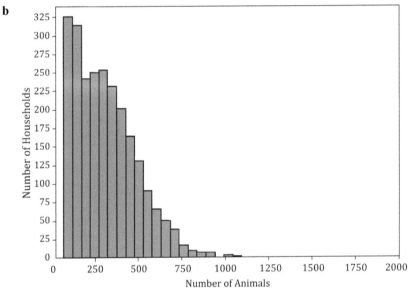

Fig. 6. (a) Reconstruction of the Narobanchin region household wealth profile based on 1950s ethnographic interviews; (b) a simulated household wealth profile (Run 90). The two bar charts have different scales because of the differences in the sources of information.

wealth distribution, with many very poor households, a few intermediate, and a very small number of wealthy—the classic many-some-rare pattern of the Pareto power-law (Cioffi-Revilla 2014: 161–168). Figure 6b illustrates the comparability of the simulation data in household wealth distributions based on a specific summer day in simulation Run 90. The simulation data has a large number of poor households, although none below 60 animals since this was the survivability threshold. Depending on exactly when a 'snapshot' of the wealth distribution in the simulation is taken, there may rarely be a few households with as many as 3,000 animals.

In the simulation, survivability of the wealthiest and the poorest quartiles was nearly the same. This is an unexpected result when compared to the ethnographic data. There are two reasons that account for this difference. First, the ethnographic data reports nuances of economic and social status and success is measured on an ordinal dichotomous scale of doing "better" or "worse". The simulation, however, uses only the crude measure of household and lineage survivability and does not include increasingly common disaster assistance that may come from outside the region. Secondly, the simulation is explicitly based only on material wealth; relational and embodied wealth are implied, but are not programmed characteristics of wealthy households. Clearly, in the real world wealthy families use social connections to their advantage.

Poor households may be more vulnerable to a specific event. However, viewed from the perspective of generational time scales and lineage survival, poorer lineages did survive in the simulation and were even able to increase their wealth. There is a practical limit to the size of herds being cared for by a single family (Vreeland 1957: 35–37). If wealthy households are near this limit (3,000 SSUs in this study) they have little room for growth, whereas poor households still have a labor surplus that permits growth, assuming sufficient grazing. Declining population in a region would undoubtedly open up mobility and new grazing opportunities for the remaining families. There is no empirical data that would allow direct comparison of long-term household survivability.

Broader Theoretical Implications

One of the most difficult aspects of simulation modeling is also its core advantage—it is truly dynamic. Time series data can be collected at scales ranging from days to millennia and from individual families to entire social groups. Rarely if ever is there access to such detailed empirical data. The richness of the simulated data makes scale recognition an unavoidable priority. For instance, should results be collected in the aggregate; if not, then where in the constantly changing passage of days and years? Herd wealth for any particular lineage fluctuates day by day, more so month by month, and year by year. Because of this variability, results are generally collected in the aggregate, but with a focus on time scales not available ethnographically (e.g., Irons 1994, Vainshtein 1980).

In situations where clans/lineages may be explicitly endogamous or exogamous, it is very unlikely that such social roles would continue in an unbroken chain for more than about 120 yr, assuming each generation is 20 yr long. These social characteristics exist in HHW and this is why lineages seldom persist for more than 100–150 yr. This is not to say that in the real world kinship never has great time depth, only that the mechanisms of fictive kinship become a factor. Royal and other elite lineages often represent an exception to the typical depth of ancestry recognition. In Mongolia for instance, ancestral ties to Chinggis Khan's (d. 1227 C.E.) lineage still play a social role and were an even more powerful connection to privilege and status in Mongolia before the communist revolution of 1921.

In 20th century Mongolia, wealthier families had more mobility options and were able to take advantage of the best grazing (Fernández-Giménez 1997, Vladimirtsov 1934). These families could afford the costs of moves and their more extensive social networks provided information and reciprocal opportunities. Poorer families could not easily afford the cost of moving and their social connections were more limited. In a recent major snow disaster in Mongolia, Fernández-Giménez et al. (2012: 7) note that

"Poor households lost a significantly higher proportion of their livestock compared to mid-level and wealthy herders....", consistent with global evidence on the positive correlation between poverty and disaster consequences. Herd losses have both short and long-term consequences for the family. It reduces the household's earning potential and consumption practices, affecting the nutritional status of the family, especially women and children. Social networks and obligations are also affected if a family becomes too poor to provide the proper gifts and honor reciprocal social expectations (Siurua and Swift 2002, Templer et al. 1993).

The maintenance of wealth across generations is a key aspect of how institutionalized inequality is developed and maintained (see Smith et al. 2010: 92). The existence of recognized elites is a key reminder that there is more to longevity of lineages than wealth on the hoof. Recognized privilege can outlive material wealth and is often the key to reestablishing fortunes. This can often be seen in dramatic fashion when a defeated royal house reemerges generations later to lead a new hegemony. Privilege does not entirely depend on recognized categories of elite group membership. Concepts like heterarchy recognize that there are socially accepted but officially unrecognized networks of leadership that may emerge from the personal abilities of individuals, such as oppositional social and political groups, subaltern groups, crime syndicates, and many others (Crumley 1995, Guha and Spivak 1988).

The institutionalization of status and the implication of greater access to wealth are likewise foundational in the complex history of early states and empires in Inner Asia. In particular Sneath questions the dichotomy of the state versus the non-state within the Inner Asian steppe, bolstering the argument that pastoralists, particularly within Mongolia, are not egalitarian (2007). This argument is highly debated among scholars of Inner Asian history and archaeology, for example Golden (2010) and Kradin and Skrynnikova (2009) heavily critique the assumption of a headless state in Mongolia. Sneath (2007: 3) argues that close examination of historical texts provides proof of the intergenerational transmission of wealth and privilege. Nineteenth-century social theory declared a distinction between the territorialized and stratified state societies and nomadic, egalitarian non-state societies. Egalitarian non-state societies became associated with 'primitive' or 'tribal' societies representing an inverted image to the modern state throughout the twentieth century. Levchine (1840), for instance, observed that the tribal model of kinship groups coming together to form larger aggregates is counter to what actually happened within the Kirghiz nation. The Kazakh polity was instead formed as administrative divisions created through a hierarchy of political administration. The formation of the polity was furthered by the emphasis on patrilineal descent groups used to structure and order society, but they were not the foundation of the process.

An example from the Ming Dynasty (1368 to 1644 C.E.) provides another instance of political strategy counter to prevailing cultural evolutionary thinking. After the fall of Mongol rule in China, patrilineal descent was used as the foundation of the taxation system. This actually forced groups to organize along kinship lines to fulfill tax obligations, rather than along the lines of their existing organization as politically segmented groups. This is an example of the state creating kin groups, rather than kin groups as precursors to the state. Further examples exist from the early 20th century under the influence of Soviet nation building (e.g., Munkh-Erdene 2006). Through historical records and accounts it can be seen how pastoralists within Inner Asia are part of a functioning aristocracy and not simply egalitarian as has often been assumed.

Conclusions

The nature of wealth transmission and inequality among mobile pastoralists is far more complex than characterized in the experiments here. Salzman, for instance, describes the political and social structure of mobile pastoralists in the Middle East and their relationship to the state (2004: 65–66). The complexities of the situation come less from internal social relations than from external political and power differentials. The simulation purposefully does not model political hierarchies beyond the lineage, nor external trade or migration, but instead seeks to focus on fundamental social and environmental relationships at the local scale.

In addition to the relationship between sustained wealth (economic advantage) and inequality there are several other sources of power used by elites in ranked societies like those being modeled here. Leadership strategies for maintaining control include genealogical seniority, supernatural authority, fictive kinship, force, sacred and long-distance knowledge, and political expertise. Ultimately, it is not just the accumulation of wealth or its sustainability over time, but how resources are translated and justified in the social web of privilege. Wealth alone is insufficient as an explanation for institutionalized inequality. However, the correlation between material wealth and forms of privilege is very strong.

In the experiments above, three aspects related to the transmission of wealth across generations were examined using the agent-based model. In Experiment 1, it was found that relatively independent households allowed greater longevity of lineages, compared to lineages with centralized control. Independent households have greater mobility which allows for better survivability in extreme weather events and generally better access to grazing. Experiment 2 dealt directly with intergenerational transfer of wealth and kinship-based sharing. Both marriage payments and bailouts in times of need were extremely important to lineage longevity. Experiment 3 examined the role of wealth in surviving extreme weather events. Surprisingly, it was found that wealthy and poor lineages suffered almost equally. Ethnographically, it is well known that poor households do less well in herd loss disasters. In the simulation both wealthy and poor lineages contain both wealthy and poor households. If the focus is on lineage survival then sharing among relatives allows survival of the social unit (e.g., Cooper 1993, 1995). More research is needed to clarify this result. However, in ethnographic findings wealthy households exercise their relational and embodied wealth to sustain their advantage. Those two features of wealth are less prominent in the agent-based model, but could be implemented in future versions.

Certain lineages grow and sustained their wealth for generations, through two simple mechanisms—marriage payment and sharing within lineages and local camps in times of crisis (bailout). A third ever present factor within the simulation was luck of access to preferred pasture. As might be expected, in some cases the actions of lineage leaders helped and sometimes their actions hindered group success. Ethnographically, the benefits of wealth are mobilized to enhance success. In the simulation the benefits of wealth work in even more subtle ways to solidify emergent patterns of grazing access. The three mechanisms of wealth differentiation provide an explanatory foundation for how social inequalities emerged and were sustained. Pastoralists have generally been considered marginal in the development of states, empires, and the earliest civilizations of the world. However, recent reanalysis of ethnographic data, and now computational modeling, have shown that simple methods of wealth transfer can sustain lineages over very long periods of time. These results support what we know from historical research including the existence of aristocratic lineages and centralized authority among pastoralists of Inner Asia.

Acknowledgements

We wish to offer a special thanks to several individuals who specifically helped in preparation of this study. We thank Meghan Mulkerin for her editorial eye and attention to detail, and Margaret Mariani for initial data collection. We thank William Honeychurch and Diimaajayn Erdenebaatar for access to ethnographic and archaeological data from the Egiin Gol region of northern Mongolia. The HouseholdsWorld model was developed with funding from the National Science Foundation (grant number BCS-0527471). Maciej Latek coded the simulation in Java using the MASON simulation toolkit developed by Sean Luke and collaborators at the EC Lab and Center for Social Complexity at George Mason University.

References Cited

Barnard, H. and W. Wendrich (eds.). 2008. The archaeology of mobility: old world and new world nomadism. Cotsen Institute of Archaeology, University of California, Los Angeles.

Batbuyan, B. 1997. Territorial organization of Mongolian pastoral livestock husbandry in the transition to a market economy. *In*: Proceedings of the Rural Development International Workshop. Food and Agriculture Organization of the United Nations (FAO), Rural Development Div., Rome, Italy. Retrieved from http://www.fao.org/docrep/w4760e/w4760e0o.htm.

Batima, P., L. Natsagdorj and N. Batnasan. 2008. Vulnerability of Mongolia's pastoralists to climate extremes and changes. pp. 67–87. *In*: N. Leary, C. Conde, J. Kulkarni, A. Nyong and J. Pulhin (eds.). Climate Change and Vulnerability. Earthscan, London.

Bedunah, D.J., D.E. McArthur and M. Fernández-Giménez (eds.). 2006. Rangelands of Central Asia: proceedings of the conference on transformations, issues, and future challenges. U.S. Department of Agriculture, Forest Service, Rocky Mountain Research Station, Fort Collins, Colorado.

Begzsuren, S., J.E. Ellis, D.S. Ojima, M.B. Coughenour and T. Chuluun. 2004. Livestock responses to droughts and severe winter weather in the Gobi Three Beauty National Park, Mongolia. Journal of Arid Environments 59: 785–796.

Bold, B.-O. 1996. Socio-economic segmentation—Khot-Ail in nomadic livestock keeping of Mongolia. Nomadic Peoples 39: 69–86.

Borgerhoff Mulder, M., S. Bowles, T. Hertz, A. Bell, J. Beise, G. Clark, I. Fazzio et al. 2009. Intergenerational transmission of wealth and the dynamics of inequality in small-scale societies. Science 326(5953): 682–688.

Borgerhoff Mulder, M., I. Fazzio, W. Irons, R.L. McElreath, S. Bowles, A. Bell, T. Hertz and L. Hazzah. 2010. Pastoralism and wealth inequality. Current Anthropology 51(1): 35–48. doi:10.1086/648561.

Bowles, S., E.A. Smith and M. Borgerhoff Mulder. 2010. The emergence and persistence of inequality in premodern societies. Current Anthropology 51(1): 7–17. doi:10.1086/649206.

Charles, K.K. and E. Hurst. 2003. The correlation of wealth across generations. Journal of Political Economy 111(6): 1155–1182. doi:10.1086/378526.

Cioffi-Revilla, C. 2005. A canonical theory of origins and development of social complexity. Journal of Mathematical Sociology 29 (April–June): 133–153.

Cioffi-Revilla, C. 2010. On the methodology of complex social simulations. Journal of Artificial Societies and Social Simulations 13(1): 7.

Cioffi-Revilla, C. 2014. Introduction to computational social science: principles and applications. Springer, London and Heidelberg.

Cioffi-Revilla, C., S. Luke, D.C. Parker, J.D. Rogers, W.W. Fitzhugh, W. Honeychurch, B. Frohlich, P. DePriest and C. Amartuvshin. 2007. Agent-based modeling simulation of social adaptation and long-term change in inner Asia. pp. 189–200. *In*: S. Takahashi, D. Sallach and J. Rouchier (eds.). Advancing Social Simulation: The First World Congress in Social Simulation. Springer, Tokyo.

Cioffi-Revilla, C., J.D. Rogers and M. Latek. 2010. The MASON Households World model of pastoral nomad societies. pp. 193–204. *In*: K. Takadama, C. Cioffi-Revilla and G. Deffaunt (eds.). Simulating Interacting Agents and Social Phenomena: The Second World Congress on Social Simulation, Vol. 7. Springer, Tokyo.

Cooper, L. 1993. Patterns of mutual assistance in the Mongolian pastoral economy. Nomadic Peoples 33: 153–162.

Cooper, L. 1995. Wealth and poverty in the mongolian pastoral economy, PALD Research Report No. 2. University of Sussex: Institute of Development Studies, Brighton.

Cribb, R. 1991. Nomads in archaeology. Cambridge University Press, Cambridge.

Crumley, C.L. 1995. Heterarchy and the analysis of complex societies. pp. 1–5. *In*: R. Ehrenreigh, C. Crumley and J. Levy (eds.). Heterarchy and the Analysis of Complex Societies, Vol. 6. American Anthropological Association, Washington, DC.

Dorfman, R. 1979. A formula for the Gini coefficient. The Review of Economics and Statistics 61(1): 146.

Erdenebaatar, D. 2009. Expedition report on ethnographic research conducted in the basin of the Egiin River in Khantai Baga, Khutag-Ondor Soum of Bulgan Aimag. Mongolian Academy of Sciences Institute of History, Mongolia.

Fernández-Giménez, M. 1997. Landscape, livestock, and livelihoods: social ecological and land-use change among the nomadic pastoralists of Mongolia. UMI, Ann Arbor.

Fernández-Giménez, M.E., B. Batkhishig and B. Batbuyan. 2012. Cross-boundary and cross-level dynamics increase vulnerability to severe winter disasters (dzud) in Mongolia. Global Environmental Change 22(4): 836–851. doi:10.1016/j.gloenvcha.2012.07.001.

Flanagan, J.G. 1989. Hierarchy in simple "egalitarian" societies. Annual Review of Anthropology 18: 245–266.

Flores Ettinger, S. and S.J. Linford. 2013. Working together: ethnography and archaeology. *In*: M. Mulkerin (ed.). Rogers Archaeology Lab. April 29. Retrieved from http://nmnh.typepad.com/rogers_archaeology_lab/2013/04/workingtogether.html.

Frachetti, M.D. 2008. Pastoralist landscapes and social interaction in bronze age Eurasia. University of California Press, Berkeley.

Galvin, K.A., R.S. Reid, R.H. Behnke Jr. and N.T. Hobbs (eds.). 2008. Fragmentation in semi-arid and arid landscapes: consequences for human and natural systems. Springer, The Netherlands.

Golden, P.B. 2010. Review of "The Headless State" by David Sneath. Journal of Asian Studies 68: 293–296.

Goody, J. 1973. Bridewealth and dowry in Africa and Eurasia. pp. 1–58. *In*: Bridewealth and Dowry (eds.). Vol. 7. Cambridge University Press, Cambridge.

Guha, R. and G.C. Spivak (eds.). 1988. Selected subaltern studies. Oxford University Press, New York.

Hansen, M., R. DeFries, J.R.G. Townshend and R. Sohlberg. 1998. UMD global land cover classification, 1 kilometer, 1.0. Department of Geography, University of Maryland, College Park.

Hansen, M., R. DeFries, J.R.G. Townshend and R. Sohlberg. 2000. Global land cover classification at 1km resolution using a decision tree classifier. International Journal of Remote Sensing 21: 1331–1365.

Humphrey, C. and D. Sneath. 1999. The end of nomadism? Society, state, and the environment in Inner Asia. Duke University Press, Durham.

Irons, W. 1994. Why are the Yomut not more stratified? pp. 275–296. *In*: C. Chang and H.A. Koster (eds.). Pastoralists at the Periphery. University of Arizona Press, Tucson.

Kawamura, K., T. Akiyama, H. Yokota, M. Tsutsumi, T. Yasuda, O. Watanabe and S. Wang. 2005. Comparing MODIS vegetation indices with AVHRR NDVI for monitoring the forage quantity and quality in inner Mongolia grassland, China. Japanese Society of Grassland Science 51: 33–40.

Kerven, C., I.I. Alimaev, R. Behnke, G. Davidson, N. Malmakov, A. Smailov and I. Wright. 2006. Fragmenting pastoral mobility: Changing grazing patterns in post-Soviet Kazakhstan. pp. 99–110. *In*: D.J. Bedunah, E.D. McArthur and M. Fernández-Giménez (eds.). Rangelands of Central Asia: Proceedings of the Conference on Transformations, Issues, and Future Challenges, Vol.

39. U.S. Department of Agriculture, Forest Service, Rocky Mountain Research Station, Fort Collins, Colorado.

Kleiber, C. and S. Kotz. 2003. Statistical size distributions in economics and actuarial sciences. Wiley, New York.

Kohler, T.A. and S.E. van der Leeuw. 2007. Introduction. pp. 1–12. *In:* T.A. Kohler and S.E. van der Leeuw (eds.). The Model-based Archaeology of Socionatural Systems. School for Advanced Research Press, Santa Fe.

Kradin, N.N. and T.D. Skrynnikova. 2009. "Stateless head": notes on revisionism in the studies of nomadic societies. Ab Imperio 4: 117–128.

Levchine, A.I. 1840. Description des hordes et des steppes des Kirghiz-Kazaks ou Kirghiz-Kaïssaks. (P.F. De Pigny, trans., C.A. Bertrand, ed.). Imprimerie royale, Paris.

Munkh-Erdene, L. 2006. The Mongolian nationality lexicon: From the Chinggisid lineage to Mongolian nationality (from the Seventeenth to the Early Twentieth Century). Inner Asia 8(1): 51–98.

National Statistical Office. 2001. Mongolian statistical yearbook 2001. National Statistics Office, Ulaanbaatar, Mongolia.

National Statistical Office. 2005. Mongolian statistical yearbook 2005. National Statistical Office, Ulaanbaatar, Mongolia.

Redding Jr., R.W. 1981. Decision making in subsistence herding of sheep and goats in the Middle East. Anthropology and Biological Sciences, The University of Michigan, Ann Arbor.

Rogers, J.D. 2012. Inner Asian states and empires: theories and synthesis. Journal of Archaeological Research 20: 205–256. doi:10.1007/s10814-011-9053-2.

Rogers, J.D. and C. Cioffi-Revilla. 2009. Expanding empires and the analysis of change. pp. 445–459. *In:* J. Bemann, H. Parzinger, E. Pohl and D. Tseveendorzh (eds.). Current Archaeological Research in Mongolia. Bonn University Press, Bonn, Germany.

Rogers, J.D., T. Nichols, T. Emmerich, M. Latek and C. Cioffi-Revilla. 2012. Modeling scale and variability in human-environmental interactions in Inner Asia. Ecological Modelling 241: 5–14.

Salzman, P.C. 2004. Pastoralists: equality, hierarchy, and the state. Westview Press, Boulder.

Siurua, H. and J. Swift. 2002. Drought and Zud but no famine (yet) in the Mongolian herding economy. Institute for Developmental Studies Bulletin 33(4): 88–97.

Smith, E.A., M. Borgerhoff Mulder, S. Bowles, M. Gurven, T. Hertz and M.K. Shenk. 2010. Production systems, inheritance, and inequality in premodern societies. Current Anthropology 51(1): 85–94. doi:10.1086/649029.

Sneath, D. 2007. The headless state: aristocratic orders, kinship society, and misrepresentations of nomadic inner asia. Columbia University Press, New York.

Takadama, K., C. Cioffi-Revilla and G. Deffuant (eds.). 2010. Simulating Interacting Agents and Social Phenomena: the Second World Congress. Springer, Tokyo.

Templer, G., J. Swift and P. Payne. 1993. The changing significance of risk in the Mongolian pastoral economy. Nomadic Peoples 33: 105–122.

Vainshtein, S. 1980. Nomads of South Siberia: the pastoral economies of Tuva. Cambridge University Press, Cambridge.

Vladimirtsov, B.Y. 1934. Obsèestvennyi stroi Mongolov, Mongol'skii koèevoi feodalizm (the social structure of the Mongols with reference to feudalism). Isdatelstvo Akademii Nauk, Liningrad.

Vreeland, H.H. 1957. Mongol community and kinship structure. Human Relations Area Files, New Haven.

24

Simulating the Emergence of Proto-Urban Centres in Ancient Southern Etruria

Federico Cecconi,[1,*] *Francesco di Gennaro,*[2] *Domenico Parisi*[3] and *Andrea Schiappelli*[4]

Computer Simulations in Archaeology and History

Computer simulations are a new way of expressing theories of human behavior and societies. A theory is no longer expressed in words or mathematical symbols but it is used as a blueprint for constructing a computational artifact. If the artifact reproduces the available empirical data, the theory is confirmed (Cangelosi and Parisi 1998, Cecconi and Parisi 1998). Furthermore, computer simulations can suggest new empirical data to look for and they can function as experimental laboratories in which the researcher varies the value of one or more variables and observes the effects of these variations.

When they are applied to archaeological and historical data, computer simulations can be used to test different reconstructions and explanations of the past proposed by different scholars (Andrighetto 2010, Bankes 2002, Berry et al. 2002, Bonabeau 2002) and, given the continuously increasing memory and computational capacities, computers can inaugurate a non-disciplinary science of history by making it possible to study the interactions between biological, cognitive, and social phenomena and to

[1] LABSS-ISTC-CNR, Rome, Italy, Via Palestro 32, 00185.
 Email: federico.cecconi@istc.cnr.it
[2] National Prehistoric and Ethnographic Museum, Rome, Italy, Piazza Guglielmo Marconi, 14, 00144.
 Email: francesco.digennaro@beniculturali.it
[3] ISTC-CNR, Rome, Italy, Via San Martino della Battaglia, 42, 00185.
 Email: domenico.parisi@istc.cnr.it
[4] Matrix96 - http://www.matrix96.it/-, Rome, Italy, Via Enrico Stevenson 26, 00162.
 Email: skya@libero.it
* Corresponding author

understand historical phenomena in which the nature of the geographical environment plays a critical role (Parisi 2001).

Two different tools have been developed for simulating social and historical phenomena: cellular automata and agent-based models. Both cellular automata and agent-based models assume the existence of a certain number of entities that receive an input from outside and generate an output which depends on both the input and the properties of the particular entity. Cellular automata are spatial models in which an entity is a cell in a grid of cells. Each cell receives an input from the adjacent cells and it responds with an output which affects the adjacent cells. Agent-based models are spatially more flexible. Each entity is an agent that can interact with other agents independently from space and an agent can receive inputs not only from other agents but also from the non-social (natural) environment (Parisi et al. 2003).

Two examples of historical simulations based on a cellular automata model concern the expansion of the Assyrian Empire in the Near East from the fourteenth to the seventh century BC (Parisi 1998) and the diffusion and differentiation of Indo-European languages in Europe as a result of the gradual expansion of agriculture in Europe (Parisi et al. 2008). In the first case the simulation replicated the available historical data but it did not reproduce the abrupt collapse of the Assyrian Empire in the last part of the seventh century BC, which indicates that other factors must be included in the simulation. In the second case, the simulation reproduces the "similarity tree" of Indo-European languages on the basis of the principle that the diversification of languages depends on the length of time elapsed since the separation of a group of people from a larger group of people who spoke the same language. One well-known example of application of the agent-based methodology is the simulation of the settlement dynamics of Anasazi Indians in the south-western United States in the early centuries of the second millennium of CE. This simulation confirmed the hypothesis that the sudden depopulation of a long inhabited valley was not due to the irreversible depletion of the soil but to endogenous reasons concerning the community itself (Dean et al. 2000).

In the present study we describe an agent-based simulation of changes in human settlements in Ancient Southern Etruria (Italy) during the second millennium BC, where an agent is a village existing in an environment that contains both other villages (other agents) and natural resources that vary in space and time. Each agent is defined by a set of rules which, given certain conditions (input), determine what the agent does (output)—for example, the agent can do nothing, it can move to another place, or it can fight with another agent. The results of the simulation are then compared with the archaeological and historical data on human settlements in this historical period in this geographical area.

Agent-based simulations are based on the idea that their results are entirely explained by their internal mechanisms, namely, the program instructions. By representing an agent as a piece of the program, it is possible to simulate an artificial world inhabited by interacting agents—individuals, communities of individuals, social institutions. The properties of each agent are specified with all the required details and these properties determine the agent's behavior and how its behavior affects other agents.

Agent-based simulations can be used for different purposes and, in particular, they can be used to test hypotheses about the emergence of social structures from the interactions of a set of individuals by finding the minimal conditions at the micro-level that are necessary to generate the structures at the macro-level. This type of agent-based simulations can contribute to understanding social systems by relating behaviors to structural and organizational properties and they can be used to integrate theories from various disciplines—cognitive psychology, sociology, economics, cultural anthropology, and history—into a general explanatory framework (Bonabeau 2002).

Though the use of simulations in anthropology and archaeology goes back to the 1960's, there is, as yet, no consensus about the direction that the simulation method should take or is likely to take in the future, and this may reflect the theoretical and methodological heterogeneity of the field. However, most simulations share a common concern with addressing research questions which are methodologically difficult, if not intractable, when approached with more traditional methods. Two examples are the demographic simulations undertaken by Nancy Howell (Howell and Lehotay 1978) to better understand the population dynamics of hunting-gathering groups and the simulations of David Thomas (Thomas 1973) aimed at modeling the distribution of paleo-Indian sites in the Great Basin of the western part of the United States. The first of these two simulations developed out of the Harvard Bushman (San) Research Program begun in 1963 among the !Kung San of Botswana and directed by Richard Lee and Irvin DeVore. The program was concerned with the biological, demographic, ecological, social, and cultural aspects of the !Kung San. The simulation made it possible to construct a demographic profile of the !Kung San notwithstanding the sampling issues that arise from both the small size of the population (only several hundred individuals) and the lack of direct information on the age of the individuals. The simulation kept track of individuals as they aged, gave birth, and died but, as only a relatively small number of anthropologists were familiar with computing methods at that time, the simulation software was not widely used beyond analysis of the demographic structure of the !Kung. However, this pioneering research has had an extensive impact as a detailed demographic study of a small-scale, non-Western society.

Thomas' simulation had a different kind of impact. While Howell's simulation addressed primarily problems that arise from stochastic effects in small populations coupled with incomplete demographic data, Thomas' simulation attempted to account for the distribution of artifacts and the spatial location of paleo-Indian sites in the Great Basin of the United States by applying then current theories about the factors that affect the spatial location of hunting and gathering groups as they underwent their yearly round. The verification problem faced by archaeologists when positing an argument for site location is substantial. Archaeologists have only artifactual remains as indicators of the locus of activities but not the nature of the activities and hence the general character of the archaeological site (e.g., permanent village, temporary camp, hunting site) are necessarily inferential. Even more problematic are attempts to infer the reasons why those activities took place at a particular geographic locality. Thomas saw simulations as providing one way to model the distribution of sites across space using an ecological model of site location. In addition, the model also made assumptions about the likelihood that artifactual materials would be discarded rather than carried away when the group moved to another locality. While the intent of the !Kung San demographic simulation was to construct a "best possible" demographic profile of the !Kund San, the importance of Thomas's simulation lay not so much in its ability to accurately model site location and spatial density of artifacts as in providing data that could be used to assess theories archaeologists were proposing about site formation processes and site location.

These two themes—simulations as tools to work out the consequences of processes already reasonably well modeled versus simulations as attempts to clarify what those processes and their consequences might have been—are both related with the problem of empirical evidence. To illustrate this problem we mention two other simulations which, though quite different in content, are both concerned with the problem of deciding whether or not a posited process accounts for a set of empirical observations. Di Piazza and Pearthree (1999) wanted to know if the process of down-the-line migration as posited by the archaeologist Patrick Kirch (1997) is consistent with the archaeological data on the migration and colonization of the Lapita peoples in the southwest portion of the Pacific which began about 3500 years ago and took place over a time period of about 500 years. For Di Piazza and Pearthree the validity of the migration model, per se, was not at issue. They were only interested in knowing if the model predicts (produces) results which are consistent with the archaeological data. They constructed a simulation which was in accordance with the posited migration process and they used a range of parameter values (growth rate, migration rate, and initial population size) estimated from historical data. What they found was that the historical pattern of colonization, measured in terms of number and population size of colonies over the time period in question, could not be duplicated except with unrealistic parameters such as a 90% migration rate. The simulation results led them to suggest a different model for the migration that removes the assumption of the sequential character of colony formation inherent in the down-the-line migration model. This assumption implies that colony formation is only weakly affected by earlier, distant colonies. For example, with a 10% migration rate from a colony, only 0.1% of the initial population would directly become part of migrants making up the next colony with down-the-line migrations. In place of the down-the-line migration model they suggest "sea-nomadism" as a way to account for both the rapid development of colonies and the archaeological evidence that suggests early settlement and approximately equal population sizes throughout the entire region colonized by the Lapita peoples.

The simulation undertaken by White (1999) was concerned with establishing if a marriage-based social network has a pattern which is only due to stochastic processes biased by demographic factors or, after removing stochastic and demographic biasing effects, the marriage structure is still patterned. In some cases there might be an a priori expectation of patterning, for example, patterning arising from marriage rules that form what Lévi-Strauss calls elementary kinship systems. In elementary kinship systems marriage rules (whether preferential or proscribed) are expressed in terms of kinship categories. According to White, previous work raised the question of whether such systems did, in fact, induce patterning, and demographic simulations suggested that, once demographic factors were taken into account, the incidence of prescribed marriages was no greater than would be expected by chance. In other societies where marriage is guided only by prohibitions such as incest taboos, patterning in marriages might arise from what White refers to as *strategies*, namely marriage decisions that take into account a variety of factors such as inheritance and prior family alliances. In both cases the methodological problem identified by White is disaggregating the social aspect of marriage choice from the demographic aspect. White does so by constructing a baseline of random marriage choices that takes into account the actual frequency of marriage at a specified level of the marriage structure. Statistical comparisons were made between the actual marriage frequency for some specified criterion (e.g., spouses have similar age) with the frequency of that kind of marriage in a distribution of marriages produced through random permutation of actual marriage partners made in accordance with whatever structural criteria are used to define the comparison level (e.g., all marriages, ignoring age). According to White, "we would expect this subset [of marriages satisfying a specified criterion] to have non-random characteristics in comparison to a uniform-probability model at that level. Marriages of similar-age spouses, for example, appear non-random against a uniform-probability model across all age groups". The method obviates the need for modeling the demographic history of the population being examined by using random permutations of actual marriage partners to generate a baseline frequency distribution against which comparisons of actual marriage frequencies can be made. The baseline model captures the patterning due to both the historical demographic facts of the population and the criteria affecting marriage choice, much in the way a regression model separates variation due to the regression model from other effects, thereby making it possible to identify any patterning in the residuals above and beyond what is expressed in the regression model. This also provides a method, as White discusses, that is sensitive to the distinction made by Lévi-Strauss between elementary and complex kinship systems.

Simulations vs. Mathematical Modeling

An interesting question is how agent-based simulations are related to equation-based or, more generally, mathematical models of social and historical phenomena. Agent-based simulations are not in contrast with mathematical models but, unlike mathematical models, they are intended to be used at a local level as analogical mappings of a real system. And from this mapping one can derive global parameters that can be incorporated into a mathematical model. In agent-based simulations, numerical data and statistics are not eliminated but they are used as evaluation procedures to compare the simulation results with data coming from the 'real' world.

An important difference between agent-based simulations and mathematical models is in the capacity of agent-based simulations to capture different stochastic aspects of the phenomena. Agent-based simulations describe stochastic fluctuations. Mathematical models describe the statistics of the fluctuations, for example, the mean and shape of the distribution. The results of the simulation may fit the analytical model and they may provide *in silico* data for developing mathematical models that make it possible to generalize the simulation results and to make predictions. Mathematical models make it possible to describe the relation between the strategies and their payoffs through closed-form equations and to generalize the simulation results, allowing accurate predictions. Agent-based simulations generate controlled data on the basis of variables, properties, local rules, and other factors decided by the modeler and these artificial data are useful to test the theory and to reduce its complexity. Mathematical models "close" the data, making it possible to falsify the theory (Cecconi et al. 2010, Gilbert and Bankes 2002).

Agent-based simulations and mathematical models can be compared in other ways. Mathematical models incorporate hypotheses on how the phenomena of interest are structured because they require specifying a certain number of variables and their range of application. Almost all mathematical models are expressed as a set of equations that define the dynamics of the phenomena and there are a number of mathematical techniques that generate indications on the phenomena's asymptotic behaviour. Agent-based simulations create artificial worlds made of a certain number of "agents" to determine if and how the behaviour of the agents generates structures and patterns. They are aimed at showing how the interactions among the agents and between the agents and the environment produce emerging phenomena which are complex and unexpected and how small variations in the initial conditions can result in very different results. They follow a "bottom up" strategy and they can be called generative models because they "generate" the phenomena that we want to explain.

Three properties characterize agent-based simulations (Epstein 1999): non-linearity, heterogeneity, and explicit extension. Non-linearity means that there is no "top-down" control of the model. What happens in the model is a result of the interactions among the agents, is not imposed from outside, and is strongly dependent on the initial conditions. Heterogeneity refers to the fact that each agent is characterized by rules which are different from the rules of other agents and this makes the interactions among the agents unique and unpredictable. Explicit extension means that the development of the model takes place in a well-defined environment which may be very complex and made of parts with different properties.

Can these properties be replicated by mathematical models? Are there phenomena that can only be modelled with agent-based simulations and are intrinsically impossible to model with an approach based on mathematical equations that describe their dynamics? Most phenomena studied by archaeologists, historians and, more generally, students of human behaviour and human societies are phenomena in which the agents are very numerous and very different from one another, the interactions among the agents play a central role, and non-linearity is the rule, not the exception. If we exclude non-linearity, which can be easily incorporated in mathematical models, the other two properties, heterogeneity and explicit extension, pose serious problems—both theoretical and practical—if we want to write a set of equations that describes the dynamics of the system.

Heterogeneity means that the agents have different properties, not the same properties with different values. The molecules of a gas are not heterogeneous because, even if they are very numerous, they are all definable in terms of a small number of state variables, which, in classical mechanics, are position and quantity of movement. All molecules possess these variables and, therefore, even if the distribution of the values of these variables can be very complicated and can actually be described only with a statistical approach, we cannot speak of heterogeneity. For explicit extension the problem is the same. If the equations must incorporate homogeneous space, models based on differential equations with partial derivates, which are at the basis, for example, of field physics, play their role egregiously. But if space becomes in some way heterogeneous, things become more complicated.

Our conclusion is that archaeological, historical and, more generally, human social phenomena possess characteristics that can be easily incorporated in agent-based simulations but not in equation-based models. In principle, it might be possible to increase the number of the variables that define the agents or the environment but how far we can go? Is it possible to imagine tens and tens of differential equations that describe the changing value of hundreds of variable?

In this chapter we first formulate a mathematical model that describes the dynamics of a population of agents (human settlements) using the concept of ZOC (zone of control). The mathematical model identifies two dynamics and we study these dynamics in a strongly heterogeneous physical environment using an agent-based model.

Proto-Urban Centers in Ancient Southern Etruria

Our simulation reproduces the phenomenon of the emergence at the beginning of the first millennium BC of proto-urban centers in Ancient Southern Etruria, an area of about 6000 square kilometers corresponding roughly to northern Lazio (a region of Italy), between the rivers Tiber and Fiora. Human communities living in this area were characterized by a growing interest for forms of settlement that ensure safety to people and probably also to resources (cattle, food, storage, etc.). This is evidenced by the steady increase in the percentage of settlements located on summit plateau heights called 'castelline', often completely isolated from the valley floor and defended by steep slopes (Patterson et al. 2000, Patterson et al. 2004, Rich and Wallace-Hadrill 2003).

At the beginning of the Bronze Age, the area was characterized by a multitude of settlements with an average size of around 5 hectares and widely distributed in the environment (Di Gennaro and Schiappelli 2002). Then, in the course of the second millennium we see a radical and widespread depopulation of these settlements and the emergence of a few centers located in more defensible positions and with a considerably higher dimensional scale: Vulci (160 hectares), Tarquinia (150 hectares), Orvieto-Volsinii (85 hectares), Cerveteri (150 hectares), Veio (185 hectares), and Bisenzio (80 hectares). These large agglomerations are called 'proto-urban centers' and they will become in later centuries the main Etruscan cities (Brown and Ellis 1995, Dyson 1982).

Structure of the Simulation

The first step of the simulation was to construct a model of the environment. The territory was divided into a grid of 90,000 square cells, with each cell representing an area of about 5 hectares. The cells are characterized by three properties: (a) soil quality and presence of natural resources, (b) existence of water courses, (c) morphology of the ground from the point of view of defensibility—and the model assumes that each settlement decides what to do on the basis of these properties. The attribution to the cells of these properties was made possible by the creation of three thematic maps: (1) a map of resources, (2) a hydrographic map which divides the streams into three classes according to the average flow rate, and (3) a map of the defensibility of places. This last map was based on identifying all areas morphologically equipped with natural defenses and on assigning to each cell one of five different levels of defensive potential. (The maps were developed by Andrea Schiappelli and they can be downloaded from www.edulabss.com.)

We also prepared historical maps of the entire area which indicate the distribution of settlements archaeologically documented in the course of the second millennium BC and the location of each settlement was assigned an index of defensive potential (Fig. 1).

These maps provide the indispensable term of reference for the simulation because the simulation can be considered as successful if it replicates the historical maps. The settlements, or village communities, are software agents, consisting of a number of individuals, N, which grows as a function of time, initially according to a constant index. Each settlement controls an area of the surrounding territory (called 'zone of control', ZOC), which is more or less large as a function of N. Each ZOC has a size and a quality which is measured in terms of the resources present in its territory. The resources collected in the ZOC are divided in equal parts among the inhabitants of the

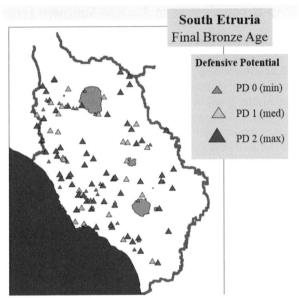

South Etruria
Final Bronze Age

Defensive Potential

△   PD 0 (min)

△   PD 1 (med)

▲   PD 2 (max)

Fig. 1. The map of defensive potential.

settlement, providing the main parameter of the simulation, the 'energy per capita'. The 'energy per capita' is consumed according to a fixed rate and, therefore, it does not increase indefinitely. For a community to settle in a well-defended position has two advantages. If the community lives in a place with high defensive potential, a bonus is applied to the productivity value of the land within its ZOC. This bonus captures two phenomena: (a) a better control of the surrounding territory, and (b) the possibility to store food and shelter cattle in places safe from external raids. Secondly, being positioned in highly defensible sites implies obvious advantages in circumstances of conflict, as, for example, in the case of a siege determined by the attack of a hostile group.

The simulation is a succession of discrete time steps and, in each time step, the condition of each settlement is updated as function of the decisions of its inhabitants. One time step is equivalent to two years and, since the simulation begins in 2300 BC and covers a period of approximately 1500 years, a community takes around 750 successive decisions.

Once an environment such as the one described is set up, we run a simulation to reproduce the process that led to the appearance towards the end of the second millennium of a few large centers in well-defended sites and numerous small settlements. To reduce more complex calculations, this preliminary simulation did not foresee the use of a hydrographic map. A partial justification for this choice is that, given the wide diffusion of water in Etruria, its supply must not have constituted a problem for proto-historic communities. Furthermore, since Etruscan waterways are numerous but not very large, they would not have represented insuperable obstacles to the movements of a group.

At the beginning of the simulation (Early Bronze Age), 250 settlements are placed in the territory, projecting over the entire area a density found in areas which are better known from the archaeological point of view. Based on archaeological data, these settlements are located in sites with low defensive potential. For each settlement the number of individuals is 80, for a total of 20,000 individuals present in Ancient Southern Etruria. During each cycle of the simulation, each settlement (a) takes into consideration its N (number of inhabitants), (b) defines its zone of control with respect to the surrounding area, and (c) calculates how many resources are available for its inhabitants, resulting in the relative value of resources per capita.

The simulation can develop according to two different types of dynamics; a positive and a negative dynamics. A positive dynamics means that the number of inhabitants of a settlement increases together with an increase in the size of its zone of control. Therefore, the available resources also increase and the value of the resources per capita remains high. The settlement is a prosperous one. On the contrary, a negative dynamics implies an increase in the number of inhabitants but not of the settlement's zone of control because of the presence of the zones of control of other settlements or because the zone of control is made of soil with low productivity. In this case, the resources per capita become insufficient, and the settlement is in trouble. When the resources do not meet the needs of a settlement's inhabitants, the number of inhabitant decreases and, if is reduced to zero, the settlement disappears.

To avoid a complete collapse, the settlement can choose among three different options: Split, Move, and Fight.

1. Split. A certain number of inhabitants leave their village and settle in other parts of the environment inside or outside Ancient Southern Etruria so that the resource requirements of the remaining inhabitants can continue to be satisfied. This is the less expensive and less painful solution and there is both archaeological and historical evidence that in some cases it was actually adopted. (In the classical era this solution was called *ver sacrum*—"sacred spring", probably the season in which mostly young people left their village to settle elsewhere.)
2. Move. The entire village moves to another place in a more productive piece of land in which the areas of control of other settlements do not prevent the expansion of the village.
3. Fight. This solution is to opt for war as a means to expand the zone of control of one's village at the expense of the neighboring villages, choosing the area productively more profitable. This is the most expensive and dangerous solution.

Analytic Model

With this agent-based simulation we can capture the relations between the properties of the environment and the behavior of the agents. In particular, we can find how the resources available to a settlement are related to its population dynamics. For each settlement we have three variables, N, ZOC, and R, where N is the number of inhabitants, ZOC is the size of the area controlled by the settlement, and R is the quantity of resources contained in the settlement's ZOC. We use these three variables to describe an equation-based model.

Equation 1.1 describes the dynamic of N:

$$\dot{N} = aN(1 - N/K)$$

which means that the size of the group depends on a logistics dynamic. K is the carrying capacity of the settlement and K is a function of ZOC and R. In fact, the number of individuals who can live in a certain area depends on the extent of their ZOC and on the resources contained in their ZOC. We rewrite Equation 1.1 as Equation 1.2:

$$\dot{N} = aN(1 - \frac{N}{(zoc \cdot R)})$$

Now, suppose that the settlement is in a potentially exploitable area, for example in a circular area surrounded by soils with higher profitability (see Fig. 2).

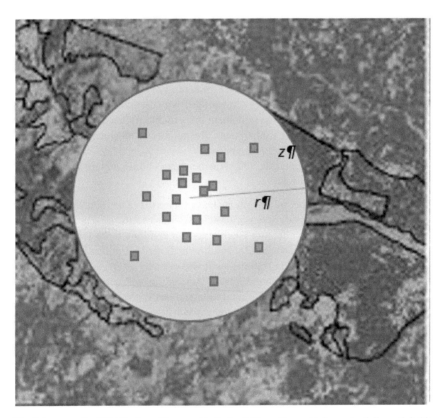

Fig. 2. An example of distribution of the agent in the land. The settlement is in a potentially exploitable area.

We can write:

$$R = f(zoc)$$

and the function f can be written easily by integrating along the radius r:

$$R = \int_0^{\sqrt{zoc/\pi}} \beta 2\pi r^2 \, dr = \frac{\beta 2\pi}{3} (zoc/_\pi)^{\frac{3}{2}}$$

This is Equation 2 and it described R as a function of ZOC.

Finally, we imagine a linear relationship between ZOC and N: ZOC grows when N grows. This is Equation 3:

$$zoc = \gamma N$$

A study of the dynamics based on Equation 1.2 and the relation between Equation 2 and Equation 3 leads to three conclusions that confirm what we have said before:

1. there are two opposite trends, one that leads to the disappearance of the settlement or to its indefinite growth
2. a, β and γ are critical
3. the linear relationship between ZOC and N is a "pretty" assumption but it's not realistic. In fact, the relationship is a function of the orographic properties which remain the main focus of the model.

By applying the rules that we have described, at the end of the simulation we find that some results replicate the archaeological evidence while other results don't. The virtual Ancient Southern Etruria appears to be divided into five main zones of control and four of these five zones of control correspond to the historical proto-urban centers of Orvieto-Volsinii, Tarquinia, Cerveteri and Veio. These four settlements have a considerable number of inhabitants and this number is expressed by the amplitude of the radius of their circle. The absence of a zone of control corresponding to the site of Vulci can be explained by the fact that a large portion of its pertaining territory lay outside the geographical boundaries of the simulation. In contrast, the absence of a zone of control relative to the center of Bisenzio can be explained by other factors since the settlement becomes extinct immediately after the archaic period.

Among the results that differ from those historically expected is that the number of settlements decreases too rapidly (Fig. 3).

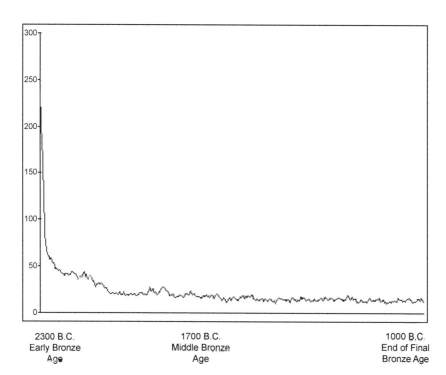

Fig. 3. The number of settlements expected from the simulation.

In the simulation, already during the virtual Early Bronze Age the system seems to undergo a collapse, going from 250 villages to about 60, while in the following centuries it remains roughly stable, with limited fluctuations. This diverges from what we know from the archaeological evidence which tells us that, after an increase between the beginning of the Early Bronze Age and Middle Bronze Age, the total number settlement remains pretty stable until, in the First Iron Age, the number of settlements is drastically reduced. This phenomenon is interpreted as due to a gradual but steady population growth but it is not captured by our simulation.

In fact, the real population growth is captured only in part by the simulation (Fig. 4). The average number of individuals for settlement increases from around 100 individuals at the beginning of the simulation to 1600 individuals to the end of the simulation—and this is in accordance with the archaeological data. However, the total size of the real population is slightly smaller and one can conclude that wars—one of the most practiced options in ancient Italy—have led to temporary reductions in the number of individuals. But the effects of wars are not reproduced in our simulation. On the contrary, the proportion of settlement in high defensive potential sites and the trend to settle down on sites with high Defensive Potential is reproduced (Fig. 5 and Fig. 6).

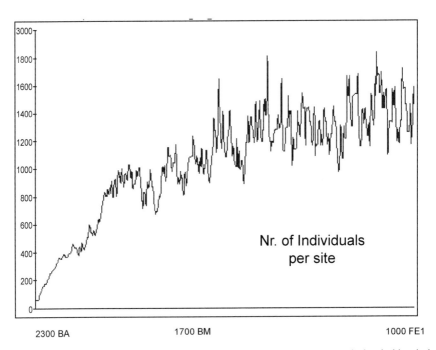

Fig. 4. The average number of individuals per settlement. This value increases (reproducing the historical data), but the total population decreases.

Conclusion

What have we learned from our simulation? By using the concept of zone of control we have obtained a distribution of proto-urban centers very similar to that observed empirically. This result may be relevant for an issue that remains open in research on the

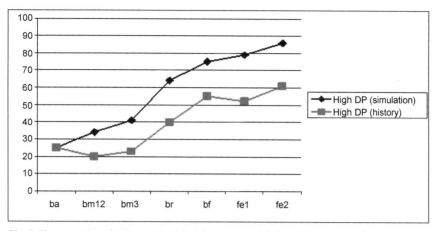

Fig. 5. The proportion of settlement in high defensive potential sites. The trend to settle down on sites with high Defensive Potential is reproduced.

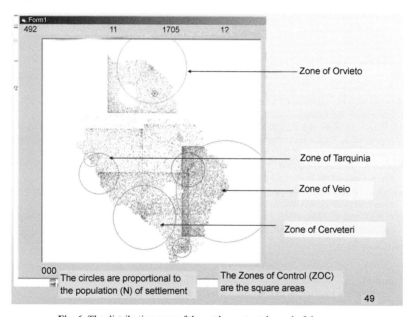

Fig. 6. The distribution map of the settlements at the end of the process.

Middle Bronze period: Is demography the variable with the greatest role in deciding the distribution of resources in a territory? The three strategies (Split, Move, and Fight) do not have any 'logistics' management of their dynamics, such as the creation of centers of assignment or routing resources. In addition, we must remember that the maps used for the calculation of the effectiveness of ZOC only indirectly affect the dynamic of the historical process. So, we have some indication that demography is the main driver for the creation of new centers.

Our simulation can profitably complement other methods of historical and archaeological research. Computer simulations are gradual and iterative. Once you have made a basic simulation and you have analyzed its results to verify your underlying model, you can make all sorts of changes to the model and add new variables so that the model/simulation better replicates the empirical data. In our case some additional variables are:

— number of calories needed for the survival of an individual
— better calculation of annual productivity, expressed in calories
— different types of land
— obstacles (mountains, lakes) and both natural (the Tiber river and its valley, the Tyrrhenian sea) and man-made communication and transportation lines (roads, boats, ports)
— level of social complexity of settlements as a function of number of individuals, with the emergence of élites which transform resource surplus into infrastructure and war
— technologies for the production of new goods such as weapons, agricultural tools, and boats. For example the factor "technology" can explain the location of new settlements on the coast of the Tyrrhenian Sea.

References Cited

www.edulabss.com.

Andrighetto, G., M. Campenni, F. Cecconi and R. Conte. 2010. The complex loop of norm emergence: A simulation model 'Simulating interacting agents and social phenomena'. Springer.

Bankes, S.C. 2002. Agent-based modeling: A revolution? Proceedings of the National Academy of Sciences of the United States of America.

Berry, B.J., L.D. Kiel and E. Elliott. 2002. Adaptive agents, intelligence, and emergent human organization: Capturing complexity through agent-based modeling. Proceedings of the National Academy of Sciences of the United States of America.

Bonabeau, E. 2002. Agent-based modeling: Methods and techniques for simulating human systems. Proceedings of the National Academy of Sciences of the United States of America.

Bourland, Lynn. 1970. Effectiveness of simulation in teaching selected principles of learning to home economics education students. MS Thesis. Texas Tech University.

Brown, A.G. and C. Ellis. 1995. People, climate and alluviation: theory, research design and new sedimentological and stratigraphic data from Etruria. Papers of the British School at Rome.

Cangelosi, A. and D. Parisi. 1998. The emergence of a 'language' in an evolving population of neural networks. Connection Science 10(2).

Cecconi, F., M. Campenni, G. Andrighetto and R. Conte. 2010. What do agent-based and equation-based modelling tell us about social conventions: the clash between ABM and EBM in a congestion game framework. Journal of Artificial Societies and Social Simulation 13(1): 6.

Cecconi, F. and D. Parisi. 1998. Individual versus social survival strategies. Journal of Artificial Societies and Social Simulation 1(2).

Dean, Jeffrey S., George J. Gumerman, Joshua M. Epstein, Robert l. Axtell, Alan C. Swedlund, Miles T. Parker and Steven McCarroll. 2000. Understanding Anasazi culture change through agent-based modeling. Dynamics in human and primate societies. Oxford University Press, Oxford pp. 179–206.

Di Gennaro, F. and A. Schiappelli. 2002. Recent research on the city and territory of Nepi (VT). Papers of the British School at Rome pp. 29–77.

Di Piazza, A. and E. Pearthree. 1999. The spread of the 'Lapita people': a demographic simulation. Journal of Artificial Societies and Social Simulation 2(3).

Dyson, S.L. 1982. Archaeological survey in the Mediterranean basin: A review of recent research. American Antiquity.

Epstein, J.M. 1999. Agent-based Computational Models and Generative Social Science. Complexity 4(5): 41–60.

Gilbert, N. and S. Bankes. 2002. Platforms and methods for agent-based modeling, Proceedings of the National Academy of Sciences of the United States of America 99(Suppl 3).

Howell, N. and V. Lehotay. 1978. AMBUSH: A Computer Program for Stochastic Microsimulation of Small Human Populations. American Anthropologist 80(4).

Parisi, D. 1998. A cellular automata model of the expansion of the Assyrian empire. Cellular Automata: Research Towards Industry. Springer London.

Parisi, D. 2001. Simulazioni: la realtà rifatta nel computer. Il mulino.

Parisi, D., F. Cecconi and F. Natale. 2003. Cultural Change in Spatial Environments. The Role of Cultural Assimilation and Internal Changes in Cultures. Journal of conflict resolution 47(2).

Parisi, D., F. Antinucci, F. Natale and F. Cecconi. 2008. Simulating the expansion of farming and the differentiation of European languages. Origin and Evolution of Languages: Approaches, Models, Paradigms.

Patterson, H., F. Di Gennaro, H. Di Giuseppe, S. Fontana, V. Gaffney, A. Harrison, S.J. Keay, M. Millett, M. Rendeli, P. Roberts and others. 2000. The Tiber Valley Project: the Tiber and Rome through two millennia. Antiquity 74(284).

Patterson, H., F. Di Gennaro, H. Di Giuseppe, S. Fontana, M. Rendeli, M. Sansoni, A. Schiapelli and R. Witcher. 2004. The re-evaluation of the South Etruria Survey: the first results from Veii.

Rich, J. and A. Wallace-Hadrill. 2003. City and country in the ancient world. Psychology Press.

The Lapita Peoples: Ancestors of the Oceanic World by Patrick Vinton Kirch and Thomas S. Dye (1997). Journal of World History 11(2): 361–364.

Thomas, D.H. 1973. An Empirical Test for Steward's Model of Great Basin Settlement Patterns. American Antiquity 38.

White, D.R. 1999. Controlled Simulation of Marriage Systems. Journal of Artificial Societies and Social Simulation 2(3).

25

The Probabilities of Prehistoric Events: A Bayesian Network Approach

Juan A. Barceló,[1,*] *Florencia Del Castillo*[1] and *Laura Mameli*[2]

Introduction

Archaeology is the science of "historical gaps"; we always deal with incomplete knowledge, with fragments and materiality left by human action. To make sense of this limitation is one, if not the most relevant of the goals of our discipline. Dealing with incomplete series of archaeological observations of broken objects and remains means dealing with uncertainty. In other words, we should assign levels of reliability to our explanations in situations of uncertainty.

Bayesian networks offer a probabilistic reasoning approach based on graph theory for the establishment of causal relationships between events, using probability theory to set levels of confidence (Cowell et al. 1999, Jensen 2001). Predicting that event X occurred there and then is difficult if we do not know or we cannot observe its consequence Y. To achieve an historical explanatory goal we should know what precipitating conditions generated at a certain moment of time and at a certain place of the world cause an increase in the probability of occurrence of an effect; consequently we can assign values to Y based on indirect observables or, as we are using in this work, based on ethnoarchaeological data. The main advantage of such an approach is that an historical explanation of archaeological observations can be expressed in terms of the historical probability of the occurrence of an event in the past, given the inversely

[1] Laboratory of Quantitative Archaeology, Universitat Autònoma de Barcelona. Spain.
  Emails: juanantonio.barcelo@uab.es; florenciadelcastillo@hotmail.com
[2] Department of Physics, Universitat de Girona. Spain.
  Email: lauramameliiriarte@gmail.com
* Corresponding author

proportional relation between the empirical evidence we have observed, the capacity of some mechanisms (social or natural) to have produced this occurrence, and the known probability that some observation in the present be in fact the material consequence of some social action in the past.

Knowledge of environmental events has been used to infer the historical occurrence of social actions. The weight of environmental conditions in archaeological explanations is still a fundamental argument of inference because it is usually assumed that ecological changes in the past were invariably followed by social consequences, according to the Humean view of a cause as "an object, followed by another, and where all the objects similar to the first, are followed by objects similar to the second" (Hume 1748, section VII). There are a number of well-known difficulties with such understanding of social dynamics in a prehistoric past.

The first difficulty is that most ecological or climatic changes were not invariably followed by changes in social organization or economic behavior. Such an imperfect regularity may arise for two different reasons. First, they may arise because of the *heterogeneity* of circumstances which the proper cause of social action generates. Second, an imperfect regularity between environmental conditions and social action may also arise because of a failure of *physical determinism*. If an event is not determined to occur, then no other event can be (or be a part of) a sufficient condition for that event.

The central idea behind our approach in this chapter is that changing environmental conditions may affect the probability of social-economic decisions; but social action may have also occurred in the absence of any environmental change. Thus the availability of resources can be considered a cause of survival in a prehistoric past, not because all human groups need resources to survive, but because survival was *more likely* to occur in a rich environment than in a poor one, but not exclusively. This is entirely consistent with there being some populations having survived in very difficult conditions (deserts, for instance), and others who died in the context of high resource availability.

If a cause raises the probability of its effect then it can be expressed formally using conditional probabilities. Wealth of resources (C) raises the probability of survival (E) just in case:

$$P(E|C) > P(E).$$

In words, the probability that a particular human group survived in Prehistory, given that there were enough resources there and then, should be higher than the unconditional probability of surviving. Alternatively, we might say that resource abundance raises the probability of surviving just in case:

$$P(E|C) > P(E|{\sim}C);$$

In words, the probability that E occurs, given that C occurs, is higher than the probability that E occurs, given that C does not occur. These two formulations turn out to be equivalent. Resource availability may raise the probability of *survival* even though instances of *rich environments* are not invariably followed by evidences of prehistoric settlements. Throughout this chapter we are going to develop this probabilistic cause-effect modeling methodology to analyse how non-linear relationships influenced hunter-gatherer survival.

Hypothesis 1: Survival as a Consequence of Resource Availability

In a hunting and gathering prehistoric society, humans survived only if they had success in acquiring subsistence available in the environment. We may build a preliminary and oversimplified model in which the occurrence of the effect, *survival* in the past, depends on the probabilities of the supposed cause: *resource availability* at that time (Fig. 1).

Thus, the probability of an archaeological observation in the present depends on the probabilities of the event having occurred in the past (Fig. 2)

In those cases, *prior probabilities* are the probabilities of survival (or archaeological observation) before any evidence is considered, and before the effects of resource availability are taken into account. Technically speaking, it is the marginal posterior probability distribution of the event in the past (or the observation in the present), given the available evidence and the factors dependencies (the direction of the arrows as depicted in the model).

For the moment, we will take into account the probabilities of the occurrence of the event (survival) in the past; therefore, we do not consider the influence of the actual observation of an archaeological site. We will come back to archaeological observation at the end of this chapter. In theory, we can know the probability of hunter-gatherer *survival* if we surmise the probability of existence of enough resources. Expressed in more formal terms, "the probability of survival given the occurrence of enough resources is equal to the probability of finding enough resources given the occurrence of survival, multiplied by the probability of survival divided by the probability of the actual existence of enough resources. This follows directly from the Bayes Theorem:

$$P(b\backslash a) = \frac{P(a\backslash b) \cdot P(b)}{P(a)}$$

Where $P(b/a)$ = the probability of survival giving n resources, $P(a)$ = probability of finding enough resources, and $P(b)$ = probability of survival. The Bayes theorem can also be read as:

$$\text{Posterior} = \frac{\text{Likelihood} \times \text{Prior}}{\text{Prob of evidence}}$$

Then, the probability that a hunter-gatherer household survived should be equal to the probability of the existence of enough resources at that time and place, conditioned on the evidence for survival, multiplied by the unconditional probability of finding enough resources in that circumstances.

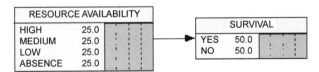

Fig. 1. An oversimplified hypothesis: survival is the direct consequence of resource availability.

Fig. 2. Archaeological observation in the present as a consequence of the probability of occurrence of events having occurred in the past.

Hypothesis 2: Survival as a Consequence of Hunting-Gathering Success and Resource Availability

The previous hypothesis is said to be *deterministic* since we assumed the occurrence of survival was determined only by resource availability at that place and moment, as if there were no element of chance involved. To avoid this determinism, we can add an intermediate factor: *hunting-gathering success,* indicating that there were other factor(s) influencing survival (Fig. 3).

Considering A to represent the probability of existence of enough resources, B as the probability of success in hunting, and C as the probability survival, the associated conditional probability can be calculated as:

$$P(C\backslash A^\wedge B) = P(C\backslash B)$$

This means that the probability of survival, given hunting success, is exactly the same as the probability of survival, given both hunting success and the availability of resources. The probability that a hunter-gatherer band survived depended directly only on whether hunting and gathering were successful at those circumstances. If we don't know whether they had actually captured enough animal preys, but we do find out there were abundance of resources in the environment during that historical circumstances, that would increase our belief both that they were successful in hunting and they survived. However, if we already knew they had success in hunting, then the availability of resources wouldn't make any difference to the probability of survival. That is, survival is conditionally independent of resource availability given that the human group succeeded when hunting on their own.

This is still a too simplistic assumption about the reasons explaining why prehistoric hunter-gatherers survived.

In an enhanced hypothetical model (Fig. 4), the "success" of an economic activity (hunting-gathering) would not depend on the availability of resources but would be the direct consequence of the amount of energy effectively acquired and transformed from the environment. We can suggest a causal chain conditioning the amount of effectively acquired energy on the energy needs (based upon the number of members the household had, and expressed in *labor units*: the higher the number of members a household has, the more energy is needed to feed all, but the easier it is to obtain such energy from the environment of the actual availability of resources), and the effective access to resources, which depends on the existence of enough labor and the proper technological efficiency of labor. In this enhanced model, the lack of an arrow from an event to its successors in the sequence (for instance, from resource availability to hunting-gathering success) implies conditional independence between the events. We see that an event, apparently independent to another may arrive to affect it at the end through intermediate nodes.

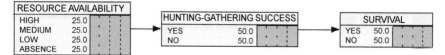

Fig. 3. This is a causal chain of three events, where resource availability may be considered as the cause of hunting-gathering success, which in turn is viewed as the cause of the actual survival of the hunter gatherer band.

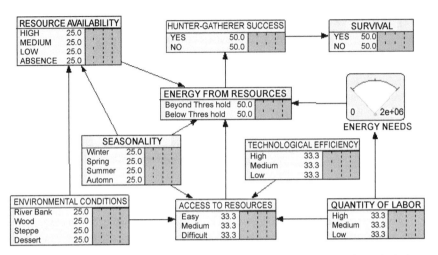

Fig. 4. An enhanced hypothetical model of the "success" of an economic activity (hunting-gathering) as a direct consequence of the amount of energy effectively acquired and transformed from the environment.

In such an enhanced hypothetical model, access to resources is a common effect of the Quantity of Labor (number of members a household has), and the technological efficiency of the instruments they have manufactured. Here, common effects (or their descendants) produce the exact opposite conditional independence structure to that of the simple causal chain we explored at the beginning. That is, the causes are marginally independent, but become dependent given information about the common effect (i.e., they are conditionally dependent):

$$P(A|C \wedge B) \neq P(A|B) \equiv A \perp\!\!\!\perp C|B$$

Before we have any evidence for the accessibility to resources, the quantity of labor and the efficiency of technology are independent. That is, changing one has no effect on the other. But as soon as we have evidence for the probability to effectively access existing resources, any change in the quantity of labor has an opposite effect in the efficiency of technology and vice versa, because they are competing explanations of the probability to access existing resources in that environment at that historical circumstances. Thus, if we observe the effect (e.g., the hunter-gatherers effectively have obtained the resource), and then we find out that one of the causes is absent (e.g., the technology was hardly efficient to do that task), this raises the probability of the other cause (e.g., that the household had a lot of members whose joint work was enough to compensate the poor technology available)—which is just the inverse of the causal chain. The same can be said about resource availability as a common effect of Seasonality and Environmental Conditions. In this case, both factors are assumed to be independent. But those factors do not only affect the amount of resources; some aspects of the effective access to resources also depend on them. In any case, access to resources is independent to the availability of resources.

Our hypothesis is now presented in form of a directed graph. We can use the following terminology (Shafer 1996, Jordan and Sejnowski 2001, Neapolitan 2004):

- v_i is a vertex (variable) in the vertex (variable) set V. *Descendants (V_i)* is the set of all the descendants of V_i in the graph—that is, all the effects of V_i in the set V.

- *Parents (V)* is the set of all the parents of V_i in the graph—that is the set of all the direct causes of V_i in the set V.
- Since *Parents (V)* is a subset of V, direct causes of V_i that are not in V will not be in *Parents (V)*.
- Directed edges from vertex v_1 to vertex v_2 in the graph represent then 'direct' causal relations between v_1 and v_2.

For instance, if there is a link in our model (Fig. 4) from node *Hunter-Gatherer Success* to node *Survival*, then we refer to node *Hunter-Gatherer Success* as the parent, and node *Survival* as the child. Of course, *Hunter-Gatherer Success* is in its turn the child of another node, *Energy from Resources*, which is the child of *Access to Resources*. Therefore, *Hunter-Gatherer Success, Energy from Resources, Energy Needs, Access to Resources, Technological Efficiency, Quantity of Labor, Seasonality, Environmental Conditions* and *Resource Availability* constitute the parents set of *Survival*. *Survival* is within the descendent set of *Resource Availability*, although there is not any direct edge between both nodes.

For each node we need to look at all the possible combinations of values of those parent nodes. Each such combination is called an instantiation of the parent set. For each distinct instantiation of the values of parent node, we need to specify the probability that the child will take each of its values. The graph indicates that probability relationships are known to exist between the states of those nodes. The direction of the link arrows suggests that the nodes where arrows begin tend to influence those where the arrows end; at least, more than the other way around. We assume that a link from *Seasonality* to *Access of Resources* indicates that the effective access to resources is causally related with the seasonality in which the economic action took place, that seasonality predisposes the effective access to resources, that the access to resources can be an imperfect observation of seasonality, that seasonality and access to resources are functionally related, or that both factors statistically correlated when we study the probabilities of survival of a hunter-gatherer household at some specific circumstances. The only constraint on the arcs allowed is that there must not be any directed cycles: an hypothetical causal factor cannot return to itself simply by following directed arcs. A sequence of vertices $\{v_1,...,v_n\}$ is a path from v_1 to v_n if and only if for all $i(i<n)$, there is a directed edge from v_i to v_{i+1} (Koller and Friedman 2009).

The Causal Markov Condition says that every variable is screened off by the set of all its parents from every other variable except its effects (Hausman and Woodward 1999). Reichenbach (1928) introduced the terminology of "screening off" to describe a particular type of probabilistic relationship. If $P(E|A \& C) = P(E|C)$, then C is said to screen A off from E. When $P(E \& C) > 0$, this equality is equivalent to $P(A \& E|C) = P(A|C)P(E|C)$; i.e., A and E are probabilistically independent conditional upon C. That is, if A and E are spuriously correlated, then A will be screened off from E by a common cause. Nancy Cartwright (1979) and Brian Skyrms (1980) at about the same time sought to rectify the problem with the requirement that causes must raise the probability of their effects in various *background contexts*. A background context is a conjunction of factors. When such a conjunction of factors is conditioned on, those factors are said to be "held fixed." Cartwright proposed the following definition:

C causes E if and only if $P(E|C \& B) > P(E|{\sim}C \& B)$ for every background context B.

Skyrms proposed a slightly weaker condition: a cause must raise the probability of its effect in at least one background context, and lower it in none.

In order to specify what the background contexts will be, we must specify what factors are to be held fixed. In our case study, we saw that the true cause of hunter-gatherer survival may be revealed when we held the level of available resources fixed, either positively or negatively. This suggests that in evaluating the causal relevance of *the availability of resources* for *survival*, we need to hold fixed other causes of such survival, either positively or negatively. This suggestion is not entirely correct, however. Let *C* and *E* be *Resources Availability* and *Survival*, respectively. There is not a direct edge connecting both, but they are not entirely independent, because *Survival* is in the descendent set of *Resource Availability* through the partial effects that the local availability of resources has on the energy taken from resources and hence on hunting and gathering success. Consider *D* is *Energy from resources* and acts as a causal intermediary. If *C* causes *E* exclusively via *D*, then *D* will screen *C* off from *E*: given the probabilities of obtaining energy beyond the subsistence threshold thanks to the joint effect of additional labor and technological efficiency on the effective access to resources, the probability of survival is not affected by the mere high or low availability of resources. Thus we will not want to hold fixed any causes of *E that are themselves caused by C*. Let us call the set of all factors that are causes of *E*, but are not caused by *C*, the set of *independent* causes of *E*. In our case, these are *Technological* Efficiency and *Quantity of* Labor. A background context for *C* and *E* will then be a maximal conjunction, each of whose conjuncts is either an independent cause of *E*, or the negation of an independent cause of *E*.

The Markov Condition states a sufficient condition but not a necessary condition for conditional probabilistic independence. As such, the Markov Condition by itself can never entail that two variables are conditionally or unconditionally dependent. The Minimality and Faithfulness Conditions are two conditions that give necessary conditions for probabilistic independence. The terminology comes from Spirtes et al. (2001). Pearl (2000) provides analogous conditions with different terminology.

i) *The Minimality Condition.* It asserts that no subgraph of **G** over **V** also satisfies the Markov Condition with respect to P. As an illustration, consider the variable set {*Resource Availability, Energy from resources*}. There is an arrow from *one* to the *other node*, but suppose that *both* are probabilistically independent of each other. In such a graph, none of the independence relations mandated by the Markov Conditions are absent (in fact, it mandates no independence relations). But this graph would violate the Minimality Condition, since the subgraph that omits the arrow from *Resource Availability* to *Energy from resources* would also satisfy the Markov Condition.

ii) *The Faithfulness Condition.* It says that all of the (conditional and unconditional) probabilistic independencies that exist among the variables in **V** are *required* by the Markov Condition. For example, suppose that **V** = {*Technological Efficiency, Quantity of Labor, Access to Resources*}. *Technological Efficiency* and *Quantity of Labor* are unconditionally independent of one another, but dependent, conditionally upon *Access to Resources*. (The other two variable pairs are dependent, both conditionally and unconditionally.)

From Evidence To Probabilities

Once the topology of the graph G on V, is specified in terms of a set of directed edges, or 'arrows', having the variables in V as their vertices encoding assumed causal relationships between factors, the next step is to quantify the relationships between connected nodes—this is done by specifying a conditional probability distribution for each node, based on available evidence. That is to say, the hypothesis is expressed at two levels, qualitative and quantitative. At the qualitative level, we have directed the acyclic graph in which nodes represent local conditions, and directed arcs to describe the conditional independence relations embedded in the model. At the quantitative level, the dependence relations will be expressed in terms of conditional probability distributions for each variable in the network.

If the variables are discrete, the probability distribution for each factor can be represented as a Conditional Probability Table (*CPT*), which lists the probability that the child node takes on each of its different values for each combination of values of its parents (Fig. 5). Technically speaking, it is the marginal posterior probability distribution of the node, given the available evidence and the factors dependencies (the arrows).

Such Conditional Probability Tables can be entered by hand based on some theoretical assumptions, or they can be learnt automatically using the appropriate algorithm based on Artificial Intelligence techniques. We may use conditional independence semantics of Bayes nets to induce models from data (Heckerman 1996, Nadkarni and Shenoy 2004, Naïm et al. 2007). The idea is to use evidence to find the *maximum likelihood* Bayesian network, which is the net that is the most likely given

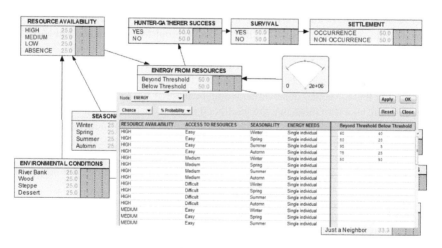

Fig. 5. A Conditional Probability Tale for the "Energy from Resources" node. On the left of the table, presented vertically, there is a subset of possible combinations of the parent states. On the top right are all possible states of Resource Availability, "High", "Medium", "Low" and "Absence". The probabilities of each combination of parent states and child state are then given at the right part. Note that the probabilities of each row in the table must sum exactly to 100 (or 1 if we express it as single probabilities). This is because each row is summarizing the probabilities of one possible world, one where the parents are in the given states. And for that possible world, the chances of the child being in any one state must sum to 100.

the data. If N is the net and D is the data, we are looking for the N which gives the highest $P(N|D)$. Using Bayes rule,

$$P(N|D) = P(D|N) \ P(N)/P(D)$$

Since $P(D)$ will be the same for all the candidate nets, we are trying to maximize $P(D|N) \ P(N)$, which is the same as maximizing its logarithm:

$$log(P(D|N)) + log(P(N))$$

Using Ethnoarchaeological Evidence to Generalize the Probabilities of Surviving in Prehistory: The Patagonian Hunter-Gatherers Case-Study

A definite finding that a node X has a particular value, x, which we write as $X = x$, is referred as a *single evidence*. In other words, it is a value for one of the nodes when it is applied to a particular situation. For instance: "High Availability of Resources", "Low Efficiency of Technology". *Positive evidence* is knowledge that some factor definitely occurred in a definite state. However, sometimes evidence is available that is not so definite. The evidence might be that a node Y has the value y_1 or y_2 (implying that all other values are impossible). Or the evidence might be that Y is not in state y_1 (but may take any of its other values); this is sometimes called a negative evidence. In fact, the new information might simply be any new probability distribution over Y. Another type of evidence is a *likelihood evidence*. In this case we receive uncertain information about the value of some discrete node. It could be from an imperfect archaeological excavation, or from a biased old ethnographic account which was not always right. Suppose, for example, that the archaeologist who has taken and analyzed the availability of resources at those specific historical and paleoclimatic conditions is uncertain. She thinks that resource availability looks positive, but is only 80% sure. Likelihood evidence consists of one probability for each state of the node, which is the probability that the actual observation would be made if the node were in that state. A common mistake is to think that the likelihood is the probability of the state given the observation made (in which case the numbers would have to add to one), but it is the other way around. Such information can be incorporated, in which case it would correspond to adopting a new posterior distribution for the node in question. In Bayesian networks this is also known as virtual evidence. Since it is handled via likelihood information, it is also known as likelihood evidence.

The set of all evidences entered into all the nodes of the model is referred to as a *case*. To enter a case in the model, we simply enter each of the evidences for each known node. Case files may consist of many cases (acting as a database, in which each case is a database *record*) following this general structure (Table 1).

We have entered data in a spreadsheet arranged with columns corresponding to factors (nodes) of which we have evidence, and each row being a case. At the intersection of each row and column is the cell that gives the value of the variable indicated by the column, for the case indicated by the row. The first row contains the names of the variables. Each will correspond to a node in the acyclic graph, although the spreadsheet file may have some variables that don't appear in the model and vice-versa. The order of columns has no relevance, given that the algorithm will use the dependence relationships expressed in terms of parent-sets and descendant-sets.

Table 1. A spreadsheet template for entering *cases* and *evidences*. Columns correspond to factors (nodes) of which we have evidence, and each row being a case. Each cell gives the value of the variable indicated by the column, for the case indicated by the row.

SETTLEMENT	OBSERVATION	RESOURCE	ENERGY	SUCCESS	SURVIVAL	SEASONALITY	ACCESS	EFFICIENCY	ENVIRONMENT

Each case provides some information about a particular instance. Case files may consist of many cases (acting as a database, in which each case is a database *record*). In hunter-gatherer research, we have the risk of generalizing a trivial conditional probability table of assumed universal validity as if all prehistoric hunter-gatherers in the world acted in the same way. To avoid this simplification, we suggest building a different model for different societies for which we have enough information, and then integrating all partial models into a single one in which each case would correspond to a society at a definite historical circumstance known through ethnological, historical or archaeological data. When we want to work with a new historical case, we save all the information gathered for the first society into a case file before removing it from the net, perhaps using the ethnic label, the archaeological site name or a reference to the historical period as a file name. When it comes time to reconsider the first case—perhaps some new ethnographic or archaeological information—we just read that society's case file.

In this chapter we have limited ourselves to building the model of a single historical case, ancient Patagonia (at the extreme south of South America). The historical process of those human groups is well known not only through archaeology but also from the ethnographical record (Barceló et al. 2009). Less than one hundred years ago, hunter-gatherer practices coexisted with a capitalist economy, about which there is enough ethnographical and historical information. To increase initial reliability of the probabilistic inference, we have created a partial model of Patagonian hunter-gatherers at the beginning of 19th century, using situations described by ethnographers. For instance, in 1865, a British explorer, George Musters, lived one year with a group of *aonikenk* people. He explained his adventures in a book (Muster 1871) in which he accounts different scenes of the daily life of this hunter-gatherer group. We have taken, separately, each description documenting the particular state of our social factors. For instance, on a particular day he witnessed a particular case of lack of success in hunting, and we can describe it using the following vector (Fig. 6). Another day he witnessed a particular case of failure in hunting (Fig. 7). As a consequence of the economic success and the possibilities of survival at those circumstances, hunters settled the place where they could hunt and/or gather, and as a consequence a hut was built and some resources were consumed at that place. However, archaeological research has not obtained any evidence of such site, given post-depositional disturbance. We can integrate all such observed cases into a single database (Fig. 8).

IDnum	SETTLEMENT	OBSERVATION	RESOURCE	ENERGY	SUCCESS	SURVIVAL	SEASONALITY	ACCESS	EFFICIENCY	THRESHOLD	LABOR	ENVIRONMENT
1	1	1	0	>3	0	1	Summer	Easy	Low	1.47E+10	High	River_Bank

Fig. 6. A single case to learn conditional probabilities from available ethnoarchaeological evidence.

IDnum	SETTLEMENT	OBSERVATION	RESOURCE	ENERGY	SUCCESS	SURVIVAL	SEASONALITY	ACCESS	EFFICIENCY	THRESHOLD	LABOR	ENVIRONMENT
1	1	1	0	>3	0	1	Summer	Easy	Low	1.47E+10	High	River_Bank
2	1	0	0	>3	1	1	Summer	Medium	Low	5.09E+10	High	Forest

Fig. 7. Two different cases to learn conditional probabilities from available ethnoarchaeological evidence.

IDnum	SETTLEMENT	OBSERVATION	RESOURCE	ENERGY	SUCCESS	SURVIVAL	SEASONALITY	ACCESS	EFFICIENCY	THRESHOLD	LABOR	ENVIRONMENT
1	1	1	0	>3	0	1	Summer	Easy	Low	1.47E+10	High	River_Bank
2	1	0	0	>3	1	1	Summer	Medium	Low	5.09E+10	High	Forest
3	1	0	50	>2	0	0	Spring	Easy	Medium	3.68E+10	Low	Forest
4	1	0	20	>2	0	0	Spring	Medium	Low	1.91E+10	Low	River_Bank
5	1	0	20	5	1	1	Autumn	Easy	High	1.60E+10	Medium	Desert

Fig. 8. A subset of ethnographical observations from Patagonia ethnographers referred to our study case.

We need three main types of algorithms for the parameter learning of the Conditional Probability Table in order to using those evidences for those cases:

- counting,
- expectation-maximization (EM), and
- gradient descent.

Of the three, "counting" is by far the fastest and simplest. It can be used whenever there are no latent variables, and not much missing data or uncertain findings for the learning nodes or their parents. The algorithm works as follows:

1. Before learning begins, the network starts off in a state of ignorance (providing there has been no previous learning or entry of probabilities by an expert). At each node, all CPT probabilities start as uniform, and each "experience" starts at its lowest value (normally 1.0). We are using here the terms used in the Netica software package (Norsys Software Corp., http://norsys.com/). Spiegelhalter et al. (1993) use the term "precision" equivalent to the way we refer to "experience".
2. For each case to be learned the following is done. Only nodes for which the case supplies a value (finding), and supplies values for all of its parents, have their experience and conditional probabilities modified (i.e., no missing data for that node). Each of these nodes is modified as follows:
   a. Only the single experience number, and the single probability vector, for the parent configuration which is consistent with the case, is modified.

   $$New\ experience\,' = old\ experience + degree$$

   where *degree* is the multiplicity of the case (set just before learning begins). It is set normally to one.
   b. Within the probability vector, the probability for the node state that is consistent with the case is changed from $prob_c$ to $prob_c'$ as follows:

   $$prob_c\,' = (prob_c * old\ experience + degree)/new\ experience$$

   The other probabilities in that vector are changed by:

   $$prob_i\,' = (prob_i * old\ experience)/new\ experience$$

   which will keep the vector normalized (old experience and experience act as the old and new normalization constants).

This learning algorithm is equivalent to a system of true Bayesian learning, under the assumptions that the conditional probabilities being learned are independent of each other, and the prior distributions are Dirichlet functions (if a node has 2 states, these are "beta functions") (D'Agostino 2003, Neapolitan 2003). Assuming the prior distributions to be Dirichlet generally does not result in a significant loss of accuracy, since precise priors aren't usually available, and Dirichlet functions can fairly flexibly fit a wide variety of simple functions (Gelman et al. 1995, Press 1989). Assuming the conditional probabilities to be independent generally results in poor performance when the number of usable cases isn't large compared to the number of parent configurations of each node to be learned (see also Korb and Nicholson 2004, Naïm et al. 2007, Welse and Wöger 1993).

If we can't use the above procedure, then we can use EM learning or gradient descent. For each application area, it is usually best to try each one to see which gives the better results. Generally speaking, EM learning is more robust (i.e., it gives good results in wide variety of situations), but sometimes gradient descent is faster. For all three algorithms, the order of the cases doesn't matter.

Both EM and gradient descent learning work by an iterative process, starting with a candidate net, reports its log likelihood, then processes the entire case set with it to find a better net. By the nature of each algorithm the log likelihood of the new net is always as good as or better than the previous. That process is repeated until the log likelihood numbers are no longer improving enough (according to a tolerance that you can specify), or the desired number of iterations has been reached (also a quantity you can specify). Briefly, EM learning repeatedly takes a Bayes net and uses it to find a better one by doing an expectation (E) step followed by a maximization (M) step. In the E step, it uses regular Bayes net inference with the existing Bayes net to compute the expected value of all the missing data, and then the M step finds the maximum likelihood Bayes net given the now extended data (i.e., original data plus expected value of missing data). Gradient descent learning searches the space of Bayes net parameters by using the negative log likelihood as an objective function it is trying to minimize. Given a Bayes net, it can find a better one by using Bayes net inference to calculate the direction of steepest gradient to know how to change the parameters (i.e., CPTs) to go in the steepest direction of the gradient (i.e., maximum improvement). A much more efficient approach that always takes the steepest path implies taking into account the previous path, which is why it's called *conjugate* gradient descent. Both algorithms can get stuck in local minima, but in actual practice do quite well, especially the EM algorithm (Jordan and Sejnowski 2001, Neapolitan 2003, Koller and Friedman 2009).

Most neural network learning algorithms (such as backpropagation and its improvements) are gradient descent algorithms (Principe et al. 2000). That invites a comparison between Bayes net learning and neural net learning, with latent variables corresponding to hidden neurons. In the case of Bayes net learning, there are generally fewer hidden nodes, the learned relationships between the nodes are generally more complex, the result of the learning has a direct physical interpretation (by probability theory) rather than just being black-box type weights, and the result of the learning is more modular (parts can be separated off and combined with other learned structures).

Hypothesis 3: Survival as a Consequence of Labour Cooperation

The current hypothesis is still too deterministic to be used as a model of survival in prehistory, even calibrating probabilities with ethnographical Patagonian data. We have stressed that the energy that can be obtained from resources is not an external factor, nor does it depend exclusively on what there is in the environment. There are social and cultural factors affecting it: Technological Efficiency and the Quantity of Labor. In this simple model we consider technological efficiency as a free parameter, to be adjusted from evidence we have of particular historical cases. The Quantity of labor, however, is a much more socially built factor. The quantity of available labor is not only a consequence of demography and biological reproduction, but an effect of social interaction among neighbours. As such, it depends on the previous organization of the social group and on the way social agents moved across a socially bounded territory.

The quantity of available labor force depends critically on the way individuals from different groups met other groups when moving and took the decision whether to cooperate or not. Cooperation in hunting is of basic importance to understand the probabilities of surviving, because in the absence of high technological efficiency, the only way to increase the probability for hunting success is by increasing the number of hunters collaborating in looking for preys and killing them. We are suggesting that the higher the cooperation, the higher the probability of hunting success. We are not arguing that the higher the number of hunters, the higher the amount of meat. In some cases, the total amount of meat per person can be lesser, but more frequent (more probable). Sharing labor and technology is a way to increase the chances of survival when the resource productivity at place (quantity of resources modulated by labor and technological efficiency) is below the survival threshold.

The question is how such cooperation may emerge? Why a social agent takes the decision of cooperating? Although the decision may seem rational, and based on the maximization of benefits and the minimization of costs, such rationality is bounded and affected by the non-monotonic and non-linear summation of social and/or political factors present or absent at such circumstances (Hamilton et al. 2007). We suggest that hunter-gatherers first decide whether neighbors are possible partners for cooperation, and second whether cooperation is really needed at the present circumstances (Barceló et al. 2014). In the first case, the decision maker should maximize the idea of *disposition* or *capability* to share labor; in the second case expected benefits from cooperating at work should be maximized in given the success of hunting and gathering until the moment of the decision.

We have enhanced our causal model by including a decision mechanism modulated by some utility parameters (Fig. 9). The new nodes in the model represent the variables that involve a decision. The possible states of a decision node represent the options for that decision (i.e., a choice set). Links between *decision* nodes are special as they do not represent cause-effect relationships. Rather, they indicate the sequence in which decisions are to be made. However, we do not assume that prehistoric hunter-gatherer were free to decide which option to choose, therefore, the initial probability distribution across states of a decision node is not uniform, but depends on a previous decision: is there in the neighborhood a suitable partner with whom labor can be shared?

In Fig. 10, an arc is drawn from the decision to accept a neighbor as a suitable partner to the decision whether to cooperate, because the disposition or capability of the neighbor to cooperate should be known at the moment the decision whether labor may be shared has to be taken. That is, the decision maker will know the probability

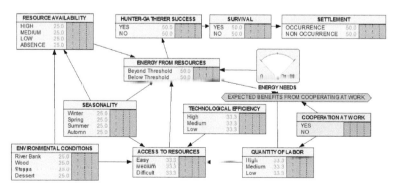

Fig. 9. A Bayesian causal model enhanced with a rational decision.

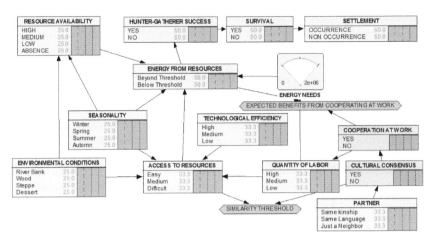

Fig. 10. A social decision affected by a previous social decision.

that there is someone in the vicinity with which sharing labor may be possible. Knowing such probability is also a decision to be taken and not an external factor. That also means that the decision maker remembers the result of a previous decision when trying to decide whether to cooperate or not. This previous decision is influenced by the intensity of the existing interaction between the decision maker and the prospective partner (same_kinship>same_language>neighborhood), but also on the actual conditions in which the need for cooperation emerges: the more needed the cooperation, the higher the tolerance to other's difference.

To be rational, any decision should imply a way to evaluate the advantages or drawbacks of each decision. This is implemented in terms of *utilities* to be maximized. A utility node is a node that allows valuating the states defined by the modality combinations of its parents. Those numerical values represent the quality or the cost of these states. According to our model, prehistoric hunter-gatherers maximized the existence of Cultural Consensus to decide with whom to share labor and they also maximized the expected benefits of sharing labor to cooperate or refuse cooperation even in presence of suitable partners (Del Castillo et al. 2014, Del Castillo and Barceló 2013).

We have defined the total (expected) utility as the sum of (expected) utilities across utility nodes and, therefore, the utilities defined for each benefit outcome should take the relative weight of the benefit variable into account. The benefit nodes included in the network intend to cover the most important considerations individuals generally have in making a choice. It is computed as:

$$EU\ (A\backslash E) = \sum P(O_i \backslash E, A)U(O_i \backslash A)$$

where:

E = available evidence of acquired energy from resources up to the moment of decision
A = the decision whether cooperate or refuse cooperation
O_i = possible decision
U = utility: expected benefits from cooperating at work.

A so-called conditional utility table (CUT) should be associated to each utility node. Rather than a probability distribution, it defines a utility value for each combined state of the parent nodes indicating how the decision maker evaluates that state (Table 2).

For each combination of values of the parents of the decision node:

a) For each action value in the decision node:
   i. Set the decision node to that value;
   ii. Calculate the posterior probabilities for the parent nodes of the utility node,
   iii. Calculate the resulting expected utility for the action.
b) Record the action with the highest expected utility in the decision table.

Table 2. A Conditional Utility Table for decision node "Cooperate At Work".

ACTION	Hunting has produced less than 2000 calories per group member until this moment (p=0.4)	Hunting has produced more than 2000 calories per group member until this moment (1–p = 0.6)
COOPERATE	30	10
REFUSE COOPERATION	−100	50

From Probabilities to Causal Inference

With these preliminaries, we are now ready to elucidate the probability that a particular human group survived at a particular place and at a particular moment in prehistory by hunting and gathering. The idea is fusing and propagating the impact of new evidence and beliefs through the graph so that each proposition eventually will be assigned a certainty measure consistent with the axioms of Bayesian probability theory (Pearl 2000). Therefore, we may assume the actual probabilities of survival in a prehistoric past can be calculated if we know some local facts and circumstances (we have archaeological evidence of that), and we may reconstruct the process of socially or politically mediated decisions on economic action. The Bayesian network we have built should be seen as a probabilistic inference system to compute the posterior probability distribution for survival in prehistory, given:

a) some structural parameters: the directed graph relates different factors, positively or negatively;
b) some metric parameters: the conditional probability tables related to each factor according to what we have learnt so far about previous known instances of survival in similar conditions;
c) archaeological, paleoecological or ethnological evidence about the actual occurrence of some factors at those local circumstances.

Bayes theorem allows us to update the probabilities of variables whose state has not been observed given some set of new observations. Bayesian networks automate this process, allowing reasoning to proceed in any direction across the network of variables. They do this by combining qualitative information about direct dependencies (perhaps causal relations) in arcs and quantitative information about the strengths of those dependencies in conditional probability distributions.

Judea Pearl (2000) has suggested the use of a sum-product message passing to calculate the marginal distribution for each unobserved node, conditional on any observed nodes. If $X = (X_v)$ is a set of discrete random variables with a joint mass function p, the marginal distribution of a single X_i is simply the summation of p over all other variables:

$$P_{X_i}(x_i) = \sum_{x': x'_i \neq x_i} p(x').$$

However this quickly becomes computationally prohibitive: if there are 100 binary variables, then one needs to sum over $2^{99} \approx 6.338 \times 10^{29}$ possible values. By exploiting the graphical structure, belief propagation allows the marginals to be computed much more efficiently. When building causal models in terms of a Bayesian Network we face the problem of determining when the model is sufficiently complete to satisfy Bayesian Network assumptions, yet compact enough to be computationally tractable. That means to assume that X causes Y, if and only if X and Y are probabilistically dependent conditional on the set of all the direct causes of X in a probability distribution generated by the given causal structure among the variables in the graph. Since the graph structure implies that the value of a particular node is conditional *only* on the values of its parent nodes, this reduces to

$$P(x_1, x_2, \ldots, x_n) = \prod_i P(x_i | Parents\,(X_i))$$
$$Parents\,(X_i) \subseteq \{x_1, \ldots, x_{i-1}\}.$$

$$P(X = pos \wedge D = T \wedge C = T \wedge P = low \wedge S = F)$$
$$= P(X = pos | D = T,\, C = T, P = lo,\, S = F)$$
$$\times P(D = T | C = T,\, P = lo,\, S = F)$$
$$\times P(C = T | P = lo,\, S = F)P(P = lo | S = F)P(S = F)$$
$$= P(X = pos | C = T)P(D = T | C = T)$$
$$\times P(C = T | P = lo,\, S = F)P(P = lo)P(S = F)$$

In a multiply connected network evidence can reach a node via multiple paths. If the model network is not too densely, connected, then we may combine nodes into clusters, propagate beliefs on clusters of nodes, and finally project resulting beliefs back to the variables in the initial network. The *junction tree algorithm* is a principled way to do this (Lauritzen and Jensen 2001). The basic idea is to transform the graph into clusters of nodes so that the graph of clusters is singly connected and has the *junction tree property* (this property ensures that evidence propagates correctly). The junction tree becomes a permanent part of the knowledge representation, and changes only if the graph changes. Since each cluster of nodes maintains its local belief table, and is proportional to joint probability of nodes in the cluster, it can be used to calculate response to query for current beliefs of nodes in the cluster; inference is performed in the junction tree using a local message-passing algorithm (Richardson and Urbanke 2008, Wainwright et al. 2005).

Conclusions: The Probabilities of Archaeological Observation

Although we have considered the example of evaluating the probabilities of survival for a particular hunter-gatherer community, based on collected evidence and the learning of general conditional probability tables, our more specific purpose is to explain the presence or absence of archaeological settlements in an area based on the concrete probability that a particular human group survived at that area given the success of their concrete actions of hunting and gathering. We assume that the *cause* of the archaeological observation lies in the probabilities of success of hunting-gathering activities in the past (and hence of survival), but also on the joint effects of human and natural disturbance processes, the probabilities of materials preservation due to the nature of sedimentary soils, and the nature of research (Fig. 11).

The notion of "immediate cause" of an archaeological actual observation is still a source of difficulty for archaeologists. Bayesian modeling allows us interpreting the classical view of Reichenbach in the sense that a non-deterministic (probabilistic) causal model, of the presence/absence of an archaeological observation at a particular place should enumerate the minimal number of processes necessary such that knowledge of the state of some (not necessary all) factors be sufficient to predict the observation in the present, provided the probabilities for that action to be performed in the past are sufficiently high. It is important to see, that the actual observation of an archaeological site is not independent of all the variables in **V** that are not depicted as direct edges to the node *archaeological observations* because our final node is in the descendent-set of all previous nodes.

In other words, what we perceive in the present at an archaeological site is simply the material effects of some action that occurred in the past. Therefore, we should analyze archaeological evidences within the context of social activity by identifying the ways people produced (and/or used) the artifact, the needs it served, and the history of its development. Any consequence of social action should be considered an *archaeological evidence* or *artefact*: the bones of a hunted animal, the bones of a buried human corpse, a territory, even an empty place is the consequence of some action (cleaning, for instance). The outcomes of social activity can be anything participating in a transformation process, including both material tools and tools for thinking (e.g., instruments, signs, procedures, machines, methods, laws, forms of work organization).

By building causal chains like (Fig. 12), we can consider archaeology as a problem solving discipline, centered on *historical* problems, whose focus is on explaining existing perceivable phenomena in terms of long past causes. The aim of using Bayesian networks

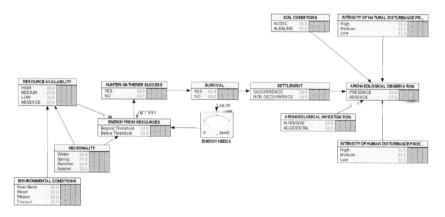

Fig. 11. A model of the probabilities for archaeological observation.

Fig. 12. Understanding why material effects of actions performed in the past can be observed in the present.

is to remark on the fact that the goal of archaeology is to study the probabilities of social action in the past and not just objects. Perceived present is the consequence of human action in the past, interacting with natural processes through time. Human action existed due to its capacity to produce and reproduce labor, goods, capital, information, and social relationships. In this situation, the obvious purpose of what we, archaeologists, "perceive" in the present is to be used as evidences for calculating the probabilities of past actions. It is something to be explained, and not something that explains social action in the past.

Network structures are a marriage between probability theory and graph theory. They provide a natural tool for dealing with two problems that occur throughout in the social sciences—uncertainty and complexity. Fundamental to the idea of a graphical model is the notion of modularity—a complex system is built by combining simpler parts. Probability theory provides the glue whereby the parts are combined, ensuring that the system as a whole is consistent, and providing ways to interface models to data. The graph theoretic side of graphical models provides both an intuitively appealing interface by which humans can model highly-interacting sets of variables as well as a data structure that lends itself naturally to the design of efficient general-purpose algorithms.

Summarizing, Bayesian Networks are an alternative for data mining, which in archaeological cases represents several advantages:

• Learn about dependency relationships and causality.
• Combine different levels of knowledge with data.
• Deal with complexity.
• Make available work with incomplete data bases.

Aknowledgments

This research has been funded by the Spanish Ministry of Science and Innovation, through Grant No.HAR2012-31036 awarded to J.A. Barceló and Project CSD2010-00034, Social and environmental transitions: Simulating the past to understand human behavior (CONSOLIDER-INGENIO 2010 program by Spanish Ministry of Science and Innovation, see: http://www.simulpast.es).

References Cited

Norsys Software Corp., http://norsys.com/.
Barceló, J.A., F. Del Castillo, L. Mameli, E. Moreno and B. Videla. 2009. Where Does the South Begin? Social Variability at the Southern Top of the World. Arctic Anthropology 46(1–2): 50–71.
Barceló, J.A., F. Del Castillo, R. Del Olmo, L. Mameli, F.J. Miguel Quesada, D. Poza and X. Vilà. 2014. Social Interaction in Hunter-Gatherer Societies: Simulating the Consequences of Cooperation and Social Aggregation. Social Science Computer Review, DOI: 10.1177/0894439313511943.

Cartwright, N. 1979. Causal laws and effective strategies. Noûs 13(4): 419–437.

Cowell, R.G., A.P. Dawid, S.L. Lauritzen and D.J. Spiegelhalter. 1999. Probabilistic Networks and Expert Systems. Springer-Verlag.

D'Agostino, G. 2003. Bayesian Reasoning in Data Analysis. A Critical Introduction. World Scientific Publishsing, London.

Del Castillo, F., J.A. Barceló, L. Mameli, F.J. Miguel Quesada and X. Vila. 2014. Modeling Mechanisms of Cultural Diversity and Ethnicity in Hunter-Gatherers. Journal of Archaeological Method and Theory. DOI: 10.1007/s10816-013-9199-.

Del Castillo, F. and J.A. Barceló. 2013. Why hunter and gatherers did not die more often? Simulating prehistoric decision making. Archaeology in the Digital Era Edited by Graeme Earl, Tim Sly, Angeliki Chrysanthi, Patricia Murrieta-Flores, Constantinos Papadopoulos, Iza Romanowska and David Wheatley. Amsterdam University Press (ISBN: 978 90 8964 663 7).

Gelman, A., J.B. Carlin, H.S. Stern and D.B. Rubin. 1995. Bayesian data analysis. Chapman & Hall.

Hamilton, M.J., B.T. Milne, R.S. Walker, O. Burger and J.H. Brown. 2007. The complex structure of hunter–gatherer social networks. Proceedings of the Royal Society 274(1622): 2195–2203.

Hausman, D.M. and J. Woodward. 1999. Independence, Invariance and the Causal Markov Condition, British Journal for the philosophy of Science 50: 521–583.

Heckerman, D. 1996. Bayesian networks for data mining. Data Mining and Knowledge Discovery 1: 79–119.

Hume.D. 178, (1748) Philosophical Essays Concerning Human Understanding. London, A. Milner (Downloaded from http://www.davidhume.org/texts/ehu.html, on March 10th., 2015).

Jordan, M.I. and T.J. Sejnowski (eds.). 2001. Graphical models: Foundations of neural computation. MIT press.

Jensen, V.F. 2001. Bayesian Networks and Decision Graphs. Springer-Verlag.

Koller, D. and N. Friedman. 2009. Probabilistic Graphical Models: Principles and Techniques. MIT Press.

Korb, K.B. and A.E. Nicholson. 2004. Bayesian artificial intelligence. Computer Science and Data Analysis. Chapman & Hall/CRC, Boca Raton.

Lauritzen, S.L. and F. Jensen. 2001. Stable local computation with conditional Gaussian distributions. Statistics and Computing 11(2): 191–203.

Musters, G.C. 1871. At home with the Patagonians: a year's wanderings over untrodden ground from the straits of Magellan to the Rio Negro. John Murray.

Naïm, P., P.H. Wuillemin, Ph. Leray, O. Pourret and A. Becker. 2007. Les Réseaux Bayésiens. Eyrolles, Paris.

Nadkarni, S. and P.P. Shenoy. 2004. A Causal Mapping Approach To Constructing Bayesian Networks, Decision Support Systems 38(2): 259–281.

Neapolitan, R. 2004. Learning Bayesian Networks. Prentice Hall, Inc. Upper Saddle River, NJ.

Pearl, J. 2000. Causality: Models, Reasoning, and Inference. Cambridge University Press, NY.

Press, S.J. 1989. Bayesian statistics: principles, models, and applications. John Wiley & Sons, NY.

Principe, J.C., N.R. Euliano and W.C. Lefebvre. 2000. Neural and adaptive systems: fundamentals through simulations. Wiley.

Reichenbach, H. 1928. Philosophie der Raum-Zeit-Lehre. English translation, 1957, The Philosophy of Space and Time. Dover.

Richardson, T.J. and R. Urbanke. 2008. Modern Coding Theory. Cambridge Univ Press, Cambridge, UK.

Shafer, G. 1996. The Art of Causal Conjecture. MIT Press, Cambridge, MA.

Skyrms, B. 1980. Causal Necessity: A Pragmatic Investigation of the Necessity of Laws. New Haven, Conn.: Yale University Press,

Spiegelhalter, D.J., A.P. Dawid, S.L. Lauritzen and R.G. Cowell. 1993. Bayesian analysis in expert systems. Statistical Science 219–247,

Spirtes, P., C. Glymour and R. Scheines. 2001. Causation, Prediction and Search, Springer, 1993, 2nd Edition MIT Press.

Wainwright, M.J., T.S. Jaakkola and A.S. Willsky. 2005. MAP estimation via agreement on trees: Message-passing and linear programming. IEEE Transactions on Information Theory 51: 3697–3717.

Weise, K. and W. Wöger. 1993. A Bayesian theory of measurement uncertainty. Measurement Science and Technology 4: 1–11.

Conclusions

26

Concluding Address: Ruminations on Mathematics in Archaeology

Keith W. Kintigh

This volume contains essays that discuss different roles for and logics underlying the use of quantitative methods in archaeology (Doran [Foreword]; Barceló [Chapter 1]; Djindjian [Chapter 2]; Nicolucci et al. [Chapter 3]; Read [Chapter 4]; and Barceló et al. [Chapter 25]), useful reviews of current technologies (Mucha et al. [Chapter 9]; Richards et al. [Chapter 12]; Bronk-Ramsey [Chapter 14]; Ducke [Chapter 18]), and descriptions and applications of novel methods (most of the remainder). Throughout his wide-ranging manifesto for digital archaeology, Barceló [Chapter 1] maintains a welcome focus on the ultimate goal of explaining social action. This introductory chapter supplies a surprising coherence to what might otherwise seem to be a collection of disparate methodological chapters. At the same time, through his essay, Barceló substantially enhances many of the individual contributions by providing needed conceptual context and background.

Both Doran's foreword and Djindjian's insightful historical chapter list quantitative topics that have continued importance in archaeology and several methods that never took off or, once popular, have receded into obscurity. This led me to wonder, in general, why some methods have become commonplace while others persist as historical footnotes, if at all. In contemplating the method-focused essays in this volume, I had the sense that some conveyed considerable potential, while others we'd hear little more about. Assuming their authors' aspirations to have an impact, several seemed notably lacking in perception of the needs of what I take to be the intended audience: the producers and higher-level consumers of quantitative and formal methods archaeology.

An obvious determinant of methods that have a major impact is the extent to which they solve pressing problems for practicing archaeologists. To take a single example,

School of Human Evolution and Social Change, Box 872402, Arizona State University, Tempe, Arizona, 85287-2402, United States.
Email: kintigh@asu.edu

seriation of grave lots or other assemblages is a common practical problem whose essence is to reduce a multidimensional similarity into a unidimensional (temporal) ordering. Manual techniques, such as the arrangement of paper strips with proportional representations of class abundance, were helpful, but did not generally result in a demonstrably optimal ordering. As Djindjian and others have documented, a number of special purpose quantitative approaches were developed. However, it is my sense that multidimensional scaling and principal components analysis fairly quickly became the standard tools, later to be largely replaced by correspondence analysis.

Here we have a well-defined and common problem. While there was a flurry of theoretical work by statisticians and archaeologists, the use of formal quantitative approaches did not become a matter of common practice until (1) we had *articles* explaining the methods and documenting applications that were understandable to a substantial fraction of the professional community (e.g., Cowgill 1972, Drennan 1976), and (2) the methods were implemented in ways that were accessible and easily usable by archaeologists, in this case through general-purpose statistical packages (such as SAS, BMDP, and Systat). In the US at least, I think the delay in adoption of correspondence analysis was mainly due to the fact that it was more recently included in the widely available statistical packages.

If practicing archaeologists lack reasonable access to a method, it is unlikely to achieve widespread use. I here use "access" in two senses, that the method can be adequately understood and that it is possible to find and use the software tools to execute it.

While all of us do not need a full understanding of all the mathematical detail, practitioners need to have a reasonable understanding of what the method does and what are its limitations. This requires understandable articles and documentation, targeted to a common level of quantitative sophistication. Formulae, of course, provide compact and precise statements of mathematical operations. However, if they aren't essential to the argument they should usually be omitted and, in all cases, plain-language statements of the main operations are needed. Use of unfamiliar and unexplained notation will stop most readers (including this one) in their tracks. Incomplete formulae (e.g., omitting the limits of a summation) and undefined terms are inexcusable. Worked out-real world examples are invaluable for understanding and acceptance.

The second "access" issue is availability of the relevant software tools. Widespread use of even basic statistical methods awaited the availability of general-purpose statistical packages. Use of these packages revolutionized the field. Similarly, general purpose GIS programs (e.g., ArcGIS and GRASS) and simulation languages (such as NetLogo http://ccl.northwestern.edu/netlogo/), have made use of GIS and, to a lesser extent simulation, everyday components of archaeological practice.

The rapidly increasing use of the open-source *R* statistical language (http://www.r-project.org/) has the potential to be truly transformative for the field because new methods easily can be made available on-line. Now, we don't need to wait for the mainstream statistical packages to implement new methods. *R* provides a powerful development platform for many kinds of statistical methods. Through the development of *R* "packages" and sharing of *R* scripts, it is easy to make new techniques available to those lacking the ability or inclination to implement a method from its description in an article or book. For example, new and promising applications of social network analysis (Mills et al. 2013, see Bordoni [Chapter 5]) have been enabled by the availability of the quantitative tools in *R*. Publication, in supplementary materials, of *R* scripts that implement new techniques is already happening (e.g., Peeples and Schachner 2012) and ought to become standard practice.

Authors of method-oriented articles should be confident that the intended reader would be able to follow the basic argument without stopping to consult outside references, and that the article should supply the reader with the necessary information (including through supplementary materials and references) to properly apply the proposed method. Some authors consistently succeed at achieving this level of clarity (To start, I'd put Cowgill, Doran, and Hodson in that category), some others, not so much. To the extent that the quantitative archaeology community is able to communicate effectively to the uninitiated, it will benefit us all. Ordinary mortals would be empowered to do a better job of archaeology, and the cognoscenti would get credit for their good works (and fewer eye rolls).

For reasons well-articulated by Doran [Foreword], Barceló [Chapter 1], Djindjian [Chapter 2], and others, modeling is essential to our progress in the field. Both agent-based (Rogers et al. [Chapter 23]; Cecconi et al. [Chapter 24] and dynamical models (Lemmen [Chapter 21]; Silva and Steele [Chapter 22]) have important roles to play. However, effective presentation of models remains problematic. The examples by Rogers et al. and Cecconi and his colleagues (this volume) seem to me the most successful. They provide accessible and reasonably rich description of the models and their results. However, with even the best presentations it is difficult to evaluate the results because it is impossible to look behind the curtain—to inspect the black box. The solution is not a detailed, component-by-component description. Even if a professional archaeologist could somehow assimilate this level of detail, one would still be unable to evaluate the model outcomes because of the complexity of the interactions. There is a role for a down-in-the-weeds model description, but it is better suited to documentation. I am not the first to remark on the difficulty of effective presentation of models and, clearly, there are no easy answers. However, part of an answer for article-length contributions might be to focus on presenting a substantive or theoretical *result*, rather than attempt to present *the model*. In this way, the discussion, including sensitivity analyses—while more limited—can be more persuasive than is probably possible for a complex model.

A closely related issue has to do with our ability to build successive generations of models directly from their ancestors. While model-building tools have improved and new models take advantage of general conceptual advances, models, small and large, are generally one-off.

There are no simple answers to this problem either. However, improved model-building toolkits and sharing of model components (e.g., through the OpenABM initiative http://www.openabm.org/), will be important, as may use of workflow management software to connect simple models into more complex ones.

The issue of data sharing plays a prominent role in several contributions. Of course, we desperately need to move both new and legacy archaeological reports and data into digital repositories such as ADS, DANS, and tDAR. Beyond the acquisition of the digital information, there are serious challenges in making the information in the repositories findable and useable. Richards et al. [Chapter 12] effectively argue that major investments in natural language processing are essential if we are to take advantage of the wealth of information we have so painstakingly documented in our innumerable reports. They also provide a summary of progress on this topic to date and some sense of future directions.

Nicolucci and his coauthors (Chapter 3) propose a general solution to data comparability based on CIDOC-CRM. While this proposal has logical appeal, it is not clear to me that this complex and, no doubt, time-consuming data documentation is really practical (or replicable) on a case-by-case basis and I am even more skeptical that this solution is scalable to the level of documenting the many tens of thousands of archaeological datasets produced each year. The problem is a real one, but it is important that the perfect not become the enemy of the good. Much can be accomplished through effective documentation of idiosyncratic recording schemes using community-developed ontologies and good data integration tools (Spielmann and Kintigh 2012, Kintigh 2013).

Finally we should return to the foreword in which James Doran reminds us of what should be archaeology's larger objective, to wit, "using such access as we have to our own past to guide our choices for our future as a species." With similar objectives in mind, with others, I recently argued (Kintigh et al. 2014) that such advances will depend heavily both on effective modeling and a new emphasis on synthetic research. Doran persuasively argues for a key role for modeling. For the needed, large-scale synthetic research, we not only must be able to discover and access to the vast amounts of legacy and newly produced data, we need powerful new tools that will empower data integration and, ultimately, synthesis. This is perhaps the most challenging new frontier for the development and application of mathematical and statistical methods in archaeology.

References Cited

Cowgill, George L. 1972. Models, methods, and techniques for seriation. pp. 381–424. *In*: David L. Clarke (ed.). Models in Archaeology. Methuen, London.

Drennan, Robert D. 1976. A refinement of chronological seriation using nonmetric multidimensional scaling. American Antiquity 290–302.

Kintigh, Keith W. 2013. Sustaining Database Semantics. *In*: F. Contreras, M. Farjas and F.J. Melero (eds.). CAA 2010: Fusion of Cultures. Proceedings of the 38th Conference on Computer Applications and Quantitative Methods in Archaeology, Granada, Spain, April 2010, BAR International Series 2494: 585–589.

Kintigh, Keith W., Jeffrey H. Altschul, Mary C. Beaudry, Robert D. Drennan, Ann P. Kinzig, Timothy A. Kohler, W. Fredrick Limp, Herbert D.G. Maschner, William K. Michener, Timothy R. Pauketat, Peter Peregrine, Jeremy A. Sabloff, Tony J. Wilkinson, Henry T. Wright and Melinda A. Zeder. 2014. Grand Challenges for Archaeology. American Antiquity 79(1): 5–24.

Mills, Barbara J., Jeffery J. Clark, Matthew A. Peeples, W.R. Haas, Jr., John M. Roberts, Jr., J. Brett Hill, Deborah L. Huntley, Lewis Borck, Ronald L. Breiger, Aaron Clauset and M. Steven Shackley. 2013. Transformation of social networks in the late pre-Hispanic US. Proceedings of the National Academy of Sciences 110(15): 5785–5790. doi:10.1073/pnas.1219966110.

Peeples, Matthew A. and Gregson Schachner. 2012. Refining correspondence analysis-based ceramic seriation of regional data sets. Journal of Archaeological Science 39(8): 2818–2827. http://dx.doi.org/10.1016/j.jas.2012.04.040.

Spielmann, Katherine A. and Keith W. Kintigh. 2012. The Digital Archaeological Record: The Potentials of Archaeozoological Data Integration through tDAR. SAA Archaeological Record 11(1): 22–25. http://digitaleditions.sheridan.com/publication/?i=58423&p=24.

27

"Mathematics and Archaeology" Rediscovered

Michael Greenacre

It was July in the year 2289 and in Gallia, the northern part of what used to be France a century earlier, they were celebrating the 500th anniversary of the French Revolution. It was also almost a hundred years since World War 4, the so-called "great technological war", which had decimated the world's scientific, technical and engineering community as well as destroying all the digital information stored in the 55 huge data centres scattered across the earth. These data centres, mostly underground, housed the "cloud", in which mankind had entrusted all its data and creative work, and were conveniently concentrated for the ensuing attacks. In a short space of time the totality of electronic records, including all books written after 2100, when paper printing was abolished, was obliterated and lost forever. All that was left of the written word were the millions of books stored in dusty warehouses, called "libraries", as well as some ancient storage media that had long fallen out of use—these were mostly disk drives with very limited storage, less than a petabyte, which had passed from one generation to another as heirlooms, usually quite deteriorated and no longer readable. Many philosophers had forecasted this disaster after the technological singularity was reached in the mid 22nd century, but nobody paid much attention to these "prophets of doom", as they called them.

I worked as an archaeolinguist at the prestigious Mediterranean Centre of Advanced Research in Barcelona. Our job was to try to understand the contents of the remnants of books discovered in various libraries across the southern Mediterranean region, from Catalonia in the west through Occitany and as far east as Lombardy and Venicia. Most books had been poorly preserved after cloud storage became universal, so we were faced with documents that were often in a highly damaged state. Nevertheless, we had excellent facilities, including robotic equipment with the new *Ocula* vision system that had been programmed to page through the documents more carefully than any human

Department of Economics and Business, Universitat Pompeu Fabra, and Barcelona Graduate School of Economics, Ramon Trias Fargas, 25-27, 08005 Barcelona, Spain.

hand and to automatically scan and interpret every letter, diagram and table that was still readable. We only had to intervene when the system stopped with a warning that some pages were stuck together, in which case we decided whether we should intervene or omit reading the hidden content, because detaching the pages might destroy them completely.

After several years working on old statistical texts, I was now working on books about archaeology itself, which then had a twofold interest for me as an archaeological linguist. I had previously reconstructed the contents of a book called "Correspondence Analysis and West Mexico Archaeology", published in 2013 by the University of New Mexico,[1] and which had been found in the Pompeu Fabra University's subterranean library in Barcelona, having been donated to the library by a statistician who worked there at that time. This book explained how artefacts could be coded and how the data on several artefacts could be compared through a visualization method called correspondence analysis. There were several chapters by another statistician named de Leeuw who explained the methodology, and having deciphered several statistical texts I could more or less follow his explanation and the mathematical calculations involved. So in my spare time I had programmed the method using the **VerbalR** language[2] and tested it out. It really was quite impressive how correspondence analysis showed the main differences between the objects as well as the features that distinguished them from one another. So I thought for my next book I would try it out, using the words that I found in the book as archaeological indicators—after all, I was in a sense excavating through these old printed artefacts ravaged by time, and the words were their features.

The next book was called "Mathematics and Archaeology", a book edited by two researchers in Barcelona at about the same time as the one I mentioned previously, i.e., early 21st century. This book was in quite bad condition, with many pages so deteriorated that they sometimes disintegrated with handling. Even the whole of Chapter 1 was missing, probably torn out secretly by some reader prior to the year 2100. Nevertheless, the contents pages were intact, and I could gather from them that the book was quite varied in its treatment of the subject, and thus a good example for my empirical exercise. The robotic scanning took a full morning, as I had to intervene several times to cope with many torn and stuck pages, but by midday I had a file of most of the text of the book stored on the atomic RAMchip that I carried permanently on my wrist.[3] I had scanned different chapters, written by different authors, separately, and soon, with a simple command to **VerbalR**, I had computed word counts of all the words in each chapter—this was similar to what I had read in the archaeology book about how quantitative archaeologists counted how many different types of artefacts

---

[1] *Correspondence Analysis and West Mexico Archaeology: Ceramics from the Long-Glasslow Collection*, edited by C. Roger Nance, Jan de Leeuw, Phil C. Weigand, Kathleen Prado and David S. Verity, University of New Mexico Press, 2013.

[2] **VerbalR** is a verbal programming language whereby the user describes the algorithm verbally to the computer, and the computer then creates the code while automatically verifying its syntax.

[3] Atomic RAMchips are now carried by every human as well as some pet animals. They carry the holder's complete personal information, including personal genome, educational and medical history, as well as a complete record of every transaction conducted by the holder (physical money was not used since the late 21st century). They were the first examples of so-called indestructible digital storage, invented as a consequence of the Great Technological War. Originally, they contained one zettabyte of memory, but nowadays you can find them with more than 100 Zb—most of this memory can be used by the holder for storing data of his or her choice.

were discovered at each site. The program also incorporated a lemmatization algorithm, grouping different forms of the same word together into a single unit.

The first thing I did was to list the most frequent words, shown in Fig. 1, which was a simple statement to **VerbalR**: "plot a horizontal barchart of the 40 most frequent words, in descending order of frequency".

The three most frequent terms were *use/d*, *archaeology/ist* and *data/base/set*. This gave me the idea that the book was practically oriented around different aspects of archaeology and that the use of data by archaeologists was a central concept. This was confirmed by other frequent words in this "top 40" list, words such as *analyse/ analytical*, *value/s*, *method/s*, *case/s*, *example/s*, etc. The word *statistical* was the 41st, just after *map/s/ping*, with 185 mentions, while the word *mathematical* was mentioned only half as much, 95 times. This led me to believe that the book was more about statistics in archaeology than mathematics. In fact, mathematics as "practised"

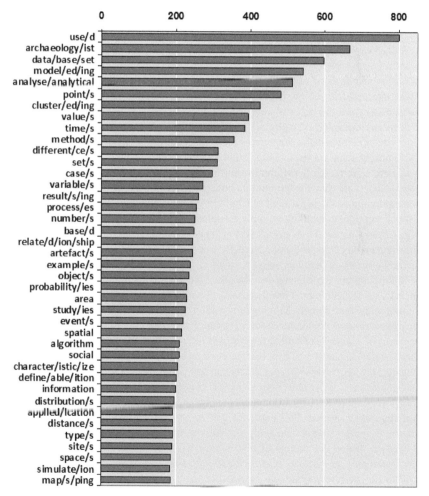

Fig. 1. Most frequent words (top 40 out of the 751 used at least twice) in the "excavation" of the text of "Mathematics and Archaeology".

in earlier centuries had now become a subject more of philosophical abstraction and esoteric debate—statisticians and data engineers had long since gleaned all the useful mathematical results they needed for their myriad of practical problems.

Correspondence analysis would show me how the distributions of words differed across the chapters, so I issued the instruction to **VerbalR** to "count the words in each chapter and perform a correspondence analysis on the table of chapters by words". In an instant the computer responded with a verbal summary and description of the results—here I transcribe the essential part:

Correspondence analysis of chapters by words

Total inertia = 2.668
Inertia dimension 1 = 0.226 (8.5%) P < 0.001
Inertia dimension 2 = 0.174 (6.5%) P < 0.001
15.0% inertia on first two dimensions, but dimensions 3, 4, 5 and 6 also non-random.
38.0% of the inertia on dimensions 1 to 6.
62.0% of the inertia can be considered random variation.

The way I understood these results is that 38% of the information in the word counts across chapters reflected real differences between the chapters, but I would need to see the results in six dimensions to fully appreciate these differences. Nevertheless, I was interested to see the best two-dimensional view, and told the computer using **VerbalR**: "plot the best two-dimensional view of the chapters and connect them in their numerical order; then plot the best two-dimensional view of the most interesting words that distinguish the chapters"—these commands gave the results in Figs. 2 and 3.

Straight away I noticed in Fig. 2 that Chapters 6 and 7, and to a certain extent 8 and 9 as well, were radically different from the remainder of the book. In Fig. 3 I further noticed that there were three important bunches of words, those pointing to bottom right corresponding to the separation of Chapters 6 and 7, and then two bunches at bottom left and top right coinciding with the diagonally oriented dispersion of the chapters in Fig. 2 from Chapter 21 (bottom left) to Chapters 4 and 9 (top right). Important words associated with Chapters 6 and 7 were, for example, *curvature, spline, images/ry, coins, convert/ed/sion, scanned/er, pixel/s, profiles, 3D* and *2D*. In fact, just from the titles of these chapters I could confirm that they were exclusively dealing with the shape analysis of archaeological artefacts, and the coding of two-dimensional images and three-dimensional objects. Moving upwards on the right hand side of Fig. 3 came many terms associated with geometry such as *triangle/s, plane/ar, cloud, circle, coordinates, geometry/ic* and *line*. The more these words moved up on the right the more they were shared with Chapters 9 and 8, in fact the word *projection* appeared as a word common to Chapters 6 to 9. The only proper nouns that appeared in Fig. 3 were the names *Fourier* and *Aitchison*. Fourier at bottom right was clearly involved in the chapters on shape analysis, while it turned out that Aitchison had been mentioned 19 times in Chapter 8 and in no other chapter. This name referred to a statistician who was the founder of a school of compositional data analysis that was much appreciated in Girona, north of Barcelona, where the authors of this chapter worked. At top right I could recognize many statistical terms such as *cluster/ed/ing, partition/s, square/s, subset/s, k-means, statistics/al, multivariate/dimensional* and *bootstrap*, which implied that Chapters 8 and 9, as well as the chapters at top right (e.g., Chapters 4, 10 and 20), were more statistically technical, especially in the area of multivariate analysis. In contrast, the chapters towards

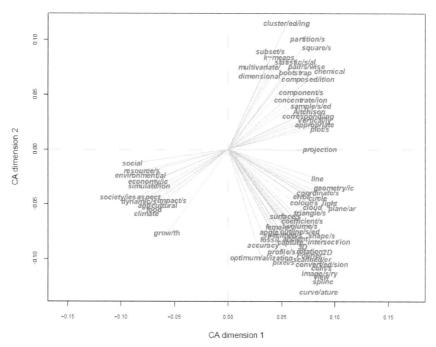

Fig. 2. Positions of chapters 3 to 25 of the book "Mathematics and Archaeology" according to first two dimensions of the correspondence analysis of their word counts.

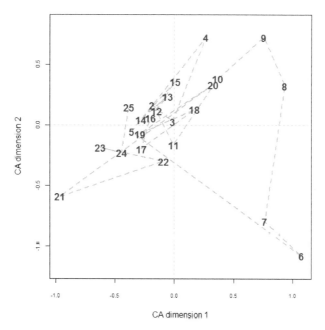

Fig. 3. Words that make the major contributions to the distinction between chapters observed in the two-dimensional solution of Fig. 27.2.

bottom left, notably Chapter 21 but also Chapters 22, 23 and 24, were characterised by non-statistical words such as *growth, climate, agricultural, society, economy/ic, environment/al, resources* and *social*. It was no coincidence then that these chapters, along with the last Chapter 25 were classified together as a separate section "Beyond Mathematics: Modeling Social Action in the Past". The position of the last Chapter 25 seemed to tend back to the top right towards the statistical terminology—looking at this chapter in more detail I confirmed that it was indeed using more technical language than the others in this section.

The correspondence analysis gave me a good overview of the book's contents and main themes, but I was rather concerned about seeing only 15% of the inertia-type measure of variance in the word counts, whereas the results pointed out that as much as 38% was non-random, and thus obviously worth exploring. The most important statistical term in the book appeared to be "cluster/ed/ing" (432 mentions—see Fig. 27.1) and I knew that clustering methods do analyse data in higher dimensions. Moreover, the term "k-means" was also important, prominent in Chapters 20, 9 and 4, and appeared to be a type of clustering adapted to large sets of objects. So, in order take all the significant dimensions into account, I asked **VisualR** to "perform a k-means clustering of the six-dimensional solution of the words in the previous correspondence analysis" and, not knowing exactly how the results might be reported, I followed this with a vague "plot some standard results". The graphical results are shown in Fig. 4, and the verbal results were simply as follows:

k-means clustering of 751 words according to six-dimensional CA coordinates
Total variance = 1.104
Number of clusters explored from 2 to 30.
Explained variance between clusters is 14.2% (3 clusters) to 82.1% (30 clusters).
Estimated number of clusters can be 8 (58.5%) or 12 (68.5%).

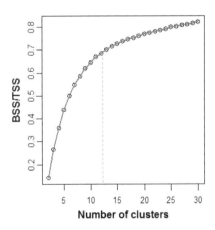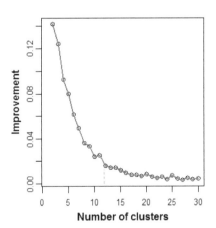

Fig. 4. (Left plot) Increasing proportion of variance explained between word clusters as number of clusters varies from 2 to 30. (Right plot) Increments in the variance explained. 12 possible clusters of words are suggested.

The method had automatically formed different numbers of clusters and proposed some possible solutions. I decided to take the larger number of clusters recommended by the program, namely 12 clusters, since it was at that point that the increments in the variance explained started to tail off, as shown in Fig. 27.4.

Now I was faced with the interpretation of these word clusters, but first I wanted to see their relationship to the chapters, using the idea of a "heat map" that was commonly used in the *DataVision* utility on my atomic RAMchip to show me graphically the history of my expenses on different items. I tested **VerbalR**'s intelligence with this instruction: "compute percentages of words in each chapter for the previous 12-cluster solution; re-order the chapters and the clusters to be to be as similar as possible to their neighbours; plot a heat map of the table of standardized percentages". There was hardly a hesitation from the program to produce the plot in Fig. 5.

I got much more than I expected, which testified to the advanced way **VerbalR** could respond to my requests. I could see more detail now why certain chapters were different from others and what characterized their vocabularies. In addition, the

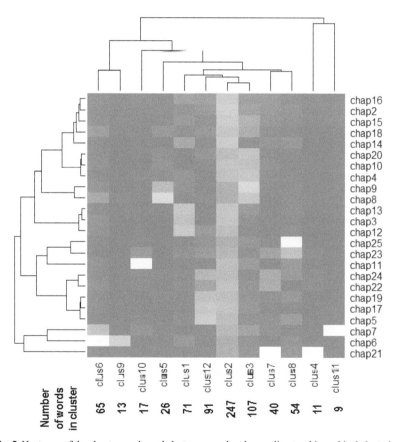

Fig. 5. Heat map of the chapters and word clusters, re-ordered according to a hierarchical clustering of each set. Colours go from black (low) through gray (medium) to white (high) in terms of standardized percentages of words.

Color image of this figure appears in the color plate section at the end of the book.

program had automatically chosen to re-order the chapters and word groups according to a hierarchical clustering of each, shown neatly in the graphical result of Fig. 5. Several associations observed in the two-dimensional correspondence analysis were similarly apparent here, for example the exceptional nature of Chapters 6 and 7. In addition, Chapter 7 was unique in having words of cluster 11, a small cluster of the nine words with *outline/s/ed* and *fossil* being the most frequent, 60 and 30 occurrences respectively, words that were hardly mentioned in the other chapters. Chapter 21 had a unique vocabulary concentrated into clusters 4 and 7: cluster 4 had only 7 words, most importantly *farmers/ing*, *agriculture/al* and *climate*, whereas cluster 7 had 40 words, most importantly *population*, *simulate/ion*, *density/ies* and *transition*. The heat map showed clearly that the words in cluster 4 were exclusive to Chapter 21 but the cluster 7 words were shared with Chapters 22, 23 and 24, all in the same section of the book mentioned above. Chapters 3, 12 and 13, each from a different section of the book, were tightly clustered due to the words in cluster 1, such as *object/s*, *event/s*, *date/s/ed/ing*, *place/s*, *domain/s* and *chronology/ies/ical*, suggesting that these chapters were dealing with space-time aspects of archaeology. Cluster 2, the biggest one with 247 words, was clearly consisting of a vocabulary common to the whole book: the most frequent words in this cluster were those dominating the word list in Fig. 1, *use/d*, *archaeology/ist*, *model/ed/ing*, *data/base/set*, *analyse/analytical*, etc. Although I could comment on many other features of this heat map, I finally mention one that was impossible to see in the correspondence analysis results. In Fig. 2, Chapter 11 was near the centre and seemed to have no special association with the words. In the heat map, however, that took into account more dimensions and formed several word clusters, Chapter 11 appeared to be exclusively associated with word cluster 10, another small cluster consisting of 17 words, the most important in terms of frequency being *tree*, *lineage/s*, *ancestor*, *hypothesis/ es/tical*, *evolve/evolution/ary* and *clade/istics*. The title of the chapter "Phylogenetic Systematics" confirmed its particular vocabulary and thus content.

Time precluded me from carrying on with my investigation of the text of "Mathematics and Archaeology"—my next job was to scan a related but older book "Mathematics in the Archaeological and Historical Sciences", which had also been found in a highly deteriorated state in the same library. My reconstruction of the text, where it had been possible, of "Mathematics and Archaeology" would now go to another section of our research group. There specialists in archaeology and mathematical modelling would try to fill in the gaps and establish as complete a text as possible, which would then be added to the archives of a subject that would otherwise have been well and truly buried in the past. Nevertheless, I felt satisfied that I could apply many of the statistical techniques that the archaeological specialists themselves had applied to their artefacts of interest, be they coins, ceramics or fossils, to the words I found in the text of this old book. In the same way I had been able to discover patterns in the vocabulary in the various chapters in order to determine the book's structure, and to re-establish content just like the archaeologists of the past. The reconstructed text will now be stored in the new indestructible atomic databases in our research centre, and a copy of all my findings will remain in the atomic RAMchip on my wrist, be buried with me one day and never be lost to future generations.

My experience of rediscovering the statistical and archaeological texts gave me much food for thought. It seemed that although archaeology dealt with physical fragments of the past, which could be inspected, described and discussed amongst experts, and conclusions drawn, it lent more credence to their research to actually code the information in the objects in some type of quantitative or qualitative format and apply mathematically-inspired statistical tools to the resultant data. Sometimes the patterns in a corpus of objects emerged naturally from the data, using methods that "handled" the data minimally, allowing for objective inference, whereas in other cases researchers had a distinct hypothesis in mind, which could then be confronted with the numerical data.

This reminded me of my holidays in Samos Island, Greece, where I had stayed several times in Pythagorion, the home village of Pythagoras. Once I met a retired mathematical philosopher there who told me that the dictum of the Pythagoreans was "*All is number*", that all things in the universe had numerical attributes that uniquely described them. And another surprising thing was that this dictum had been passed along generations from ancient times to the 23rd century! Word-of-mouth, story-telling, the process of teacher-student communication, all these ways of transferring knowledge had become absolutely crucial after the technological war, when it was realized that knowledge was "perdurant" and not "endurant". The solid objects studied by archaeologists—ceramics, coins and fossils, amongst others—were enduring reminders of past societies. The written word itself had proved to be as ephemeral as the nebulous cloud that used to house it, and even the Rosetta stone and its various copies had outlived most of the writings of the succeeding millennia.

Apart from the verbal transfer of knowledge, the remnants of books remained as artefacts of mankind's scientific progress. Understanding this progress required a particular technology for the "data" collection and—crucially, when the books had extensively deteriorated, as in the case of "Mathematics and Archaeology"—statistical tools to help reconstruct content and form. Thanks to the work of archaeolinguistic groups like ours and the application of these statistical methodologies, we were painstakingly piecing together the lost knowledge of the past.

Index

Color Plate Section

Chapter 1

Fig. 1. A General Schema for Archaeological Problems.

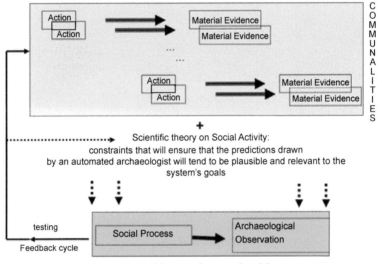

Fig. 2. From Observed Instance to Inferred causes. A schema for archaeological problem-solving.

Chapter 9

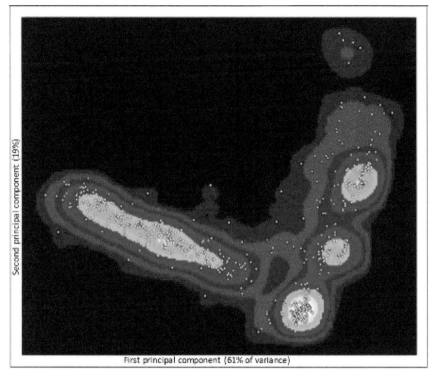

Fig. 7. The archaeological objects (points) and several cuts of the bivariate density in the plane of the first two principal components.

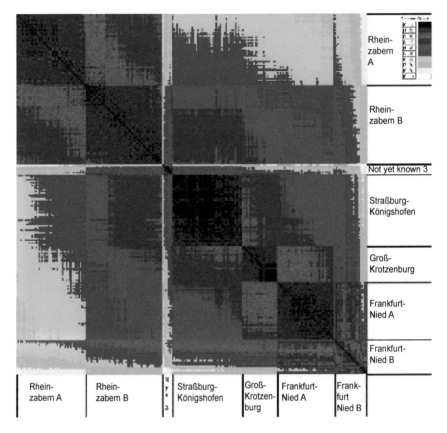

Fig. 9. Heatplot of the ordered distance matrix.

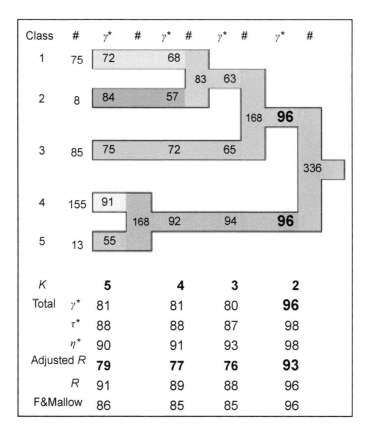

Fig. 12. Informative dendrogram: a binary tree with results of validation of hierarchical clustering.

Chapter 10

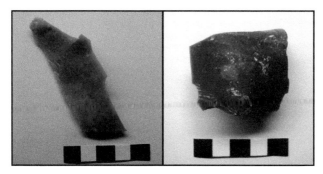

Fig. 3. Blades resulting of the flint knapping experience with soft hammer (left) and hard one (right).

Chapter 11

Fig. 3. Examples of Paleoindian fluted projectile points from North America: (a) Clovis (Logan Co., Kentucky); (b) Cumberland (Colbert Co., Alabama); (c) Crowfield (Addison Co., Vermont); (d) Dalton (Lyon Co., Kentucky); (e) Gainey/Bull Brook (Essex Co., Massachusetts); (f) Suwannee (Santa Fe River, Florida).

Chapter 20

Fig. 3. 1949 aerial photograph of the Puerco Mesa area in Malibu, California. Large ellipse indicates approximate location of LAn-19 and smaller ellipse the location of LAn-803. LAn-19 was probably first occupied after 4000 B.C. and LAn-803 was probably used somewhat earlier (ca. 5500–4000 B.C.). Photograph provided by Dr. Chester King.

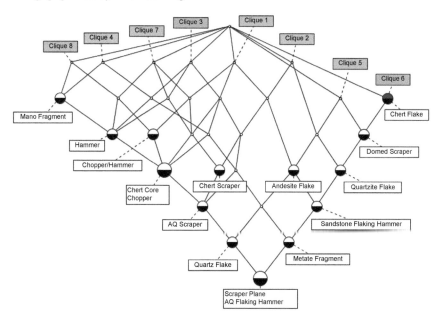

Fig. 6. Minimal representation of the lattice generated from the presence/absence matrix in Table 4. This simplified representation helps to visually identify the six distinct and complete artifact type chains in the lattice (Table 5).

Chapter 27

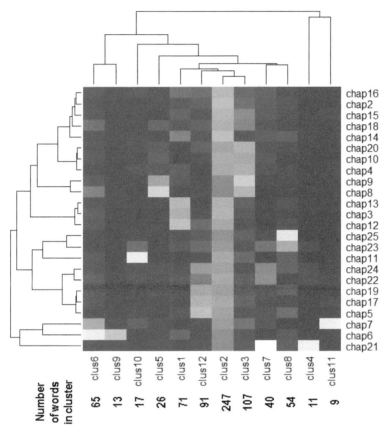

Fig. 5. Heat map of the chapters and word clusters, re-ordered according to a hierarchical clustering of each set. Colours go from red (low) through yellow (medium) to white (high) in terms of standardized percentages of words.